SECOND EDITION

# EXPLORATION
## STRATIGRAPHY

SECOND EDITION

# EXPLORATION STRATIGRAPHY

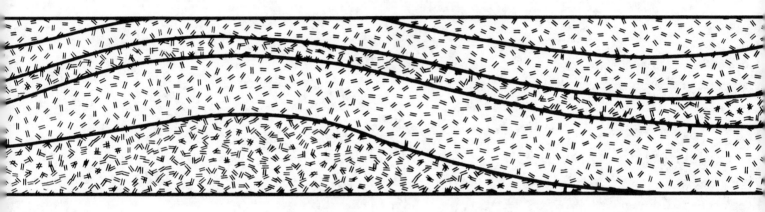

GLENN S.
VISHER

# PennWell Books

PennWell Publishing Company
Tulsa, Oklahoma

Copyright © 1990 by
PennWell Publishing Company
1421 South Sheridan/P.O. Box 1260
Tulsa, Oklahoma 74101

Visher, Glenn S.
  Exploration stratigraphy/Glenn S. Visher—2nd ed. p. cm.
  Bibliography: p.
  Includes index.
  ISBN 0-87814-342-4
  1. Petroleum—Prospecting.  2. Geology, Stratigraphic.  I. Title.
TN271.P4V57 1990
622'. 18282—dc20
89-35422  CIP

Printed in the United States of America
1  2  3  4  5  94  93  92  91  90

During the writing of this book, I have been constantly

reminded of those that preceeded me. Without systematic

observation and thought there would be little to

understand and interpret. The discipline taught to

me by my professors at the University of Cincinnati—

most importantly John L. Rich—and at Northwestern

University—notably Arthur Howland, Larry Sloss, and

Bill Krumbein—provided the basis and the need for

objective analysis. Always a historical precedent must be

found, new data must be carefully collected, and a rig-

orous analysis developed. If this book has merit, much

must be credited to their scientific discipline and example.

# CONTENTS

2nd EDITION

Five years ago I believed the need was for a book that outlined the principle new stratigraphic concepts useful for the exploration for hydrocarbon accumulations. Little consensus existed with regard to time, unconformities, correlation, and sequences in the interpretation of the stratigraphic record. Source rocks, expulsion of hydrocarbons, migration, and the correlation of hydrocarbon extracts to reservoired hydrocarbons were suggested for only a few basins. Seismic stratigraphy was based upon a very generalized unconformity history, and the interpretation of the causality of changes in wavelets due to facies changes was still a major area of research. Similarly the understanding of response patterns as a product of sedimentary processes was not often applied to the interpretation of the origin of stratigraphic sequences.

The thrust of this current research was on developing sedimentary models, useful only for the identification of depositional environments, not on depositional systems and their resultant architecture. Basinal history and architecture was principally focused on provenance, the geometry of depositional areas, and paleogeographic reconstruction was limited to the recognition of shallower to deeper water patterns, continental to coastal patterns, and, in a few instances, to shelf sequences. Little was said about the causality for a particular stratigraphic sequence, cycles of deposition, or the boundary conditions that would make the genetic interpretation of stratigraphic patterns possible.

This second edition builds upon the model framework, environmental interpretations, and mapping of facies patterns, but adds concepts useful for predicting the time and place for the occurrence of particular stratigraphic response patterns. With new precision for time-rock stratigraphy, on a scale in some instances of less than 250,000 years, it is now possible to understand the importance of climatic, impact, eustatic, and plate tectonic events. Opening of ocean basins, changes in patterns of oceanic circulation, and eustatic variations on scales of millions of years, and smaller secular variations in solar flux on scales of 21,000, 40,000, and 100,000 years suggest the causality for biogenic extinction events, the formation of anoxic events in ocean basins, regressive-transgressive depositional cycles, and the resultant architecture of depositional systems.

In an attempt to use these exciting new themes as tools for stratigraphic interpretation, I have attempted in the second edition to synthesize complex response patterns into simple depositional systems. Books are now available outlining the details of depositional processes, for example: *Sedimentary Environments and Facies, 2nd Edition; Tide-Influenced Sedimentary Environments and Facies; Shelf Sands and Sandstones;* and *Recent Developments in Fluvial Sedimentology.* The challenge today is to synthesize this wealth of observational detail into easily identifiable depositional systems. Such a synthesis has only recently been possible because of these new books, and by our ability to separate stratigraphic units into genetic sequences reflecting short time intervals.

Much new material has been added to the second edition to illustrate these stratigraphic and depositional themes. Since this book is a synthesis of accepted stratigraphic and environmental principles, it should be useful as a textbook for training petroleum explorationists, and should also be useful as a textbook for an undergraduate course in stratigraphy and sedimentation.

The expanded observational base from studies of Modern depositional processes, with new surface and subsurface information, has made it possible now to present larger integrating themes. This new information base provided the basis for the interpretation of depositional patterns, analysis of non-reccurring events, and for the interpretation of basinal and plate tectonic histories.

For those interested in using this book as a textbook, at the end of each chapter are suggested readings for further study. It is hoped that this book, with its extensive bibliography and many illustrations, will provide the needed creative synthesis to obtain additional insights and understanding of the complex interplay of boundary conditions, depositional processes, and themes of sedimentation and stratigraphy.

April, 1989
Glenn S. Visher

## 1st EDITION

This book is an outgrowth of a continuing education course, *Stratigraphic Controls for Hydrocarbon Accumulations*. It reflects a synthesis of information, concepts, and ideas developed to communicate a rapidly developing understanding of the stratigraphic record. This record is a product of interpretable response patterns. Utilizing the concepts of causality, process-response models, developmental history, and most importantly, conceptual models, it has become possible to understand the origin of specific response patterns and to predict stratigraphic patterns from a limited observational base.

Of necessity a synthesis must be brief and explicit. The concepts are simply stated, and only sufficient examples are described to illustrate the pattern to be interpreted. The complexity of processes, boundary conditions, and non-recurring events makes each example slightly different. It is only the illustration of the integrating theme that is emphasized. This approach has allowed the book to be short, without the overwhelming detail that flaws most comprehensive treatises. Description is not the scientific basis applied to the formulation of stratigraphic models or the prediction of patterns. Those wishing this detail must provide their own observations as the basis for stratigraphic analysis. It is not my purpose to ignore careful observation, but rather to illustrate developmental patterns and models.

Concepts or models are useful tools or devices to structure observation; they cannot be the final word in the developing an understanding of causality in the stratigraphic sciences. It has been my purpose to illustrate what we do know, not to dwell on paterns or observations that are poorly understood. These may become the bases for new syntheses, but I believe it timely to present this book as a possible basis for developing broader syntheses and understanding the origin of the stratigraphic record. At this juncture it is possible to use what is known to enhance the predictability of reservoir patterns and hydrocarbon accumulations. With this as the goal, the book should be useful.

April, 1984

G.S. Visher

# THE STRATIGRAPHIC FRAMEWORK

Our estimate . . . of the value of all geological evidence . . . must depend entirely on the degree

of confidence which we feel in regard to the permanency of the laws of nature. Their immutable

constancy alone can enable us to reason from analogy by the strict rules of induction, respecting

the events of former ages. **C. Lyell** *Principles of Geology,* **1830**

## CONCEPTS

The stratigraphic record represents the history of processes and events that occurred at the surface of the earth. The time-rock framework, periods and areas of missing geologic history, patterns of systematic depositional changes, and lithologic response to processes and events are truly the bases for stratigraphic analysis.

If the goal is to predict the pattern of a specific lithology that may have economic significance, then observations relating to boundary conditions during deposition have particular importance. The array of processes involved in development of any stratigraphic unit can produce similar-appearing patterns. If processes can be separated, their contributions can be analyzed and related to their specific responses and aspects of the developmental history for the stratigraphic unit can be isolated.

### Time-Stratigraphic Patterns

If the stratigraphic record represents a continuum that is interrupted by process-related responses to events or systematic changes in earth history, then identification of these changes can provide the basis for correlating strata and the identification of genetic units. The unrelated event has little meaning in the historical interpretation of a stratigraphic section.

The identification of processes lies at the base of stratigraphic interpretation, for if there is no theme, there will be no pattern, and the mapping of stratigraphic units has no purpose other than description of discontinuous formations and random events. Responses to developmental patterns must be the basis for interpretation, and these patterns are limited in number. Time-stratigraphic patterns can be identified and used as a basis for stratigraphic analysis.

## A HISTORICAL RESUME

### Naturalism of the Eighteenth Century

Geology as a predictive science may find its roots in the naturalism of the eighteenth century. The need for a naturalistic philosophy was developed due to the discovery of the predictability of the motion of bodies, useful in discovering the array of order throughout the solar system. The works of Boyle and Newton, the spacetime teleology of Barrow, and the naturalistic philosophies of Descartes and Hobbes were the philosophical foundation for science. The discovery of order led to mathematical laws, to experimental verification, and to the concept that cause and effect was the basis for science.

Application of these thoughts to geology led to the recognition of the similarity between observed characteristics in the strata exposed on the earth's surface and those that could be studied in modern observations of nature. The great contribution of Hutton (1788) cannot be overstated. The belief that natural causes produced these strata formed the basis for his synthesis:

> We must read the transactions of time past in the present state of natural bodies, and, for the reading of this character, we have nothing but the laws of nature, established in the science of man by his inductive reasoning. For man is not satisfied in seeing things which are; he seeks to know how things have been, and what they are to be.

Mapping of strata led to the need for scientific theories to explain uplift, erosion, unconformities, a pattern of change of fossils found in the strata, and the countless detailed observations based upon the stratigraphic record, which became observations requiring interpretation. Word on mapping stratigraphic units by Guettard (1757) and William Smith beginning in the 1790's (Smith 1815a,b) was a methodology new to science, and provided the basis for nineteenth century insights.

The developmental basis for natural laws was principally inductive, a discovery process, and first causes were not entertained. Abraham Gottlieb Werner, a popular professor at the Freiberg Mining Academy during the eighteenth and early nineteenth centuries, formulated a causal basis for the formation of stratigraphic units. A five-fold classification of deposits from and within a primitive ocean was envisioned. This "Neptunist" theory was important in the history of stratigraphic thought. Catastrophism and uniformitarianism were, and in some quarters still are, hotly debated. This controversy led to new observations and subdivision of rock units.

## Rationalism and Geological Theories of the Nineteenth Century

The transition from inductive reasoning and the discovery of the laws of nature to a deductive scientific approach to the synthesis of observation was crucial to geologic thought. Herschel (1830) suggested that in addition to inductive reasoning, synthesis can be created by "bold hypotheses that establish an interrelation of previously unconnected laws". He further suggested that "a meticulous inductive ascent, and a wild guess are on the same footing if their deductive consequences are confirmed by observation" (Lohsee 1972).

Whewell went beyond this position and suggested that: "Induction is a term applied to describe the process of true Colligation of Facts by means of an exact and appropriate Conception" (Whewell 1859, 70). A corollary to this is that inductive inference is always something more than a mere collection of facts.

> Whewell stated that "the facts are not only brought together, but seen in a new point of view. A new mental element is superinduced; and a peculiar constitution and discipline of mind are requisite in order to make this induction" (Whewell 1858, 71).

This extension of inductive discovery may lead to a broader synthesis, and this synthesis can be tested against facts (observation) and "appropriateness."

Fundamental to the history of geological thought was the understanding of the age of the earth. Early in the eighteenth century, the significance of fossils in stratigraphic units, the mapping of strata across Europe, and the recognition of raised beaches and wave-cut terraces above the present shorelines were compelling evidences for a much longer geological history than that suggested by biblical scholars. De Maillet (in Telliamed) suggested the age of the earth to be millions or billions of years (Albritton 1980). Similarly, Albritton documents the importance of work by Buffon (1778) during the latter part of the eighteenth century. His "Des Epoques de la Nature" appeared in his fifth supplementary volume published in 1778. From these beginnings it was possible for Hutton to conceive that continuing processes, over extended periods of time, could account for stratigraphic patterns. From these beginnings it was Lyell that recognized that the antiquity of the earth was essential for the interpretation of stratigraphic patterns. Haber (1959) suggests that: "although no work was more decisive on the course of geological thought than Lyell's *Principles of Geology*, its assimilation and acceptance took time".

At this same time other observers were attempting to classify stratigraphic units on the basis of lithologic characteristics, remains of ancient life, and major discordant spatial relationships. From this work by Murchison, Sedgwich, and Ulrich and many others, a sequential ranking of clearly identifiable stratigraphic units emerged. This work made it possible to map units across areas of differing elevation, across structural discordances, and even across non-contiguous geographic areas, such as areas across the English Channel and across the Atlantic Ocean. The ability to make geologic maps, which represented not only patterns of particular lithologic units but also patterns in relation to relative time, was the first and possibly the most important contribution of stratigraphy to understanding the history of the earth.

Parallel to the work of the lithostratigraphers and those geologic mapmakers was the work done by naturalists concerned with the nature of the life forms found in the strata. Their usefulness was not envisaged until concepts of organic evolution, the sequential pattern of stratigraphic units, and the interpretation of lithologic units as the normal product of natural processes were clearly established. The role of the biostratigrapher is one of subdividing stratigraphic units on the basis of relative age and temporal equivalence of stratigraphic units throughout the world.

Lyell's *Principles of Geology* (1830), followed by contributions of Agassiz (1842) and Darwin (1859), changed the nature of scientific "discovery." By deduction, analogy, and creative insight they produced syntheses that could be tested by prediction. The physical scientists were busy "discovering" laws that interrelated observations. Those scientists involved in the interpretation of the history of the earth required new methods of synthesis. Lyell (1830) could only conceive of the interpretation of the origin of strata in terms of causality:

Geology is the science which investigates the successive changes that have taken place in the organic and inorganic kingdoms of nature; it inquires into the causes of these changes, and the influence which they have exerted in modifying the surface and external structure of our planet.

Louis Agassiz fundamentally changed the approach for demonstrating the validity of hypotheses. He recognized that uniformitarianism provided the approach for interpreting the origin in the past. Later Agassiz wrote:

In the early part of the summer of 1840, I started from Switzerland for England with the express object of finding traces of glaciers in Great Britain....Inexperienced as I then was, and ignorant of modes by which new views, if founded on the truth, commend themselves gradually to general acceptation. I was often deeply depressed by the scepticism of men whose scientific position gave them a right to condemn the views of younger and less experienced students.....It is but lately, that, in turning over the leaves of a journal, published some twelve or fifteen years ago...I was amused to find a formal announcement, under the signature of the greatest geologist of Europe, of the demise of the glacial theory (Agassiz 1896, 1–2).

It was through Agassiz's example that careful observation and comparison made it possible to develop new understandings in the natural history of the earth (Agassiz 1842). The historical patterns reflected in the stratigraphic record thus could be used to gain insights into scientific theories.

Darwin could then conceive of the developmental history of forms of life. Each taxa could be studied in terms of its variation and its development could be inferred. Gould (1986) suggests that Darwin's principle scientific contributions were: (1) the uniformitarian argument that one should work by extrapolating from small-scale phenomena that can be seen and investigated, and (2) the establishment of a graded set of methods for inferring history when only large-scale results are available for study.

In summary, Gould (1986) states that *The Origin of Species* achieves its conceptual power by using:

uniformitarianism in extrapolating the observed results of artificial selection by breeders and farmers; inference of history from temporal ordering of coexisting phenomena (in constructing, for example, a sequence leading from variation within a population, to small-scale geographic differentiation of races, to separate species, to the origin of major groups and key innovations in morphology); and imperfection(s) (vestigial organs, odd biogeographic distributions made sensible only as products of history...).

These contributions essentially created a hypothetical-deductive basis for historical science.

The hierarchy of fossil zones allowed the geologist to construct geologic maps showing patterns of rock and time-rock units. Structural patterns with the identification of folds, faults, monoclines, arches, basins, domes, and even geosynclines were interpreted from this data base. In the search for minerals, the prediction of positions of individual stratigraphic units or veins, and more recently the position of structural highs, has required precision. The idea of a single value—elevation based upon topographic surveys—was applied to the relation of a rock unit or datum to sea level. The structural map was a useful tool in prediction and became the most significant map in the interpretation of the pattern of units on and beneath the earth's surface. This map was an objective interpretation based upon easily measured data, but it required the finding of a suitable marker. In most areas this was done simply by choosing a thin limestone or another distinctive lithology. Structure maps became the basis for interpreting positions of coal beds, mineral deposits, and oil fields.

A second single-valued map was derived from structural and stratigraphic data. This map, the thickness or isopachous map, was first used to determine the volume of a coal bed or a limestone unit, but soon it was used to interpret the history of structural features, such as basins, arches, geosynclines, and erosional intervals.

## Unconformity

The concept of the unconformity was developed early in the history of stratigraphy. Direct observation of differing attitudes of beds above and below a surface was a source of confusion. By understanding the role of faulting, folding, and the concept of superposition of younger on older units, it was possible to see that these surfaces were important discontinuities in the history of the earth and they represented catastrophic events that were useful in subdividing stratigraphic sections. Of all the observations concerning stratigraphic sections, probably the most troublesome was the interpretation of unconformities. The belief that they were synchronous and distributed throughout the world was widely held. Discontinuities in the evolutionary record, changes in thickness of strata, missing units, sea-level changes, glaciation, and countless other changes were universally attributed to the presence of unconformities. A simple classification of the geometric and stratigraphic relationships of unconformities did provide important insights: the angular unconformity representing conformable and progressive patterns of onlap of younger units across a surface of either rocks of nonsedimentary origin or of strongly contrasting physical characteristics, the erosional unconformity that showed an obvious truncation and fill pattern discernible in surface exposures, and the surface of apparent conformity that had strongly differing biologic content. These relationships were described, but their interpretation required the collection of additional data.

## Early Interpretations of Data

Early in the analysis of sedimentary strata, observers found patterns or cycles of units. These cycles were variously interpreted. A European school of stratigraphy led by Johannes Walther suggested that the principle of the present as a key to the past indicated these patterns represented a migration of biologic or lithologic units on the surface of deposition. A second school suggested these patterns represented systematic changes in the factors that controlled deposition, for example, sea level, mountain building, and tectonics. New constructs, based upon a wider range of observations and utilizing tools of synthesis beyond those available, were needed to resolve the problem. Other repeating patterns were observed: the worldwide lithologic similarity of units of the same age, the recognition of a broad tectonically based history of sedimentation through a complete tectonic cycle, and local repeating patterns of successive depositional units.

The interpretation of these patterns required a frame of reference or a scientific construct that was unavailable to the observer of outcropping sequences.

The drilling of wells for economic gain, e.g. minerals and petroleum, provided new information to be interpreted. New types of data were collected, for example, samples from wells, physical measurements of changes in stratigraphic units, techniques of mineralogic analysis, more detailed microscopic analysis of lithologies, and the recognition of biostratigraphic zones. From this expanded data base came the recognition of incremental changes in lithologic characteristics, and the concept of sedimentary facies was developed to explain the temporal and areal lithologic variations seen in many stratigraphic sections.

With more data available, it was possible to represent changes in a quantitative manner. Not only could structural and isopachous maps be drawn, but maps representing lithology could be constructed. The problem in presenting these data was that for them to be interpretable, some larger basis of integration was required. The basis found useful for geologic maps was time, and time was also the basis for integration of the wider range of stratigraphic data. The range of problems and possibilities that an insight into aspects of absolute and relative time provided was the basis for developing more general concepts of stratigraphic analysis.

The three-dimensional representations of time-rock units became the basis for paleogeographic and paleotectonic patterns, the analysis of sequential patterns, and the interpretation of biostratigraphic and lithostratigraphic successions. The problem in this type of analysis was providing a rigorous time framework. We have seen books, papers, and many symposia concerning time and stratigraphy. The paleontologists provided well-documented time-rock correlations and constructed maps to show the areal pattern of stratigraphic units. Throughout the Mesozoic in Europe, and the Tertiary in the United States, time-rock subdivisions were produced that illustrated the potential of this type of stratigraphic integration. This method of integration of stratigraphic data was valuable but was severely limited due to fossil preservation, ecologic changes, and the availability of a worldwide evolutionary sequence.

Using time-rock correlations, a wide range of data derived from independent observational bases can be integrated. The concepts of evolution, eustatic changes, interdependence of sediment supply and position of the strandline, climate, tectonic patterns, position on cratonic plates and many other factors must be included in a historical analysis.

New, or at least redefined, scientific constructs had to be developed to take diverse data with little internal relationship and construct unifying principles. If we accept that the origin of the stratigraphic record is largely deterministic, then we can focus on causal factors. The unifying principle will be the process or the association of processes that were operative during a short period of geologic history. The recognition of these processes and the evaluation of their effects both in relation to time and stratigraphic patterns are the bases for predicting the distribution of hydrocarbons.

## STRATIGRAPHIC PALEONTOLOGY

Stratigraphic paleontology has been an evolving stratigraphic science. It was not presented as one of the founding principles of stratigraphy. Originally the work of William Smith in 1790 suggested that stratigraphic units could be identified by their contained fauna. The empirical association of the fauna and flora to specific stratigraphic units was one of the first bases for their recognition, and by using the principle of superposition, the relative ages of stratigraphic units could be determined. The significance of the correlation was not formulated until 1859 when Darwin recognized that species were a part of a developmental history of life. This concept revolutionized concepts of time, the significance of fossils, the approach to stratigraphic correlations, and the basis for determining relative ages of stratigraphic units.

The uses of preserved fossils can be summarized into three different applications and stages:
- The identity of fossils in a stratigraphic unit,
- The empirical correlation of fossil patterns and sequences to stratigraphic units, and
- The dynamic formulation as seen in evolutionary paleontology.

Each of these approaches was successfully applied to the interpretation of the stratigraphic record, but confusion concerning aspects of the nature of the changes in the biota preserved in the stratigraphic record has existed since Darwin. Stratigraphic paleontologists still discuss (1) index fossils for a stratigraphic unit in the way "Strata" Smith did; (2) the grade of evolution in terms of abrupt and catastrophic changes, as did Lyell; or (3) as a continuing developmental system as suggested by Shaw (1964) and Valentine (1973).

### Index Fossils

The fundamental basis for determining the relative ages of stratigraphic units is by reference to the established sequence of index fossils. This approach is simple, direct, and unambiguous in its application but requires that the absolute stratigraphic range for index fossils be known or assumed. Unfortunately, this never was the case. The system works well when local areas are examined and where superposed stratigraphic units are demonstrably different with no recurrence of similar ecologic conditions to provide homotaxial ambiguities. The basis for use of index fossils is primarily negative: it is the absence of a particular form, either above or below a specific stratigraphic unit. As is the case for other historical sciences, a single observation is often not sufficient to test a hypothesis: it is the association of many observations that increases confidence in the conclusion. The presence of any index fossils in a stratigraphic unit or concurrent ranges of several taxa permits an ordering in a historical sense.

The importance of index or guide fossils to the interpretation of stratigraphic relationships was evident. Faunal lists were judged by the number of species included and, more importantly, by the differences that could be identified from other formations. The more different species present, the better the paleontologist could identify a particular formation and place it in its historical framework. Subdivision of faunal and floral elements into new species, and in a parallel fashion, subdivision of the stratigraphic continuum into smaller increments were often the goals. This approach provided the basis for correlation, mapping, and historical changes in the biota and was the basis for determining the history of the earth. Paleogeographic maps, concepts of tectonic history, and biologic evolution are in part based upon this observational base.

### Catastrophism and Index Fossils

Rarely in the stratigraphic record does the sequence of stratigraphic units record a continuum of change. Bedding planes, hiatuses, unconformities, and ecologic changes produce a discontinuous fossil record. The principle of uniformity was recognized, but the abrupt nature of change, both in the physical and the faunal and floral records, suggested that both sedimentation and faunal changes also were abrupt or occurred in response to non-systematic events.

Concepts pertaining to the continuity of the fossil record have been a subject of detailed analysis. Raup and Sepkoski (1986) reported on their studies of patterns of mass extictions in the stratigraphic record. The data set includes the number of genera and families that became extinct during each stage and the percentage of living genera that became extinct (Fig. 1–1). These data require a reexamination of the concept of phyletic gradualism. Concepts of clad diversity and extinction, need to explained in terms of both speciation and extinction (Sepkoski and Hulver 1985) (Fig. 1–2).

Causal mechanisms need to be considered. Few

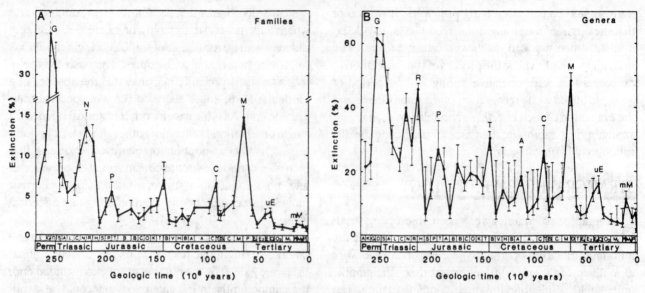

**Fig. 1–1.** Percentage of extinction for *A,* marine animal families and *B,* genera in stratigraphic intervals between the mid-Permian and the Recent. Letters along the abscissa denote standard stratigraphic units. From Raup and Sepkoski, 1986.

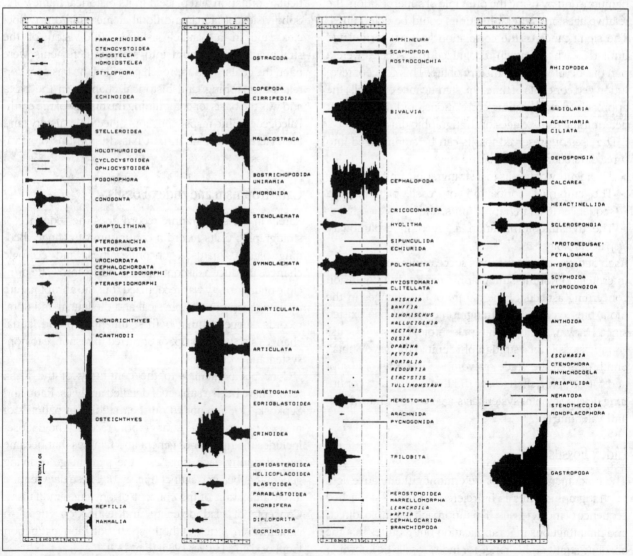

**Fig. 1–2.** The Marine Fossil Record. From Sepkoski and Hulver, 1985.

attempts at synthesizing these mechanisms have been attempted, but Gould (1985) has suggested that both speciation and extinction may occur at three differing levels or tiers of time. Tier 1 would be related to the Darwinian mode of gradualism, natural selection, and competition. Tier 2 would be related to punctuated equilibria related to cladogenesis, reflecting nonequilibrium changes in ecology, changes involving geographic or spatial segregation of ancestral stocks, defined in terms of allopatric, parapatric, or sympatric modes of separation. If speciation represents extrapolated adaptive struggles of organisms, the change would accumulate fitfully through time (Gould 1985). Tier 3 represents the observational base of family and species terminations evidenced by mass extinctions. These events could not be controlling in patterns of cladogenesis: they would be irregular and dramatic changes in the environment. Changes in temperature, composition of the atmosphere, and oceanic circulation patterns are only a few of the possible causal mechanisms for extinction events. After major extinction events, not only did the physical environment change, but the ecology reflected an entirely different controlling structure. New phyletic branching rates would appear to drive adaptive strategies to fill ecologic voids.

These concepts form the basis for refining the time scale of paleontologic subdivision of the stratigraphic record. The discontinuous nature of the fossil record has been recognized as the basis of subdivision of the stratigraphic record, and used to recognize periods, epochs, and ages. Is it now possible to recognize smaller episodic events utilizing concepts of punctuated evolutionary change, or smaller scale extinction events. These factors may be the only limits for subdividing the stratigraphic record. New meaning may be placed upon the index fossil, since its temporal range may be narrowly defined, based upon punctuated evolution and abrupt extinction. The temporal scale of stratigraphic subdivision may be more precise than the rates of phyletic evolution, where species persist for more than a million years, but genera and families have evolved in periods of only a few million years (Stanley 1985, 14).

The stratigraphic approach is the basis for most age determinations throughout the world. The relative grade or complexity of the biotic elements, the multiplicity of taxa and overlapping zones, and the rates of faunal development and distribution all make it possible to identify faunal zones throughout the world and to establish a worldwide stratigraphic subdivision at the level of epochs. A discussion of the methods and precision of biostratigraphic correlation is developed based upon the organizing principle of discrete faunal changes.

## Evolutionary Paleontology

The implication of a dynamic, evolving biologic system was recognized by Huxley. In keeping with Darwin's concept of natural selections, he visualized that evolution may have a direction related to environmental factors and the tenet, "The survival of the fittest." He suggested that a similar biota may be preserved in the stratigraphic record at differing places and times due to homotaxis. This potential for confusion was accepted by many stratigraphic paleontologists, and the proven principles of index fossils and biozones were retained as the basis for stratigraphic correlation and subdivision.

After Darwin new significance was placed upon paleontology as recording a dynamic, evolving developmental history. This approach was fundamentally different from Lyell's perodic change or special creation approach. Today there appears to be some convergence of these two approaches. Paleontologists studying systematics and documenting the paleontologic record are interested in defining, as carefully as possible, phyletic lineages, but others are interested in the causality of evolution and extinction, and aspects of *Paleobiology*, including paleobiogeography, evolutionary paleoecology, and evolutionary morphology. These latter aspects have great significance in reconstructing environmental perturbations, plate tectonic patterns, and environmental changes for biofacies analysis. These can lead to an historical interpretation of the stratigraphic record on a scale perhaps of tens or hundreds of thousands of years, rather than on a scale of five hundred thousand or a few millions of years. Even with phyletic lineages, following Brinkman (1929), subdivision of the Jurassic into time-rock intervals based upon faunal zones is not much better than five to ten million years (Harland et al. 1982).

The modern synthesis of phyletic gradualism also has its roots in Brinkmann's 1929-study of ammonite lineages. The fossil record is the only source for recognizing patterns of evolution and extinction. The evolutionary tree has definable aspects and possibly from observation of the stratigraphic record some aspects of causality can be inferred (Raup 1985) (Fig. 1–3). Application of this theme has provided the first unique basis for determining the time and place of evolutionary events (van Hinte 1976b). From these data precision for stratigraphic subdivision was claimed to be on the order of one million years (Fig. 1–4).

A new field of stratigraphic paleontology, paleoecology, was developed to aid in the interpretation of stratigraphic units and biofacies patterns (Ager 1981). Also, biologic concepts concerning the nature of genes, gene pools, taxanomic diversity, and morphologic adaptations modified the principle of natural selection and, consequently, bios-

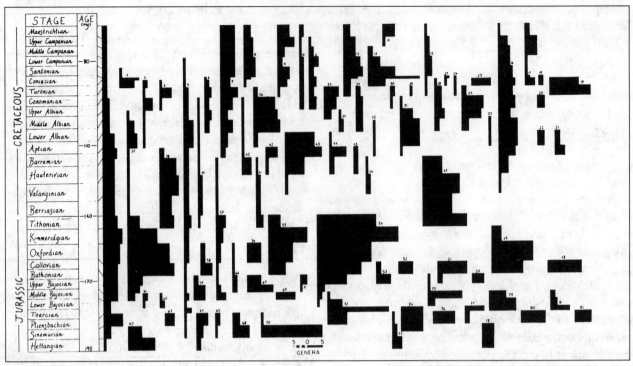

**Fig. 1–3.** From Raup, 1985.

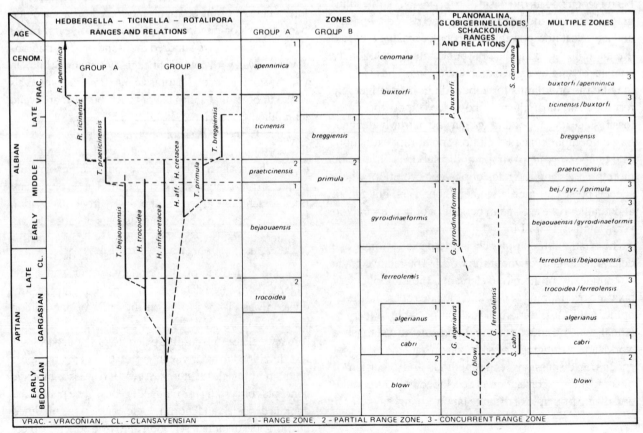

**Fig. 1–4.** Cretaceous pelagic foraminiferal evolutionary lineages. From van Hinte, 1976b, reprinted by permission.

tratigraphic interpretations based upon it. The end product of a biostratigraphic collection does not have to be a faunal list. It can be interpreted as a response to evolutionary development, ecology, faunal diversity, significant morphologic adaptations, and physical and chemical aspects of the environment.

# BIOSTRATIGRAPHY

Increased four dimensional resolution of stratigraphic record utilizing cores from the "Deep Sea Drilling Program", magnetic stratigraphy, increased seismic resolution, and the measurement of temperature changes on the scale of tens of thousands of years has pointed out the discontinuous, episodic nature of the stratigraphic record. It has become possible to address rates of evolution and extinction, ecologic changes, and faunal diversity across provincial boundaries and within ecologic communities. Elredge and Gould (1972) and Gould and Elredge (1977) with the formulation of the concept of a Punctuated Equilibria in evolution have changed the way we think about the biostratigraphic record (Vrba 1980) (Fig. 1–5). Causality for evolutionary changes can be placed in the context of environmental changes, and related to specific historic events. Various models for evolutionary change have been suggested, including Wright's Adaptive Landscape (Lewin 1986) (Fig. 1–6). For more than 60 years Wright has been writing about genetics and evolution (Wright 1988). Still little is understood about the causality of evolution. The relative roles of adaptive radiation, biogeography, adjustment to environmental change, and sociobiology are subjects of continuing discussion and controversy (Newman 1985; Lande 1985).

No biostratigrapher claimed it was possible to subdivide the stratigraphic record into absolute time zones. The goal was to establish relative ages for stratigraphic units on a worldwide basis. The development of a new form was a unique occurrence. The place and time for this event could be precisely defined and the new form obviously was a basis for relative age dating of a rock unit.

Certainly the presence of a specific new form in any stratum any place in the world must indicate that the stratum was younger than the evolutionary event that produced the new form. From this the useful concepts of biozones were developed and applied to the stratigraphic record (Mallory 1970, 559, 563).

Types of biozones were the subject of study, and their definition and significance were summarized in *An International Guide to Stratigraphic Classification, Terminology, and Usage* (Hedburg 1972, 24–27):

▸ **Lineage zone:** . . . a type of range zone consisting of the body of strata containing specimens of a bioseries

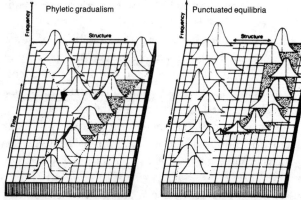

**Fig. 1–5.** Alternative models of evolutionary change. From Vrba, 1980.

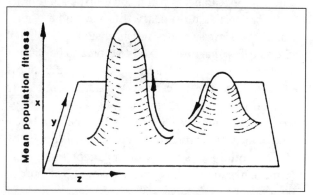

**Fig. 1–6.** Under a given set of environmental conditions, each combination of characters y and z has a certain fitness (x), some of which will be greater than others. The result is an adaptive landscape, with one or more fitness optima representing possible morphologies. Transition from one optimum to another by genetic drift alsays occurs rapidly. From Lewin, 1986.

representing the evolutionary or developmental line or trend of a taxon, or biologic group, or any segment of such, defined above and below by features of the line or trend. Lineage zones are considered to be particularly valuable guides to time correlation.

▸ **Taxon-range zone:** . . . is the body of strata representing the total range of occurrence (horizontal and vertical) of specimens of a taxon (species, genus, family, etc.). Taxon-range zones are useful as indicators of the age of the constituent strata and their environment of deposition. There can be as many taxon-range zones as there are taxons, and they do not lend themselves to any complete and systematic partitioning of all strata into a single set of units without gaps and overlaps.

▸ **Assemblage zone:** . . . is a body of strata whose content of fossils, or of fossils of a certain kind, taken in its entirety constitutes a natural assemblage or association which distinguished it in biostratigraphic character from adjacent strata. Assemblage zones are particularly significant as indicators of past environments. They also usually have considerable age significance and value in correlation.

▸ **Concurrent-range zone:** . . . is a body of strata defined by those parts of the ranges of two or more selected taxons which are concurrent or coincident. The definitive taxons of a concurrent-range zone are usually selected in order to give the zone maximum time significance and it therefore serves primarily as an aid to time correlation and chronostratigraphic classification, although it is not in itself a chronostratigraphic unit.

▸ **Oppel zone:** . . . has been applied to a less strictly defined form of concurrent-range zone in which first and last appearances of individual taxons and other paleontologic features believed to have time significance also serve as guides and in which the number of the so-called diagnostic taxons required to be present at any one place to establish the presence of the zone is discretionary. The oppel zone is broader, more readily usable, less precise relative to the concurrent-range zone, and is employed principally for time correlation and as an aid to chronostratigraphic classification.

▸ **Acme zone:** . . . is a body of strata representing the acme or maximum development of some taxon but not its total range.

▸ **Interval zone:** . . . comprises the strata between two distinctive biohorizons but does not itself necessarily represent any distinctive biostratigraphic range or even any particularly distinctive biostratigraphic assemblage. The interval zone is in common use for correlation purposes. These zones are subject to varying precision in their ability to predict relative ages of stratigraphic units. A single discovery in strata that is interpreted to be younger requires the re-evaluation of all taxon-range zones present in intervening strata to determine if they contain newly developed taxa. The problems of preservation, species diversity, faunal distribution, and ecology all can cause the apparent absence of a particular taxon in a correlative time-stratigraphic unit.

# BIOSTRATIGRAPHIC TIME-STRATIGRAPHIC CORRELATION

It is the evolutionary change of all the taxa and their association areally and vertically that is the basis for time-stratigraphic subdivision (Mallory 1970, 554). The array of overlapping developmental ranges of taxa is the basis for relative age determination.

Rates of evolution were indicated to be in the order of 500,000 to a few million years (Weller 1960, 559). If the grade of evolution recognized in stratigraphic sections from many parts of the world is used, the general level of

precision possible is approximately 1 to 20 million years (Henbest 1952, 311). Subdivision of stratigraphic units into zones useful for paleogeographic or basin analysis is at best at the level of Series. In basins where an evolutionary scheme developed and the same depositional environments are repeated in successive stratigraphic intervals it is possible to obtain local time-rock zonations useful in establishing the pattern of migration of time-rock units.

## Biofacies Patterns

Ecologic controls can only be demonstrated in the areal variation of physical and chemical aspects of the environment (Berry 1960, 18). Progressive changes in water turbidity, depth, salinity, energy, or some other ecologic aspect can be recognized in the biotic community. The formation of new taxa, changes in populations, loss of particular forms, or systematic adaptation of morphologic changes in one or more taxa all are modern bases for ecologic interpretation (Bretsky and Lorenz 1970, 540). Systematic changes in the community can be related to sequential patterns of regression and transgression, tectonic patterns producing changes in water depth, barriers to faunal migration, and changes in rates of supply of detritus.

**Table 1-1**
**Unit Terms and Hierarchies in Stratigraphic Classifications**

| Categories | Principal Unit Terms | |
|---|---|---|
| Lithostratigraphic | Group<br>  Formation<br>    Member<br>      Bed | |
| Biostratigraphic | Biozones<br>  Assemblage zones<br>  Range zones (various kinds)<br>  Acme zones<br>  Interval zones<br>  Other kinds of biozones | |
| Chronostratigraphic | Eonothem<br>  Erathem<br>    System<br>      Series<br>        Stage<br>    ....<br>        Chronozone | Equivalent time terms (Geochronologic)<br>Eon<br>  Era<br>    Period<br>      Epoch<br>        Age<br>  ....<br>        Chronozone |
| Other stratigraphic categories (Mineralogic, environmental, seismic, magnetic, etc.) | Zone | |

Prefixes *sub* and *super* may be used with unit terms when appropriate if additional ranks are needed. *Zone* is a general term that may be used in any kind of stratigraphic classification. It should be prefixed to indicate the kind of zone (e.g., lithozone, biozone, chronozone, mineralzone, and range zone) if this is not otherwise clear. *Marker-horizons* of various sorts may be designated as lithohorizons, biohorizons, chronohorizons, etc. From Hedburg, 1972.

**Table 1-2**
**The Chief Methods of Radiometric Age Determination**

| Parent Nuclide | Half-life (years) | Useful Age Ranges (years B.P.) | Daughter Nuclide | Minerals and Rocks Commonly Dated |
|---|---|---|---|---|
| Carbon-14 | 5,730 | < 25,00 | Carbon-12 | Wood, peat, CaCO$_3$ |
| Uranium-235 | — | < 150,000 | Protactinium-231[1] | Aragonite corals Deep-sea sediment |
| Uranium-234 | — | < 250,000 | Thorium-230[1] | Aragonite corals Deep-sea sediment |
| Uranium series | — | 200,000 to tens of millions | Helium 4 | Aragonite corals |
| Uranium-238 | 4,510 million | > 5 million[2] | Lead-206 | Zircon Uraninite Pitchblende |
| Uranium-235 | 713 million | > 60 million[2] | Lead-207 | Zircon Uraninite Pitchblende |
| Potassium-40 | 1,300 million | > 50,000[2] | Argon-40 | Muscovite Biotite Hornblende Glauconite Sanidine Whole volcanic rock |
| Rubidium-87 | 47,000 million | > 5 million[2] | Strontium-87 | Muscovite Biotite Lepidolite Microcline Glauconite Whole metamorphic rock |

(1) Intermediate daughter products.
(2) These methods may give useful information below the minimum indicated, but the methods become increasingly subject to serious error with decreasing age. From Faul, 1966.

## Migration of Stratigraphic Units

The three-dimensional pattern of lithologic and biologic units is the observational basis for stratigraphy. To obtain this basis, some method of relating stratigraphic units to time must be found. Much effort has been expended to find an absolute time scale. These were either based upon rates and systematic changes in the evolution of the biota or on systematic rates of degradation of radioactive isotopes of commonly occurring elements. Both of these methods were applied in determining a time scale and subdivision of the stratigraphic record (Table 1-1). The level of precision each method offers depends upon the nature of the materials and observations that are available.

## Measurement of Absolute Time

The basis for a time scale useful for measuring absolute time is the decay of particular isotopes (Faul 1966) (Table 1-2). Decay rates and variation in the products of decay have been studied in great detail, and the probabilities of these occurrences have been established. These represent the theoretical models, and measurement represents the verification of the theoretical construct. However, other variables stand in the way of application to stratigraphic sections. Problems of measurement, preservation of all related elements within a mineral from the time of its formation, presence of appropriate mineral phases in stratigraphic sections, and equilibrium concentrations of all, or at least the important, isotopes in each mineral phase at the time of formation were found to affect the precision of an age measurement (Faul 1966, 61). The ability to measure absolute time must be tempered by a statement of the probable variance. This variance was found to be significant and is larger the older the stratigraphic unit. A precision of plus or minus 10% is the probable limit of variance (Eicher, 1968; Harland, et al. 1982).

As in similar problems in science, it is not the single measurement that is the most important; it is the association of measurements from differing systems that allows the verification. If several differing methods of isotopic dating are applied to the same stratigraphic unit, more understanding and a closer measure of the most probable age can be made. The range of methods applicable to dating a rock does allow this type of synthesis (Table 1–2). This table illustrates the many approaches to radiometric age dating. Often subjective interpretations must be made concerning the formation of the mineral phase that is to be dated. Application of several methods can provide information on relative ages. These may be based upon relation to igneous intrusions, synsedimentary authigenesis, post-sedimentary diagenetic changes, and determination of a detrital or an authigenic origin for a mineral phase. Evaluating all of these aspects allows determination of ages that are useful in stratigraphic correlation.

Radiometric age dates are important in stratigraphy and an absolute time scale for the ages of the periods (Table 1–3) has been developed. Table 1–4 (Palmer 1983) indicates in more detail absolute ages and probable errors for all Ages in the Phanerozoic Era. This is useful in analysis of rates of biologic evolution, sedimentation, and the spans of time reflected in orogenic events. Little understanding of any of these aspects was possible prior to the development of an absolute time scale. If the stratigraphic goal is to evaluate these variables, then this is the method of choice; if the goal is to subdivide the stratigraphic section into units representing historical development, then some other tool is required.

## Time-Stratigraphic Correlation

Correlation of stratigraphic sections is different than zonation. A single section represents a specific history of development represented by changes in biota and lithology. In a three-dimensional reconstruction, variation includes systematic changes of both lithology and biota with respect to each other and to time. There is little evidence that every section to be correlated represents the same intervals of history, the same patterns of response on a vertical or two-dimensional level, or the same pattern of faunal or lithologic elements. Certainly, a bed-by-bed or zone-by-zone two-dimensional correlation is subject to gross errors if that is all the lithologic or biozone data available. There is no rational basis to say that a bed-by-bed or zone-by-zone correlation in any way relates to time. Certainly, if lithosomes and biozones are diachronous, as they are at any time surface, then no temporal basis for a correlation exists. If the goal is to develop a history of sedimentation, then the areal and historical pattern of lithosomes or biozones must be established. The only common ground lies in a process control that interrelates deposition at individual locations, and this process control must reflect systematic changes in positions of lithosomes and biozones.

**Table 1-3**
**Major Units of Standard Global Chronostratigraphic (Geochronologic) Scale (Palmer, 1983)**

| Erathems and Eras | Systems and Periods | Radiometric Dating (millions of years)* | |
| --- | --- | --- | --- |
| | | Duration of Unit | Age of Beginning of Unit |
| Cenozoic | Quaternary | 1.6 | 1.6 |
| | Tertiary | 64.8 | 66.4 |
| Mesozoic | Cretaceous | 77.6 | 144 |
| | Jurassic | 64 | 208 |
| | Triassic | 37 | 245 |
| Paleozoic | Permian | 41 | 286 |
| | Carboniferous | 74 | 360 |
| | Devonian | 48 | 408 |
| | Silurian | 30 | 438 |
| | Ordovician | 67 | 505 |
| | Cambrian | 65 | 570 |
| Archeozoic Precambrian | | 3,230 + | 3,800 + |

## Identification of Time-Stratigraphic Surfaces

The key to understanding a section is to find an array of objective criteria that approaches a parallelism to a time surface. These can include the following:

▸ Parallelism or identity with an evolutionary event. The formation of a new species, the acme of development of a taxon, the unique assemblage of a group of taxa, or a zone representing a unique overlap of taxa with well-defined ranges.

▸ Parallelism to a single event that is represented by a characteristic aspect over a large portion of the area. The event may represent a unique climatic condition, chemical change, tectonic occurrence (earthquake, landslide, volcanic eruption, etc.), or biologic change (radiolaria or dinoflagellate bloom, etc.).

▸ A normal pattern of deposition and nondeposition produces an array of bedding planes. Each bedding plane is a local time parallel surface, and an infinite array of these three-dimensionally distributed bedding plane surfaces is parallel to a time surface.

▸ Depositional surfaces parallel or subparallel to a horizontal datum (sea level) should be parallel to a time surface. Hence, deposition at the strandline or on a shelf surface with a slope of less than 2 meters per kilometer may be considered as parallel to a time surface. This would include transgressive shoreline deposits or widespread blanket shelf stratigraphic units.

▸ A surface developed by mapping apexes of systematic changes in the surface of deposition at or below sea level (bathymetric cycles). If aspects of the strati-

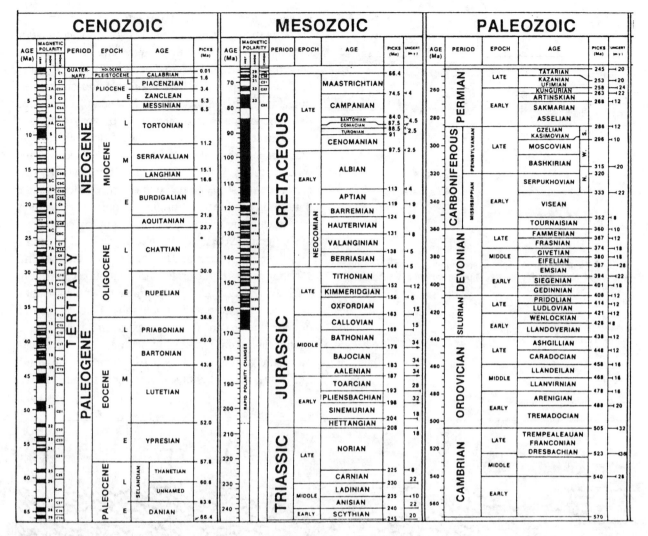

**Table 1-4. From Palmer, 1983.**

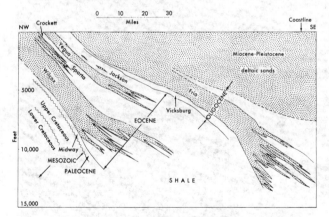

**Fig. 1-7.** Cross section of the sediments of the Gulf Coastal Plain along the Colorado River, Texas. The alternation of sandstone and shale units is illustrated, as is the overall spread with time of the sand facies toward the Gulf. The stippled areas are interbedded sandstone and shale. From Clark and Stearns, 1968.

graphic section can be related to depositional water depth, removal of this depth produces a surface parallel to sea level and to a time surface.

▶ Relation of successive depositional units to a surface of erosion or nondeposition provides a relationship to a time surface. Successive surfaces of nondeposition or erosion may have similar form and be parallel to a time surface.

The establishing of a time-rock correlation framework is the single most important aspect in the construction of stratigraphic history. Without the scale of time, there is no basis for comparing stratigraphic sections or understanding the flow of events that produced even a single stratigraphic section. Each of the above is a basis for establishing such a correlation, but rarely are data sufficient to demonstrate that a unit is truly parallel to a time surface. It is only in the integration of all of these objective criteria with additional information that confidence can be established in a correlation pattern (Fig. 1-7).

Mapping of a correlation datum should show a coherent interrelationship among all the data. If one or a group of points shows no rational pattern, either the rational basis was not found or there was a mistake in the correlation pattern. In either case more data are required

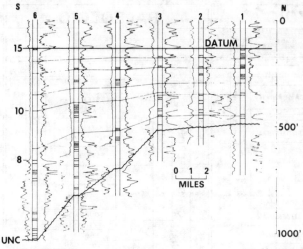

**Fig. 1-8.** Geologic cross section showing electric log correlations. Tertiary example, South America. Stippled pattern in wellbore represents sandstone. Key horizons indicated on left. From Vail, Todd, and Sangree, 1977, reprinted by permission.

to determine the validity of the correlation.

The mapping of an interval between two correlation datums developed within a stratigraphic continuum should show an interpretable pattern. Tectonic features, such as growth faults, basins, arches, and faulted sections, may be reflected in such a map; in addition depositional patterns related to topography, lithology, and source areas may be identified. These are changes that can be verified by comparison to other sections and other portions of the same section. For example, succeeding time-stratigraphic units should show similar patterns in relation to continuing tectonic or depositional processes. Long seismic sections integrate large volumes of stratigraphic data. They reflect the regional or continuing theme or process, and the local event or stratigraphic variation is lost in the array of discontinuous reflections (Fig. 1-8 and 1-9). One of the problems with correlation and the recognition of stratigraphic patterns is that unifying themes are difficult to extract from the mass of stratigraphic data. The seismic method acts as a filter with only stratigraphic units with three-dimensional continuity, on a scale of kilometers, recognizable in the data

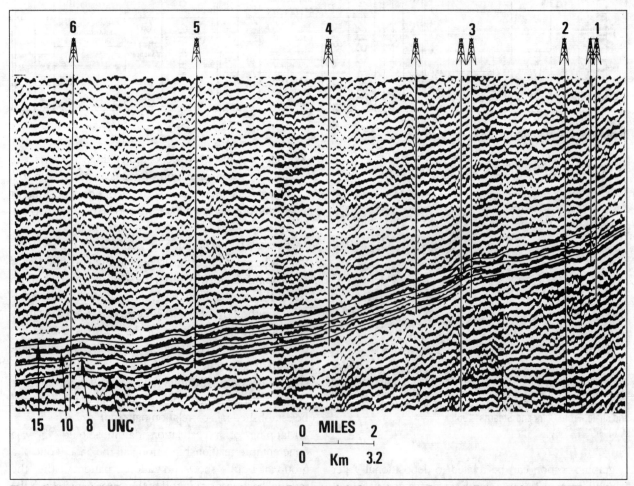

**Fig. 1-9.** Single-fold VDF magnetic tape seismic section from a Tertiary basin in South America. From Vail, Todd, and Sangree, 1977, reprinted by permission.

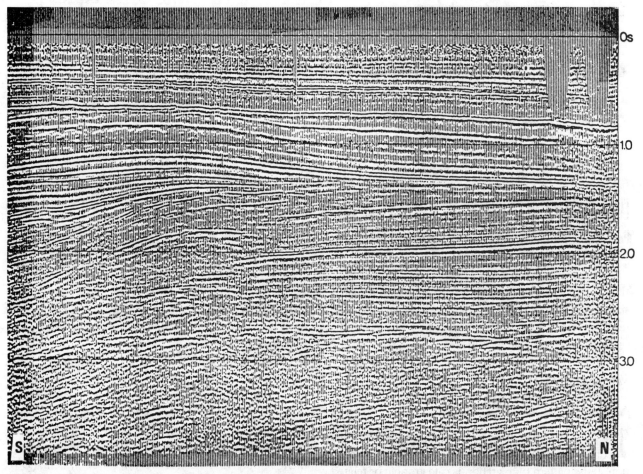

**Fig. 1–10.** Seismic cross section, Oldenburg overthrust zone, Northwest Germany. Reprinted courtesy Prakla Geophysical Co.

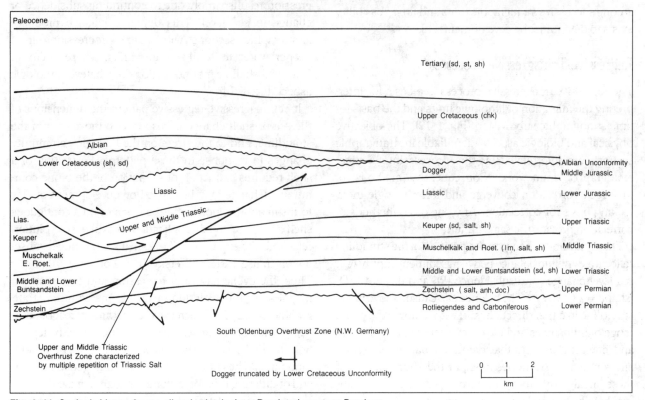

**Fig. 1–11.** Geologic history from well and seismic data. Reprinted courtesy Deminex.

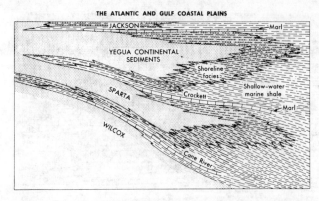

**Fig. 1-12.** Cross section showing the facies changes in the Eocene Claiborne Group of central Louisiana. The Sparta and Yegua sands are regressive deposits separated by marine shales that were laid down during transgressions of the sea. From Clark and Stearns, 1968.

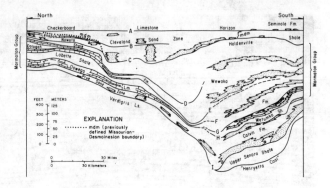

**Fig. 1-13.** Basin to shelf correlation is impossible if there is no theme. From Krumme, 1981.

(Fig. 1-10 and 1-11). Systematic variations of this magnitude are produced by regressive-transgressive, unconformity, and regional tectonic responses to specific unifying developmental histories.

# MIGRATION OF TIME-STRATIGRAPHIC UNITS

The systematic migration of lithologic and biologic zones is in response to continuing patterns of deposition. If the pattern of migration is catastrophic or related to nonrecurring events, there is no rational relationship among successive stratigraphic units and little possibility exists to determine the developmental history.

## Regressive-Transgressive Processes

The regressive-transgressive process is a basis for interpreting the migration of lithologic units and the basis for time-stratigraphic subdivision (Fig. 1-12). The objective physical and biologic aspects identifiable in stratigraphic sections are interrelated by this process into a time-stratigraphic framework. Consequently, methods of time-rock correlation converge and form a single interlocking basis for correlation. Any one or all methods of correlation provide the same pattern.

The continuous change in the position of the strandline reflects a dynamic balance between sediment supply ($Q$), sea-level changes and subsidence ($R$), and dispersal ($D$) [shape $= f(Q, R, D)$]. The balance can cause a move in the strandline in either direction, and with it moves the entire range of lithosomes and biozones related to the physical and chemical aspects that define all paralic and shelf environments. By the response of the shoreline, all of these areally distributed environments can be stratigraphically interrelated. Boundary conditions for a particular shelf or interior sea often do not change rap-

idly, and water depths, wave energy, supply of detritus, tidal effects, climate, ocean circulation, and subsidence or tectonic activity are relatively constant or change in a systematic manner.

Systematic patterns of subsidence, change in water depth, uplift in mountain areas, or climate changes can produce cyclic patterns of regression and transgression (Fig. 1-13). A delta builds outward (regression of the strandline) into progressively deeper water until it reaches an equilibrium related to receptor capacity, sediment supply and destructive forces. A systematic or ongoing process reflects a changed response, and a transgression results simply due to continuing subsidence or change in sea level. This dynamic balance produces stratigraphic sections with many progressions from deeper water to shallow-water units. These occur in nearly all shallow marine sections of whatever lithologic aspect (Fig. 1-14).

In a transgressive-regressive pattern the dimensions of the sea or shelf change in response to movement of the strandline, and to the stress on the biologic community changes in response to these patterns. Restriction of a shelf requires extinction or migration of the biotic community. Change may be caused by development of new forms in another portion of the basin or on a contiguous shelf area or by the diversification or explosive evolution related to the expansion of the shelf area during a subsequent transgression (Bretsky and Lorenz 1970, 544). Thus, the cyclic nature of regression and transgression may lead to the evolution of new forms within a stratigraphic section. Transgressive units are in response to subsidence or rising sea level catching up with deposition; thus, the entire shelf or basin area may be rapidly inundated. These units tend to be homogeneous, thin, and of differing lithologic character than regressive units. Their contained fauna also should have similar aspects and provide an opportunity to recognize new taxonomic

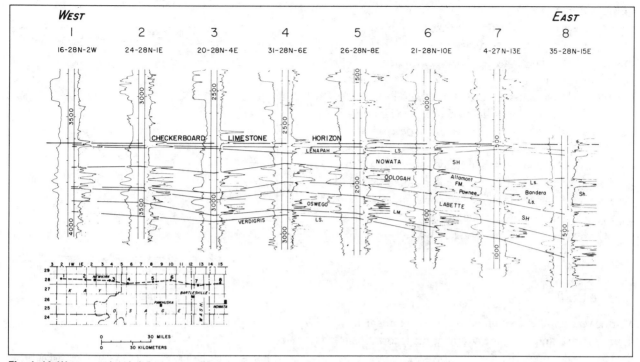

**Fig. 1-14.** West-east electric log cross section across northern platform, showing proximity of Checkerboard and Lenapah Limestones. From Krumme, 1981.

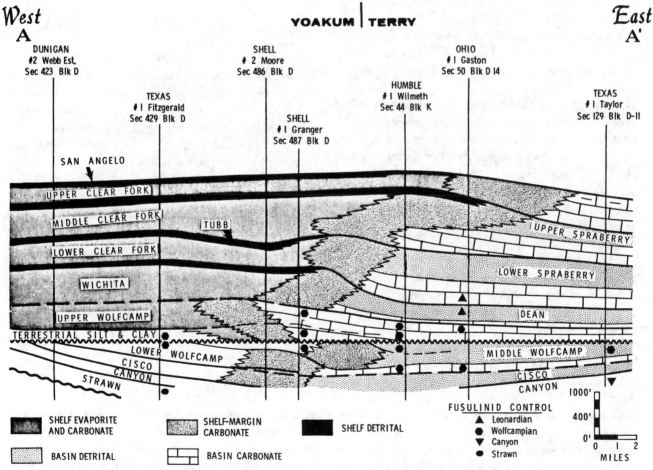

**Fig. 1-15.** Cross section depicting shelf to basin transition, northwest margin of the Midland basin. The basic stratigraphic generalities are explained by a two-stage multimodel. Facies (1), (3), and (4) are associated with high stands of the sea; facies (2) and (5) are associated with the intervening lowstands of the sea. Sediment types are as follows: (1) shelf evaporites and carbonates, (2) basin clastics, (3) shelf-margin carbonates, (4) basin carbonates, (5) shelf clastics. From Silver and Todd, 1969, reprinted by permission.

forms within a single rock-stratigraphic unit, thus providing a basis for time-stratigraphic correlation.

The regressive-transgressive pattern is only one type of migration of time-stratigraphic units, albeit a very common one in many stratigraphic sections. This pattern illustrates the process control for methods of time-stratigraphic correlation. The object of correlation is to place lithologic and biologic zones into a developmental framework. If the pattern can be recognized, then each transgressive break represents the boundary to another genetic unit and each unit represents a single period of development. Carbonate, delta, shoreline, and shelf units so divided represent decipherable periods of the earth's history (Silver and Todd 1969) (Fig. 1–15).

## Marine Units

Marine sections that contain no key beds or transgressive intervals are common in the stratigraphic record. If similar lithologic units form an unbroken stratigraphic sequence, as in sheet sandstones, alluvial regimes, or flysch basins, correlation on a local scale is often impossible. However, the entire unit can be treated as a whole, and areal changes provide the insight into paleogeography and to the unifying theme of deposition.

Of particular importance in these homogeneous units is the determination of the boundary conditions. These provide the framework for the processes that produce the depositional unit. The geometry of the depositional site controls in large measure time-stratigraphic, lithologic, and biostratigraphic patterns. Identifiable responses to changes in depth suggest the form of time stratigraphic subdivision of such units and the pattern of time parallel surfaces. Depth indicators include biofacies, physical aspects of the sedimentary unit, chemical responses, and patterns of bedding planes.

If bounding units can be established, the internal geometry of bedding planes may indicate the form of time-stratigraphic correlations. Bedding planes represent lithologic transitions, and as indicated, an array of subparallel bedding planes may approximate time parallelism (Fig. 1–16). This type of information is often available from reflection seismic sections. If the form of thickening and thinning in relation to time-stratigraphic datums can be ascertained, then maps can be constructed from seismic and log or outcrop information relating patterns of thickening of time-stratigraphic units to lithologic changes (Figs. 1–17 through 1–20). From this information, a basis may be derived for determining the historical development of the unit and the pattern of migration of time-stratigraphic units.

Stratigraphic associations that fit this particular pattern include the following:

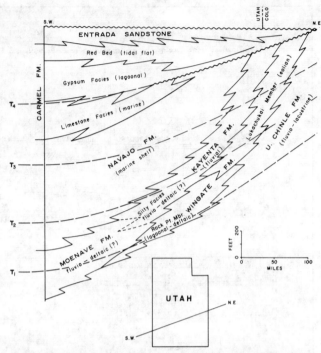

**Fig. 1–16.** Cross section showing stratal relationships and environmental facies changes within the Glen Canyon Group and Carmel-Navajo unconformity. From Freeman and Visher, 1975.

- Slope and fan deposits related to continental margins, marginal basins, and to deep cratonic basins;
- Volcanic arc-trench systems that include sections of volcanoclastics, submarine volcanics, chemical sediments, and canyon-continental rise associations;
- Evaporite sections dominated by a single mineral, for example, anhydrite and halite; and
- Blanket units (nearly continental in dimension) dominated by a single lithology, for example, sheet sandstone, black shales, and certain carbonate units.

These associations are developed in response to a continuing set of conditions where shorelines are not a part of the depositional pattern. Their absence may be due to marine transgression that nearly inundates emergent continental areas, water depths too great to respond to cyclic patterns of fill, or to a dynamic equilibrium that maintains the depositional pattern through an extended period of time.

## Continental Units

A differing set of conditions and processes prevails in the continental regime. Bounding surfaces are not nearly as important a control for interpretation and correlation. The interrelationship of eustatic changes, climate, and tectonics provides the basis for the depositional themes or patterns.

Limits of continental strata may be defined sequentially or areally, by unconformities or erosional intervals, or by

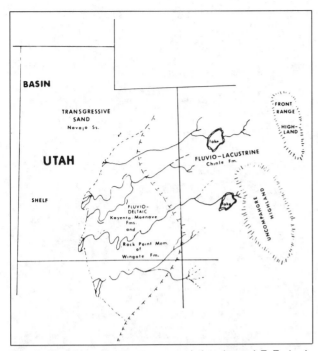

**Fig. 1–17.** Paleoenvironment: assumed time interval T₁-T₂ (early stage of transgression). From Freeman and Visher, 1975.

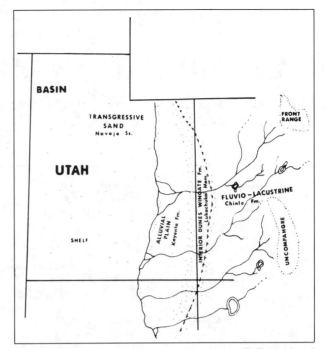

**Fig. 1–19.** Paleoenvironment: assumed time interval T₃-T₄ (continued transgression). From Freeman and Visher, 1975.

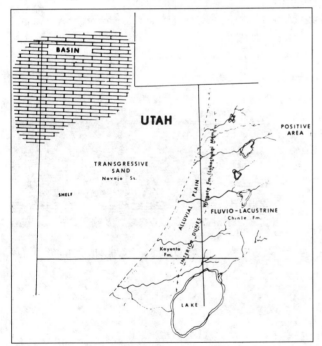

**Fig. 1–18.** Paleoenvironment: assumed time interval T₂-T₃ (continued transgression). From Freeman and Visher, 1975.

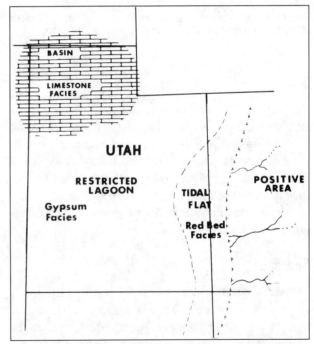

**Fig. 1–20.** Paleoenvironment: assumed time surface T₄ plus (time of deposition of Carmel Formation) (advanced stage of transgression). From Freeman and Visher, 1975.

marine transgressive units. In these cases the limits are not only defined geographically but temporally. Thus, the time-stratigraphic subdivision usually must start with the bounding units. Neither one of the bounding planes or units need necessarily be a time datum, and the recognition of such surfaces within a continental unit is usually impossible (Fig. 1–21).

Bedding planes do not offer a basis for identifying time datums; the areal variability of lithofacies is so great that lithologic transitions across bedding planes do not have a systematic regional pattern. Lithologic changes occur at a scale of less than a kilometer, and no topographic or depositional theme interrelates individual transitions.

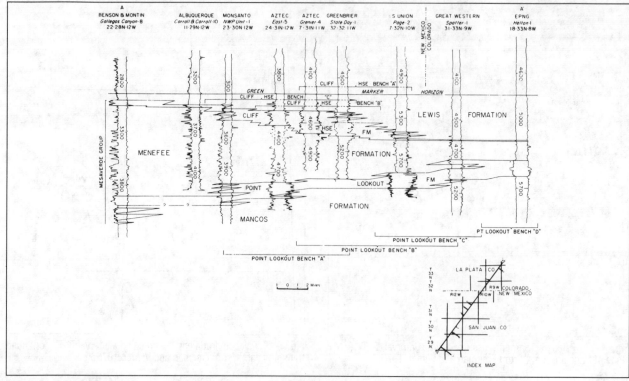

**Fig. 1–21.** Southwest-northwest cross section A-A¹ through the central part of the San Juan basin, showing stratigraphic relations of various Mesaverde sandstone units. From Hollenshead and Pritchard, 1961, reprinted by permission.

Therefore, the interpretation of a continental association requires an understanding of the dynamic relationship of the sedimentation to the structural or tectonic framework. The internal subdivision of these units on a time-stratigraphic basis is often not possible, but their three-dimensional distribution in relation to the tectonics provides the framework for a genetic interpretation (Figs. 1–22 and 1–23).

## Summary

Time-rock patterns are not simply based upon stratigraphic observations. The development of more and more detail in description does not suggest the framework for the areal or sequential patterns of stratigraphic units. Even at the level of observation and classification, the relationships within the data are not always obvious. Larger levels of data synthesis are required to determine the significance of individual observations. The basis of time correlation for any stratigraphic association may change as the boundary conditions and depositional processes change. Simply listing methods of correlation does not provide the insight necessary to develop a time-stratigraphic framework. However, if the processes or patterns of development are provided, then each observation of the biologic, lithologic, or sequential pattern is significant. Synthesis does not require greater levels of

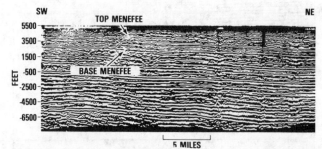

**Fig. 1–22.** Seismic section. From Vail, Todd, and Sangree, 1977, reprinted by permission.

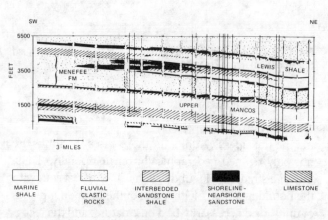

**Fig. 1–23.** Geologic section based on electric logs from the San Juan basin, New Mexico. From Vail, Todd, and Sangree, 1977, reprinted by permission.

description, but only a basis for the integration of observed facts.

Time relations in stratigraphy must necessarily be based upon interpretation, and this requires a dimension that is not available to the observer and recorder of local observations. If patterns can be developed that are useful in the interrelation of observations, then interpretation is possible; if no basis can be found for interrelating observations, a time basis for historical development cannot be constructed. In those cases each observation, measured section, or mapped unit must rest on an empirical basis, and a single observation can cast doubt on an interpretation. This is not the case if a developmental pattern, based on converging lines of evidence, is the basis for interpretation.

## UNCONFORMITIES

With the added precision in measurement of relative time from biostratigraphy, and measurement of absolute time based upon increasing numbers of radiometric age dates of stratigraphic events, discontinuities can be established with increasing areal and temporal precision. As the number of tie points increases, comparisons to magnetic reversals, events related to extinctions, and first and last floral and faunal occurrences, sea-level falls (Vail and Hardenbohl 1979) (and more recently, Haq et al. 1987), geologic events, and temperature changes measured in the marine biosphere (Fischer 1981) have led to increased understanding of the continuity of the stratigraphic record (Harland et al. 1982) (Tables 1–5 and 1–6).

Precision of time events is variable, with high confidence for much of the Cretaceous and Tertiary, but showing increased probable errors through the balance of the Phanerozoic Eon. Assumptions for some dates are related to rates of evolution, sea floor spreading, or sedimentation, while others are firmly established by multiple worldwide radiomentric dates. Some variance therefore is based upon assumptions, see, for example, Haq and Eysinga (1987), DNAG (Palmer 1983), and various authors using glauconites (Odin 1982). Correlation of varying types of events into a rigorous chronological series of stratigraphic ages is still not possible, but convergence in dating is steadily increasing.

From observations now available it can be seen that the stratigraphic record contains many small and large scale discontinuities on both a local and worldwide scales. Many authors are now attempting to place a periodicity into the various type events. Raup and Sepkoski (1986) suggest that extinction events have a periodicity of 26 million years. Rampino and Stothers (1984b), however, see

extinction events related to cometary or asteroid impact events at a periodicity approximating 32 million years (Fig. 1–24). They also suggest periodicities of 33 million years, including sea-level falls, sea floor spreading discontinuities, tectonic episodes, and impact craters greater than 10 kms in diameter (Rampino and Stothers 1984b) (Figs. 1–24, 1–25, 1–26).

Major stratigraphic discontinuities (see Table 1–5 and 1–6) do not appear to correlate well with the geologic record of impacts greater than 10 kilometers in diameter (Grieve 1982). However a complete record of impacts would not be recorded in the stratigraphic record, and the nature of their worldwide effects is still being debated (Waldrop 1988).

In this author's opinion it does not appear reasonable that all stratigraphic events can be interrelated into a grand scheme of catastrophism. Many events are very poorly dated, and some events result in complex stratigraphic responses, such as back-stripping of deposited stratigraphic intervals and periods of nondeposition. These may preclude the obtaining of precise temporal correlation of a specific event. Other stratigraphic discordances possibly may be related to sea floor spreading, plate tectonics, or glaciation, and would be developed over a period of millions of years rather than resulting from a specific event, such as a meteorite shower, or the solar system passing through the galactic plane (Rampino and Stothers 1984a). Certainly increased temporal resolution and more detailed analysis of unconformities in the stratigraphic record may indicate that periodicities do exist in the stratigraphic record, and new bases for the understanding of causality for unconformities can be entertained.

Unconformities have been classified in a descriptive manner, and it is useful to present the terminology to avoid confusion in use of terms (Table 1–7). This classification is not genetic, but the terms can be used to imply genesis in the framework of units presented if scales of area and time are added (Fig. 1–27). This table represents a possible descriptive framework and is useful in suggesting the nature and scale of observations required to identify unconformity surfaces. It is not single observations at one point in space or time that is the basis for analysis; it is the flow of changes represented by patterns of biofacies and lithofacies, evolution, erosion, nondeposition, specific lithosomes, structural relationships, and stratigraphic responses to continuing processes in underlying and overlying units that must be interrelated.

The analysis of the stratigraphic responses requires a detailed time-stratigraphic framework. The presence of an unconformity can only be defined by either missing time from an absolute or evolutionary reference or by

**Table 1-5**
**Uncomformity Chart, With Age Dates**

| Time M.Y. | Unconformity History | Harland Age[1] | Haq Age[2] | S. Level Falls[3] | Orogenies[4] | Flood Basalts[5] | Impacts >10kms[6] | Rapino & Stothers[7] | Raup and Sepkoski[8] | Fischer & Arthur[9] | Plate Tectonics |
|---|---|---|---|---|---|---|---|---|---|---|---|
| 0 | B. Messian | 6.3 | 6.3 | | | | 1.3±.2 3.5±.5 4.5±.5 | | | | |
| 10 | B. Tortonian | 11.3 | 10.2 | 10.5 | Styrian 10-14 | | 10 14.8±.7 | 11.0 | 11.3 | | Orogeny |
|  | B. Burdegalian | 19 | 20 | 21 | | 17±1 | 15 | | | | |
| 20 | | | | | | | | | | | |
| 30 | B. Chattian | 32.8 | 30 | 30 | | 35±2 | 38.4± 39±9 | 37 | 38 | 30 | Spreading |
| 40 | B.U. Eocene | 42 | 39.4 | 39.5 | Pyrean 40-44 | | | | | | Orogeny |
| 50 | Post Ypresian | 50.5 | 49 | 49.5 | | | 57 | | | | Spreading |
| 60 | B. Thanetian | 60.2 | 60.2 | 58.5 | Laramide 60-65 | 62±2 | 65 | 66 | 65 | 62 | Orogeny |
|  | T. Cretaceous | 65 | 66.5 | 67.5 | | 66±2 | <70 | | | | |
| 70 | B. Maastrich. | 73 | 74 | | | | 77±4 | | | | Spreading |
| 80 | M. Campanian | | | 80 | | | | | | | |
| 90 | U. Turonian | | 98 | 90 | Austrian Oregonian 91-96 | | 95±7 | 91 | 91 | 94 | Orogeny |
|  | M. Cenomanian | | | 94 | | | | | | | |
| 100 | U. Albian | | | | | | 100 100±5 100±20 100±50 | | | | |
| 110 | L. Aptian | | 112 | 112 | | 110±5 | 100 100±5 100±20 100±50 | | 113 | | Spreading |
| 120 | L. Valanginian | 138 | 128 | 126 | | 130±5 | <120 | | 125? | 126 | |
|  | B. Valanginian | | 136 | 128.5 | | 135±5 | 130±6 | | | | |
|  | B. Portlandian | | | 136&138 | | | | | | | |
| 140 | U. Kimmeridg. U. Bathonian | | | 158.5 | L. Kimmerian 140-145 | | | 144 | 144 | 158 | Orogeny Spreading |
| 160 | L. Bajocian | | | 169 | M. Kimmerian 170-175 | 170±5 | 160±5 | 176 | 163? 175? | | Orogeny |
| 180 | L. Aalenian T. Sinnemurian | 200±10 | 194 | 177 | E. Kimmerian | 190±5 | 183±3 | 193 | 194 | 190 | Spreading Orogeny |
| 200 | B. Sinnemurian | 206 | 201 | 202 | 195-200 Hardgessen | 200±5 | 200±100 | 217 | 213 | | Spreading |
|  | B. Rhetian | 219 | 215 | 215 | | | 210±4 | | 219 | | |
| 220 | B. Norian | 225 | 223 | 224 | 219-225 | | 230 | | 225 | 222 | Orogeny Spreading |
|  | B. Carnian | 231 | 231 | 232 | | | | | | | |
|  | L. Anisian | | 239 | 237 | | | | | | | |
| 240 | B. Tatarian | 253 | 255 | 255- | 246-248 Palatinian Appalachian | 250±10 | <250 | 245 | 243 248-253 | | Orogeny |
| 260 | | | | 260 | | | | | | | Spreading |

References:

1) Harland, 1982
2) Haq & Van Eysinga, 1987
3) Haq, et al, 1987
4) Stille, 1936
5) Rapino and Stothers, 1988
6) Grieve, 1982
7) Rapino and Stothers, 1984b
8) Raup and Sepkoski, 1986
9) Fischer and Arthur, 1977

**Table 1-6**

| Time M.Y. | Unconformity | Harland | S.L falls | Orogenies | Impacts[6] | Extinctions | Plates |
|---|---|---|---|---|---|---|---|
| 250 | Pre-Late Permian | 258 | 255-260 | Hercynian 275-0280 | | 248-253 | Orogeny Spread. |
| | Mid.-Wolf. | 275-280 | | Asturian 286-295 | 290 20 300 | 286-295 | Orogeny |
| 290 | Post-Atoka<br>B. Penna. | 315<br>320 | 310<br>320-5 | Erzgebir. | | | Spread.<br>Orogeny |
| 330 | | | 330-5 | Sudetian 345 | 350 | | Spread. |
| | B. Carbon. | 360 | 360- | Acadian | 360 10 | 360 | Orogeny |
| 370 | | | 365 | Antler<br>Erian<br>395 | 360 25<br>365 7 | | Spread.<br>Orogeny |
| | B. Emsian | | 394 | 395 | | | |
| 410 | | | | | | | Spread. |
| | B. Silurian | 438 | 435-440 | Taconic 435 | | 438 | Orogeny |
| 450 | B. Mid. Ord. | 480- | 470 | Sardinian 485 50<br>470 | 450 | Spread. | Spread |
| | | | 480- | | 485 50 | | Orogeny<br>Spread. |
| 490 | | | 490 | | | 505 | |

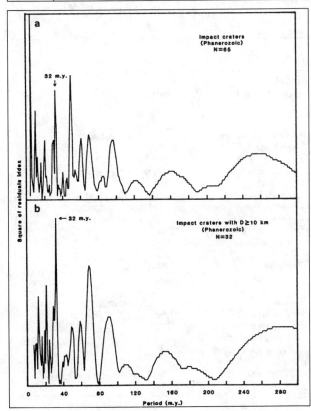

**Fig. 1–24.** Terrestrial impact craters: the square of the residuals index as a measure of goodness of fit for various trial periods. The unmarked high spectral peaks at periods of 5, 10, and 50 m.y. are artifacts due to rounding of the ages of many of the older craters. *a,* Impact craters; *b,* Impact craters with D > 10 km. From Rampino and Stothers, 1984b.

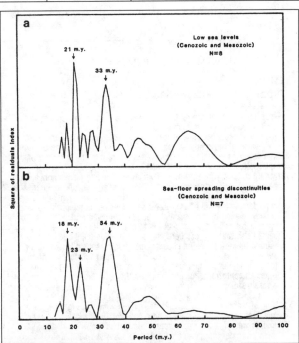

**Fig. 1–25.** Marine phenomena: the square of the residuals index as a measure of goodness of fit for various trial periods. *a,* Low sea levels; *b,* Sea-floor spreading discontinuities. From Rampino and Stothers, 1984b.

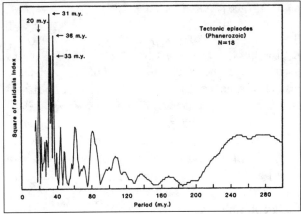

**Fig. 1–26.** Tectonic episodes: the square of the residuals index as a measure of goodness of fit for various trial periods. From Rampino and Stothers, 1984b.

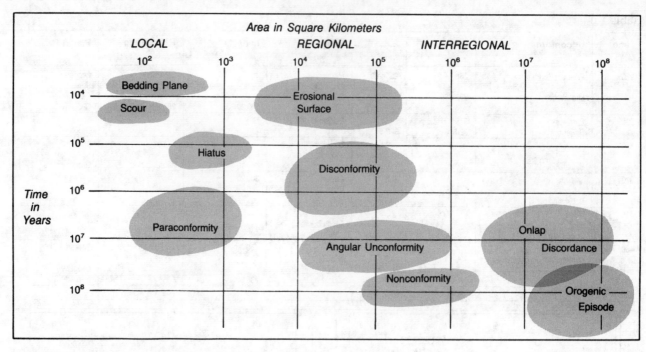

**Fig. 1–27.** Unconformities—Scale of Area and Time

**Table 1-7**
**Types of Unconformities**

**Hiatus** (a) A break or interruption in the continuity of the geologic records, such as the absence in a stratigraphic sequence of rocks that would normally be present but either were never deposited or were eroded before deposition of the overlying beds. (b) A lapse in time, such as the time interval not represented by rocks at an unconformity; the *time value* of an episode of nondeposition or of nondeposition and erosion together.

**Disconformity** An unconformity in which the bedding planes above and below the break are essentially parallel, indicating a significant interruption in the orderly sequence of sedimentary rocks, generally by a considerable interval of erosion (or sometimes of nondeposition), and usually marked by a visible and irregular or uneven erosion surface of appreciable relief. The tendency is to apply the term to breaks represented elsewhere by rock units of at least formation rank.

**Erosional Unconformity** An unconformity made manifest by erosion or a surface that separates older rocks that have been subjected to erosion from younger sediments that cover them; specifically *disconformity*.

**Angular Unconformity** An unconformity between two groups of rocks whose bedding planes are not parallel or in which the older, underlying rocks dip at a different angle (usually steeper) than the younger, overlying strata; specifically an unconformity in which younger sediments rest upon the eroded surface of tilted or folded older rocks.

**Nonconformity** (a) An unconformity developed between sedimentary rocks and older rocks (plutonic igneous or massive metamorphic rocks) that had been exposed to erosion before the overlying sediments covered them.

**Onlap Unconformity** A surface of nondeposition that is progressively onlapped by successively younger stratigraphic units.

*Glossary of Geologic Terms, Bates, 1987

**Table 1-8**
**Modern Sedimentation Rates**

| | μmm/year |
|---|---|
| Lake Vierwaldstätter[1] | 3,500-5,000 freshwater |
| Lake Lunz[1] | 1,800 |
| Rhone delta | 700 delta |
| Nile delta | 660 |
| Clyde Sea[2] (shallow) | 5,000 largely terrigenous |
| Norwegian fjord[2] | 1,500 |
| Gulf of California[2] | 1,000 |
| Moluccas[2] (volcanic ash) | 700 |
| Tyrrhenian Sea[2] | 100-500 inland seas |
| Black Sea | 200 |
| Bahamas[3] | 33.8 carbonate environments |
| Florida Keys[3] | 80 |
| Florida inner reef tract[3] (contaminated by terrigenous material) | 220 |
| Globigerina ooze[2] | 8-14 deep sea |
| Red clay[2] | 7-13 |

(1) Quoted after Schwarzacher (1946)  (3) Stockmann et al. (1967)
(2) Quoted after Kuenen (1950)       From Schwarzacher, 1979

missing time-stratigraphic datums as defined by temporal events. If the absolute or relative time scale available is very coarse, the detail of the unconformity analysis must also be general and the local or short-term disruption in the stratigraphic record cannot be recognized. In many stratigraphic sequences the pattern of sedimentation allows for a time-rock subdivision on a level of hundreds or possibly tens of thousands of years. Where this is possible, the detailed analysis of unconformity surfaces provides a tool useful in developing a dynamic stratigraphic history.

## Stratigraphic Framework

Sedimentation rates and both areal and temporal dimensions are important in interpreting unconformities within the stratigraphic record (Table 1–8). Analysis of these aspects provides an insight into stratigraphic processes and responses represented by discontinuity surfaces. These boundary conditions are developed and applied to differing paleogeographic settings as a basis for stratigraphic interpretations.

**Table 1-9**
**Ancient Sedimentation Rates**

µmm/year

| | Duration (m.y.) | Maximal America | Maximal Europe | Effective maximum | Shelf |
|---|---|---|---|---|---|
| Cambrian | 100 | 86 | 55 | 40 | 15 |
| Ordovician | 60 | 147 | 77 | 66 | 25 |
| Silurian | 40 | 49 | 154 | 113 | 30 |
| Devonian | 50 | 78 | 314* | 160 | 32 |
| Lower Carboniferous | 40 | 51 | 88 | 50 | 13 |
| Upper Carboniferous | 40 | 188 | 210 | 150 | 13 |
| Permian | 45 | 62 | 224* | 100 | 23 |
| Triassic | 45 | 178 | 140 | 67 | 43 |
| Jurassic | 45 | 152 | 111 | 67 | 33 |
| Cretaceous | 65 | 354 | 230* | 230 | 43 |
| Paleogene | 45 | 236 | 224 | 133 | 31 |
| Neogene | 23 | 533 | 533* | 226 | 47 |
| Weighted average | | 160.9 | 171.5 | 108.9 | 27.7 |

(1) Caledonian, Variscian, and Alpine orogenic periods
From Schwarzacher, 1979

## Rates of Sedimentation

The preserved stratigraphic record, representing deposition on continental shelves and in epeiric seas in the Holocene, is often in orders of magnitude thicker than for similar environmental conditions from ancient stratigraphic units (Schwarzacher 1975, Table 1–9). Comparable units sedimentation rates are found only in marginal basins, "geosynclines," and rarely in cratonic basins. The discrepancy must lie in the presence of unconformities in the shelf and epeiric sea sections. Studies of these stratigraphic units indicate that the evolutionary record of the biota is fragmentary, with few genetic lineages preserved (Hedburg 1961, 503). This suggests that more time is missing in a section than is preserved (Shaw 1964). Also, lithofacies analysis, or migration of lithostratigraphic units within extended spans of time (tens of millions of years), shows coherent lateral and vertical patterns (Krumbein and Sloss 1963, 299). From these two observations, it can be concluded that a unifying theme or process is responsible for the development of the stratigraphic section, but a periodicity in sedimentation also is required. Erosion can account for missing time-stratigraphic units, but this requires resedimentation with no obvious depositional site available in shelf and epeiric sea environments. Non-uniformitarian concepts may be invoked to explain differences, but observation of systematic stratal relationships suggests a process control for sedimentation.

The single most important aspect of Holocene sedimentary environments is the variability in sedimentation rates. In clastic regimes, regressions and transgressions are documented that interrelate areas of no sediment accumulation with those of rapid deposition. Sediment is trapped near the shoreline, and thousands of square kilometers of shelf receive little or no sediment. Thus, the simple process of regression across the shelf produces a blanket of neoshore marine sediments. In contrast, little sedimentation occurs in areas and during periods of open shelf development (Curray 1960; Schwarzacher 1975). Cycles of high and low rates of sediment accumulation result, and the dynamic process of regression and transgression produces both rational lithologic patterns and gaps in the time-stratigraphic record. Similar patterns develop in carbonate terrains with the added possibility of solution and removal of strata during subaerial exposure and by abrasion and removal of clastic particles from the shelf environment (Ginsburg 1971).

Regional unconformities are identified by mapping three-dimensional discontinuity surfaces and demonstrating missing time-stratigraphic units and events. If major stratigraphic relationships are involved, at least regional continuity of the unconformity surface must be demonstrated. If the scale of missing time is measured in millions of years, then that missing interval must be accounted for by a historic pattern or event.

Surfaces reflecting shorter periods of missing stratigraphic history or of local extent (Table 1–5) can only be recognized by cross sections, detailed stratigraphic maps, or disruption in the sequence of depositional units.

## Summary

Boundary conditions with respect to rates of sedimentation and areal and temporal patterns illustrate the importance of unconformities in the stratigraphic record. The depositional history can only be inferred from the preserved stratigraphic record and missing periods must be identified from aspects of that record before a developmental analysis is possible. Time is continuous, and for history to be complete, a stratigraphic continuity must be developed in terms of events, rates, and processes.

# STRATIGRAPHIC APPLICATIONS

A hierarchical classification of possible successively larger levels of disruption of the stratigraphic continuum can be used as a basis for discussion of unconformities. Such a sequence includes:

▸ Episodic events,
▸ Depositional cycles,
▸ Stratigraphic lithozones,
▸ Onlap discordances, and
▸ Worldwide tectonic events.

At each level differing measurements need to be applied, boundary conditions defined, and processes evaluated. Thus, observation can be used to suggest both the level of discordance and the process relationship. Also, the nature of an unconformity surface may be evaluated with reference to the classification of unconformity types (Fig. 1–27). From these two differing aspects, one descriptive of time and area and the other related to stratigraphic units bounded by the surface of discontinuity, a genetic approach to unconformity analysis is possible.

## Episodic Events

It is possible that as much time is represented by local discontinuities as by the stratigraphic record. As the time stratigraphic framework is refined, additional insights concerning rates of evolution, sedimentation, and erosion can be obtained to interpret these local discontinuities.

## Depositional Cycles

Discontinuity surfaces can be established by changes in systematic lithologic sequences. Many rhythmic or identifiable lithologic patterns may be identified both areally and vertically. They represent a response to changing boundary or local process controls. Sedimentary sequences are composed of dynamically interrelated lithologic units, and the bounding surfaces represent the close of a continuing depositional process. Changes in geometry of the depositional site (water depth, restriction, etc.), provenance, water chemistry, or some other active process control produce the break in sedimentation.

Areal units of the dimension indicated are suggested by the classification (Fig. 1–27) to be bounded by surfaces representing missing time of from 10,000 to several hundred thousand years. A time dimension of this magnitude does not allow the use of paleontologic evolutionary zones, but stratigraphic patterns are present, and missing units can be mapped.

## Stratigraphic Lithozones

Widespread lithozones or patterns bounded by key beds or datums are easily identified and correlated. They may be seen on seismic profiles and petrophysical log sections, and represent patterns of regression and transgression in shelf areas and depositional cycles in marine basins and are identifiable in continental regimes by progressive changes in lithologic or even biologic aspect (Thayer 1974, 141) (Figs. 1–28, 1–29, and 1–30). They represent stratigraphic responses to tectonic controls. The bounding surfaces are formed in response to tectonic or eustatic controls. Units are measured in hun-

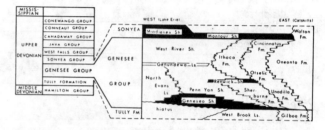

**Fig. 1–28.** Stratigraphic units of the New York Upper Devonian. Vertical axis is time. From Thayer, 1974.

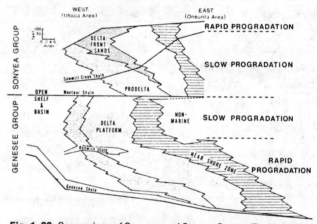

**Fig. 1–29.** Comparison of Genesee and Sonyea Groups/Environment. From Thayer, 1974.

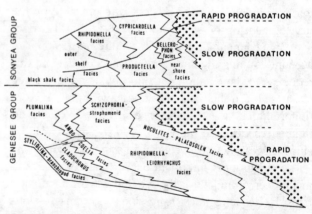

**Fig. 1–30.** Faunal facies. Dashed lines on right mark periods of transgression. Datum is base of Montour shale. From Thayer, 1974.

dreds of meters and represent complex interrelationships between depositional environments and processes, dispersal centers and patterns, and tectonic elements.

Bounding erosional or paraconformable surfaces are formed in response to tectonic or eustatic controls (Figs. 1–10 and 1–11). They are widespread with limits defined by the tectonic or physical boundaries of the area. Stratigraphic relations involving onlap or offlap, convergence or divergence, or erosional discordance define these surfaces. Angular relationships may be seen on a regional basis, and locally significant differences in faunal and floral content may suggest "missing time." The classification suggests "missing time" from possibly 1 to 10 million years (Fig. 1–27).

## Onlap Discordances

The next level of unconformities separates the largest level of coherent depositional patterns. The single most important process identifiable in the stratigraphic record is the response of sedimentation to systematic eustatic or tectonic patterns extending over periods ranging from 10 to 100 million years. The pattern most often exemplified is continent-wide patterns of offlap and onlap. The stratigraphic response is identified by large-scale change of units from marine to continental, followed by erosion and then by systematic marine onlap across exposed cratonic surfaces. Changes in continental relief, provenance area, lithology, water depth, and biota are identifiable. Stratigraphic aspects are treated in detail in the section on sequences.

Unconformities represent areas from 100,000 to possibly 10 million square kilometers. The smaller area represents onlap across an orogenic zone, and the latter represents a craton-wide unconformity and onlap. Relief on the unconformity surface may range from a few hundred meters to possibly 1,000 meters. The sedimentation pattern is controlled by the topographic and tectonic aspects reflected at the unconformity surface. Source areas, dispersal patterns, lithologic characteristics, thicknesses, and climatic changes are responses to both boundary conditions and systematic changes in sea level and tectonism.

## Worldwide Tectonic Events

Radiometric ages for mountain-building periods and major unconformity surfaces have a general correlation throughout the world. Certainly orogenic epochs, as evidenced by the differing ages of granites, thrusting, and mountain building, often result in nearly continuous tectonic activity, but the work by paleontologists in separating the eras, the pattern of onlap unconformities, and

newer concepts of plate tectonics are useful in subdividing the stratigraphic record.

## Unconformity Analysis

Relationships between unconformity surfaces and overlying and underlying stratigraphic units are useful in interpreting depositional history. Maps showing areal and temporal patterns of stratigraphic units can indicate topography, tectonic events, and offlap-onlap relationships. These aspects can be determined by specific time-stratigraphic maps. For example:

- ▸ Directions and magnitude of onlap are related to time-stratigraphic datums overlying unconformity surfaces;
- ▸ Topography on an unconformity surface can be determined by thickness variations between an overlying time-rock datum and an unconformity;
- ▸ Paleogeographic patterns can be related to lithology of onlapping strata;
- ▸ Time-stratigraphic relationships can be determined by offlap and onlap time-rock datums;
- ▸ Regional tectonic patterns may be determined by the relationship of thickness variations between the unconformity surface and both underlying and overlying time-stratigraphic datums;
- ▸ Tectonic history is related to the pattern of time-stratigraphic units below the unconformity surface; and
- ▸ Erosion and tectonism are related to thickness variations between the unconformity surface and a lower time-stratigraphic datum.

Direct observation provides the basis for these maps, but their interpretation requires the understanding of specific process-response patterns. Unconformity analysis may be subdivided into three general categories: (1) relationships between the unconformity and overlying units; (2) stratigraphic relationships across an unconformity surface; and (3) time-stratigraphic relationships between the unconformity surface and underlying units.

Paleogeography also may be interpreted from lithologic aspects of units overlying an unconformity surface. Unusual lithologic units developed in relation to unconformity surfaces reflect possible soil zones, paraconformities, unusual environments, missing time, and unique historical events. In addition, lithologic maps of units overlying an unconformity are useful tools in stratigraphic analysis. These all are interpretable in terms of paleogeographic history of an unconformity surface.

## Paleogeographic Aspects

The unconformity surface provides a reference datum for reconstructing paleogeographic patterns (Andressen

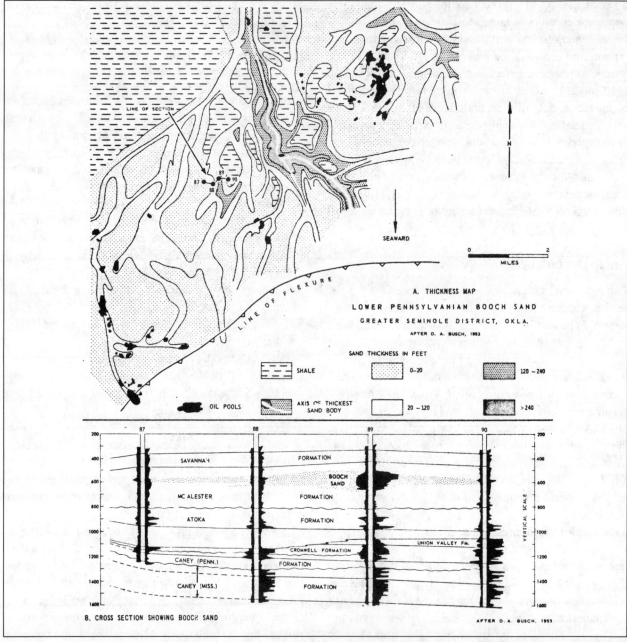

**Fig. 1-31.** *a*, thickness of the Booch sandstone; *b*, cross section of related Pennsylvania strata in central Oklahoma. From Busch, 1953, reprinted by permission.

1962; Martin 1966; Busch 1974) (Fig. 1-31). Topography on an unconformity surface can be determined if a horizontal time-rock datum can be identified close to the unconformity surface. The identification of a horizontal datum is often difficult, but the shoreline represents a horizontal datum, and marine depositional events are often subparallel to the sea surface or ancient shorelines. Also, if thin transgressive lithologic units are present, these often are subparallel to a horizontal datum (Busch 1974).

Rates of transgression and the regional shape of the unconformity surface may be established by mapping successive shoreline positions (Melton 1968; Chenoweth 1967). The identification of time-rock datums and mapping of their termination at a shoreline suggests a process response to eustatic and subsidence variation (Figs. 1-32, 1-33, 1-34, and 1-35).

## Temporal Patterns

The temporal relationship of the unconformity surface to the depositional history can be interpreted by combined lithostratigraphic and biostratigraphic sections (Figs. 1-29 and 1-30). Integration of both types of information

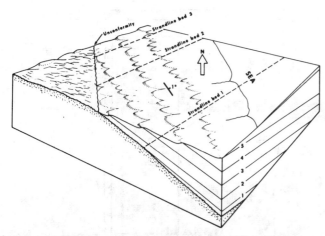

**Fig. 1-32.** This represents block diagram after a differential tilt, sea withdrawal, and some erosion. Tilting, as shown by dip-and-strike symbol, was toward far right corner of block. Erosion has exposed all key beds except number 1, but outcrops in this case intersect unconformity at small angle rather than paralleling it. Strandlines of key beds, established during marine transgression, parallel left and right edges of block and are shown partially by dashed lines. From Chenoweth, 1967, reprinted by permission.

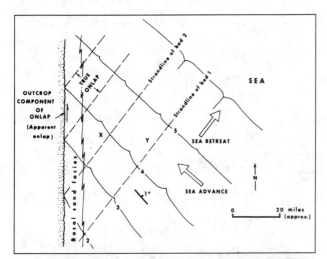

**Fig. 1-33.** Geologic map and "worm's eye" map of block diagram. Wavy lines indicate outcrop of key beds; dashed lines show shoreward limits at unconformity. Outcrops diverge slightly basinward (toward lower right) and cross facies boundaries landward. True direction and magnitude of onlap, shown in subsurface by double-headed arrow, are toward upper left. A component of onlap appears at unconformity where successively younger strata onlap toward top of map. From Chenoweth, 1967, reprinted by permission.

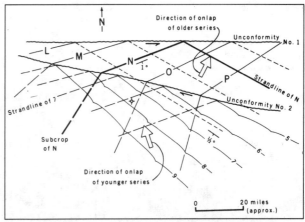

**Fig. 1-34.** Analysis of two overlapping series. Older strata, indicated by letter lines *L-P* were deposited by sea advancing northeast across area, onlapping unconformity *No. 1*. These beds were tilted southeast and eroded to sea level. A second transgression followed with sea advancing northwest. Numbered beds overstep unconformity *No. 2*, which was developed on truncated edges of older series. Later tilting southwest and erosion produced present pattern. Shown are: strandlines of each series on two unconformities ("worm's eye" part of the map), subcrops of older lettered series beneath unconformity *No. 2* (subcrop part), and outcrops of both series (geologic map part). Dashes are used for all subsurface lines. Shown also are directions of onlap and outcrop components of onlap (half arrows). From Chenoweth, 1967, reprinted by permission.

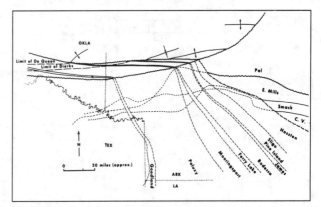

**Fig. 1-35.** From Chenoweth, 1967, reprinted by permission.

allows the interpretation of the developmental history. These independent lines of observation are correlated, and three-dimensional observational patterns interrelate the data in a dynamic process. The most useful method for this analysis is the determination of the position where the unconformity surface disappears into a depositional continuum and the tracing of the unconformity from this line progressively into areas of more profound discordance or missing stratigraphic intervals.

Rarely in the stratigraphic literature have unconformity surfaces been studied with regard to the various mappable aspects. For this reason few unconformities

reported in stratigraphic sections have interpretable eustatic, tectonic, and paleogeographic conditions prior to, during, and after their development. A case in point is the analysis of unconformities in the Silurian-Devonian section of central and southern Oklahoma (Shannon 1962; Amsden 1962; Boucot 1962) (Figs. 1-36, 1-37, and 1-38). This interval has been particularly well studied because of the availability of paleontologic, outcrop, and subsurface information. The subsurface Hunton Formation is an important economic objective in the exploration for hydrocarbons, and the relationship of the stratigraphic units to unconformity surfaces appears to be an impor-

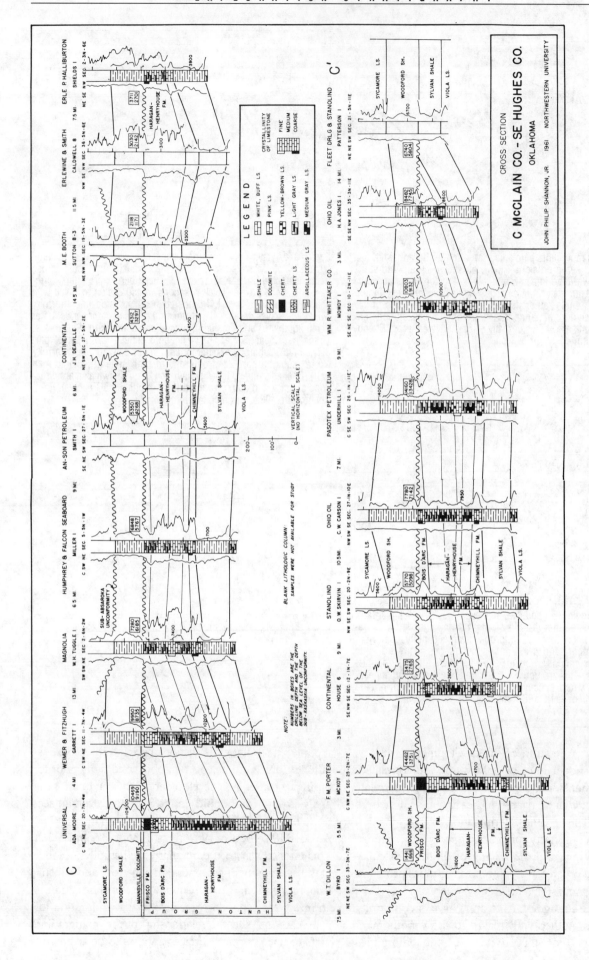

**Fig. 1-36.** Cross section *C-C'* of central McClain and southeast Hughes counties. From Shannon, 1962, reprinted by permission.

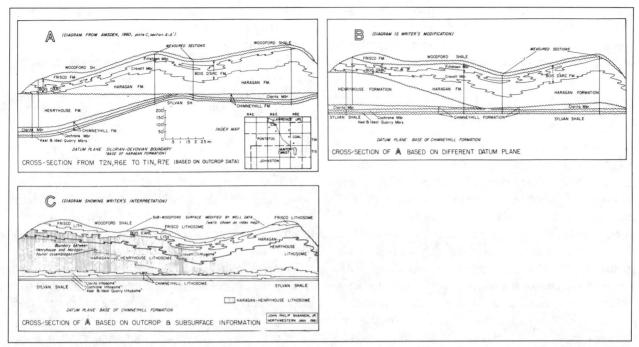

**Fig. 1–37.** Comparison of Amsden's and Shannon's interpretation of stratigraphic relations of Hunton Group in outcrop. From Shannon, 1962, reprinted by permission.

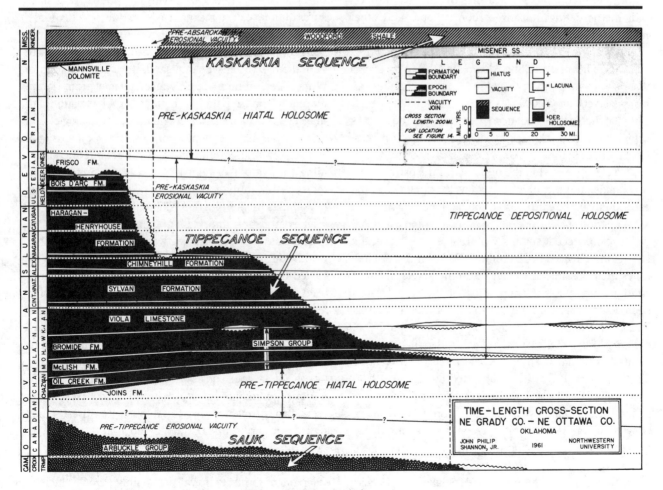

**Fig. 1–38.** Unconformity analysis. From Shannon, 1962, reprinted by permission.

tant control for the development of porosity. If all the stratigraphic approaches to unconformity analysis were applied to this problem, a single objective interpretation using all lines of evidence might resolve the stratigraphic controversy.

# SEQUENCES

Unconformities occur at many differing temporal and areal scales. Causality for these stratigraphic breaks has been the focus of continuing research. Worldwide disruptions recognized by sea level falls, faunal and floral extinctions, and depositional events have been suggested to be related to plate tectonics, and possibly periodic catastrophies. Within basins a differing temporal scale appears to control stratigraphic response patterns. Disconformities, erosional surfaces, litho- and bio-facies patterns, and bedding sequences show remarkable periodicities. Some of these may be related to sea-level changes measured on a scale of hundreds of thousands or millions of years, but others appear to reflect cycles that can be measured on a scale of tens of thousands of years.

Increased temporal precision related to magnetostratigraphy, and more precise measures of time, based upon radiometric dating utilizing carbon isotopes and other radionuclides, has been correlated to oxygen isotope changes in the tests of calcareous and siliceous microfossils (Hays, Imbrie, and Shackleton 1976). Results of this work show that periodicities of temperature changes can be observed at 21,000, 41,000, and 100,000 years. Milankovitch suggested from variations in the earth's orbit, that the 21,000 year cycle correlated with precession changes in the earth's rotational axis, that the 41,000 year cycle correlated to the obliquity of the earth's rotational axis, and the 100,000 year cycle correlated to the eccentricity of the earth's orbit. Interactions may produce longer period cycles, for example a 135,000 and a 400,000 year cycle. A recent summary of these events within the stratigraphic record by Arthur and Garrison (1986) illustrates their possible causal effects in the development of stratigraphic cycles. Confirmation of these cycles has been demonstrated for the late Triassic and early Jurassic lacustrine sequences of the Newark supergroup of eastern North America (Olsen 1986), and for marine sediments of a similar age from the south coast of England, a 40,000 year cycle has been suggested by House (1986). Grotzinger (1986) suggests similar cycles may be identified in rocks 2 billion years old from northwest Canada. How temperature changes may produce stratigraphic cycles in widely contrasting sedimentary environments is unknown, but possibly storm cycles may change rates of sediment dispersion, supply, and sea or lake levels in near shore depositional systems, or relate to rates of biogenic productivity in deeper water systems.

The identification of possible causal mechanisms for stratigraphic cycles has resulted in increased emphasis on allocyclic interpretation of stratigraphic responses. A functional relationship between sedimentation rates, controls related to dispersion of sediment, and to the interrelation of sea level changes, erosion, and uplift or subsidence, can also be invoked to produce repetitive autocyclic stratigraphic sequences.

Objective study of carbonate cycles can be related to tidal range, shelf slopes, rates of sedimentation, and rates of progradation (Bova and Read 1987). Sequences related to varying amplitudes of sea-level changes can be correlated to described stratigraphic sequences. Computer modeling of repetitive sequences requires assumptions relating to rates of subsidence, and the amplitude of sea-level oscillations (Read, Grotzinger, Bova, and Koerschner 1986). These a priori assumptions are predicated on an allocyclic control, but similar patterns are possible utilizing differing controls for time periods of disconformities, rates of relative sea level rise, and sedimentation rates. Computer modeling may match observational patterns, but without independent measures of time, causal mechanisms are only speculation.

Areal and temporal scales for cycles have been demonstrated on basinal and worldwide scales, and with periodicities of tens of thousands to a few millions of years. Determining causal mechanisms from stratigraphic response patterns requires the observation and interrelation of climatic, geochemical, mineralogic, and many other independent variables. Four dimensional analysis of stratigraphic intervals are required to determine the pattern of depositional discontinuities. With this type of information it may be possible to test various hypotheses to interpret depositional patterns.

One of the most important unifying principles in stratigraphy has involved the objective identification of similar worldwide time-stratigraphic patterns. Ager (1981) presented many of the stratigraphic observations that have been unrelated to a unifying concept or stratigraphic responses. Worldwide similarities of Cambrian sandstones, Ordovician and Silurian carbonates, Devonian reefs, Carboniferous limestone transitional into fluvial and deltaic sandstones, Triassic redbeds, Jurassic limestones, Cretaceous shales and chalk, and Tertiary clastic associations require an explanation (Table 1–10). It has been demonstrated in North America that these lithologic associations are bounded by unconformities (Sloss 1949; Sloss 1963; Wheeler 1963). These North American unconformities were related to a worldwide pattern of onlap unconformities by Vail and Wilbur (1966). Much of

**Table 1-10**
**Vertical Sections of Lithologies**

| | Cratonic | | Shelf | | Intrashelf Basin | | Deep Water Basins | |
|---|---|---|---|---|---|---|---|---|
| | Clastic Supply: | | Clastic Supply: | | Clastic Supply: | | Clastic Supply: | |
| | High | Low | High | Low | High | Low | High | Low |
| Limit of transgression | Carbonaceous shales | Playa carbonates | Black shale | Sabkha | Biogenic shales | Chert chalk | Biogenic shale | Chert |
| Onlapping units | Riverine deltas | Molluscan carbonates Dolomites | Deltas Algal carbonates | Bioherms Molluscan carbonates | Submarine fans | Reefs Argillaceous micrite | Volcano-clastic turbidites | Carbonate gravity flows |
| Basal units | Stacked fluvial channels | Silty dolomite | Strike valley sandstones | Sheet sandstones | Submarine fans | Debris Flows | Turbidity current deposits | Carbonate breccias |
| Rising Sea Level | Lakes red siltstones | Lagoonal evaporites | Marsh | Nodular anhydrite | Red shale | Basin center evaporites | Debris flows | Hard ground |

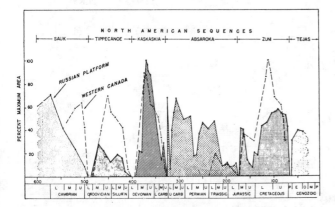

**Fig. 1-39.** Areas of preservation of units in western Canada and on the Russian Platform plotted as a function of geologic time. Major time-stratigraphic subdivisions are shown, as are the limits of cratonic sequences recognized in North America. From Sloss, 1972, reprinted by permission.

this information was developed from seismic sections and detailed micropaleontologic analysis of cores from marginal basins (Vail et al. 1977a). Worldwide onlap was independently confirmed by Sloss (1972) (Fig. 1–39), Holmgren et al. (1975), and by Zaaza and Visher (1975), and the historical basis was illustrated by Wise (1974) (Fig. 1–40).

## Seismic Evidences

The integration of both paleontology and stratigraphic relationships can be accomplished within the framework of seismic reflection profiles. Continuity of depositional events, patterns of plate tectonic history, and the preservation of a continuous historical record in marginal basins are recorded in continuous seismic profiles. The areal and temporal patterns are interrelated, and paleontological analysis can provide the basis for historical

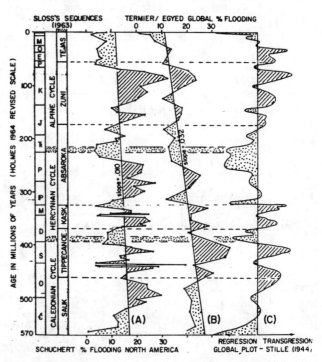

**Fig. 1-40.** Sequence boundaries and onlapping stratigraphic units. From Wise, 1974.

interpretations (Fig. 1–41). Seismic sections are particularly complete for Mesozoic and Cenozoic sedimentary sections, and wells through these sections provide the samples required for paleontological reconstruction (Vail et al. 1977). Much of this information is available through the search for hydrocarbons by major petroleum companies.

The degree of precision possible in identifying sequences in shelf and marginal basins depends upon the ability to correlate paleontological data, depositional events, unconformity surfaces, and onlap relationships

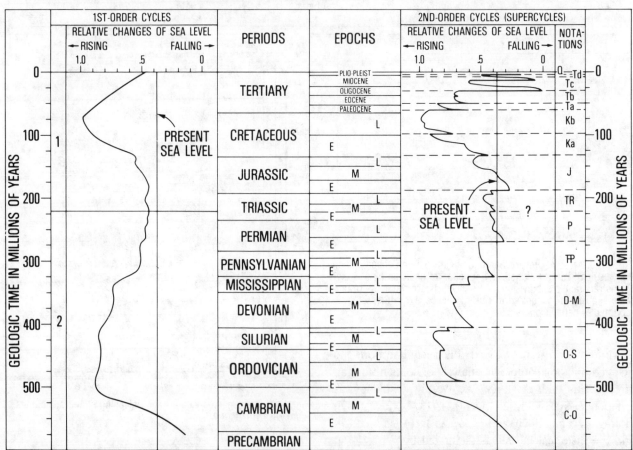

**Fig. 1-41.** Correlation of regional cycles of relative change of sea level from three continents and averaging to construct global cycles. From Vail, Mitchum, and Thompson, 1977, reprinted by permission.

**Fig. 1-42.** First and second order global cycles of relative change of sea level during Phanerozoic time. From Vail, Mitchum, and Thompson, 1977, reprinted by permission.

on a worldwide basis. The amount and differing types of data required in this analysis are overwhelming, but if the sequence concept is utilized, the unifying stratigraphic sequence framework can be observed.

The sea-level curve appears to be a derivative of the curve prepared by Pitman (1978), but additional inflection points can be interpreted as the result of loading and unloading related to sea-level changes. Transitions

appear to be abrupt, which are related to truncation both by subaerial and submarine processes (Fig. 1-42).

Seismic observation has increased in resolution. It is now possible to observe stratigraphic terminations on a scale as little as 10 meters. With the increased biostratigraphic and absolute time determinations from radiometric, magnetostratigraphy, and sedimentologic and subsidence patterns, it is now possible to examine stratigraphic patterns on a scale of hundreds of thousands to a few million years (Vail et al. 1977). This synthesis is shown by work at Exxon, where Haq, Hardenbol, and Vail (1987) have suggested three levels of unconformity surfaces ranging from 1 million to 30 million years. Their work reflects a precision at all levels of stratigraphic analysis. Utilizing depositional patterns, magnetostratigapy, biostratigraphy, and the most precise radiometric age dating, their synthesis can be correlated with worldwide stratigraphic events (Haq et al. 1987). Causality is still in question, but the scientific test is in prediction, and data are available to confirm a worldwide synchroneity of stratigraphic patterns.

Examination of seismic record sections illustrates the depositional patterns which may be recognized on a worldwide scale (Bally 1986). Correlations of events need to be related to many causal mechanisms (Tables 1–5, 1–6), and related to extinction, plate-tectonic, and worldwide unconformities recorded in the stratigraphic record. The interrelation of plate tectonic and biostratigraphic patterns was suggested by Valentine and Moores (1972). (Fig. 1–43). A combination of observation is required to produce a synthesis, but we are forced to believe that worldwide patterns are reflected in the stratigraphic record. The interrelation of a general sea level history, to depositional sequences appears to be related to a multiplicity of causes. (Tables 1–5 and 1–6) The temporal and areal scale of depositional events requires more data than is presently available. Data analysis indicates that magnitudes of sea-level change, the time duration of each event, and the stratigraphic response patterns related to tectonics and sedimentation, change from basin to basin, and from continent to continent. Periodicity of events must of necessity contain a great deal of variance, and from more detailed stratigraphic observations, based upon an enlarged data base a more complete understanding of causality and predictability may be possible.

***Paleontological Evidence.*** The paleontologic response to worldwide sequence patterns was outlined by Valentine and Moores (1972) (Fig. 1–43). Additional local work by paleontologists on sequence unconformities suggests a relationship to tectonic events. This evidence is only fragmentary, but diversity patterns, explosive evolution, and extinction events are identifiable.

***Worldwide Evidences.*** If sequences represent systematic changes in the sea level, and if this is the major control, each cratonic plate throughout the world should be similarly affected. Since sequences represent the largest systematic stratigraphic pattern, their identification must be possible in nearly all sedimentary basins throughout the world.

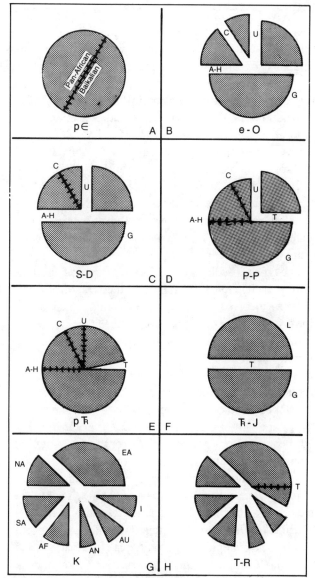

**Fig. 1–43.** Detailed paleontological analysis is hampered by these changing evolutionary patterns developed in response to plate tectonic spreading and suturing patterns. From Valentine and Moores, 1972.

## Descriptive Evidences

The integration of areal lithologic, biologic, and stratigraphic patterns in a time framework provides the observational basis for the recognition of sequences.

## Unconformity Surfaces

Detailed stratigraphic cross sections that are continuous into basinal areas are needed to determine unconformity surfaces of the sequence dimension. Recognition of onlap and downlap relationships is easier on cratonic margins where continuous seismic profiles can be used in a manner similar to continuous stratigraphic outcrop sections.

## Lithologic Pattern

Plate tectonics is the theme that interrelates biogeography, sequence unconformities, and stratigraphic onlap. The manner in which these responses are related to plate tectonics and to orogenic episodes on cratonic plates is not clearly understood. Periods of active plate separation and oceanic intrusive and volcanic activity appear to correlate with periods of onlap. Granite age dates and continental uplift appear to correlate with offlap and sequence unconformities. Continental breakup, rifting, and formation of marginal basins also appear to correlate with the first stages of onlap. The sedimentary response may be reflected in the formation of extensive evaporite units, lacustrine basins, and volcanic activity.

The types of sedimentary responses depends upon the nature of the cratonic margin and the type of orogenic process involved. Three types of continental margins have been defined: (1) passive, with either narrow or broad shelf surfaces, and well-defined sequence unconformities and onlap; (2) block faulted, with local basins and volcanic events and little evidence of continuity of the sequence unconformity, but with thick stratigraphic units developed in response to onlap; and (3) compressional zones and suturing of cratonic plates or linear trenches adjacent to cratonic plates, with discontinuous sequence unconformities and little evidence of onlap (Dewey and Bird 1970). These patterns developed at the same time in differing parts of the world in response to the same plate tectonic event. Thus, the sequence response is variable and must be interpreted in relation to larger paleogeographic and cratonic tectonic patterns. These patterns are identifiable from stratigraphic responses and may be useful in determining the developmental history. Additional stratigraphic aspects are required to predict the pattern of deposition and the controls for individual stratigraphic sections, but the determination of magnitudes of onlap, patterns of paleogeography, nature of cratonic margins, and plate distribution is possible. These can be used for local developmental interpretations.

## Stratigraphic Patterns

Worldwide stratigraphic patterns are reflected by the uniformity of sedimentary responses discussed by Ager (1981) and in the introduction of this section. Similarities in stratigraphic responses to onlap and climatic conditions are controlled by the local paleogeographic setting, but in comparable tectonic and environmental regimes, a worldwide stratigraphic pattern is developed for each sequence.

Lithologic responses may be reflected in many ways. Examples would include the following:

► Transgression across a cratonic surface characterized by second-cycle sandstones overlying a low relief surface dominated by small rivers and aeolian processes. Low rates of tectonic activity and sea-level rise produce blanket sandstones, thickening to the edge of the continental shelf. Overlying units represent carbonate and evaporite units developed in response to diminished source area. Examples include basal Cambrian (Lockman-Balk 1970), lower Middle Ordovician (Dapples 1955), Lower Permian (Rascoe and Baars 1972), and basal Jurassic sandstones with overlying onlapping clastic carbonate units (Freeman and Visher 1975).

► Transgression across a cratonic surface characterized by heterogeneous lithologic units with moderate relief. Erosional processes are dominated by integrated rivers, moderate to high rainfall, and a combination of chemical and mechanical weathering. The response would be development of regressive-transgressive delta patterns with the interrelation of paralic and shelf sedimentary environments. Examples include Upper Carboniferous (Visher et al. 1971), Lower Cretaceous (Weimer 1960), and Paleocene, Oligocene, and Miocene (Vail et al. 1977b). These units may be followed by thick shales or locally by shale and shelf carbonate units.

► A transgression across a cratonic surface characterized by lithologic units with little potential for clastic provenance produces differing patterns. Climatic conditions may result in low runoff or the paleogeography precludes supply of clastic detrititus to the area (Sheldon 1963). The sedimentary unit reflects the dominant shelf processes and the interrelation of shorelines and shelf geometry to chemical aspects of sedimentation (Bissell 1964). Regressive-transgressive shoreline deposition dominated by paralic, shelf, or lagoonal sedimentary and biologic controls are reflected. Massive carbonate units related to reefs, tidal flats, or shelf environments developed (Ginsburg and James 1974). These units are followed by restricted environments and red-bed or evaporite associations (Irwin 1965).

## Stratigraphic Responses

Four tectono-environmental depositional areas are useful for developing characteristic sequence models: (1) cratonic continental sections, (2) marine continental shelves and epeiric seas, (3) intrashelf basins, and (4) deep water basin (Table 1–10).

*Cratonic Continental Sections.* Onlap may be indicated by accumulation of thick clastic sections and widespread distribution of a single clastic unit followed by abrupt change in depositional pattern.

*Continental Shelves.* The progressive vertical and landward stacking of regressive-transgressive depositional cycles is characteristic. Continental shelf models may be in dynamic equilibrium with sediment supply and subsidence and produce a single transgressive sheet sandstone unit in response to increasing sea level and subsidence. Continental shelves devoid of clastic provenance also reflect systematic changes in subsidence and/or sea-level rise. Recognition of stratigraphic relations requires careful correlation and detailed paleogeographic reconstruction of the depositional environment from process-related sedimentologic analysis. Mapping of these patterns by use of biofacies and lithofacies data and by isopach and interpretive tectono-environmental maps allows the recognition of sequences.

▸ Transgression across a low-relief surface with low potential for clastic provenance may provide a homogeneous blanket carbonate or clastic unit with little variation.

▸ Increasing sea level may be reflected by blanket black shale lithologies (Rich 1951). Sequences of this type are difficult to identify. Local tectonic events or paleogeographic patterns may alter both the stratigraphic pattern and the sedimentologic response.

*Mobile Belts.* Sequence patterns influence sedimentation within mobile belts. Recognition of clastic fill related to dispersal points, canyons, and fans and to periods of cratonic emergence is interpretable within these stratigraphic sections.

## Summary

Sequences were recognized by objective criteria in local sections and over a single cratonic area. Also, evidence suggesting a worldwide history of sequences was presented to interrelate objective data from many sections throughout the world. Finally, sequences were shown to leave an indelible record in the stratigraphic record due to a systematic response in the developmental history. Of all the stratigraphic responses to a continuing pattern, the sequence appears to be unique in its ability to interrelate worldwide stratigraphic patterns.

The problem with the sequence as a model for stratigraphic interpretation is the need for observation from so many disciplines. Rarely does one stratigrapher understand the tectonic, biologic, sedimentologic, and historic responses that can be obtained from the stratigraphic record. Synthesis requires the integration of diverse types of data from many parts of the world. The application of the sequence model is only possible because it provides the framework for interpreting and organizing observations and also provides the basis for asking important questions. With all of this, it can be used as a tool for interpreting and predicting the developmental history of individual and collective stratigraphic sections from around the world.

## SEDIMENTARY FACIES

The stratigraphic record represents a developmental pattern of areal and vertical change of lithologic and biologic aspects. The history represented in the stratigraphic record is recorded in a continuum of variation. Analysis of this four-dimensional flow is the purpose of facies mapping. Variation in aspects within the stratigraphic record is in response to continuing processes. The identification of these processes within the natural limits of a stratigraphic unit requires the use of stratigraphic response patterns as a guide to observation. The diversity represented by a single stratigraphic section, not to mention the variation represented both horizontally and vertically, requires the understanding of the dominant sedimentary controls.

The facies approach allows integration of many types of observation and presents data in a manner that illustrates the interrelationships between the changing variables. This summation separates the rationally varying aspects from those that represent local events that do not reflect a sedimentary process control. Mapping a stratigraphic unit provides the basis for comparison with other mapped intervals and for interpreting boundary conditions and sedimentary processes. Quantitative data presentation provides the objective basis for interpretation, which is useful for predicting patterns and measuring responses to specific process controls.

The design criteria of facies maps include selection of the area to be mapped, the unit thickness, the time interval represented, the method of data presentation, and the function used to represent variation. The nature of the stratigraphic controls that are important in the development of a unit must be evaluated prior to construction of a series of maps. Many techniques are available for mapping stratigraphic units, and the ones selected should most clearly reveal the nature of the stratigraphic response to developmental processes. The final step in the study of a stratigraphic unit is the interpretation of the depositional history based upon the mapped variables. Thus, three aspects: (1) determining stratigraphic controls, (2) designing facies maps, and (3) interpreting provide the basis for stratigraphic analysis and for presenting information on sedimentary facies.

## Stratigraphic Controls

The pattern of a stratigraphic unit, both areally and temporally and in relation to patterns of provenance and tectonism, provides the framework for interpreting the developmental history. The definition of these boundary conditions is required for any facies study. Simply stated, a facies map represents the areal variation of one or more measurable aspects of a stratigraphic unit. A facies map can represent three-dimensional changes in biota, lithology, or any other aspect that may be quantified and presented by a map. However, if the map is to be used for historical interpretation, a time dimension must be placed as one of the boundary conditions. This requires recognition of time-rock units and correlation of the interval of time represented by the facies map. Only with a temporal basis can the map be used for interpreting rates of change and for prediction and interpretation of sedimentary process controls.

## Interval Thickness and Time

The thickness of a unit is unimportant in terms of interval. Units many hundreds of meters thick may show a systematic relationship to a continuing process, and a unit only a few meters in thickness and representing a single depositional event may reveal an interpretable pattern. However, if a process control does not interrelate depositional units within the interval, no matter what the thickness of the unit, an interpretable pattern may not be recognizable.

If the stratigraphic unit is defined in terms of time, it represents a period of history and can be so interpreted. If the unit is defined by a lithologic or biologic aspect and the mapped variable is another measure, the map may be useful for predicting change within the stratigraphic unit, but the developmental history of the unit or the measured aspect is not interpretable. Within the stratigraphic framework, facies maps that are not defined in a temporal manner have little historical value.

The interval of time represented by a stratigraphic unit can vary from a few tens of thousands of years, bounded by time-rock datums, or may include an entire sequence, representing possibly 30 million years, bounded by unconformity surfaces. The span of time represented is not critical, but the unit should represent the deposition of a unit genetically related or controlled by one or more continuing sedimentary processes.

## Interpreting Facies Patterns

Changes in thickness or time may be important responses to continuing processes, and maps showing these two variables are important in the interpretation of a stratigraphic unit. Relationships between onlap and erosion within a time framework were found to be useful in the analysis of unconformities, sequences, and the pattern of time-stratigraphic datums. These data, combined with facies information, provide the basis for interpreting tectonic controls and many depositional patterns including regression and transgression.

## Correlation

Interpretation of facies maps is based upon understanding the nature of the stratigraphic unit. Relationships concerning the bounding contacts of a stratigraphic unit to time-rock datums and the determination of time equivalency throughout the distribution of a unit are important considerations. Correlation requires establishment of the framework of deposition before analysis of changing lithologic and biologic aspects within the unit can be made. With this defined framework, it is possible to determine the relationship of changing biologic and lithologic aspects to the boundaries of the unit and to changes in thickness and to tectonic patterns. If the correlation is based upon time-stratigraphic relationships, then the interpretation represents the integration of observed aspects during a period of history.

The bases for correlation (as outlined in a previous section) are varied, and differing techniques and methods are required for delineating stratigraphic units for facies study. The selection of a basis for correlation is the first and probably most important step in constructing facies maps. Measurement of absolute time is rarely possible at the scale used in defining mappable stratigraphic units. Thus, the determining of relative time equivalence is a critical aspect in the design of a facies map. Rates of regression and transgression, duration of hiatuses and paraconformities, rates of sedimentation throughout the mapped area, and rates of climatic or tectonic change are variables that may be ascertained by correlation. Only after determining these sedimentary controls is it possible to select either the rock unit or time interval to be mapped.

If the purpose of constructing a facies map is to determine the distribution of porosity in a laterally continuous lithologic unit, the basis for correlation does not require precision in determining temporal relationships. If the purpose is to show patterns of changing provenance and to evaluate tectonic and paleogeographic patterns, then the unit must be carefully defined by both upper and lower time-stratigraphic datums. The types of stratigraphic units, the method of correlation for each type of unit, and their possible interpretation are presented in Table 1–11. This table summarizes the changing basis for correlation in the construction of facies maps and sug-

**Table 1-11**
**Types of Mappable Units**

| Types of Units | Methods of Correlation | Significance |
| --- | --- | --- |
| Slice—arbitrary thickness or percentage | Subdivide the unit parallel to one or two stratigraphic markers | Shows lithologic and biologic responses to provenance and deposition |
| Lithosome or biozone | Determine the three-dimensional continuity of a measurable aspect | Shows changes in depositional process |
| Regressive-transgressive cycle | Determine continuity of bounding lithologic units | Shows depositional process and local tectonic controls |
| Erosional surface down to a time-rock datum | Determine onlap of time-rock datums across an erosion surface | Reflects post-depositional history |
| Time-rock datum down to an unconformity | Determine continuity of a marker and onlap across an erosion surface | Shows paleogeography and history of transgression |
| Time-rock datum to time-rock datum | | |
| Chronozone | Define and trace time-rock markers | Shows tectonic and depositional history |
| Stage-Series | Relate biozones to time-rock datums | Reflects tectonic and paleogeographic history |
| Sequence | Relate biozones to onlap-offlap patterns of time-rock datums | Shows tectonic and paleogeographic framework |

gests the importance of correlation in the initial design stage. Correlation may require much more time than any other aspect of the study, but without the understanding of temporal and/or lithologic equivalence, there is little basis for interpreting mapped aspects.

The bases for correlating stratigraphic units for facies mapping vary with the type of unit, and the type of unit used varies with the purpose for constructing the map. If a correlation cannot be constructed on one basis, another basis may be chosen that provides part of the required information. Only if a particular aspect is interpretable from the type of stratigraphic unit chosen can provenance, local or regional tectonic events or patterns, paleogeography, depositional process, or history be determined.

## Boundary Conditions

The interpretation of variation in some aspect within one of the above described stratigraphic units requires understanding rates of change. Specifically, rates of uplift or subsidence, eustatic change, deposition, and topography of the depositional surface are reflected in changing patterns of lithology, thickness, and biologic aspect. Information concerning these boundary conditions is necessary to the interpretation of sedimentary facies. Much of this information is available from other sources relating to unconformities, to continuing processes reflected in sedimentary features, and to the tectonic framework. These features can be determined by comparing measurable aspects in single or correlative stratigraphic sections.

Unconformity surfaces do reflect the topography and the rate or pattern of onlap across the surface. Also, tectonic history may be determined by isopachous maps and paleogeographic maps relative to the unconformity. These can be used specifically to determine rates of uplift, eustatic change, and topographic boundary conditions. The scale of the unconformity in terms of time is important in determining the nature of the continuing process that might be revealed within stratigraphic units containing unconformity surfaces.

Indicators of depositional environments, water depths, and local depositional processes also are available to aid in the design of facies maps and to the interpretation of these maps. Local aspects may be interpretable within the facies framework and may provide additional information concerning stratigraphic responses to specific processes. The facies map then becomes a tool for the genetic interpretation of individual stratigraphic sections.

The most important boundary condition is the tectonic framework. This controls the sedimentary response and also patterns of unconformities, provenance, and rates of change. Information concerning the tectonic framework is available from continuing processes of sedimentation reflected in associated stratigraphic units.

It is this interrelationship of all the data into a continuum of change with the development of a consistent, recognizable, and rational pattern that is useful.

## Provenance

Sedimentary responses are not merely controlled by time, thickness, unconformities, or the boundary conditions; they are also controlled by the latitude, position on the craton, relation to land areas, relief, and the types of sediment available in a source area. The response to provenance is identifiable, but predicting the pattern of lithofacies and biofacies is not always possible. The stratigraphic record reveals the developmental history, but it does not predict succeeding events and patterns unless a unifying process or theme can be identified.

## MAPPING STRATIGRAPHIC INTERVALS

Mapping sedimentary facies may be accomplished with a wide range of objective and interpretive maps. The principal requirements are mapping an equivalent interval, quantifying the mapped parameter, standardizing operational definitions, and selecting mapped aspects that have a rational pattern of variation and data points distributed throughout the area of investigation. If these characteristics can be met, a map showing areal variation can be constructed. A wide range of measurable aspects meets all of these requirements. The diversity of observational characteristics requires that decisions must be made concerning the following factors: (1) the selection of the unit (thickness, area, and time interval); (2) variables to be mapped and in what manner they are to be presented (number, type, and scale); (3) how the data will be presented (ratio, percentage, and vertical position); (4) which maps are to be compared or presented together (combined ratio, isopach, and structure); and (5) what derived maps can be used for interpretation (rates of change, trend surface, variance, factors, or discriminant functions).

## Selecting a Unit

The thickness and area of a stratigraphic unit to be mapped depends upon the nature of the control, the ability to correlate, and the purpose for the map. Table 1-11 illustrates the types of interpretation possible for each stratigraphic unit, and this may be used as the basis for selecting a map interval. The higher the rate of change of thickness and lithology, the closer the spacing of the control is required.

The environment of deposition reflects whether sedimentary cycles or continuous depositional patterns are developed. The stratigraphic response to continuing processes may occur on a scale of a few kilometers in area and a few meters in stratigraphic interval or may represent thousands of kilometers and hundreds of meters of stratigraphic interval. The intertonguing relationships of the Mesaverde of southwestern Colorado and Utah can be subdivided into small units with process-related responses identifiable for stratigraphic intervals measured in a few tens of meters (Sears et al. 1941; Speiker 1949; Weimer 1960) (Fig. 1–44).

Of particular importance in selection of units for sedimentary facies analysis is the determination of unconformity surfaces. The facies pattern may be strongly affected by unconformities either overlying or underlying the stratigraphic unit. In each case either onlap or erosion may change the time interval across the mapped area, remove correlative units, or reflect periods and areas of nondeposition. Identification of these responses is necessary to the interpretation of the stratigraphic unit.

The purpose of the stratigraphic study must be understood prior to the selection of a stratigraphic unit for facies analysis. Facies maps are models for synthesizing and presenting stratigraphic data. They should be used to communicate information, not as a rigid procedure to reveal hidden and unknown stratigraphic patterns. Certainly, the synthesis of stratigraphic information into a facies map may suggest processes and patterns that were not originally obvious, but then the procedure is to develop new correlations, objective maps, and methods of data presentation that will illustrate these stratigraphic relationships.

## Selecting Map Variables

Selecting a map variable is similar to selecting a unit for facies analysis. Understanding rates of change of the variable within the stratigraphic unit must be understood prior to designing a series of stratigraphic maps to illustrate changes in sedimentary facies. The number of variables, the types of information to be mapped, the scale of the map, the contour interval, and the pattern of control points must be considered in the design of a facies map. With thin stratigraphic intervals or very dense control, it is possible to map a single variable and show detailed changes in a particular aspect. Conversely, with thick stratigraphic units and widely spaced control, a combination of several variables is necessary to illustrate a rationally changing pattern.

This principle must be modified if the variable critical for interpretation is not interrelated with other aspects or if the mapped aspect is not simply related to other measurable aspects within the stratigraphic unit. If the

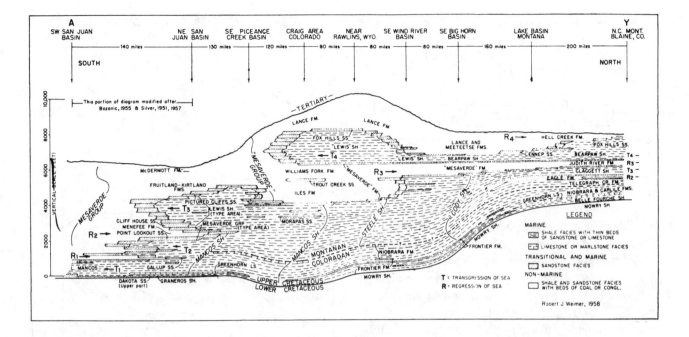

**Fig. 1–44.** Diagrammatic restored section of Upper Cretaceous rocks extending from north-central Montana to southwest flank of San Juan Basin, New Mexico. Tertiary rocks regionally cover erosional surface of Cretaceous strata. Diagram structurally distorts this surface. From Weimer, 1960.

distribution of petroleum is to be predicted, there may be no simple relationship to any stratigraphic aspect, and a map showing the distribution of petroleum in the section best suggests the basis for predicting its occurrence. Such a single variable map may be correlated with other maps, and a stratigraphic control may then be observed. A single map or a combination of additional maps can be constructed to predict the pattern of petroleum occurrence.

The types of variables encompass all measurable aspects that can be determined at all control points. These include biologic, lithologic, and thickness aspects, and their position or distribution within the stratigraphic section. Typically, a combination of lithologic and thickness measurements characterizes the response to tectonic, environmental, and depositional processes and events. For thick widespread units, the thickness of sand, shale, carbonate, and evaporite provides sufficient information for predicting the sedimentary controls. For thin units of local distribution, detailed aspects may be used, including shale color, carbonate types and texture, sand composition and texture, evaporite mineralogy, and possibly other quantifiable aspects.

In a similar fashion biofacies maps may be based upon the number of species, number of forms, proportion of differing types of taxa, and the distribution of each of the above criteria in differing lithologies. Biofacies allow the determination of properties of the depositional environment, and they are particularly useful in thin units of local distribution.

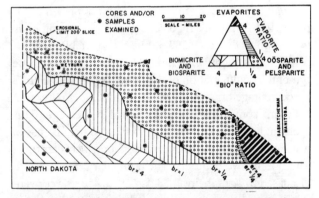

**Fig. 1–45.** A 200-foot slice map at the top of the Mission Canyon Formation (Mississippian) in southern Saskatchewan. The map shows the distribution of lithologic types in a carbonate bank. From Krumbein and Sloss, 1963.

## Presentation of Data

The purpose of facies maps is to communicate information about a stratigraphic unit. The design of a facies map must present data efficiently to allow communication of as much information as possible in a single map. The power of a facies map is the ability to present, in an integrated form, a great range of observational data. A range of data makes possible the development of the interrelationships between control points, changing patterns of depositional units, and the response to ongoing depositional processes (Fig. 1–45).

Univariant maps dealing with a single variable over the map area are used most often. The maps show structure, thickness of a unit, percentage of a particular lithology or aspect within the unit, or some other absolute single-valued measure. These maps are important as a part of the process-related interpretation for the genesis of the unit, but rarely do they provide sufficient information for interpretation. These maps are particularly useful in restricted areas or in thin stratigraphic units where a single process may be responsible for the observed variation. In these cases with an understanding of the process causing the variation, the map is interpretable and can be used as a basis for predicting changes between control points and in regions outside the area of control.

The data may be interrelated by use of patterns or by contours. The contours may have a simple linear spacing, a percentage or ratio distribution, or some other function. The effects on the contour spacing are significant (Fig. 1–46), and the selection of the nature of data presentation may be related to distribution of the data, to easy conveyance of information to another investigator, and to rates of change in the data set. No single method is always the most useful. This selection is a part of the design problem in facies map construction.

Various methods are available to present more than one aspect on a map. A ratio is a measure of the proportion of two varying aspects, and a single value represents this proportion. Similarly, a percentage is a proportion of all the variables that total 100%, but each variable must be contoured separately if more than two variables are present. The illustration of two, three, or more variables by percentage on a map leads to a confusing pattern of contours connecting points of equal percentage. This problem is avoided in the three-variable system by use of two ratios, which represent the variation between three variables by two contours. The triangular facies diagram discussed by Krumbein and Sloss (1963, 458) is a convenient method of representing the relationships (Fig. 1–47).

The value in this type of presentation is in the continuity of facies patterns. The three variables are represented by two ratios that vary from zero to infinity and, therefore, include all possible values within a stratigraphic unit. The continuum of sedimentation is represented by a continuum within the triangular diagram. The pattern of stratigraphic change shown by the map is analogous to the pattern of change within the triangular diagram. A progression of lithologic change can be mapped by dividing the triangle into areas, identifying these areas with a unique pattern, and then applying this pattern to areas bounded by similar ratios on the map. By this technique the combined aspects represented by two ratios vary continuously across the map and allow for interpolation between control points (Fig. 1–45). The power of this type of map is similar to that available for

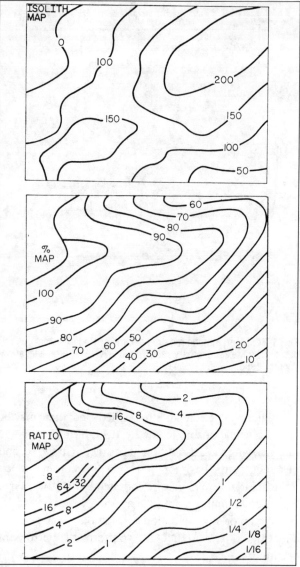

**Fig. 1–46.** Isolith, percentage, and ratio maps of a hypothetical stratigraphic unit. Contours on the isolith map are in feet. Note the similarity in the form of percentage and ratio maps, with crowding of the ratio contours as they approach infinite values. The percentage contours tend to "bunch up" in the vicinity of 50%. From Krumbein and Sloss, 1963.

mapping two aspects or a single-valued aspect, which allows every data point to be used in the construction of an integrated pattern of change similar to the natural variation.

An additional value to this type of mapping is that areas may be represented by a pattern, and contours may be added to illustrate additional aspects of the stratigraphic unit. Of particular importance in the interpretation of a stratigraphic unit is the change of thickness of the unit, which is often a response to tectonic processes operating during its deposition. Other aspects, such as paleocurrent directions, textural characteristics, and composition may be added in a similar manner. Thus, a large body of

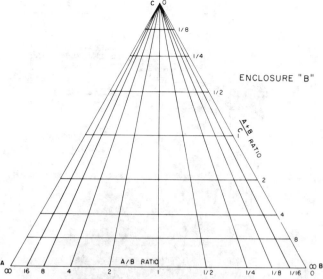

**Fig. 1–47.** Standard facies triangle shows continuous variation in three end members. Also provides a geometric progression to illustrate equivalent areas. ¼ CR ratio is equivalent to 8 CR x ⅛ ss/sh R, ¼ CR x 1 ss/sh R x 1 CR is equivalent to 8 CR x 1 ss/sh R. Provides equal areas on most facies maps. From Krumbein and Sloss, 1963.

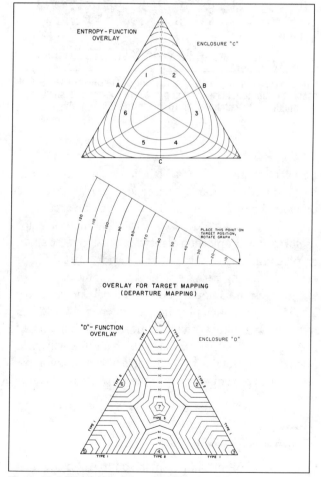

**Fig. 1–48.** (a) Entropy-function maps, (b) classifying-function maps, and (c) distance distribution maps describe the relationship among three variables. These are simple area subdivisions of three component data, plotted on triangular graph paper. They represent methods of emphasizing some particular aspect of percentage data. From Forgotson, 1966, reprinted by permission.

the available stratigraphic information can be mapped, and combining it into a single map provides a basis for interpretation, prediction, and stratigraphic analysis.

Other types of multivariant presentation of data include the entropy function, classifying function, and distance distribution maps. These maps describe the relationship among three variables and are simple areal subdivision of three component data plotted on triangular graph paper. They represent methods of emphasizing particular aspects of percentage data (Fig. 1–48). Selection of one of these maps is based upon design criteria and is useful to illustrate particular stratigraphic relationships. However, care must be taken that interpretation of the map is based upon mapped relationships reflecting changes in the map function. The entropy map emphasizes end members, while the classifying or D function map simply shows the pattern of seven combinations derived from the percentage data. For either of these maps, it is difficult to show the continuum of change illustrated by the conventional facies triangle. This is also true with the facies tetrahedron, which illustrates discrete changes but with four end members (Krumbein and Sloss 1963, 471–473).

The vertical variability map illustrates the flexibility of mapping techniques. The measure of the distribution of lithologies within a single stratigraphic section can be shown either by the center of gravity of the distribution or by the standard deviation (Krumbein and Libby 1957) (Figs. 1–49, 1–50). These maps are useful in mapping patterns of change both areally and vertically within a stratigraphic section. Modification of these functions and use of similar functions are possible, but care must be exercised that the resulting map can be easily interpreted. The goal is communication of information, not constructing maps that are confusing or do not present additional information.

Additional methods for presenting stratigraphic information can be grouped under statistical methods. These include factor and discriminant function analysis (Fig. 1–51), trend surface analysis (Figs. 1–52, 1–53, 1–54, 1–55, and 1–56), and various functions related to patterns of change. The latter category could include log or exponential patterns, values related to the normal distribution curve for example, the mean, standard deviation or a higher moment measure or a vectorial measure (Figs. 1–57 and 1–58). The representation of multicomponent stratigraphic aspects can be presented by any of these measures, and they may have value in the interpretation of specific processes related to sedimentation of a stratigraphic unit.

***Map Combinations.*** The combination of any of the above maps may serve as a useful purpose in the communication of stratigraphic data.

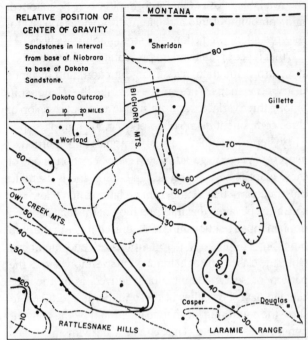

Fig. 1-49. Vertical variability. From Krumbein and Libby, 1957, reprinted by permission.

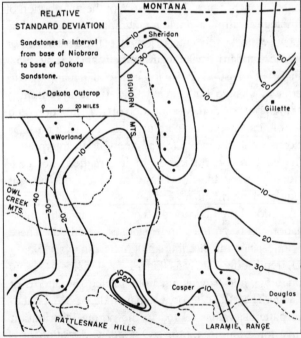

Fig. 1-50. Vertical variability. From Krumbein and Libby, 1957, reprinted by permission.

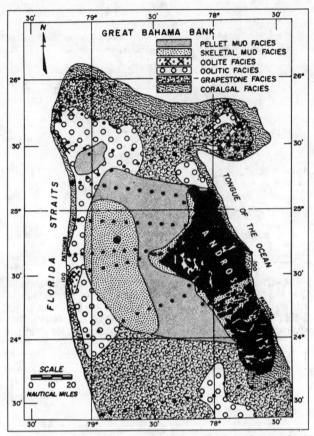

Fig. 1-51. Facies map of the Andros lobe of the Great Bahama Bank based on optimum classification procedure. Heavy dots show location of 200 sample stations. From Imbrie and Purdy, 1962.

Placing stratigraphic data into computer storage allows for sequential analysis, testing differing combinations of the data, and development of derived maps. Differing combinations of maps may be produced, and mapping programs may be prepared for automatic plotters to present this type of data. This flexibility allows observation of the variation of many aspects in differing associations and functions. Many map combinations can be produced in order to select a particular map that provides the basis for predicting hydrocarbon accumulations. There is an added advantage: the map or maps useful for communication of information about the stratigraphic unit may be selected from a wide range of maps developed during analysis of the stratigraphic data.

The types and amount of observational data available in any single stratigraphic unit are often overwhelming, but the possibilities provided by the combination of maps helps cut a path through the data that will yield interpretable patterns within the data from a stratigraphic unit. These patterns may form the basis for facies analysis.

## FACIES ANALYSIS

The nature of the stratigraphic controls and the map design provides the basis for interpretation of facies maps. As indicated by Table 1-11, the type of unit and the correlation framework can be used to suggest the significance of each type of map. Interpretations and examples for each of the classes or types of units indicated in Table 1-11 are used as the basis for analysis of facies maps.

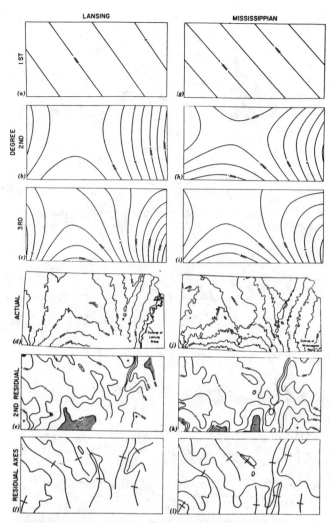

**Fig. 1-52.** Series of contour maps that portray fitting of trend surfaces to four structural horizons, top of Lansing Group, top of Mississippian rocks over entire state of Kansas. Maps include first, second, and third-degree trend surfaces, actual structural surfaces, residuals from second-degree trend surfaces, and axes of residual highs and lows. From Merriam and Harbaugh, 1964.

## Slice Maps

In areas where correlation is difficult or where the subdivision of the section cannot be based upon time-stratigraphic datums, it is often possible to approximate a time-rock unit by an arbitrary subdivision. An unconformity surface or a time-rock datum may be used as a guide; then the section is subdivided into thickness units above and below the surface. If two surfaces can be defined, an approximation of a time-rock unit is represented by a percentage of the interval.

A slice map effectively demonstrates provenance and depositional variation throughout an area and is especially valuable when the boundaries of the unit are close

**Fig. 1-53.** Facies map, lower part of Winchell formation. From Wermund and Jenkins, 1970.

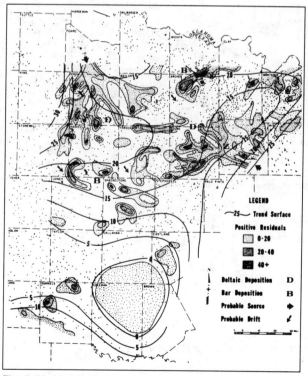

**Fig. 1-54.** Map of fourth-order trend surface and resulting positive residuals from surface fitting to percentage of sandstone, lower part of Winchell formation. From Wermund and Jenkins, 1970.

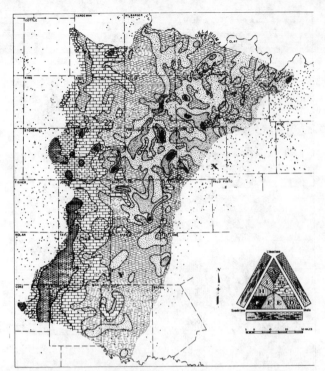

**Fig. 1-55.** Three-component facies map of approximate interval of the Black Ranch formation. From Wermund and Jenkins, 1970.

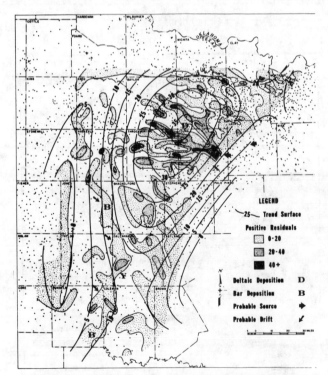

**Fig. 1-56.** Map of fourth-order trend surface and resulting positive residuals from surface fitting to percentage of sandstone. From Wermund and Jenkins, 1970.

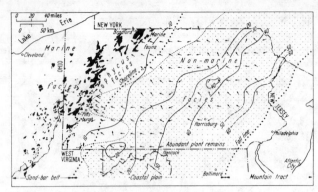

**Fig. 1-57.** Sedimentary framework, Pocono formation, (Mississippian) in central Appalachians. After Yeakel, from Potter and Pettijohn, 1965.

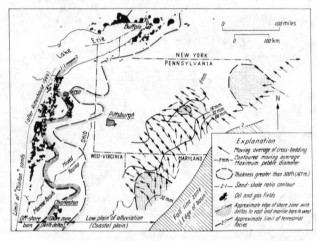

**Fig. 1-58.** Sedimentary framework, Tuscarora quartzite (Silurian) in central Appalachians. After Pelletier, from Potter and Pettijohn, 1965.

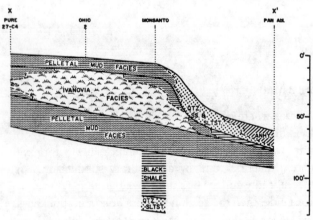

**Fig. 1-59.** Shoal to basin cross section of cycle 1 Ismay zone, illustrating facies changes. From Elias, 1963.

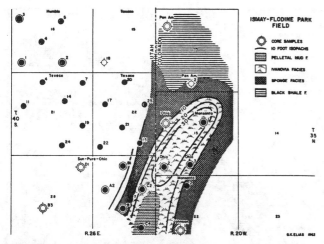

**Fig. 1–60.** Isopach map and distribution of Ivanovia facies and contemporaneous facies at end of transgressive phase of cycle 1 Ismay zone. The Ivanovia plants grew only on carbonate-mud substrates located on earlier formed ridges in the shoal environment. The best inter-particle primary porosity, formed by winnowing of the Ivanovia facies, is commonly found near the channels and at the basinal edge of the bioherm. From Elias, 1963.

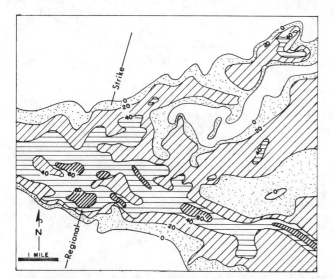

**Fig. 1–61.** Isopach map of total sand thickness. From Nanz, 1954, reprinted by permission.

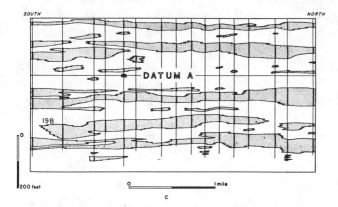

**Fig. 1–62.** Preparation of vertically exaggerated cross sections by use of correlations in shale intervals. From Nanz, 1954, reprinted by permission.

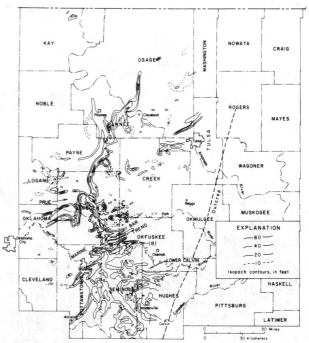

**Fig. 1–63.** Isopach map of Prue sandstone and lower Calvin sandstone in eastern Oklahoma. From Krumme, 1981.

to time-rock datums (Figs. 1–59 and 1–60). These maps are constructed to demonstrate local variations within a short time span. Maps showing a single aspect can provide insight into depositional processes. They may reveal energy within the depositional environment, ecological controls, or a lithologic response to some aspect of provenance (Figs. 1–61 and 1–62). A series of slice maps, especially if they approximate time-rock units, can show the developmental history of the units and may reflect paleogeographic change.

Since changes in thickness are arbitrarily established, these maps have little meaning with respect to rates of sedimentation and tectonism. Also, since boundaries are not established with respect to time-rock datums, their relationship to unconformities, onlap and offlap, or to historic patterns and events cannot be demonstrated. The lithologic response they reflect may have a process control, but a dynamic interrelationship between process and rates of change cannot be determined.

## Lithosome or Biozone Maps

A facies map of a lithologic unit or formation or strata encompassing the local occurrence of a taxon may have little relation to a time-stratigraphic unit. Variation of either a biologic or lithologic aspect with respect to thickness can be meaningful to depositional processes. The geometry of a lithosome may suggest a process of deposition and be useful in predicting the occurrence of other aspects, such as petroleum occurrence or paleogeography (Fig. 1–63).

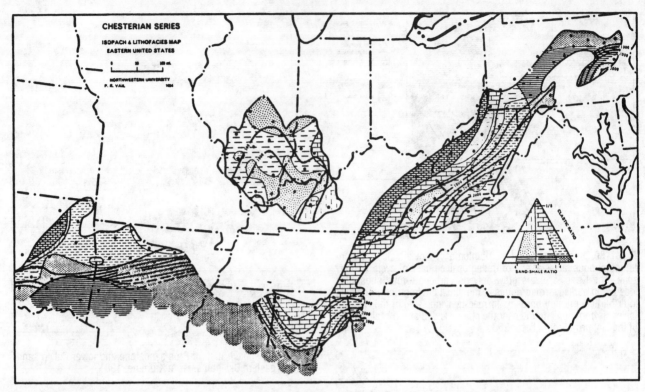

**Fig. 1–64.** Isopach-lithofacies map from Chesterian Series in the eastern U.S. From Sloss, Dapples, and Krumbein, 1960.

Maps based upon the geometry and areal distribution of lithologic or biologic units are important in the analysis of responses to depositional processes.

The thickness aspect of these maps is useful for determining the interrelationship between processes and the depositional setting. The relation to base level, systematic eustatic changes, or continuing tectonic processes is reflected. Other controls are possible, such as erosion, bypassing, paleogeography, and environmental aspects. These must be considered in any interpretation of the geometry of a unit.

## Regressive-Transgressive Cycles

Correlation of a single regressive-transgressive cycle allows the mapping of a single time-stratigraphic unit. A time-stratigraphic framework provides the basis for interpretation of changes in thickness and lithologic pattern. From this type of interval, it is possible to determine the effects of local depositional processes and tectonic controls (Visher et al. 1971).

Time-stratigraphic datums subdivide these units, and their relationship to continuing processes of deposition provides the basis for interpretation. The subdivision is not based upon biologic control, but the physical processes allow recognition of genetic sedimentary patterns. From these units, an understanding of the interrelationship of local tectonic and depositional process controls can be determined. These units are the lowest level of

time-stratigraphic units available for analysis.

From these data interpretations concerning the boundary conditions, the paleogeographic framework and environments of deposition and the depositional history of the unit can be determined. This information is useful in predicting the pattern of lithologic and biologic units.

## Erosion Surface to a Time-Rock Datum

The subsequent erosion of a time-rock unit strongly modifies the mapped pattern of a genetic depositional cycle. The pattern of truncation primarily reflects postdepositional tectonic processes. The geometry of the erosion surface may reflect regional patterns of offlap or effects of local tectonic uplift. In either case the resulting pattern is not genetically related to the depositional process. However, the erosional pattern is a part of the developmental history of a stratigraphic section, and it may reveal important aspects not identifiable within a time-rock unit.

Thickness patterns often reflect the erosional period and not deposition, but the preserved pattern of lithology may be a basis for interpreting tectonic and paleogeographic controls of sedimentation. The interpretation is complicated, but an example illustrates the value of these maps (Fig. 1–64). The pattern of truncation suggests the relative position of cratonic and basinal areas. The relationship of thickness and lithofacies contours also

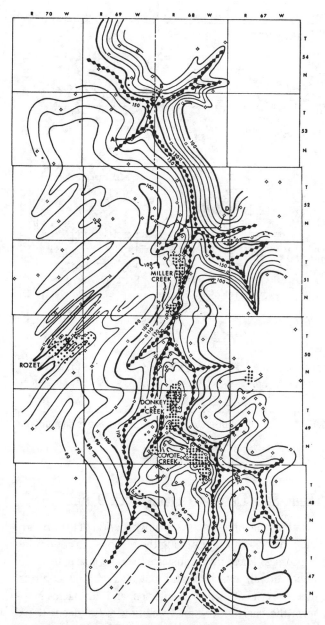

**Fig. 1-65.** Isopach map of genetic increment of strata between bentonite marker bed and unconformity at base of Muddy sandstone zone. From Busch, 1974, reprinted by permission.

suggests the zero edge of the unit to be the result of truncation and not deposition related to onlap. These patterns determine the tectonic elements and patterns related to later periods of uplift and thus allow analysis of a facies pattern with these effects to be removed. Facies maps of a unit beneath an unconformity may provide some information on the history or events not recorded in the stratigraphic record.

## Time-Rock Datum to an Unconformity

Time-stratigraphic units that overlie an unconformity surface reflect the history of onlap. The geometry and thickness of the onlapping unit reflects the paleogeo-

graphy of the unconformity surface (Fig. 1-65). Detailed correlations within the onlapping unit may suggest the history of the transgression and illustrate the relation between the unconformity surface and deposition within the stratigraphic unit. The facies pattern also may reflect the systematic change in lithology represented by onlap. If the mapped unit extends to the limit of transgression or onlap, the thickening wedge of sediment suggests the response to changing depositional environments.

The thickness pattern indicates the topography present on the unconformity surface, and the facies relationships to this isopach pattern reflect paleogeographic sedimentary controls. The facies response is controlled primarily by paleogeography and the transgression. Since facies patterns often parallel isolith thicknesses, as seen by the Morrowan onlap in Anadarko basin of Oklahoma (Forgotson and Statler 1966), these relationships can be interpreted from maps showing sandstone distribution above the unconformity (Fig. 1-66). The zero edge is recognizable by the association of a progression of depositional environments and by their developmental association.

A particularly significant problem in the interpretation of depositional history is the determination of time-stratigraphic datums within onlapping units. Onlap may result in parallel lithologic units with time datums at an angle to lithologic units or local regressive-transgressive patterns with time datums parallel to lithologic units (Masroua 1973). The interpretation of the resulting facies patterns requires integration of other depositional aspects including biofacies, environmental patterns, and the response of sedimentation to topography and provenance. Other types of depositional patterns may be developed in response to a continuing process, and the recognition of time-stratigraphic datums is not possible as suggested for the onlapping Navajo sandstone (Freeman and Visher 1975) (Fig. 1-16).

## Unconformity to Unconformity

Interpretation of depositional units bounded by unconformities is particularly difficult. Aspects of tectonism before and after the deposition of the unit and depositional response to eustatic changes and provenance may control the facies pattern. Separation of these effects is often impossible. The combination of an isopach map and a facies map does allow the recognition of the tectonic control for sedimentation.

Analysis of thin units bounded by unconformities that represent changes in depositional or tectonic processes mostly reflect pre- and post-depositional controls. Facies and isopach maps of thicker units may suggest a response to depositional processes (Figs. 1-67 and 1-68).

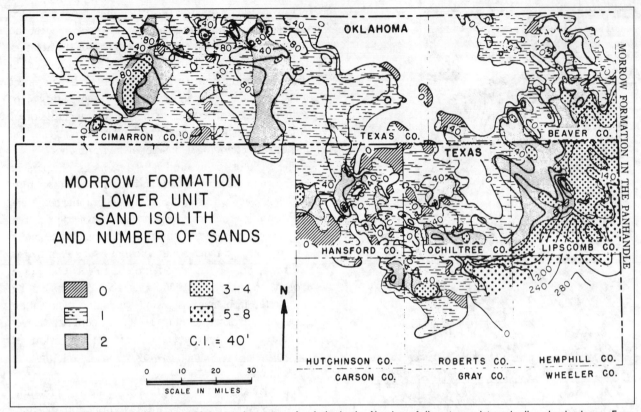

**Fig. 1-66.** Sandstone isolith map, lower Morrow unit, western Anadarko basin. Number of discrete sandstone bodies also is shown. From Forgotson, Statler, and David, 1966, reprinted by permission.

**Fig. 1-67.** Isopach-lithofacies map of the Tensleep sandstone. From Sloss, Dapples, and Krumbein, 1960.

## Time-Rock Datum to a Time-Rock Datum

Stratigraphic units encompassing a period of history provide the basis for interpreting the response to continuing processes. These units illustrate the response to tectonic, depositional, eustatic, and paleogeographic controls. The patterns of isopach, biofacies, and lithofacies maps form the basis for the interpretation of the depositional history. Units bounded by time-rock datums or markers, onlap-offlap cycles, or widespread faunal zones are recognizable. Facies maps illustrating variations in aspects reflect changing process controls:

▸ Small units, or lithozones, reflect changes in local tectonic and depositional history;

▸ Larger units (stages or series) defined by biozones and time-rock datums illustrate a longer period of history related to paleogeographic and tectonic controls; and

▸ The largest definable stratigraphic unit, the sequence, shows the history of deposition within the framework of tectonism and paleogeography.

Each complete time-rock unit is useful for determining aspects of the depositional history, and interpretation is related to temporal and scalar dimensions of the unit.

A thin time-stratigraphic unit representing a few or possibly hundreds of meters illustrates a single cycle of

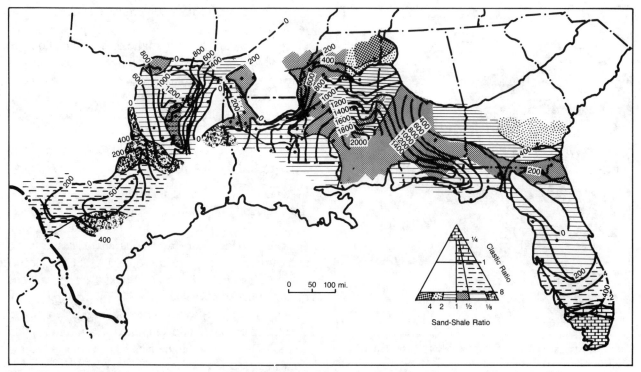

**Fig. 1–68.** Isopach-lithofacies map, Cretaceous, Woodbine Group, Gulf Coast area. From Sloss, Dapples, and Krumbein, 1960.

deposition. Local tectonic responses may be seen in the isopach map, and the facies pattern may reflect paleogeography, provenance, and the pattern of depositional environments. These maps are useful for interpreting the genesis of depositional units (Fig. 1–69).

Larger intervals representing a stage or series represent an average of many depositional events. The effects of continuing tectonic, eustatic, or paleogeographic processes may be revealed by facies and isopach maps. Each unit may represent many cycles or the continuing response to a longer period of change. The design of maps requires that broader scale variations be illustrated, and typically a facies map with three end members combined with an isopach map is the most useful (Fig. 1–70). Depositional limits of a unit can be recognized, and continuing dispersal centers or depositional responses to paleogeography can be interpreted. At this level it is possible to interpret basinal patterns, tectonic controls, and changes in paleogeography during the deposition of the unit.

The largest time-stratigraphic unit, or sequence, is bounded by unconformities, but it represents a sufficiently long span of history in which fundamental depositional controls may be recognized. Specifically, patterns of onlap and offlap are represented, and these may be related to the tectonic and paleogeographic framework. The relationships between ocean basins, cratonic areas, and basins filled with sediments are

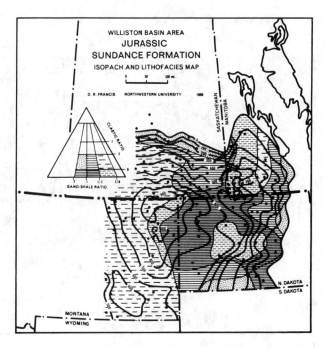

**Fig. 1–69.** Isopach-lithofacies map, Jurassic Sundance formation, Williston Basin Area. From Sloss, Dapples and Krumbein, 1960.

identifiable. Stratigraphic units of this dimension are useful in interpreting the controls to developmental history illustrated by smaller units. An example illustrates this interpretation (Fig. 1–71). The isopach map is not as useful in many respects as the facies pattern. Thickness

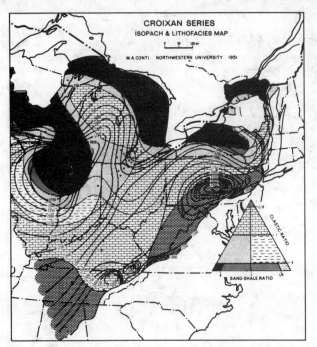

**Fig. 1–70.** Isopach-lithofacies map, Croxian Series. From Sloss, Dapples, and Krumbein, 1960.

variations reflect in part unconformities, but thicker units may show process-controlled patterns of facies. Maps integrating many types of lithologic information are required to show the response to long period tectonic and paleogeographic controls in the developmental history. These maps help establish a framework for the interpretation of maps from units of smaller areal or temporal dimension.

## Interpreting Facies Maps

Systematic patterns exist for very large areas and time units. Figure 1–72 shows the relationship between tectonism, source areas, and depositional patterns. Typically, relations of isopachous and facies patterns indicate transgressive and erosional areas. Basins, geosynclines, shelf areas, arches, and uplifts are shown. The transcontinental arch, the large NE-SW trending area of nonpreserved rocks, reflects both transgression due to parallelism of clastic ratio to uplifted area and truncation due to the local high angle facies lines adjusted to the truncated area.

The influence of subsidence rates or structure on deposition is reflected in the Missourian pattern of the southern midcontinent area (Fig. 1–73). This map reveals subsidence rates, position of source areas, directions of clastic transport, and nature of zero edges.

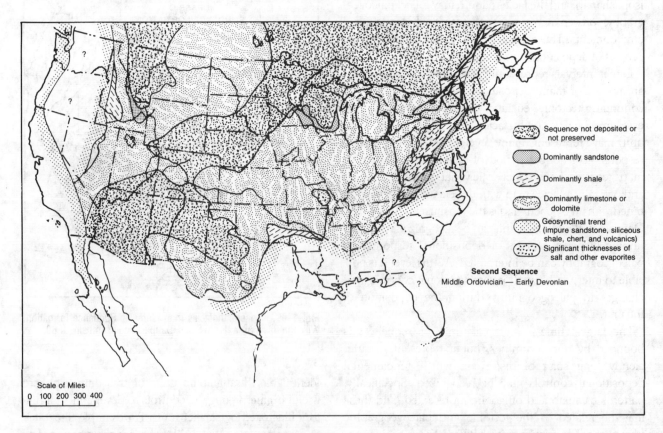

**Fig. 1–71.** Lithologic pattern of second sequence, Early-Middle Ordovician to Early Devonian. Reprinted courtesy L.L. Sloss.

The Desmoinesian and Atokan facies map illustrates the problem when two units are combined with an intervening unconformity surface (Fig. 1–74). Facies patterns are not coherent, patterns of source areas are difficult to interpret, and the paleogeographic pattern is not suggested.

A simple shale color map may indicate source areas, directions of regression, and presence of basins (Fig. 1–75). Black shale to gray shale may be a sign of restriction. Deltaic lobes and position of strand lines are identified by the clastic pattern.

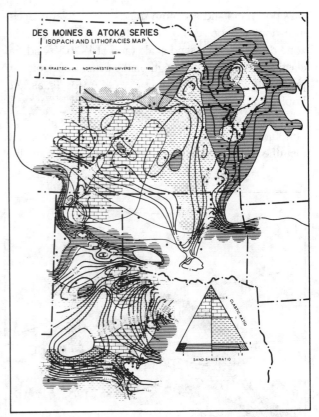

**Fig. 1–74.** Isopach-lithofacies map, Des Moinesian and Atokan Series. From Sloss, Dapples, and Krumbein, 1960.

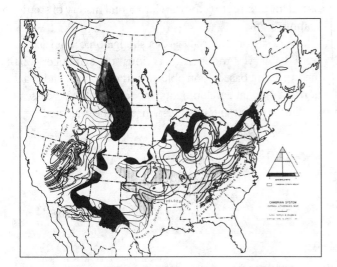

**Fig. 1–72.** Isopach-lithofacies map, Cambrian System. From Sloss, Dapples, and Krumbein, 1960.

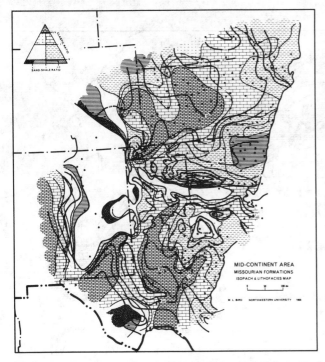

**Fig. 1–73.** Isopach-lithofacies map, Mid-Continent area, Missourian formation. From Sloss, Dapples, and Krumbein, 1960.

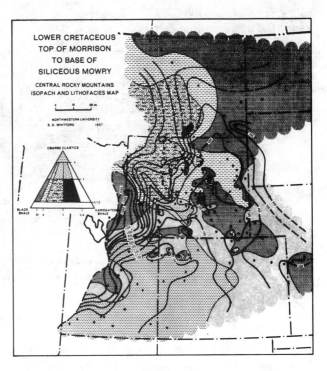

**Fig. 1–75.** Isopach-lithofacies map, Lower Cretaceous, top of Morrison to base of siliceaus Mowry, central Rocky Mountains. From Sloss, Dapples, and Krumbein, 1960.

Depocenters and optimum trends for petroleum accumulations can be seen over wide areas (Fig. 1–76). The zone defined by the sand/shale ratio of ⅛ to ½ and the clastic rates greater than eight is the primary petroleum-producing zone in the Claiborne Group of Texas and Louisiana.

Positions of intrashelf basins, shelf-edge carbonate buildups, and areas of high clastic deposition are easily identified. Regional facies trends are particularly valuable for structuring exploration programs and determining areas of high petroleum potential (Fig. 1–77).

## Summary

Facies analysis requires careful correlation, recognition of the boundary conditions, and determination of the purpose for constructing a map or series of maps. The most important aspect in analysis is the design of the map. There are few rules but many guidelines applicable to the construction and analysis of facies maps. The comparison of maps of differing aspects and from differing areas aids in the development of facies models, which are useful for interpreting the developmental history of stratigraphic units.

If every step in the analysis of a stratigraphic unit is supported by (1) independent lines of evidence, (2) measurement based upon objective criteria, and (3) its relation to a rational developmental model, then the processes relating to the stratigraphic response can be evaluated.

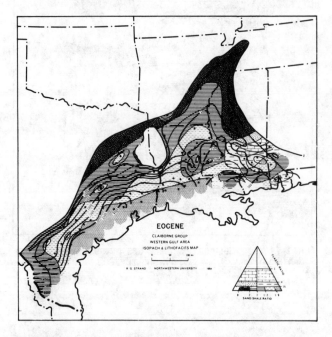

**Fig. 1–76.** Isopach-lithofacies map, Eocene Claiborne Group, western Gulf area. From Sloss, Dapples and Krumbein, 1960.

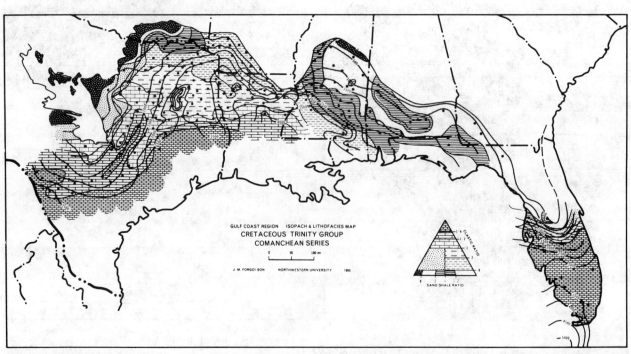

**Fig. 1–77.** Isopach-lithofacies map, Cretaceous Trinity Group, Comanchean Series, Gulf Coast area. From Sloss, Dapples, and Krumbein, 1960.

# REFERENCES FOR FURTHER STUDY — CHAPTER 1

Ager, D.V., 1981, The Nature of the Stratigraphic Record, 2nd ed.: New York, John Wiley & Sons, 122 p.

Arthur, M.A., and R.E. Garrison, eds., 1986, Milankovitch cycles through geologic time: Paleo-oceanogray, v. 1, p. 369–586.

Gould, S.J., 1985, The paradox of the first tier an agenda for paleobiology: Paleobiology, v. 11, p. 2–12.

Gould, S.J., and Eldridge, 1977, Punctuated equilibria the tempo and mode of evolution reconsidered Paleobiology, v. 3, p. 115–155.

Haq, B.U., J. Hardenbol, and P.R. Vail, 1987, Chronology of fluctuating sea levels since the Triassic: Science, v. 235, p. 1156–1167.

Haq, B.U., and F.W. van Eysinga, 1987, Geological Time Table, 4th revised edition: New York, Elsevier Science Pub. Co., Inc.

Harland, W.B., A.V. Cox, P.C. Llewellyn, C.A.G. Picton, A.G. Smith, and R. Walters, 1982, A geologic time scale: Cambridge, Cambridge Univ. Press, 131 p.

Raup, D.M., 1985, Mathematical models of cladogenesis: Paleobiology, v. 11, p. 42–52.

Read, J.F. J.P. Grotzinger, J.A. Bova, and W.F. Koerschner, 1986, Models for generation of carbonate cycles: Geology, 14, 107–110.

Schoch, R.M., 1986, Phylogeny Reconstuction in Paleontology: New York, Van Nostrand Reinhold Co., 353 p.

Stanley, S.M., 1985, Rates of evolution: Paleobiology, v. 11, p. 13–26.

Valentine, J.M., 1985, Phanerozoic Diversity Patterns: Princeton, N.J. and San Francisco, CA, Princeton Univ. Press and Pacific Division, Am. Assoc. for Advan. of Science.

# 2 EXPLORATION TOOLS

... scientific knowledge is more than a collection of experientia. Rather, science

is, to appropriate a term from psychology, a structured collection of experiences,

and any inquiry into the nature of science is really an examination of this

structure and its making. **R.C. Lewotin**
*Models, Mathematics and Metaphors,* 1963

## INTRODUCTION

Stratigraphy is the integrating framework for interpreting petroleum accumulations. Tectonic controls, plate tectonic history, basin development and fill, reservoir patterns, source rocks, and patterns of fluid migration all interrelate in the developmental history of stratigraphic units. Physical, biological, and chemical processes are reflected in identifiable stratigraphic responses and provide the data for determining the developmental history of stratigraphic sections.

The significance of many sedimentary aspects has been unknown due to the lack of understanding of process controls. Recent research in the stratigraphic sciences emphasized the study of plate tectonic, sedimentary, organic, and fluid processes. It was not a single new concept that made it possible to predict stratigraphic patterns; rather, it was the synthesis of information that allowed understanding.

The days when a stratigrapher was a "sample runner," a paleontologist was a "bug picker," a petroleum geologist mapped "net pay," a geophysicist "picked records," a subsurface geologist "slipped logs," and field geologists

"plane tabled" are long since past. We talk now of seismic stratigraphy, biostratigraphy, stratigraphic models, synsedimentary tectonics, and basin models. Explorationists use stratigraphy to interpret the developmental history illustrated by the rock record.

## STRATIGRAPHIC CONTROLS FOR HYDROCARBONS

Petroleum exploration cannot be profitably based upon the tenet, "Oil is where you find it." The unpredictability that characterized exploration reflects the lack of understanding of the controls. These can now be identified, and exploration programs designed to define the stratigraphic pattern, and can now be used to explain petroleum accumulations. These include the following:
- Distribution and tectonic history of source rocks,
- Pressure gradients to cause the expulsion and migration of generated hydrocarbons,
- Source rocks in close proximity to permeable pathways for migration,
- Permeable pathways interconnecting source and reservoir rocks,

▸ Porous and permeable reservoir rocks along the migration path, and

▸ Changes in relative permeability to separate hydrocarbons from migrating fluids.

Each of these aspects can be predicted by examining the pattern and developmental history of stratigraphic units. The problem has been that few stratigraphic techniques were available to interpret the pattern of these variables. The prediction of petroleum occurrence requires mapping these aspects and determining their rate and pattern of change. This requires an understanding of the boundary conditions, the significance of stratigraphic responses, and the quantification of observations. Aspects, such as rate of subsidence or tectonic uplift, sea-level change, sediment supply, paleogeography of the depositional area, biologic productivity, and evaporation balance, produce identifiable stratigraphic responses.

## Determining the Stratigraphic Controls of Petroleum

Research interest into the orgin of petroleum has been documented for more than 55 years. John Rich was publishing on the origin, migration, and accumulation of petroleum in 1934, and the American Petroleum Institute funded research by Parker Trask (Trask and Patnode 1942) to study a possible organic origin for petroleum. They demonstrated a correlation between percentage of

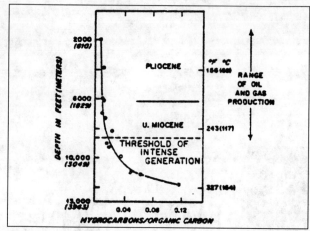

**Fig. 2–2.** Hydrocarbon generation curve for the Los Angeles basin. From Philippi, 1965.

organic carbon and the occurrence of hydrocarbons in fine grained sediments (Fig. 2–1). Research by Philippi (1965) records research carried on by Shell Oil Company during the later portion of the 1950s. This research on petroleum generation in the Los Angeles basin indicated that the generation of hydrocarbons from organic material (kerogen) was principally a thermochemical process, and that significant hydrocarbon generation did not commence until sediments reached a temperature of more than 70° Celsius (Fig. 2–2). This pioneering work set the stage for academic and industrial research during the latter portions of the 1960s and the 1970s.

**Fig. 2–1.** Organic carbon distribution in the Los Angeles basin. From Trask and Patnode, 1942, reprinted by permission.

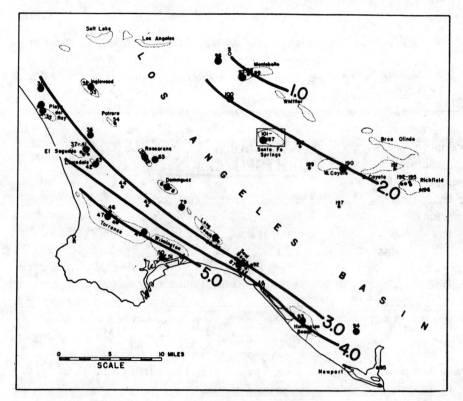

**Table 2-1**

| Anoxic Basin Type | Paleogeographic Setting | Stratigraphic Distribution of Anoxic Sediments |
|---|---|---|
| Anoxic Lakes | Equable, warm, rainy. Early rifts. Intermountain basins. | Continuous within same Anoxic Lake System. |
| Anoxic silled basins | Temperate to warm, rainy. Intracratonic seas. Also Anoxic pockets on shelves. | Variable. Tends to be richest at bottom of basin or pocket. |
| Anoxic layers w/ upwellings | Oceanic shelves at low latitudes. West side of continents. | Often narrow trends. But can be widespread. Phosphorites. Diatomites. |
| Anoxic open ocean | Best developed at times of global warm-ups & major transgressions. | Very widespread w/little variation. Often synchronous worldwide. |

**Table 2-1.** Classification of anoxic depositional models. Certain settings in marine realm may combine two or even three anoxic basin types; for example, local anoxic layers caused by upwelling may reinforce anoxic open-ocean oxygen-minimum layer impinging upon shelf with silled depressions. From Demaison and Moore, 1980, reprinted by permission.

## Source Rocks

Recent research has emphasized the role of regressive-transgressive patterns of sedimentation, changing geometries of ocean basins and cratonic plate configurations, and changing relative sea levels as important controls for the formation of source rocks. These aspects have had a major control on the worldwide patterns of organic productivity and on the time and place for the occurrence of restricted sedimentary basins. During a few periods of geologic history, stratified water columns developed, and anoxic conditions were present at the sediment/water interface. Vail (1987) (Fig. 2-53) has indicated that condensed sections, reflecting periods of transgression during the upper portion of seismic sequences, have an enhanced potential for the deposition and accumulation of organic rich sediments in cratonic marginal basins. Oceanic basin circulation patterns also led to the development of areas of enhanced upwelling, and higher rates of organic productivity (Emery 1969) (Fig. 6-1). In combination, only during a few times in the geologic record, and in a few restricted anoxic depocenters, has there been significant accumulation of hydrocarbon producing source rocks (Demaison and Moore 1980) (Table 2-1).

The distribution of major petroleum provinces recorded by Bois et al. (1982) for the Jurassic and Cretaceous (Figs. 2-3 and 2-4) illustrate the importance of marginal and foreland basins for the development of significant volumes of source rock. These basins contain

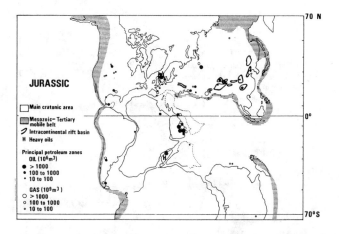

**Fig. 2-3.** World distribution of petroleum zones (or groups of petroleum zones) in Jurassic reservoirs. Size of the circles is related to the importance of total recoverable reserves. From Bois et al., 1982, reprinted by permission.

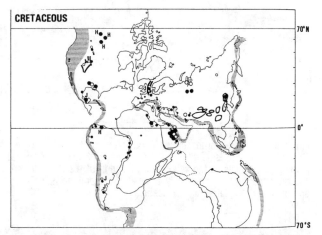

**Fig. 2-4.** World distribution of petroleum zones (or groups of petroleum zones) in Cretaceous reservoirs. Size of the circles is related to the importance of total recoverable reserves. From Bois et al., 1982, reprinted by permission.

many major transgressive sequences, and their development is related to major restructuring of the areal configuration of cratonic plates. During the Jurassic, the opening of the North Sea, the Gulf of Mexico, and subsidence of an intrashelf basin in the area of the Arabian-Iranian Gulf produced major centers for hydrocarbon accumulations. Large Cretaceous accumulations associated with rift basins in Libya and China, foreland basins in Russia, Alaska and Alberta, and intrashelf basins in the Gulf of Mexico, eastern Venezuela, and the Arabian-Iranian Gulf, illustrate the importance of sequence stratigraphy and paleogeography to the formation of source rocks.

Very special conditions are required for source rock development. An example of phosphorite deposition on the east coast of North America illustrates the association of conditions requisite to the development of an anoxic water layer at the sediment/water interface. An intrashelf

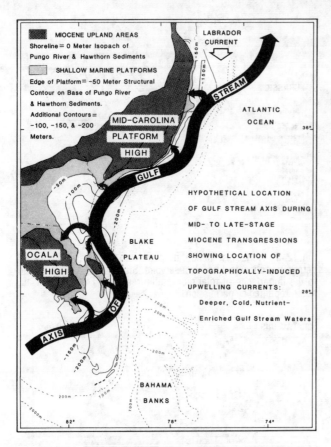

**Fig. 2-5.** Map of the southeastern United States showing the relationship of first- and second-order structures and major Miocene phosphate deposits to the hypothetical location of the Gulf Stream and the resulting topographically induced meanders and upwellings during a mid- to late-stage Miocene transgression. From Riggs, 1984.

**Fig. 2-6.** *B,* Mid- to late-stage transgression, temperate climate; phosphate deposition optimum on mid and inner shelf. *C,* Sea-level maximum (interglacials), warm climatic; shelf dominated by subtropical carbonate deposition. From Riggs, 1984.

basin between structural highs is illustrated by Riggs (1984) (Fig. 2-5). This portion of the continental shelf is marginal to the Gulf Stream, a major oceanic current system present since the Miocene. Sequence stratigraphy during the lower and middle Miocene indicates a second order supercycle (Haq et al. 1987). During mid- to late-stage transgression, phosphorite was deposited on the middle to outer shelf in an intrashelf anoxic basin (Fig. 2-6). Similar patterns of shelf and slope basin deposition appear to be important throughout the geologic record for the accumulation of significant amounts of sediment with a high percentage of organic carbon.

Evidence has developed that a wide range of types of organic matter may be able to produce commercial accumulations of hydrocarbons. Data by Hunt (1976) suggested that differing types of organic matter had differing potential for conversion to hydrocarbons (Table 2-2). This table also indicates that large differences occur in the percentage of organic carbon preserved in deposited sediments. Analysis of lithology in relation to conversion factors of organic carbon to hydrocarbon has been shown to be related to the chemical nature of deposited organic carbon (Baker 1962) (Fig. 2-7). Three general chemical types of organic material have been identified: (1) fresh water alganites with a high H/C atomic ratio; (2) marine phytoplankton, zooplankton and microorganisms (bacteria) with a lower H/C ratio; and (3) continental plants with identifiable vegetal debris and low H/C ratios and a high O/C ratio (Tissot and Welte 1984, 151–154). Examples of these types of source rocks and their evolution with burial are shown by Figure 2-8 (Tissot et al. 1984). The evolution of hydrocarbons from these precursors is one of bacterial alteration, condensation, and polymerization, leading to the formation of fulvic and humic acids as schematically illustrated by Figure 2-9 (Tissot and Welte 1984). The end product of this process is "kerogen," an organic complex insoluable in normal organic solvents. Diagenesis of kerogen leads to the formation of crude oil composed of n-alkanes, cycloalkanes, and aromatics (with nitrogen, sulfur and oxygen compounds) (Tissot and Welte 1984) (Fig. 2-10). With increasing depth (temperature) changes in the proportion and molecular weights of these compounds have been characterized by Tissot and Welte (1984) (Fig. 2-11). This figure compares to Figure 2-12, which illustrates that the path of diagenesis and catagenesis varies with the type of organic material deposited (Tissot and Welte 1984). Changing H/C and C/O ratios require differing processes of condensation and loss of carbon dioxide, which results in differing efficiencies of hydrocarbon generation. These figures illustrate that diagenetic alteration leads to differing products from differing precursors. The stage of the alteration may lead to differing molecular

**Table 2-2**
**Typical Source Rocks and Percentage of Hydrocarbon in Organic Matter**

| Rock Type | Hydrocarbons, ppm | Organic Matter weight % | Hydrocarbons in Organic Matter, % |
|---|---|---|---|
| **Shales** | | | |
| Wilcox, Louisiana | 180 | 1.0 | 1.80 |
| Frontier, Wyoming | 300 | 1.5 | 2.00 |
| Springer, Oklahoma | 400 | 1.7 | 2.35 |
| Monterey, California | 500 | 2.2 | 2.27 |
| Woodford, Oklahoma | 3,000 | 5.4 | 5.56 |
| **Limestones and Dolomites** | | | |
| Mission Cyn. Limestone, Montana | 67 | 0.11 | 6.09 |
| Ireton Limestone, Alta. | 106 | 0.28 | 3.79 |
| Madison Dolomite, Montana | 243 | 0.13 | 16.7 |
| Charles Limestone, Montana | 271 | 0.32 | 8.47 |
| Zechstein Dolomite, Denmark | 310 | 0.47 | 6.60 |
| Banff Limestone, North Dakota | 530 | 0.47 | 11.3 |
| **Calcareous Shales** | | | |
| Niobrara, Wyoming | 1,100 | 3.6 | 3.06 |
| Antrim, Michigan | 2,400 | 6.7 | 3.58 |
| Duvernay, Alta. | 3,300 | 7.9 | 4.18 |
| Nordegg, Alta. | 3,800 | 12.6 | 3.03 |
| Mean | | | 5.4 |
| Mean less two high samples | | | 4.1 |

HYDROCARBON VS. ORGANIC CARBON

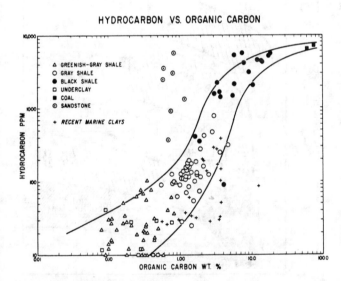

**Fig. 2–7.** Hydrocarbons vs organic carbon for principal rock types of Cherokee Group and for recent marine clays. From Baker, 1962, reprinted by permission.

weight hydrocarbon fractions and to differing proportions of waxes, asphaltines, n-alkanes, and cycloalkanes (Fig. 2–10). Of particular importance is that the response of the organic fraction to increased burial (temperature) leads to chemical changes in their structure.

Identification of these changes has been a principle area of research during the past fifteen years. Indices of these changes have been shown to relate to increased C=C bonds, and change in color of organic mascerals (TAI), vitrinite reflectance ($R_0$), and fixed carbon ratios of

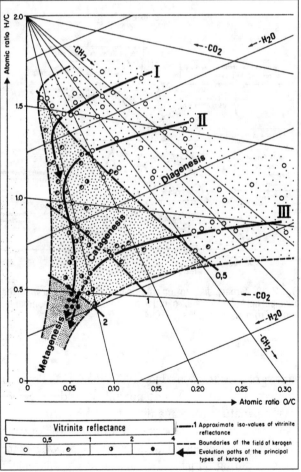

**Fig. 2–8.** General scheme of kerogen evolution from diagenesis to metagenesis in the van Krevelen diagram. Approximate values of vitrinite reflectance are shown for comparison. From Tissot and Welte, 1984.

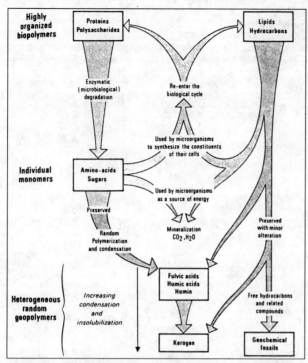

**Fig. 2-9.** Fate of organic material during sedimentation and diagenesis, two main organic fractions: kerogen and geochemical fossils. From Tissot and Welte, 1984.

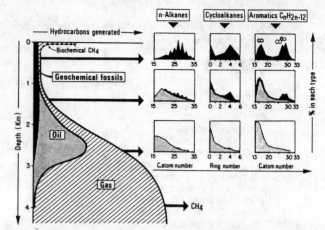

**Fig. 2-11.** General scheme of hydrocarbon formation as a function of burial of the source rock. The evolution of the hydrocarbon composition is shown in insets for three structural types. Depths are only indicative and correspond to an average on Mesozoic and Paleozoic source rocks. Actual depths vary according to the particular geological conditions: type of kerogen, burial history, geothermal gradient. (Modified after Tissot et al., 1974). This figure can be compared with other diagrams proposed by Sokolov (in Kartsev et al., 1971), Hedberg (1974), and Tissot and Welte, 1984.

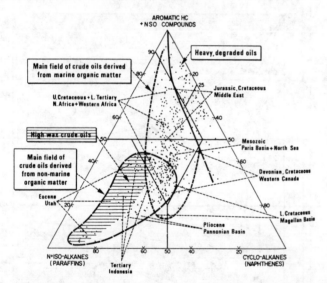

**Fig. 2-10.** Ternary diagram of crude oil composition, showing the principal field of occurrence of crude oils from marine and nonmarine origin. From Tissot and Welte, 1984.

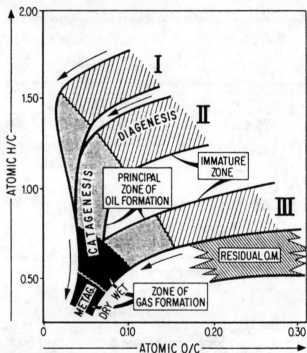

**Fig. 2-12.** From Tissot and Welte, 1984.

various organic fractions (Dow 1979) (Fig. 2-13). Physical changes outlined by Dow do not fully consider the contrasting nature of organic source materials, the transportational and depositional history of the organic fraction, or diagenetic changes in mineral or water composition during burial and compaction. The diagenetic changes in the composition of the organic fraction is illustrated by Tissot et al. (1974), following van Krevelen, and

shows the path of maturation for differing kerogen types (Fig. 2-14). These data may be compared to laboratory pyrolysis of kerogen at high temperature (Tissot et al. 1987) (Fig. 2-15). An indication of the rate, depth (temperature), and the amount of the kerogen converted to hydrocarbons, for differing types of kerogen, can be determined by a combination of pyrolysis and extract analysis of source rocks (Fig. 2-16). A more difficult

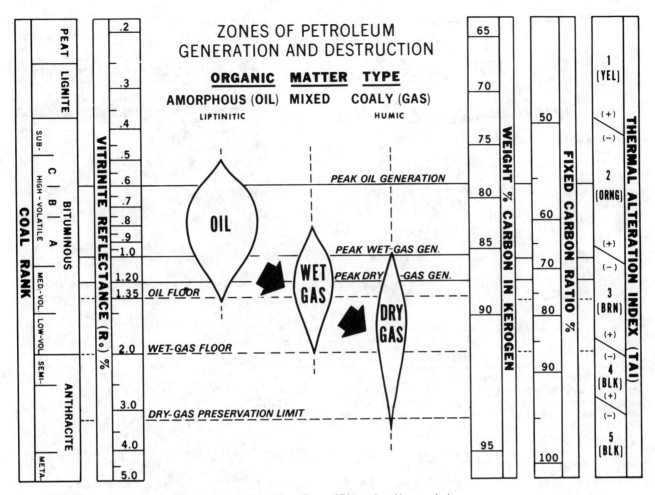

**Fig. 2-13.** Zones of petroleum generation and destruction. From Dow, 1979, reprinted by permission.

problem is adjusting laboratory data to reaction rates related to geologic time. In physiochemical rate equations effects of temperature are exponential and effects of time are linear (Fig. 2-17). A difference of only 10° Celsius may correspond to a maturation time period of 40 million years. With sufficient time, however, some stratigraphic intervals may reflect some hydrocarbon generation over an extended period of time. Waples (1980), following Lopatin (1971), suggested that both time and temperature are important in rate equations, and by calibrating reaction rates against physical and chemical indices of maturation, a Time-Temperature Index (TTI) could be devised. With knowldge of the history of the geothermal gradient, and the subsidence history of a basin, it would be possible to determine the amount and timing of petroleum generation (Fig. 2-18). This approach would be useful, especially if there is insufficient detailed geochemical data on the nature of the organic fraction, source and transportational history of the sediment, and the diagenetic history of both sediment and fluids.

Temperature is a principle control to the conversion of kerogen to hydrocarbons. Geothermal gradients are

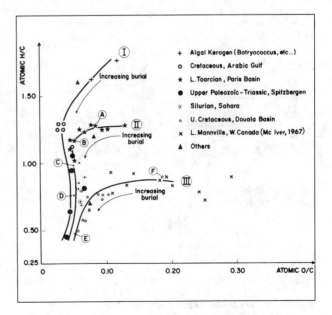

**Fig. 2-14.** Examples of kerogen evolution patterns. *Path I* includes algal kerogen excellent source rocks from Middle East; *Path II* includes good source rocks from Northern Africa and other basins; *Path III* corresponds to less oil productive organic matter but may include gas source rocks. Evolution of kerogen composition with depth is marked by arrow along each particular path. From Tissot, Durand, Espitalie, and Combaz, 1974, reprinted by permission.

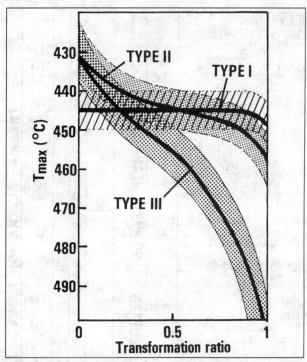

**Fig. 2-15.** Relationship between maturation index $T_{max}$ and transformation ratio of the main kerogen types, based on data form Rock-Eval pyrolysis of subsurface samples. Patterns show the range of data, with solid lines being the best fit. From Tissot et al., 1987, reprinted by permission.

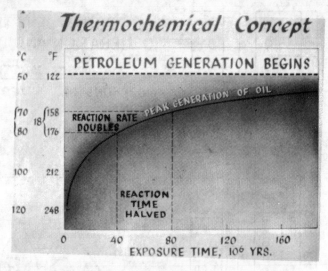

**Fig. 2-17.** Thermochemical concept. Reprinted courtesy Amoco.

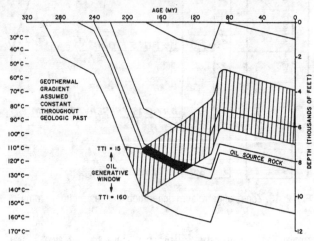

**Fig. 2-18.** Iso-TTI lines on geologic model. From Waples, 1980, reprinted by permission.

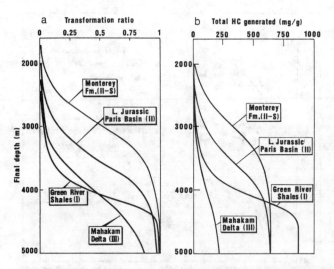

**Fig. 2-16.** Comparison of *a,* transformation ratio vs. depth for four different kerogen types subjected to the same thermal history. Note large difference of burial necessary to reach onset of oil generation between typical deltaic source (*Mahakam delta,* kerogen type *II*) and high-sulfur kerogen (*Monterey,* type *II-S*), or even between normal (type *II*) and high-sulfur (type *II-S*) kerogen. Furthermore, if we consider *b,* quantity of hydrocarbon generated per gram of kerogen, the important quantitative difference is due to respective hydrocarbon generative potential of various kerogens. Simple maturation indices obviously can not account for such diversity. From Tissot et al., 1987, reprinted by permission.

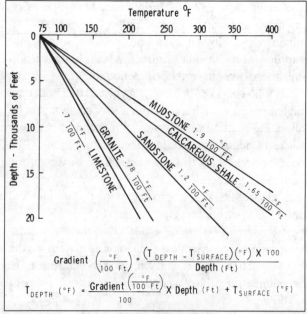

$$\text{Gradient}\left(\frac{°F}{100\ Ft}\right) = \frac{(T_{DEPTH} - T_{SURFACE})(°F) \times 100}{\text{Depth}_{(Ft)}}$$

$$T_{DEPTH}\ (°F) = \frac{\text{Gradient}\left(\frac{°F}{100\ Ft}\right)}{100} \times \text{Depth}_{(Ft)} + T_{SURFACE}\ (°F)$$

**Fig. 2-19.** Temperature gradients from wells for various rock types. From Bradley, 1975, reprinted by permission.

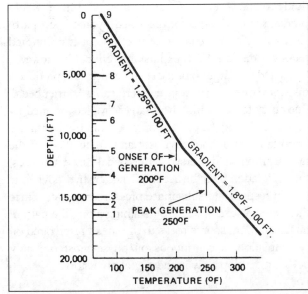

**Fig. 2-20.** Present depth-temperature profile for a well in the test area. Present depths to geologic units 1 through 9 also are shown. From Bonham, 1980, reprinted by permission.

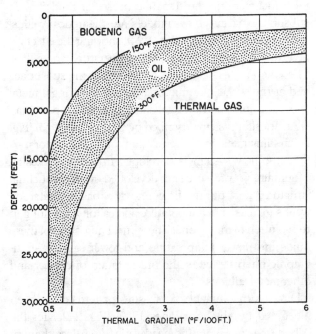

**Fig. 2-21.** The generation and destruction of crude oil related to depth, age, and geothermal gradient. The top surface corresponds to oil generation and the bottom surface to destruction. Neither should be taken as sharp boundaries but rather as gradational changes. From Barker, 1979, reprinted by permission.

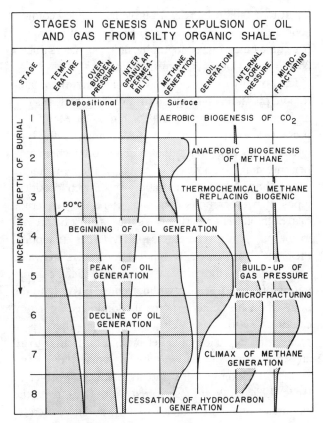

**Fig. 2-22.** Stages in genesis and expulsion of oil and gas from a silty organic-rich mud or shale. From Hedburg, 1980.

important in determining the range of depths where petroleum generation can proceed. Temperature gradients are strongly controlled by the physical and mineralogic characteristics of the stratigraphic intervals (Bradley 1975) (Fig. 2-19). Temperature at specific depths is controlled by the lithology, compaction, patterns of fluid flow, and proximity to low or highly conductive diapiric features. A common temperature effect is the

presence of thick overpressured shale intervals. Low permeability mudstones and calcareous shale stratigraphic intervals show steeper geothermal gradients due to their relatively low thermal conductivity (Bonham 1980, 78) (Fig. 2-20).

The concept of the oil window has developed and suggests that the beginning of significant hydrocarbon generation occurs close to 70° Celsius, and due to thermogenic destruction of the oil and formation of methane, there is no significant occurrence of oil above 150° Celsius (Barker 1979) (Fig. 2-21). A summary figure by Hedburg (1974) shows the interrelation of temperature, pressure, and permeability with the generation of gas and oil, and illustrates the relation of pore pressure and the development of microfracturing (Fig. 2-22).

The conversion of organic material to kerogen, and then to hydrocarbons is an inefficient process, and consequently the volumes of source rock required to generate large volumes of oil require special tectonic and depositional controls. If a source bed contains 2% organic carbon, and if the transformation ratio of kerogen to oil is about 20% (Table 2-1), then approximately 4 kilograms, or 400 parts per million by weight, of oil is generated per ton of rock (This would represent a very rich source rock). In terms of volumes, a cubic mile of oil is approximately equal to 3 billion barrels of oil, which would require a source rock volume, with 100% expulsion efficiency, of 300 cubic miles, or a stratigraphic unit 450 meters thick x 35 kilometers long x 2 kilometers wide.

The porosity of source rocks is complexly interrelated to pore pressure, compaction history, diagenetic history, and expulsion history (Hunt 1976) (Fig. 2-23). Many over-pressured source rocks may retain abnormally high (greater than 20%) porosity at burial depths where temperature may be above 70° Celsius, or the temperature where thermochemical alteration of kerogen produces types of hydrocarbons dominant in reservoired crude oils. Both oil wet and water wet surfaces must be present in the pores, with complex capillary forces developed. Preferential movement of smaller molecules of water, carbon dioxide, and methane should lead to the concentration of larger hydrocarbon molecules, possibly to the point where the center of pores are dominated by hydrocarbons (Tissot and Welte 1984, 336). There is a preferential depletion of larger molecules, from 3 to 10 angstroms in diameter, and highly polar NSO compounds, including resins and asphaltines (Tissot and Welte 1984, 332) in reservoired crude oils over the amounts present in source rocks. In addition there is apparently no relation to solubility of hydrocarbons between petroleum extracts from source rocks and reservoired crude oils. These considerations suggest that oil is expelled from source rocks as a discrete phase.

The size of pores and fractures is unknown, but estimates range from a few micrometers up to .1 millimeter. Clay particles are typically less than a few micrometers in size, and at the lower range of their size are similar in size to colloidal particles. Many hydrocaron molecules have dimensions similar to those of colloidal particles, and montmorillonite crystals range in size from a few thousandths to a few hundredths of a micrometer. Consequently, hydrocarbons expelled from source rocks must reasonably be as very small colloidal particles, not much larger than a few tens of molecular diameters in size. This would preserve compositional similarity to reservoired hydrocarbons and preferential restriction of asphaltine expulsion, and would indicate that the pore size distribution and capillary forces were sufficient to retain water, methane, and carbon dioxide in the pores. Consequently, expulsion of hydrocarbons would be associated with the simultaneous expulsion of methane, carbon dioxide, and water from the source rock.

The buildup of pressure is continuous, but the release of pressure is related to expulsion of a specific fluid volume. Expulsion of small volumes of fluid, less than 3%, will physically produce a significant pressure drop. If hydrocarbon volumes in source rocks are small, and periodic expulsion events are infrequent and reflect small fluid volumes, even a continuous oil phase would produce bulk volumes of dispersed hydrocarbon particles of only a few tens, or at most one or two hundred parts per million. Each expulsion event could, however, be related to a large volume of source rock, and produce a dispersed oil phase in adjacent permeable "carrier beds." Buoyancy forces could effect larger droplets, formed by accretion during the process of expulsion, but its relative importance is unknown (Tissot and Welte 1984). If the hydrocarbon is dispersed as colloidal sized particles, then migration must be accomplished by hydrodynamic flow in response to subsurface pressure gradients. Since there are few examples of oil staining along the path of migration from source rocks to the site of reservoired oil accumulation, a continuous oil phase most probably does not develop early in the migration history.

## Petroleum Migration

There are many lines of evidence and innumerable examples documenting the dynamic nature of fluid flow in subsurface basins. Flow models are theoretically illustrated by Roberts (1980) (Fig. 2-24), showing the effects of flow to produce pressure and temperature anomalies. These anomalies have been confirmed in the subsurface and at the surface by microseeps, saline springs, metal concentrations, and surficial changes in Redox potential.

An artesian system described by Germanov (1963) (Fig. 2-25) illustrates the interrelation of these factors. A dynamic flow system is developed from the Turkestan Mountains. A pressure sink developed over an anticlinal structure, and during flow into the basin, meteoric waters increased in temperature and salinity, resulting in the solution of metal ions. The vertical migration of these fluids into lower temperature and lower salinity stratigraphic intervals led to the precipitation of copper and other metal sulfides.

Direct measurement of pressure patterns have been illustrated by Hitchon (Jones 1980) (Fig. 2-26) for the Alberta basin. Hydraulic head varies from less than 30 meters to more than 70 meters above a datum. This contrast illustrates changed patterns of permeability, the importance of the basal Cretaceous unconformity as a permeable pathway for fluid migration, and the presence of a pressure sink at the Peace River drainage. The importance of drainages as pressure sinks is illustrated by the accumulation of hydrocarbons adjacent to the Athabasca River drainage (Jones 1980) (Fig. 2-27).

Subsurface pressure maps are rarely constructed, and their patterns have not been fully understood. Artesian systems, or systems associated with hydraulic head, have been documented from many basins throughout the world (Toth 1980) (Fig. 2-28). Pressure patterns in basins

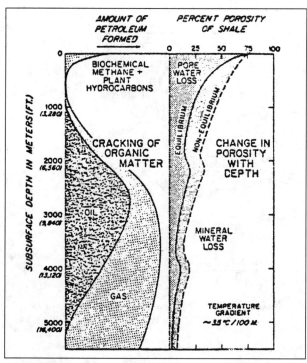

**Fig. 2-23.** Pattern of petroleum generation with depth and compaction compiled from many basins. From Hunt, 1976, reprinted by permission.

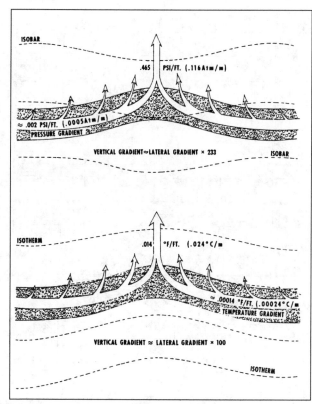

**Fig. 2-24.** *A:* Pressure change imposed on convergent, upward-moving waters and their contents. *B:* Temperature change imposed on convergent, upward-moving waters and their contents. From Roberts, 1980, reprinted by permission.

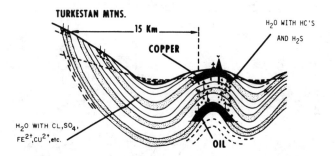

**Fig. 2-25.** Oil and copper ore trapped due to deepwater discharge in the same anticline at Ferghana USSR. Note oxidated waters outside trap and reduced waters inside trap (after Germanov) 1963. From Roberts, 1980, reprinted by permission.

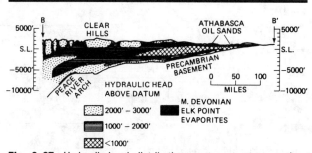

**Fig. 2-26.** Hydraulic-head cross section. Peace River oil-bearing sandstone deposit. After Hitchon, 1974, from Jones, 1980, reprinted by permission.

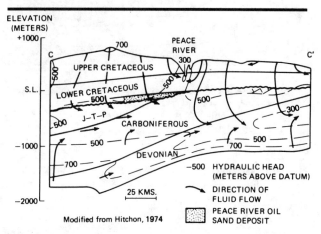

**Fig. 2-27.** Hydraulic-head distribution, east-west cross section, northern Alberta. After Hitchon 1969, from Jones, 1980, reprinted by permission.

which contain underpressured and overpressured intervals, not related to artesian pressures or to hydraulic head, have been described (Masroua 1973) (Fig. 2-29). The cause of mapped pressure patterns appears to be related to patterns of hydrocarbon generation and migration, and changes in history of permeability along the basal Pennsylvanian unconformity surface (Fig. 2-30). Individual reservoirs are filled with differing fluids at differing pressures (Fig. 2-31).

Indirect evidence for fluid migration can be shown in relation to solution of evaporite strata. From west Texas,

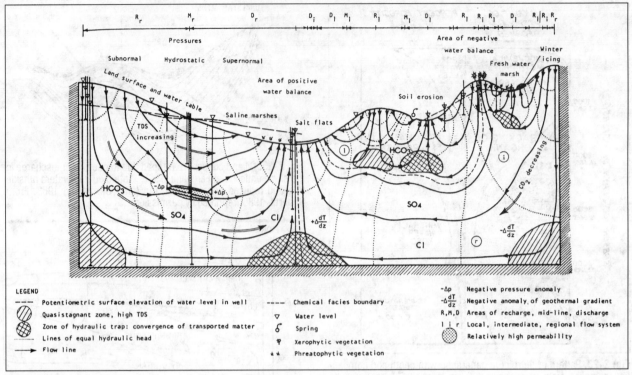

**Fig. 2–28.** Diagrammatic summary of properties and manifestations of regionally unconfined groundwater flow in simple and complex drainage basins. From Toth, 1980, reprinted by permission.

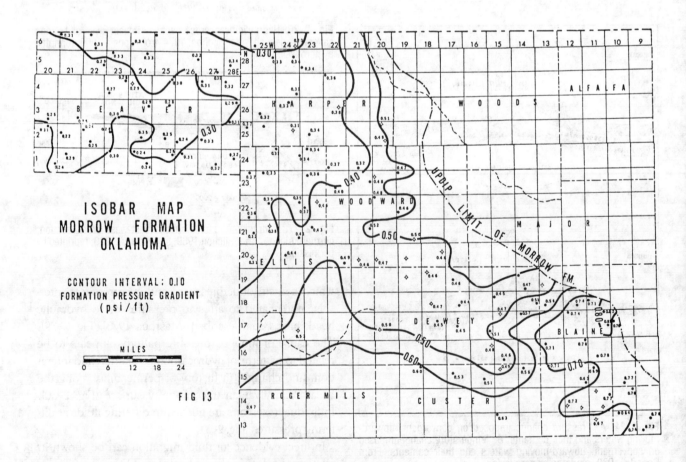

**Fig. 2–29.** Isobar map, Morrow formation, Oklahoma. Reprinted courtesy L. F. Masroua, 1973.

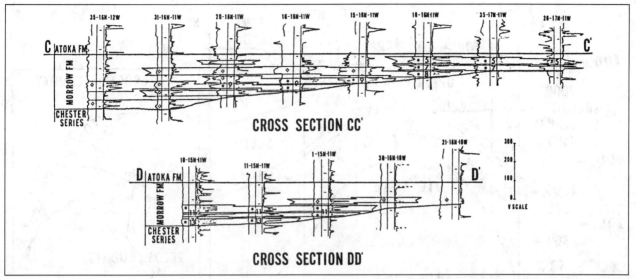

**Fig. 2-30.** Northeast-southwest stratigraphic cross sections showing the correlation of reservoirs 5 to 13. Reprinted courtesy of L. F. Masroua, 1973.

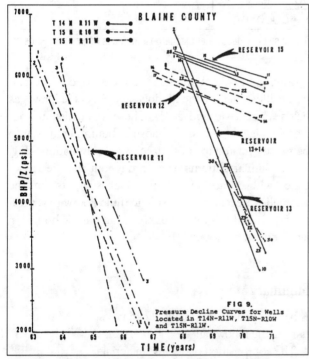

**Fig. 2-31.** Pressure decline curves for wells located in T14N-R11W, T15N-R10W and T15N-R11W. Reprinted courtesy of L. F. Masroua, 1973.

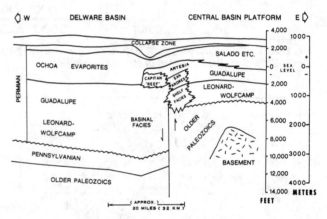

**Fig. 2-32.** Profile of westside, central (Permian) basin platform; showing removal of Permian ochoan evaporites and overburden collapse caused by deepwater discharge augmented by faulting and high-permeability shelf aquifers. From Roberts, 1980, reprinted by permission.

examples of solution of upper Permian evaporites produce thinned intervals above a faulted basin margin, resulting in the collapse of overlying strata (Roberts 1980) (Fig. 2-32), and a solution face of evaporite strata occurs below the Pecos River pressure sink (McCoy 1934) (Fig. 2-33). The patterns of salinity changes in subsurface also indicate the history of fluid migration. The Dakota sandstone from the Powder River basin illustrates the strong contrast of formation water salinity reflecting deep basin

connate waters partially flushed by meteoric water invading channel sandstone sequences (Jones 1980) (Fig. 2-34). Oil fields are associated with structural-stratigraphic traps where flushing has not occurred. Similarly, from the Delaware basin in west Texas, connate waters derived from the center of the basin with salinities of more than 250,000 milligrams/liter, are displaced by meteoric waters with salinities of less than 50,000 milligrams/liter (Collins 1975) (Fig. 2-35). Permeability traps are developed at updip pinchouts of sandstone intervals deposited in submarine channels.

Tracing of fluid flow patterns may be associated with both salinity anomalies and temperature anomalies. In southern Oklahoma overlying the Healdton field (an anticlinal trap), resistivity ($R_w$) ($R_w$ is the water resistivity, which may be measured or calculated from petrophysical logs) is contoured in ohm-meters at two depths: 9 meters

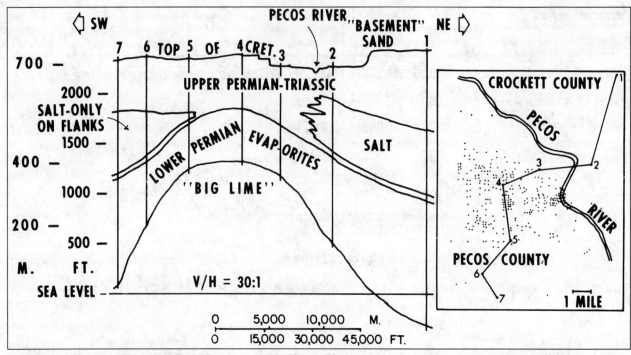

**Fig. 2-33.** Profile of Yates field, West Texas, showing removal of Permo-Triassic salts by deepwater discharge into Pecos River. From Roberts, 1980, after McCoy, 1934, reprinted by permission.

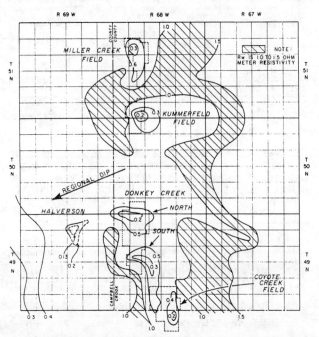

**Fig. 2-34.** Regional Dakota water resistivity pattern. From Jones, 1980, reprinted by permission.

geothermal gradient anomalies are developed. The pattern is illustrated by the Madden and Frenchie Draw gas fields (Figs. 2-40 and 2-41). These anomalies are not closely related to structural closure, but rather appear to be mostly related to subsurface gas accumulations (Fig. 2-42). Similar patterns exist throughout the area associated with oil accumulations, which suggests that the anomalies are related to subsurface fluid flow with vertical migration of fluids into shallower stratigraphic horizons, as described by Roberts (1980).

## Summary

A basin's petroleum potential can be estimated by determining the following:
- The composition of organic materials,
- The temperature history of the basin, and
- The volume of rock that could be the source of hydrocarbons.

These are measurable aspects that can and should be evaluated in every well that is drilled in a basin. Total organic content, soluble hydrocarbons, chromatographic analysis and vitrinite reflectance are only a few of the modern tools available to determine the source rock potential of a basin.

The hydrodynamic history of a basin is revealed in the patterns of salinity, pressure, and shale density. These aspects are easily measured by logging techniques, sample examinations, formation testing, and locally by seismic methods.

and 370 meters (McConnell 1985) (Figs. 2-36 and 2-37). In addition, the geothermal gradient was measured over the field area (Fig. 2-38). Both sets of data indicate a high correlation with the field outline. Other examples where temperature anomalies can be correlated with hydrocarbon accumulations have been described from the Rocky Mountain region (Meyer and McGee 1985) (Fig. 2-39). Overlying both structural and stratigraphic traps,

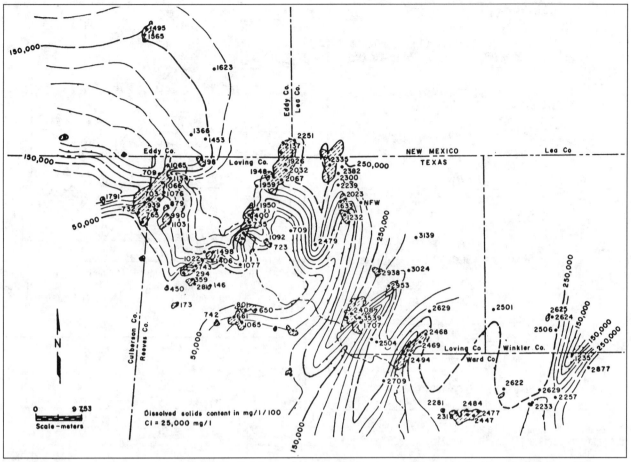

**Fig. 2-35.** Map of the dissolved solids content of the Bell Canyon formation waters, eastern Delaware basin, Texas. From Collins, 1975.

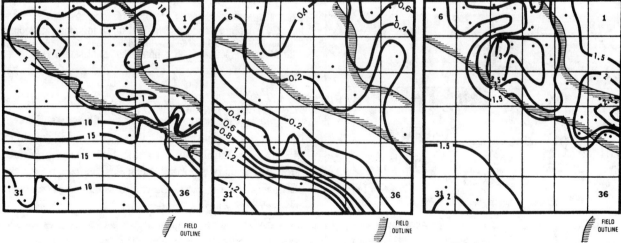

**Fig. 2-36.** Water resistivity ($R_w$) of Healdton area (T4S, R3W), at depth of 300 ft (90 m). Resistivity values are in ohm-meters ($\Omega$ m) and are SP-log derived. Dots are control wells. McConnell, 1985, reprinted by permission.

**Fig. 2-37.** Water resistivity ($R_w$) of Healdton area (T4S, R3W), at a depth of 1,200 ft (370 m). Resistivity values are in ohm-meters ($\Omega$ m) and are SP-log derived. Dots are control wells. From McConnell, 1985, reprinted by permission.

**Fig. 2-38.** Geothermal gradient map of the Healdton area (T4S, R3W). Gradient values are in °F/100 ft and were measured at about 1,200 ft (370 m). Dots are control wells. From McConnell, 1985, reprinted by permission.

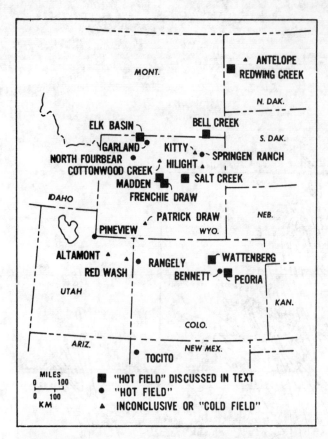

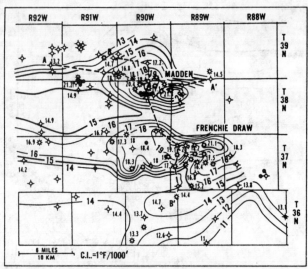

**Fig. 2-39.** Rocky Mountain oil and gas fields examined for temperature gradient anomalies. From Meyer and McGee, 1985, reprinted by permission.

**Fig. 2-41.** Temperature gradient map in Madden and Frenchie Draw field area, Wyoming. Gradient shown here is calculated from producing depth to ground surface. From Meyer and McGee, 1985, reprinted by permission.

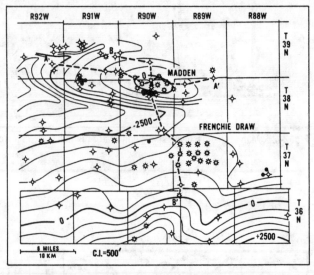

**Fig. 2-42.** Top of Paleocene lower Fort Union formation in Madden and Frenchie Draw field area in Wind River basin, central Wyoming. From Myer and McGee, 1985, reprinted by permission.

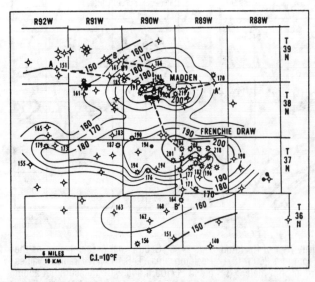

**Fig. 2-40.** Temperature map at 2,500 ft (762 m) subsea level in Madden and Frenchie Draw field area, Wyoming. Temperature values interpolated from individual well geothermal gradients. From Meyer and McGee, 1985, reprinted by permission.

Accumulation and entrapment of hydrocarbons requires the displacement of water by oil and gas. Water is the continuous phase in sedimentary rocks and is held in the pores by capillary forces and hydrostatic pressure. Globules of oil or bubbles of gas can displace this water only to the extent that the upward buoyant force can overcome capillary forces. The smaller the pores, the greater the capillary force. Migrating petroleum occurs as dispersed globules and moves with the flow of connate waters. Water continues to move through sediments long after initial compaction, and petroleum is continually concentrated at boundaries where capillary forces are higher than the force produced by the difference in density between oil and water.

Avenues of migration must be a continuous chain of large pores or fractures. Petroleum accumulation can only occur at the upward terminus of these pathways or at a change in relative permeability. This can be a fault gouge, a fine-grained shale or silt, a chemical sediment, or some other type of capillary pressure barrier. These barriers can be determined by mapping salinity and pressure patterns. Major changes in either of these is suggestive of a barrier to migration and thus a potential area for petroleum accumulation.

## TIME-STRATIGRAPHIC MAPS

Time-stratigraphic maps can be developed to illustrate the following:

▸ Geometry of the basin,
▸ Pattern of shoreline migration,
▸ Paleogeography of the depositional area, and
▸ Provenance areas.

Mapping reservoir rocks is the function of petroleum geologists. The prediction of the geometry of a reservoir unit necessitated an understanding of its origin and controls for sedimentary patterns. These are interpreted by understanding sedimentary processes. Stratigraphic patterns are the result of these processes, and by interpreting boundary conditions and by understanding these processes, reservoir patterns can be predicted. This requires interpreting water depths, synsedimentary tectonic processes, relation to ocean basins, and boundary conditions reflected in stratigraphic aspects.

Of all the maps available to the petroleum explorationist, the facies pattern, illustrated by time-rock units bounded by time datums, is the most important. These maps, at whatever scale or time interval, show both tectonic and depositional history. The larger the interval and area included, the more insight can be obtained into the controls for stratigraphic responses. These maps reflect synsedimentary tectonism, provenance, direction and pattern of shoreline shift, and the paleogeography that controls the sedimentary pattern. The Upper Cretaceous from the Rocky Mountain area illustrates the principles (Sloss et al. 1960) (Fig. 2–42). Most areas of significant hydrocarbon accumulation can be predicted by the area defined by the 1/2 to 1/4 sand-shale ratio. Areas of deltas, directions of clastic source, position of greatest subsidence, and water depths are indicated.

The chronozone map is useful to determine the areal variation of any aspect other than that chosen as the mapped aspect. It is useful for determining depositional patterns of potential reservoir rocks. Prediction can be based upon a developmental pattern (Krumme 1980) (Fig. 2–43). In conjunction with a structure map, it is a useful tool for exploration.

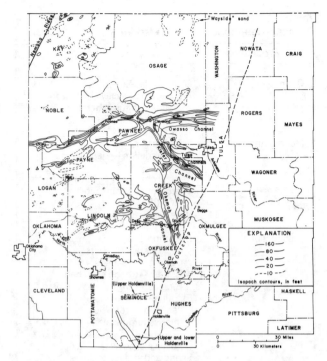

**Fig. 2–43.** Isopach map of Cleveland sandstone in eastern Oklahoma. From Krumme, 1981.

## Sedimentary Models

Of similar importance to other stratigraphic aspects is the prediction of patterns of reservoir and source rocks. Since the vertical succession of stratigraphic units is fundamentally related to the paleogeography of the depositional site, it is possible to study Holocene sedimentary patterns and use these to predict the geometry of ancient stratigraphic units. The interpretation of sedimentary processes, and therefore stratigraphic patterns, has been the basis for interpreting the origin of stratigraphic units. Whether a unit is fluvial, deltaic, shoreline, shelf, or deeper marine is important if the petroleum geologist is to predict the reservoir pattern, heterogeneities, reservoir performance, and pattern of source rock. If the proper structural, temperature, fluid dynamic, and migration history is present, then the depositional pattern of the reservoir is the basis for exploration.

Identification of the reservoir pattern from limited stratigraphic data is now possible. Core and well log data can be used to interpret sedimentary structures, textures, trace fossil assemblages, mineralogy, and other aspects formed in response to sedimentary processes. The depositional process of nearly all sedimentary units can be interpreted. The organizing principle useful in this type of interpretation is the process-response pattern. A good example of this approach is shown by the fluvial valley-fill sedimentary pattern (Table 2–3).

**Table 2-3**
**Process-Response Valley-Fill Model**

| Processes | Responses |
|---|---|
| Channeling | Scour surfaces<br>Lag deposits — clay chips, wood, coarse detritus concretions, reworked fauna |
| Meandering | Point bars<br>Depth variation — depositional surface<br>Vertical changes in grain size<br>Vertical changes in amplitude of structures<br>Valley fill |
| Unidirectional flow | Bedforms<br>  Sand waves and bars — planar crossbeds<br>  Dunes — trough crossbeds<br>  Ripples — lingoid and climbing<br>Entrainment of detritus<br>  "saltation" 100-750$\mu$<br>  Intense turbulence — suspension<br>  Positive skewness — fair sorting<br>Fabric and surface textures<br>  Alignment of particles<br>  Few impact marks<br>  Equant to rod-shaped particles |
| Seasonal discharge peaks | Sand unit as thick as channel depth<br>Rapid deposition after peak flow<br>  Separation of channel and floodplain deposits |
| Velocity pulsations | Lamination<br>Sediment accretion |
| Chemical — water<br>pH~6.8<br>Eh~ + 0.4<br>T.S.<5,000 ppm | Dispersion of clays<br>Elimination of detrital ferrous iron compounds<br>Hydrolysis of potassium and magnesium<br>Elimination of illite and chlorite<br>Colloids of silica, alumina, iron, and organic<br>Organisms restricted |

All sedimentary regimes can be characterized by similar types of associations. Since the stratigraphic responses reflect these processes, the origin of stratigraphic sequences can be interpreted. The mapping of reservoir patterns can be based upon direct comparison to Holocene depositional systems. The bases for comparison are the sedimentary responses. Whether it is a log shape, a specific sedimentary structure or texture, or the paleogeography of sedimentary units, the prediction of the reservoir pattern requires understanding the depositional process.

Data needed for interpretation are provided by plate tectonic patterns, regional geophysical surveys (gravity and seismic), detailed analysis of well cuttings, cores, and logs, sedimentary facies patterns, and the interpretation of sedimentary processes. These all can be related to the basin framework to provide an understanding of the developmental history of the stratigraphic fill. With this understanding, prediction of patterns of petroleum accumulations is possible.

# STRATIGRAPHIC PATTERNS FROM SEISMIC DATA

Stratigraphy is a four dimensional science, and the construction of such a stratigraphic framework requires the use of information on the areal variation of stratigraphic units in both space and time. Data relating to geometry of stratigraphic intervals, relative time relationships, and the application of a rigorous absolute time scale is needed to suggest causal stratigraphic patterns. Patterns derived from seismic data can provide information on all of these stratigraphic measures.

Seismic reflections are produced by the constructive interference of a wave train with an array of impedence contrasts (velocity-density) within the earth. Impedence contrasts are related to lithologic and fluid changes within a stratigraphic interval which are parallel or sub-parallel to depositional or time-rock surfaces. Thus the seismic record can portray, by the pattern of reflections, the time-rock architecture of a stratigraphic unit. Amplitude of reflections is related to the impedence contrast between lithologic units, to the areal continuity of reflecting surfaces, to the number and spacing of reflecting surfaces, and to the frequency of the seismic wave train that interacts with these surfaces (Table 2–4). Other aspects including geophone spread length, rate of change of reflection surfaces, tuning frequency, and lat-

**Table 2-4**
**Geologic Interpretation of Seismic Facies Parameters**

| Seismic Facies Parameters | Geologic Interpretation |
|---|---|
| Reflection configuration | • Bedding patterns<br>• Depositional processes<br>• Erosion and paleotopography<br>• Fluid contacts |
| Reflection continuity | • Bedding continuity<br>• Depositional processes |
| Reflection amplitude | • Velocity-density contrast<br>• Bed spacing<br>• Fluid content |
| Reflection frequency | • Bed thickness<br>• Fluid content |
| Interval velocity | • Estimation of lithology<br>• Estimation of porosity<br>• Fluid content |
| External form and areal association of seismic facies units | • Gross depositional environment<br>• Sediment source<br>• Geologic setting |

Mitchum, Vail, & Sangree; 1977.

eral changes in interval velocity also affect the continuity, polarity, and amplitude of a reflection. The most probable continuous reflections are those that are parallel to the depositional surface, or bedding planes, which are related to time-rock depositional surfaces. These surfaces reflect the geometry and history of deposition of a stratigraphic unit. Since seismic lines reflect a continuity of change, and an intersecting grid of the lines can be used to correlate a three-dimensional array of reflectors, it is possible to map a three dimensional array of time-rock surfaces.

As described in Chapter 1, the stratigraphic record is subdivided into stratigraphic units related to sedimentary, geostrophic, tectonic, extraterrestrial, and plate tectonic events. The occurrence of these events is reflected in the stratigraphic (historic) record of the earth by unconformity surfaces. These surfaces often are widely distributed bedding planes, and often reflect patterns of erosion, topographic changes, changes in sea level, and changes in the geometry of depositional units. Such changes in bedding patterns are demonstrated by changed patterns of seismic reflections. Events recorded by seismic reflection patterns may be local, basinal, or worldwide in their distribution, and on time scales of a 20,000 to 10 million years. Correlation of these "unconformity producing events" can establish local relative time scales, can be related to paleontologic evolutionary history and events, or can be related to a worldwide array of absolute age measurements (Tables 1–5 and 1–6). Thus the pattern of seismic reflections can be correlated to both local and to worldwide time scales, producing a framework for four-dimensional stratigraphic analysis.

Subsurface geologists have understood that the character of log and lithologic responses revealed in both outcrops and subsurface well data could provide the necessary information to construct four dimensional time-stratigraphic patterns. Detailed correlation of surface and subsurface log data, construction of a grid of cross sections produced the information necessary for this analysis (Mitchum et al. 1977a) (Fig. 2–44). Work of this type required the careful analysis of many wells, utilizing paleotologic, lithologic, and an understanding of local tectonic and depositional controls.

The acquiring of these data necessitated the expenditure of a great deal of time by professional geologists, and the availability of closely spaced well or outcrop sections. Inferences concerning time-rock correlations were based upon the construction of paleogeologic, facies, and iso-

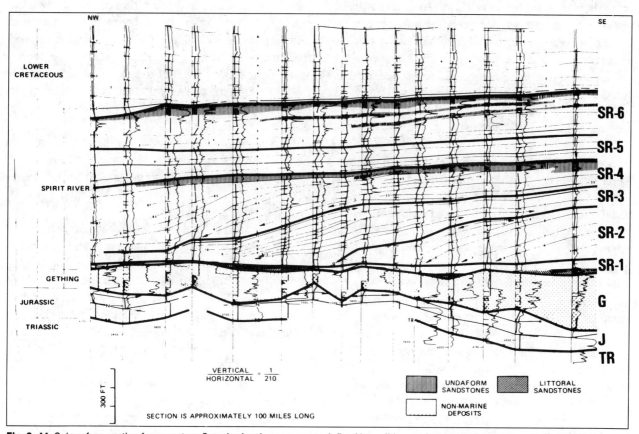

**Fig. 2–44.** Subsurface section from western Canada showing sequences defined by well-log market correlation (prepared by G.T. McCallum and provided by Imperial Oil Ltd). Market correlation of closely spaced well logs displays depositional patterns of Lower Cretaceous strata in northwestern British Columbia. Erosion truncation of Jurassic beds (*surface J*), onlap of lowermost Spirit River onto Gething (*surface G*), and progressive downlap and minor onlap of prograding upper Spirit River beds are used to determine several sequence boundaries. From Mitchum, Vail, and Thompson, 1977, reprinted by permission.

**Table 2-5**
**Geologic Interpretation of Seismic Facies Parameters**

| Reflection Terminations at Sequence Boundaries | Reflection Configurations within Sequences | | External Forms of Sequences and Seismic Facies Units |
|---|---|---|---|
| Lapout | Principal stratal configuration | | Sheet |
|   Baselap |   Parallel | | Sheet drape |
|     Onlap |   Subparallel | | Wedge |
|     Downlap |   Divergent | | Bank |
|   Toplap |   Prograding clinoforms | | Lens |
| Truncation |     Sigmoid | | Mound |
|   Erosional |     Oblique | | Fill |
|   Structural |     Complex sigmoid-oblique | | |
| Concordance |     Shingled | | |
|   No termination |     Hummocky clinoform | | |
| |   Chaotic | | |
| |   Reflection-free | | |
| | Modifying terms | | |
| |   Even | Hummocky | |
| |   Wavy | Lenticular | |
| |   Regular | Disrupted | |
| |   Irregular | Contorted | |
| |   Uniform | | |
| |   Variable | | |

From Mitchum, Vail, and Sangree; 1977.

pachous maps. Rarely was it possible to place a high degree of confidence in the paleogeographic interpretations derived from this work. Environmental analysis, reservoir characterization, and depositional history were all the subject of intense debate among geological scientists.

The understanding that seismic reflection sections could be used to establish genetic depositional units and to define the depositional patterns (bedding planes) by analysis of internal and external reflection surfaces provided the explorationist with the tool needed to reconstruct depositional history. Reflection geometries, isopachous maps of genetic units, maps showing patterns of reflection terminations, and maps of velocity variations within genetic units became the most powerful stratigraphic tool available to the explorationist. From limited data it was possible to predict patterns of stratigraphic variation, to determine facies patterns, and to interpret structural and stratigraphic depositional histories. (Mitchum et al. 1977) (Fig. 2–45) shows the nature of seismic data available for interpretation. Analysis of bedding plane terminations produces a pattern of topographic, structural, and depositional history that can be used to interpret facies patterns (Mitchum and Vail 1977) (Figs. 2–46 and 2–47). The four dimensional pattern of genetic units may be derived from a grid of seismic sections where each terminating reflection event can be correlated throughout the area of study, and the areal variation in the internal patterns of reflections can be used to develop a paleogeographic analysis of successive depositional units.

Such an analysis has been illustrated by Vail and his co-workers (Vail et al. 1977a) (Fig. 2–48). Bounding surface have been correlated to the empirical sealevel curve (Haq et al. 1987) (Tables 1–5 and 1–6). The vertical alignment of terminations, the absolute ages of sea-level events, and the internal bedding plane patterns allow the reconstructing of depositional history and pattern of facies changes within the cross section, and when correlated to other seismic sections, the four dimensional stratigraphic pattern can be mapped.

A terminology of bedding plane terminations has been developed for ease of communication (Fig. 2–49a), and the paleogeologic relationships suggested by topographic, termination, and internal reflection patterns (Vail 1987) (Fig. 2–49b).

Detailed analysis of these relationships, in both stratigraphic and inferred depositional depth, and in relation to geologic time, allows the stratigrapher to construct a subsurface seismic sequence analysis (Vail 1987, Mitchum et al. 1977) (Fig. 2–50 and Table 2–5). As indicated in Chapter 1, there is a genetic relationship between the internal and external shape of a genetic stratigraphic unit related to the functional interelation of f(Receptor capacity, $R$; Dispersive energy, $D$; and rate of supply, $Q$.: $[S = f(R,D,Q)]$. In addition, climate, latitude, and rates of subsidence and sea-level rise have been shown to be important in the architecture of a genetic sequence (Vail et al. 1977) (Figs. 2–51 and 2–52) to relative rise in sea level with coastal onlap (Fig. 2–52).

▸ Fall in sea level resulting in changes in degree of truncation, or the position of deposition of the next higher

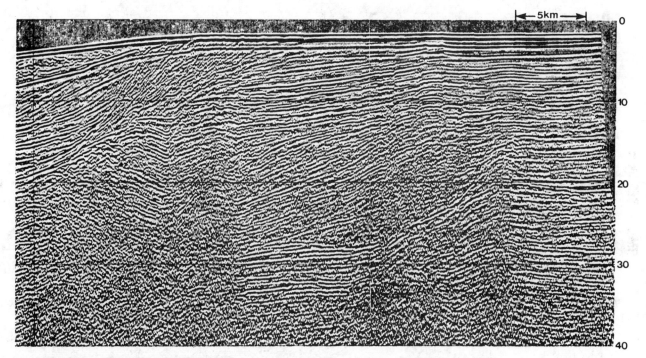

**Fig. 2–45.** Section offshore West Africa. From Mitchum, Vail, and Thompson, 1977, reprinted by permission.

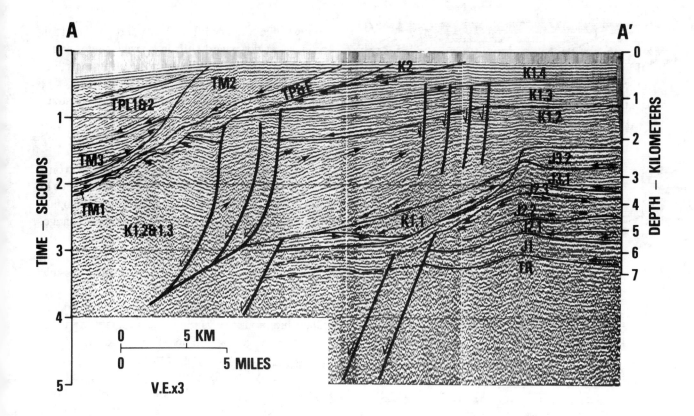

**Fig. 2–46.** Seismic sequences from offshore northwestern Africa showing sequences defined by seismic reflections. Systematic reflection termination patterns used to determine downlap, onlap, truncation, and toplap of strata G and sequence boundaries. Geologic ages of sequences determined from wells on section and in surrounding area. From Mitchum and Vail, 1977, reprinted by permission.

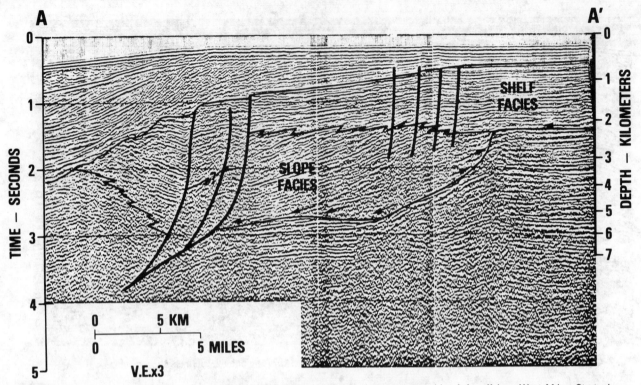

**Fig. 2-47.** Generalized seismic facies, Lower Cretaceous Valanginian through Aptian sequences, Line A-A¹, offshore West Africa. Strata deposited in shelf depositional environment prograde right to left over deposits of slope environment. From Mitchum and Vail, 1977, reprinted by permission.

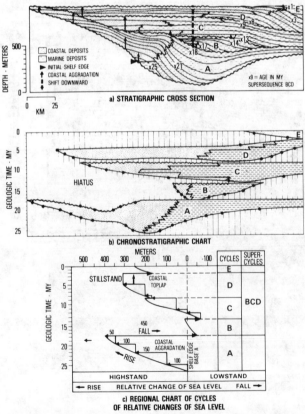

**Fig. 2-48.** Procedure for constructing regional chart of cycles of relative changes at sea level. From Vail, Mitchum, and Thompson, 1977, reprinted by permission.

stratigraphic sequence (Fig. 2-52a and b). Rapid fall produces a downward shift in coastal onlap, and a gradual fall produces a downward shift in the clinoform pattern with resulting truncation.

▶ The geometry of the progradational chronostratigraphic markers, whether parallel, oblique, or sigmoidal, is in response to the balance between rates of relative rise in sea level and sedimentation.

Due to changes in time and place of sedimentation on a cratonic or basin margin, there are relative differences in the rates of sea level rise and subsidence. Due to changing position of the shoreline, erosion, base level change, effects of loading, and bypassing of sediments, the sequence is composed of contrasting internal and external geometries. Figure 2-53 illustrates a possible sequence and depositional system pattern (Vail 1987). With linear subsidence the eustatic sea-level curve will oscillate and produce a varying paleogeographic pattern in depth and geologic time. The thickness of the sequence is primarily controlled by subsidence, and the depositional system is primarily controlled by the receptor capacity and the relative position of the shoreline. Four depositional system tracts can be defined, which include 1) the Lowstand System Tract, 2) the Transgressive System Tract, 3) the Highstand System Tract, and 4) the Shelf Margin System Tract. These systems then can

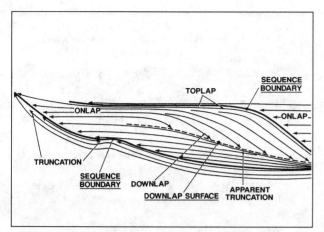

**Fig. 2–49a.** Diagram showing reflection termination patterns and types of discontinuities. Discontinuity names are underlined.

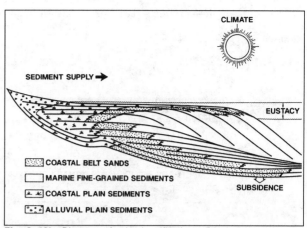

**Fig. 2–49b.** Diagram of seismic reflections, lithofacies, and major variables affecting stratigraphy. Vail, 1987, reprinted by permission.

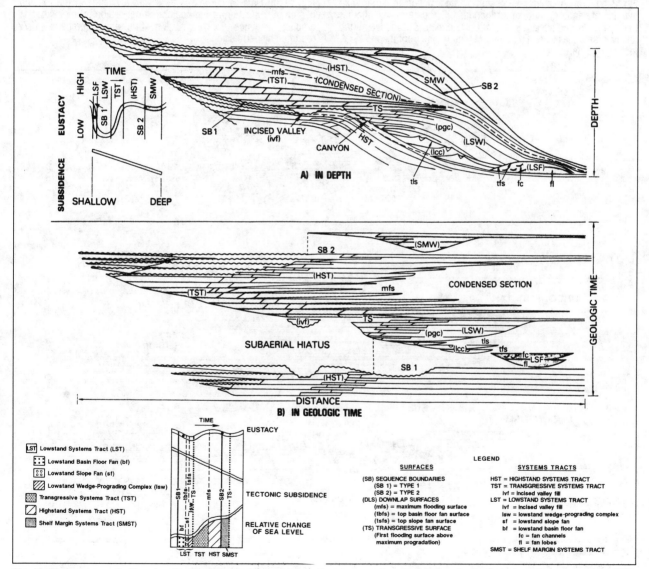

**Fig. 2–50.** Sequence stratigraphy diagrammatic section, showing sequences and systems tracts in depth and geologic time. Vail, 1987, reprinted by permission.

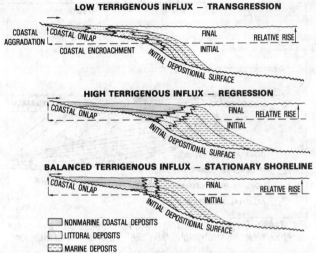

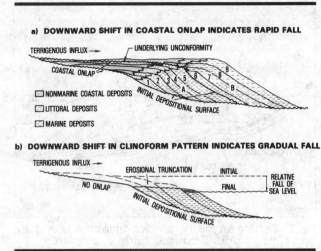

**Fig. 2-51.** Transgression, regression, and coastal onlap during relative rise of sea level. Rate of terrigenous influx determines whether transgression, regression, or stationary shoreline is produced during relative rise of sea level. From Vail, Mitchum, and Thompson, 1977, reprinted by permission.

**Fig. 2-52.** Rate of fall in sea level with a change in the resulting degree of truncation or the position of deposition of the next stratigraphic sequence. From Vail, Mitchum, and Thompson, 1977, reprinted by permission.

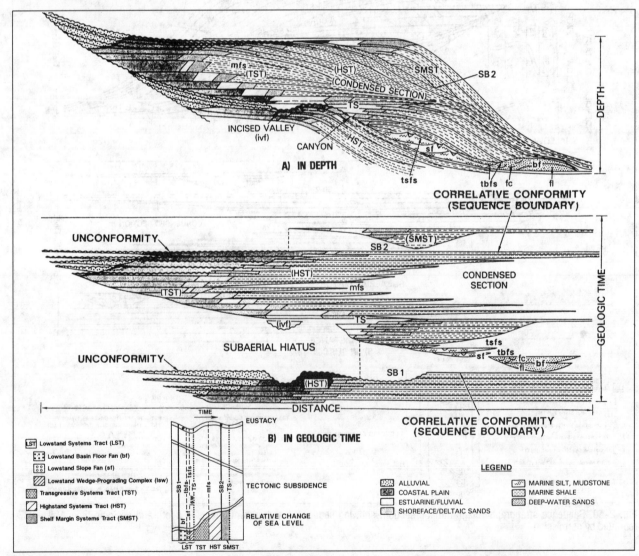

**Fig. 2-53.** Sequence stratigraphy diagrammatic siliciclastic sediments within sequences. From Vail, 1987, reprinted by permission.

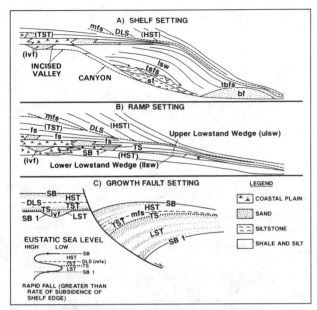

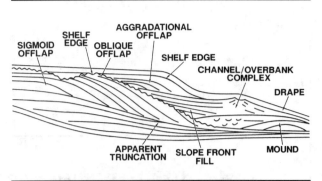

**Fig. 2-56.** From Vail, 1987, reprinted by permission.

**Fig. 2-54.** Sequence stratigraphy diagrammatic section, showing lowstand systems tracts in shelf, ramp and growth fault settings. From Vail, 1987, reprinted by permission.

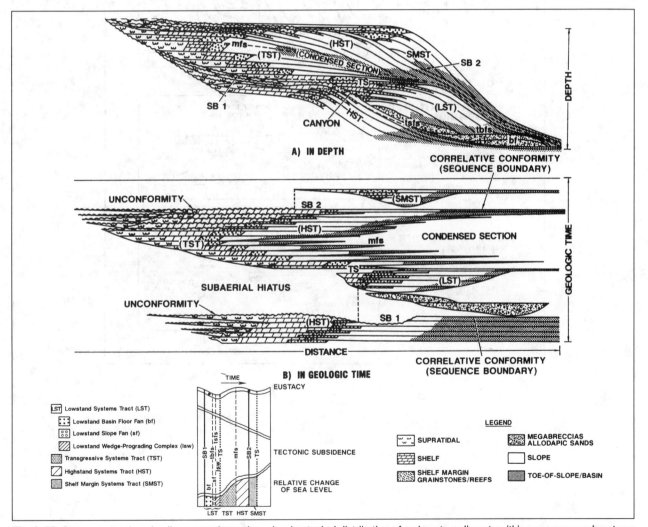

**Fig. 2-55.** Sequence stratigraphy diagrammatic section, showing typical distribution of carbonate sediments within sequences and systems tracts in depth and geologic time. From Vail, 1987, reprinted by permission.

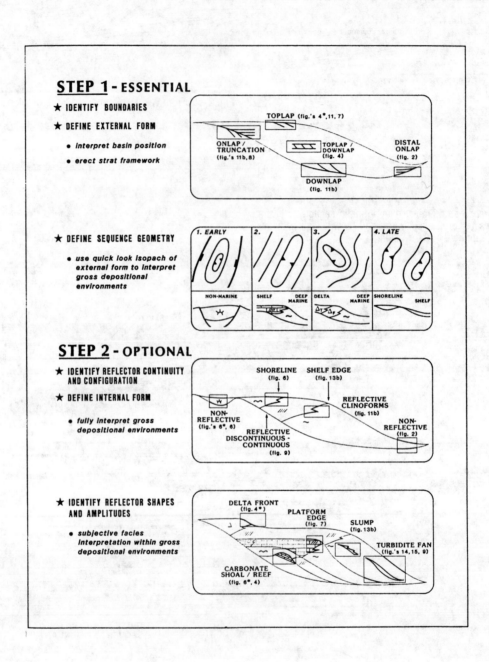

**Fig. 2-57.** From Hubbard et al., 1985, reprinted by permission.

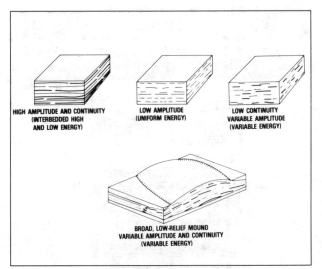

**Fig. 2–58.** Types of shelf seismic facies. From Sangree and Widmier, 1977, reprinted by permission.

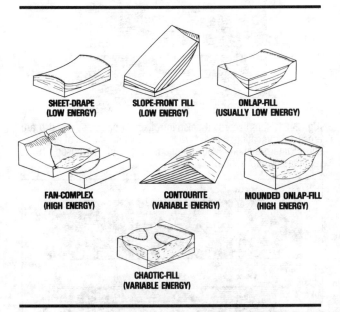

**Fig. 2–59.** Types of basin slope and floor seismic facies. From Sangree and Widmier, 1977, reprinted by permission.

be related to depositional processes and sedimetary facies patterns.

Three siliciclastic sequence frameworks have been illustrated by Vail, which include a Shelf Margin Setting, a Ramp Setting, and a Growth Fault Setting (Vail 1987) (Fig. 2–54). These are common paleogeographic patterns and many stratigraphic units show facies patterns similar to those described by Vail. Work by Brown and Fisher (1977; 1980), and Brown (1984) show other types of paleogeographic frameworks, processes, and depositional systems. The addition of controlling factors such as shelf

topography and width, basin depth, rates of sediment supply, subsidence, sea-level rise, latitude, and controlling depositional themes such as tidal, wave, and current systems will change paleogeographic patterns and the nature of depositional systems.

Sequence analysis may be developed in a similar fashion for carbonate depositional systems (Fig. 2–55). In contrast to siliciclastic systems, carbonate systems are strongly influenced by the interrelations of many variables. The receptor capacity is a more complex interrelation between patterns of subsidence, rates of sedimentation, and sea-level rise; dispersive energy is more controlled by rates of sedimentation and shelf geometry; rates of sedimentation are more controlled by oceanic and shelf circulation patterns rather than supply of sediment to the shelf. Facies patterns can be complex with the synsedimentary development of intrashelf basins, either shelf edge or basin margin biohermal buildups, a depositional ramp, and the shelf margin system illustrated by Vail (1987) (Fig. 2–55).

## SUBSURFACE APPLICATIONS

### Interpretation of Bedding Patterns

Subsurface facies analysis of seismic sequences is based upon detailed analysis of the geometry of reflection patterns (Fig. 2–56). A detailed listing of terms has been developed by Mitchum et al. (1977) (Table 2–5). Two and three dimensional aspects of bedding patterns reflect depositional processes, and can be used to interpret internal and external geometries within seismic sequences (Hubbard et al. 1985) (Fig. 2–57).

Reflector continuity and amplitude are important aspects to suggest bedding continuity and lithologic transitions. Identifying these patterns allows the suggestion of depositional processes and paleogeographic position within the stratigraphic framework (Figs. 2–58 and 2–59) (Sangree and Widmier 1977). Many of these patterns were not recognizable from conventional stratigraphic observation. The seismic tool is useful for defining bedding patterns on the scale of tens of meters vertically and hundreds of meters areally. These features often have not been identified utilizing subsurface well data, or from measured outcrop sections. The external form of features that extend over areas of many square kilometers have required innovative interpretation of subsurface data, often based upon dipmeter, detailed correlation of depositional events, and the preparation of isopachous maps between closely spaced event markers (Mitchum et al. 1977) (Fig. 2–60). Careful analysis of seismic reflection patterns are required to recognize these important depositional patterns.

Bedding plane patterns illustrated by seismic sections require definition by identifying sequence boundaries and internal bedding patterns. Marking of bedding terminations is useful to characterize both internal and external geometries (Mitchum et al. 1977) (Fig. 2–61). Internal bedding patterns defined as sigmoidal, oblique, parallel oblique, and complex sigmoid-oblique have required new stratigraphic observation to understand and interpret their origin (Mitchum et al. 1977) (Fig. 2–62). Seismic facies analysis of these complex patterns has been correlated to well data and specific depositional histories suggested for the Brazilian passive continental margin (Brown and Fisher 1977) (Fig. 2–63).

## Development of Subsurface Models

Vail and co-workers have developed a terminology to define seismic sequence types and their subsets (Van Wagoner et al. 1987). Figure 2–64 illustrates the relation of bedding geometry, sequence boundaries, erosional surfaces, parasequences, and log response patterns. This synthesis provides the basis for recognizing depositional environments, and sedimentary facies, within a genetic time-rock interval. Depositional history relating the boundary conditions described by the functional relations of $[S = f(R, D, Q)]$ is illustrated by the vertical and areal stacking of parasequences.

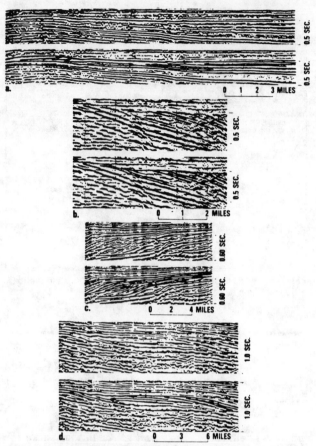

**Fig. 2–61.** Top-discordant seismic reflection patterns; *a* and *b* are erosional truncation; *c* and *d* are toplap. Second section of each pair shows interpretation. From Mitchum, Vail, and Thompson, 1977, reprinted by permission.

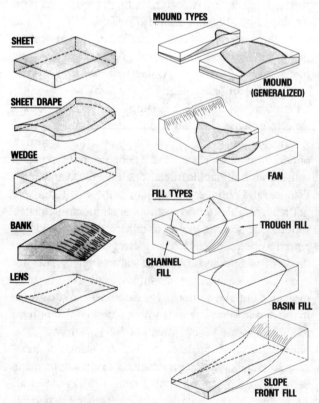

**Fig. 2–60.** External forms of some seismic facies units. From Mitchum, Vail, and Thompson, 1977, reprinted by permission.

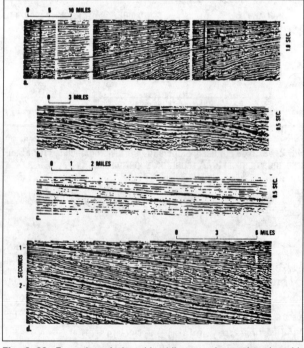

**Fig. 2–62.** Examples of sigmoid, oblique, and complex sigmoid-oblique seismic reflection configurations: *a* is sigmoid; *b* is mostly tangential oblique with some sigmoid; *c* is mostly parallel oblique; and *d* is complex sigmoid-oblique. From Mitchum, Vail, and Thompson, 1977, reprinted by permission.

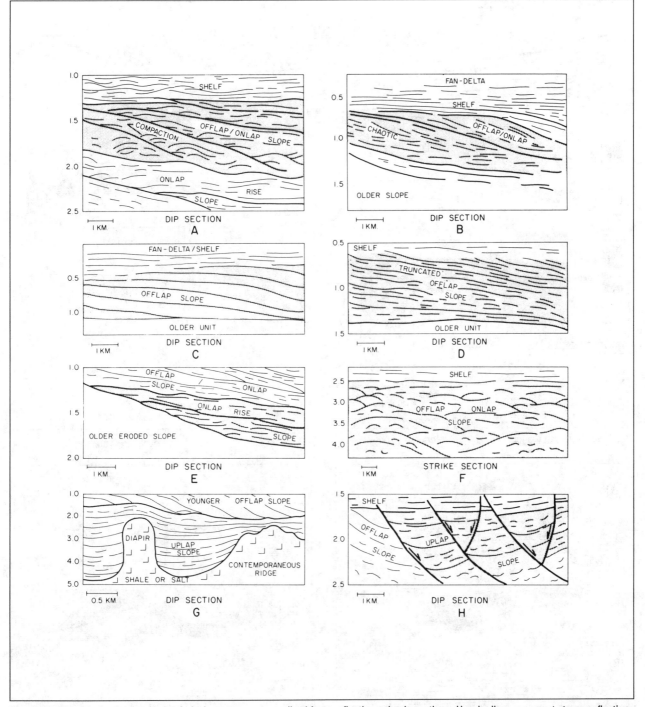

**Fig. 2–63.** Slope and associated seismic facies patterns generalized from reflection seismic sections. Heavier lines represent strong reflections and/or discontinuities. Vertical scale in seconds (two-way travel time): *A,* Dip section of a slope complex composed of older onlap rise facies and younger offlap/onlap slope facies. Onlap reflections terminate along strong local or regional reflections and/or discontinuities in reflection attitudes. *B,* Dip section through slope composed of chaotic reflections enveloped in sigmoid-shape offlap units. Complex offlap-onlap reflections shown in section *A* may represent discrete, alternating episodes of offlap deposition followed by episodes of shelf-edge, slope erosion and extensive onlap. The sigmoid-shaped units composed of onlap and chaotic reflections may, on the other hand, reflect rapid offlap, oversteepening of slopes, extensive slumping, and local onlap of submarine fan and proximal slope deposits. *C,* Dip section of slope system showing relatively uniform continuity of offlap reflections (commonly calcareous/terrigeneous, hemipelagic, upper slope facies) representing relatively slow rates of deposition. *D,* Dip section of slope system showing relatively uniform, but truncated (sometimes called oblique), offlap reflections indicating relatively rapid progradation and minimum subsidence. *E,* Dip section through thick onlap slope units representing extensive continental rise deposition; internal reflections terminate along strong reflection discontinuities. An offlap/onlap complex overlies the rise deposits representing another depositional episode. *F,* Strike section of offlap/onlap slope system characterized by convex-upward reflectors representing hemipelagic blankets draping large slope fan complexes (or cones). *G,* Dip section through a slope system characterized by uplap reflections caused by rapid subsidence of salt basins. *H,* Dip section through a slope system composed of uplap reflections resulting from subsidence of minor fault block. From Brown and Fisher, 1977, reprinted by permission.

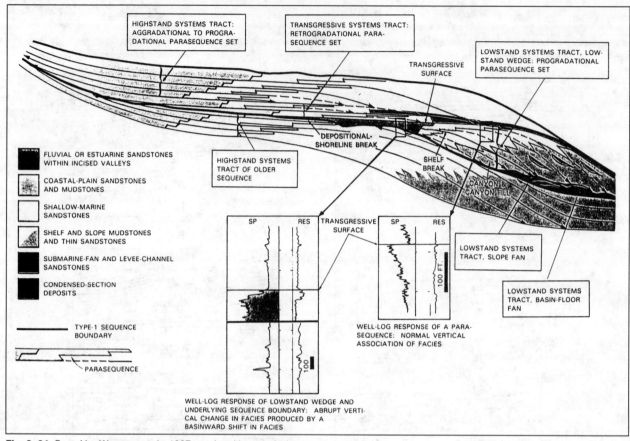

**Fig. 2-64.** From Van Wagoner et al., 1987, reprinted by permission.

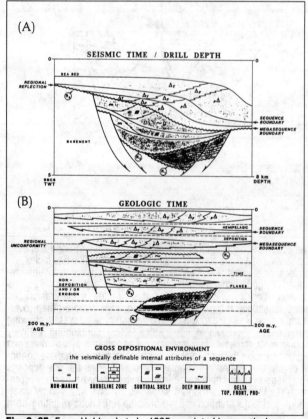

**Fig. 2-65.** From Hubbard et al., 1985, reprinted by permission.

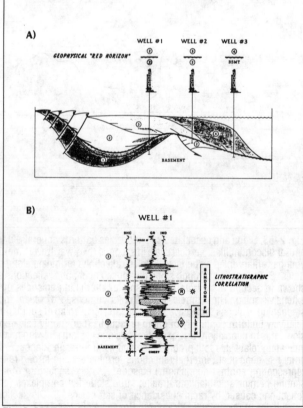

**Fig. 2-66.** From Hubbard et al., 1985, reprinted by permission.

The interpretation of seismic sections in both seismic time/drill depth, and geologic time is required to recognize sequence boundaries and to interpret the history of facies patterns (Hubbard et al. 1985) (Fig. 2–65). Thus sequences may be defined as megasequences, on a scale of tens of millions of years; sequences, on a scale of millions of years; and within sequences; parasequences, on a scale of a million years or less. The purpose of this type of analysis is to determine the scale and pattern of erosional or depositional surfaces bounding sequences (Fig. 2–66). The well section suggests that the sandstones could all be grouped into a single formation, but they actually have very different ages, and were deposited in contrasting and unrelated depositional environments. Similarly the lower shale interval of sequence 1 is unrelated to shales of sequence 3. In this example, the age of strata underlying and overlying the unconformity can be of widely differing ages, and the lithostratigraphic correlation of "formations" is equally fallacious. Internal patterns of bedding plane terminations, and thickness maps are needed to define time-rock intervals. With this type of information the analytical scheme outlined in Figure 2–57 by Hubbard et al. (1985) can be usefully applied.

Vail et al. (1977), as modified by Brown and Fisher

**Table 2-6**
**Summary of Seismic Facies Characterized by Mounded and Draped Reflection Configuration**

| Properties of Seismic Facies | Depositional Environments/Settings | | |
|---|---|---|---|
| | Reefs and Banks: shelf/platform margin, back shelf patch reefs and pinnacle/barrier reefs | Submarine Canyon and Lower Slope: proximal turbidities, slumped clastics | Hemipelagic Clastics: proximal basin and lower slope |
| Reflection configuration | Mounded, chaotic, or reflector-free; pull-up or pull-down common | Mounded; complex and variable | Parallel; mirrors underlying surface |
| Lithofacies and composition | Shallow-water carbonate biogenic buildups; may or may not exhibit reef-forming framework | Sand and shale submarine fans; complex gravity-failure fans or mounds; turbidity flow; other grain flows, submarine landslides/debris flows; clinoform/fondoform deposits | Terrigenous and calcareous clays (commonly alternating); pelagic oozes; deposition from suspension plumes and nepheloid clouds; fondoform deposits |
| Geometry and structure | Elongate lens-shaped (shelf/platform edge and barrier reefs); elongate to subcircular lens-shaped (patch and pinnacle reefs/banks); form on stable structural elements | Irregular fan-shaped to mounded geometry; common but not restricted to unstable basins | Sheet to blanket geometry exhibiting drape over underlying surface; common in deep, subsiding basins |
| Lateral relationships | Shelf/platform edge facies grade updip into parallel/divergent shelf/platform facies; grade downdip into talus and sigmoid clinoform facies; patch reef/bank facies grade updip and downdip into parallel/divergent shelf/platform facies; pinnacle and barrier facies grade downdip into talus clinoforms and to basinal plain (fondoform) facies | May grade shelfward into progradational clinoforms (normally oblique), canyon onlap fill, or pinch out against eroded slope; may grade basinward and laterally into basinal plain (fondoform); onlap fills or drapes | Commonly grades laterally or basinward into basinal plain (fondoform) facies; may grade shelfward into submarine canyon onlap fill; may onlap eroded slope |
| Nature of upper and lower boundaries | Upper surface concordant or may be onlapped by flank reflections; basal surface concordant, baselapping, or may overlie clinoform toplap; pull-up or pull-down of basal surface common | Upper surface commonly erosional and onlapped, baselapped, or concordant (with drape); basal surface irregularly baselapping; may appear concordant (low resolution) or may onlap (mounded onlap fill) | Upper surface commonly concordant, but may be onlapped or baselapped; basal surface generally concordant but may onlap eroded mound or slope |
| Amplitude | High along boundaries; may be moderate to low internally, commonly reflector-free | Variable; generally low; some higher internal amplitudes may be thin hemipelagic drapes | Low to moderate; some high amplitude reflections (well defined on high-frequency, shallow data) |
| Continuity | High along boundaries; internally discontinuous to reflector-free | Discontinuous to chaotic | High |
| Frequency (cycle breadth) | Broad; cycle may diverge into massively bedded buildup | Highly variable; commonly narrow | Narrow, uniform |

From Brown and Fisher; reprinted courtesy AAPG

(1977), have developed tables reflecting seismic patterns for a number of depositional settings (Tables 2–6, 2–7, 2–8, and 2–9). These patterns are based upon their collective experience and may be used as guide to detailed facies analysis of seismic sequences defined by the methods outlined above. It is particularly important to understand the position of the interval within an overall basinal setting, to recognize climatic and provenance controls, and to apply the analysis to a related series of seismic sequences. These facies models, when correlated with structural and isopachous maps, well log sections, and paleontologic information may provide the explorationist with the necessary information to interpret basinal history.

**Table 2-7**
**Summary of Seismic Facies Characterized by Progradational Reflection Configurations**

| *Properties of Seismic Facies* | Depositional Environments/Settings | |
|---|---|---|
| | *Slope: associated with prograding shelf/platform* | *Prodelta/Slope: associated with prograding shelf delta or shelf-margin delta; or Slope; associated with prograding neritic shelf supplied periodically by shelf delta/fan delta* |
| Reflection configuration | Sigmoid clinoforms<br><br>Progradational in dip profile; parallel to disrupted and mounded in strike profile | Oblique clinoforms<br><br>Progradational in dip profile; hummocky, progradational to mounded in strike profile; mounds more common in deepwater slope than in prodelta/slope on shelf |
| Lithofacies and composition | Hemipelagic slope facies in upper/mid-cliniform; submarine fans common in lower clinoform; generally calcareous clay, silt, and some sand (base of clinoform); clinoform deposited in deep water beyond shelf edge | *On shelf:* prodelta (upper) and shallow slope facies (mid-clinoform and lower clinoform); deposited on submerged shelf; composition generally terrigenous clay, silt, and sand; sand concentrated in submarine fans at base of clinoform<br><br>*Beyond shelf edge:* (1) prodelta and deepwater slope associated with shelf-margin delta; may be growth-faulted; clay, silt, and sand (in basal submarine fans); and (2) deepwater slope associated with prograding neritic shelf supplied periodically by shelf deltas/fan deltas; clay, silt, and sand (in basal submarine fans) |
| Geometry and structure | Lens-shaped slope system; poorly defined individual submarine fans and point sources; strike profile may intersect facies to define parallel to slightly mounded configurations; rarely affected by growth faults; represents low rate of sedimentation under relatively uniform sea-level rise and/or subsidence rate | Comlex fan geometry with apices at shelf-edge point sources; each submarine fan resembles a bisected cone; total slope system lens- to wedge-shaped; strike profiles intersect fans or cones to display complex mounds; seismic facies deposited rapidly relative to subsidence and/or sea-level rise; highly unstable slopes associated with deepwater deltas (growth faults, roll-over anticlines) |
| Lateral relationships | Grades updip through shelf/platform edge facies into parallel divergent shelf/platform (undaform) reflections; may grade downdip into basinal plain (fondoform) or mound/drape seismic facies; grades along strike to similar facies; may change landward to oblique facies | Terminates updip against base of delta platform or shelf/platform (undaform) facies and may grade downdip into basinal plain (fondoform), or mound/drape facies; may change basinward into sigmoid facies; grade along strike into mounded facies and locally submarine canyon-fill facies |
| Nature of upper/lower boundaries | Generally concordant at top and downlap (baselap) terminations at base; upper surface of outer or distal sigmoids may be eroded by submarine erosion and submarine canyons; eroded surface commonly onlapped by continental rise facies | Toplap termination at top and downlap (baselap) termination at base; may contain local or minor submarine erosion/onlap sequences; outer or distal oblique clinoforms commonly eroded by submarine erosion and submarine canyon cutting; eroded surface generally onlapped by continental rise facies |
| Amplitude | Moderate to high; uniform | Moderate to high in upper clinoform; moderate to low in lower clinoform; highly variable |
| Continuity | Generally continuous | Generally continuous in upper clinoform; discontinuous in mid-clinoform and lower cliniform; may exhibit better continuity near base |
| Frequency (cycle breadth) | Broadest in mid-clinoform where beds thickest; uniform along strike | Broadest at top and generally decreases downdip as beds thin; variable along strike |

From Brown and Fisher; reprinted courtesy AAPG

**Table 2-8**
**Summary of Seismic Facies Characterized by Parallel and Divergent Reflection Configurations**

| Properties of Seismic Facies | Depositional Environments/Setting | | | |
|---|---|---|---|---|
| | Shelf/Platform | Delta Platform: delta front/delta plain | Alluvial Plain/ Distal Fan Delta | Basinal Plain |
| Reflection configuration | Parallel/slightly divergent; highly divergent near rare growth faults | Parallel/slightly divergent on shelf; highly divergent near growth faults in deepwater deltas | Parallel, generally grades basinward into delta plain or into shelf/platform facies | Parallel/slightly divergent; may grade laterally into divergent fills or mounds |
| Lithofacies and composition | Alternating neritic limestone and shale; rare sandstone; undaform deposits | Shallow marine delta front sandstone/shale grading upward into subaerial delta plain shale, coal, sandstone channels; prodelta facies excluded except where toplap is absent; undaform deposits | Meanderbelt and channel-fill sandstone and flood-basin mudstone; marine reworked fan delta sandstones/pro-fan shale; undaform deposits | Alternating hemipelagic clays and siltstone; calcareous and terrigenous composition; fondoform deposits |
| Geometry and structure | Sheet-like to wedge shaped or tabular; very stable setting; uniform subsidence | Sheet-like to wedge shaped or tabular on shelf — prismatic to lenticular basinward of subjacent shelf edge with growth faults and rollover anticlines; relatively stable, uniform subsidence on shelf; rapid subsidence and faulting in deepwater delta | Sheet-like to wedge shaped (individually elongated ribbons or lobes), commonly tilted and eroded | Sheet-like to wedge shaped; may be slightly wavy or draped over subjacent mounds; generally stable to uniform subsidence; may grade laterally into active structural areas |
| Lateral relationships | May grade landward into coastal facies and basinward into shelf-margin carbonate facies; local carbonate mounds | May grade landward into alluvial systems and basinward into prodelta/slope clinoforms (on shelf) or growth-faulted prodelta/slope facies (deepwater setting) | Grade landward into reflection-free, high sandstone facies; alluvial facies grade basinward into upper delta plain; fan delta facies grade basinward into shelf/platform or into slope clinoforms | Commonly grades shelfward into mounded turbidites or slope clinoforms; may grade laterally into deep-water mounds or fills |
| Nature of upper/lower boundaries | Concordant coastal onlap and/or baselap over upper surface; upper surface may be eroded by submarine canyons; basal surface concordant, low-angle baselap or (rare) toplapped by subjacent clinoforms | Normally concordant at top but may be rarely onlapped or baselapped; upper surface may be eroded by submarine canyons; basal surface generally toplapped by prodelta/slope clinoforms (on shelf); rarely concordant with prodelta on shelf but common in deepwater, rollover anticlines | Upper surface may be onlapped by coastal facies; top may be angular unconformity; base is generally concordant; fan deltas rarely overlie clinoforms (toplap) | Generally concordant at top and base; may onlap eroded slope clinoforms or eroded mounds; upper surface rarely eroded |
| Amplitude | High | High is delta front and coal/lignite or marine transgressive facies within delta plain; low/moderate in most delta plain and in prodelta where in continuity with delta front | Variable — low/high | Low to moderate |
| Continuity | High | High in delta front, coal/lignite and marine transgressive facies; low/moderate in remainder of delta plain and prodelta where in lateral continuity with delta front | Discontinuous; continuity decreases landward | High |
| Frequency (cycle breadth) | Broad or moderate; little variability | Variable; broader in delta front; coal/lignite and marine transgressive facies moderate; narrower in other delta plain and prodelta where in continuity with delta front | Variable; generally narrower cycles than shelf/platform | Generally narrower than shelf/platform; commonly very uniform breadth throughout |

From Brown and Fisher; reprinted courtesy AAPG

## Examples of Subsurface Models

Application of the principles developed for sequence analysis to a complex Cretaceous depositional history from the Texas Gulf Coast illustrates the power of these techniques. A field seismic section, showing the stratal relationships for the Woodbine and Wilcox formations is illustrated (Vail et al. 1976) (Fig. 2–67). Subsequently, Vail and his co-workers developed a subsurface example of sequence analysis (Vail et al. 1977). The stratigraphic framework is illustrated in Figure 2–68, which delineates the unconformity pattern below the prograding

**Table 2-9**
**Summary of Seismic Facies Characterized by Onlap and Fill Reflection Configurations**

| Properties of Seismic Facies | Depositional Environments/Settings | | | |
|---|---|---|---|---|
| | Coastal (Paralic) Onlap Facies | Continental Rise: slope-front fill and onlap clastics | Submarine Canyon-Fill Deposits | Other Deepwater Fill Deposits: mounded, chaotic structurally active basins |
| Reflection configuration | Parallel; coastal onlap | Parallel/divergent; platform or shelfward onlap | Parallel/divergent; landward and lateral onlap | Parallel/divergent; chaotic mounded onlap |
| Lithofacies and composition | Delta/alluvial plain and medial fan delta sands and shales; supratidal clastic/carbonate facies; rarely beach/shoreface clastic facies | Sand and shale deposited in submarine fans by turbidity flows; hemipelagic terrigenous/ calcareous clays; distal pelagic oozes | Sand and shale deposited by turbidity flow in submarine fans near base; hemipelagic and neritic shale/calcareous clays in middle and upper sequence respectively; locally may contain coarse proximal turbidites | Sand and shale deposited in turbidity flow in submarine fans; hemipelagic terrigenous/ calcareous clays; pelagic clays; pelagic oozes; locally proximal turbidites |
| Geometry and structure | Sheet-like or tabular; uniform subsidence during deposition; periodic tilting and erosion; deposited near basinal hinge-line during subsidence and/or sea-level rise | Wedge-shaped lens; may be fan-shaped or lobate in plan view; slow subsidence | Elongate; lens-shaped in transverse section; may bifurcate updip; pinches out updip; slow subsidence | Variable lens-shaped; commonly irregular; reflects bathymetric configuration of structural depression; slow to rapid subsidence |
| Lateral relationships | Pinches out landward; grades basinward into lower delta plain, distal fan-delta, or shelf/ platform facies; may grade laterally into marine embay-ment facies | Pinches out updip; grades basinward into basinal plain or hemipelagic drape facies; continuous laterally for tens of kilometers | Pinches out updip and laterally; grades downdip into continental rise, mounded turbidites, or large submarine fans | Pinches out in every direction |
| Nature of upper/lower boundaries | Upper surface commonly tilted, eroded, and onlapped by similar deposits; base of facies onlaps unconformity, commonly angular | Upper surface commonly baselapped by prograding clinoforms; basal surface onlaps updip against eroded slope (and commonly outer shelf); may show baselap basinward against mounds or bathymetric highs | Upper surface may be concor-dant with overlying shelf or platform reflections or com-monly baselapped by prograding prodelta and slope facies; basal surface onlaps updip and laterally; baselap onto basin floor rarely observed | Upper surface may be concor-dant with hemipelagic drape or baselapped by prograding clinoforms; basal surface onlaps in all directions |
| Amplitude | Variable; locally high but normally low to moderate | Variable; hemipelagic facies moderate to high; clastics low to moderate | Variable; generally low to moderate | Variable; generally low to moderate |
| Continuity | Low in clastics; higher in carbonate facies; decreases landward | Moderate to high; continuous reflections in response to hemipelagic facies | Variable; generally low to moderate | Variable; poor in chaotic or mounded fill; high in low-density turbidites and hemipelagics |
| Frequency (cycle breadth) | Variable; generally moderate to narrow | Narrow; uniform | Variable but generally narrow | Variable; commonly narrow; may increase breadth toward axis of fill |

From Brown and Fisher; reprinted courtesy AAPG

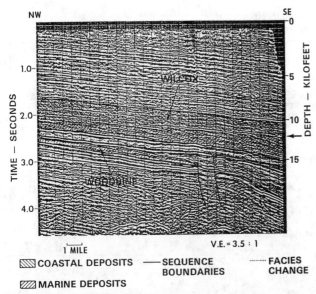

Fig. 2-67. Seismic (dip) profile, East Texas, showing Cretaceous Woodbine and Eocene lower Wilcox delta systems. Reflections document basinward (right) progradation. Higher amplitude reflections at the toplap boundary in the Wilcox system are in response to sand-rich delta-front facies; overlying delta plain/alluvial plain is represented by lower amplitude, discontinuous seismic facies. From Vail et al., 1976.

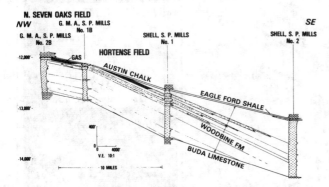

Fig. 2-68. Electric-log cross section showing distribution and geometry of Woodbine deltaic sandstone beds, Polk County, Texas. From Vail, Todd, and Sangree, 1977, reprinted by permission.

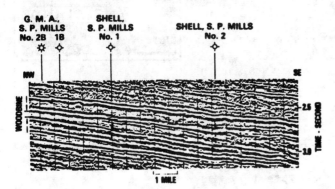

Fig. 2-69. Seismic section showing reflection pattern details of Woodbine deltaic wedge. From Vail, Todd, and Sangree, 1977, reprinted by permission.

interbedded sandstone and shale clinoform units of the Woodbine formation. The regional seismic section (Fig. 2-67), which is blown up in Figure 2-69, illustrates the onlap of the Eagleford shale as shown by bedding plane terminations. From the well data synthetic seismograms can be developed to illustrate the reflection patterns (Fig. 2-70), and correlated to a seismic record section (Fig. 2-69). Mapping of reflection termination patterns can be developed following a table of patterns of above and below internal bedding configurations (A – B/C) (Table 2-10). Figure 2-70a (Ramsayer 1979) converts the bedding plane patterns and terminations into a map form. An environmental interpretation can be suggested by examining the history of progradation within the mapped area (Fig. 2-70b). This pattern can be related to well logs, paleontology, and sample examination to confirm this interpretation.

An example from New Zealand in *Atlas of Seismic Stratigraphy* (1987), edited by A. W. Bally, is useful in that it illustrates the methods of sequence identification, facies analysis, and paleogeographic analysis from patterns of bedding plane terminations. The area studied (Fig. 2-71) shows very limited well control, the presence of a giant gas accumulation, and the line of a seismic section that was used to illustrate the depositional history. A well is included in the section to tie to sequence stratigraphy (Figs. 2-72a and 2-72b). Paleontology and an understanding of the worldwide sequence unconformity patterns allows the constuction of a time-rock, facies framework for analysis (Figs. 2-73a and 2-73b).

Lower sequences, 5-8, are mostly continental except in the very northwest portion of the area where continuous parallel low amplitude reflection configurations are developed (Figs. 2-72a and 2-72b). These discontinuous reflection patterns are difficult to separate, but patterns of onlap and convergence can be seen to the east. Sequence analysis is well illustrated by examining intervals 2-4. A basal transgressive sand of Eocene-Oligocene age is not discernable by seismic sequence analysis, and the overlying sequence number 4 contains contrasting lithofacies, with marine deep water deposits below a prograding pattern of clinoform deposition (Figs. 2-74 and 2-75). This contact has not been indicated on Figure 2-71, but can be observed just below the number 4. This can correlate with a upper Oligocene-lower Miocene deeper water sequence, overlain by a progradational middle Miocene sequence (Figs. 2-72a and 2-72b), and is consistent with worldwide sequence patterns described by Vail (1987). Middle Miocene and upper Miocene-lowest Pliocene paleogeographic maps (Figs. 2-76 and 2-77) show the progradational patterns reflected in Figures 2-72a and 2-72b. These patterns are based on the

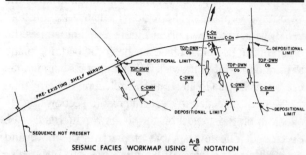

**Fig. 2–70a.** Reconstructing depositional environments by mapping bedding geometries obtained from seismic sections. From Ramsayer, 1979.

**Table 2-10**
**Internal Configuration**

| Above | Below |
|---|---|
| Erosional truncation (Te) | Downlap (Dn) |
| Toplap (Tp) | Onlap (On) |
| Concordance (C) | Concordance (C) |
| *Within* | |
| Parallel (P) | Prograding clinoforms (PC) |
| Subparallel (SP) | Sigmoid (S) |
| Divergent (D) | Oblique (O) |
| Chaotic (C) | Complex sigmoid-oblique (SO) |
| Reflection-free (RF) | Shingled (Sh) |
| Mounded (M) | Hummocky clinoforms |

Mitchum, Vail, & Sangree; courtesy AAPG.

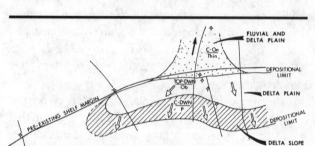

**Fig. 2–70b.** From Ramsayer, 1979.

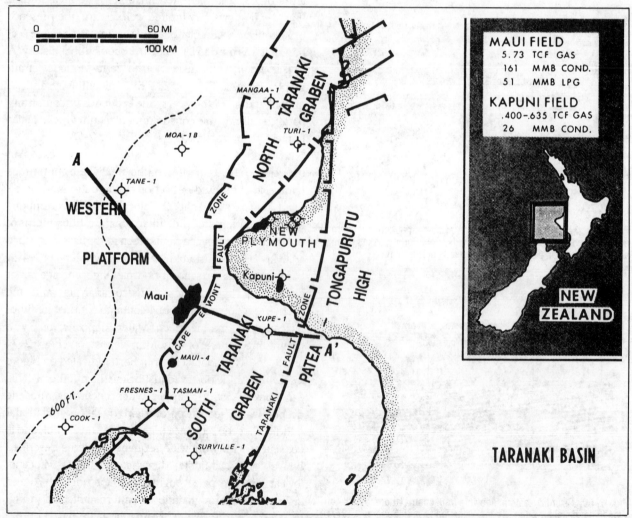

**Fig. 2–71.** Tectonic and index maps of the Taranaki basin, including location of exploration wells, gas-condensate fields and structural cross section A-A[1]. From Bally, 1987, reprinted by permission.

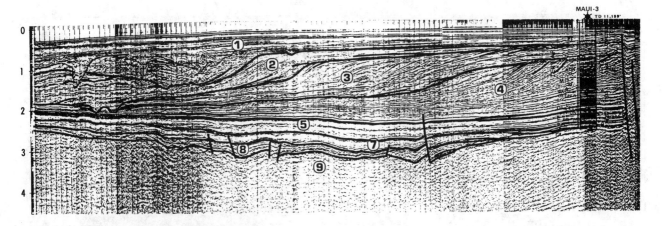

**Fig. 2–72a.** From Bally, 1987, reprinted by permission.

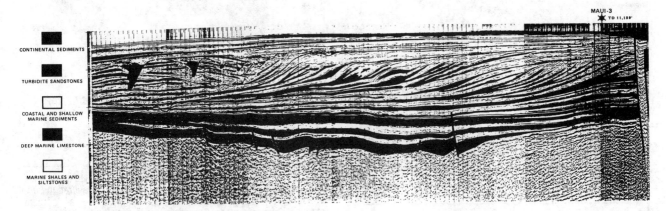

**Fig. 2–72b.** From Bally, 1987, reprinted by permission.

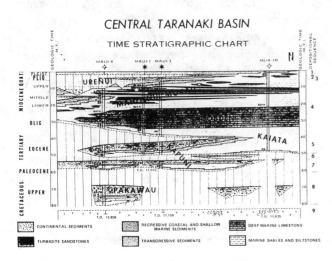

**Fig. 2–73a.** Time stratigraphic or chronostratigraphic correlation chart parallel to basin depositional strike. Note well control, different facies and environments and hiatal gaps along unconformities. From Bally, 1987, reprinted by permission.

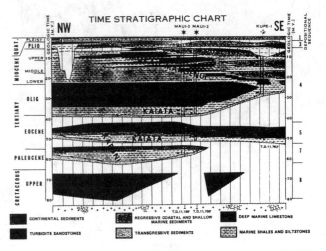

**Fig. 2–73b.** Time stratigraphic or chronostratigraphic correlation chart parallel to seismic line in 2–72a. Note the different facies and environments, hiatal gaps along unconformities and well control.

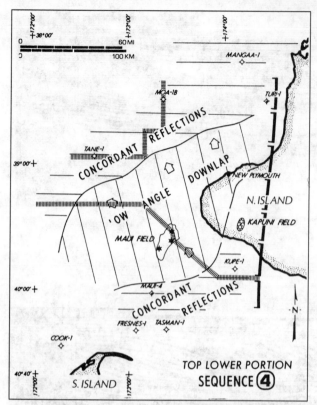

**Fig. 2-74.** Seismic facies map showing the general distribution of the different reflection configuration patterns for the lower portion of sequence-4. From Bally, 1987, reprinted by permission.

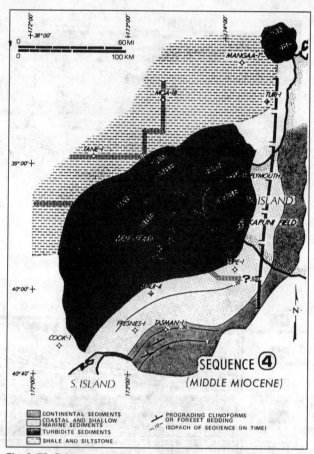

**Fig. 2-76.** Paleogeographic and isochron map. This map depicts the different environments, lithofacies and thickness (measured in two-way time) for the lower portion of sequence-4. Note the development of two deep sea fan areas. From Bally, 1987, reprinted by permission.

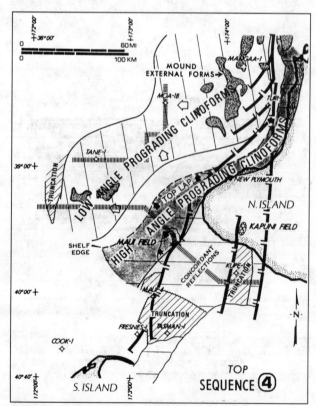

**Fig. 2-75.** Seismic facies map showing the areal distribution of the different reflection configuration patterns for the upper portion of sequence-4. From Bally, 1987, reprinted by permission.

patterns of reflector terminations, internal seismic-time/depth seismic lines, and the limited well control available. Similarly the Pliocene sequence 3, shows the pattern of reflection terminations and a reconstructed paleogeographic map (Figs. 2-78, 2-79, and 2-80).

Pliocene-Pleistocene sequences 1 and 2 are also described in the depositional reconstruction, but are more complex due to more complex topography, rapidity of sea-level changes, and to rapid changes in rates of sedimentation. Complex sequence patterns possibly exist due to the history of Pleistocene sea-level changes (Figs. 2-72a and 2-72b).

## DATA-PROCESSING TECHNIQUES

### Wavelet Processing

With the use of spiking deconvolution, minimum phase, and zero phase filtering, it is possible to produce a wavelet of the desired amplitude, frequency, and symmetry to allow analysis of specific reflections.

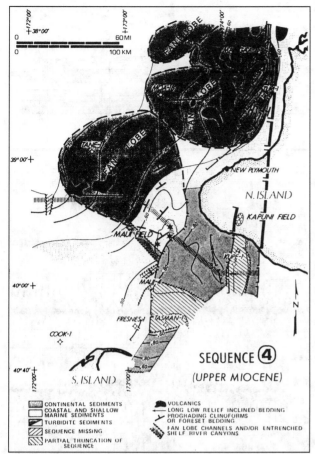

**Fig. 2-77.** Paleogeographic and isochron map. This map depicts the different environments, lithofacies and thickness (measured in two-way time) for the upper portion of seuqence-4. Note the change in the deep sea fan development. From Bally, 1987, reprinted by permission.

The following two sections illustrate the apparent increase in resolution (Fig. 2–81). It must be remembered that a reflection is often not the product of a single bedding interface and mapping of a single reflection is simply a useful device for examining patterns of bedding planes and sedimentary units.

With this in mind the modeling of complex interbedding of lithologies is both a complex sedimentary and seismic problem. To understand these relationships better, a theoretical model may be used to interpret these patterns (Figs. 2–82a and 2–82b). The difference in reflection position and configuration is well illustrated. More resolution appears in the section using a symmetrical wavelet.

## Modeling Depositional Patterns

Synthetic seismic signal responses using variable impedance and frequencies, but with a symmetrical wavelet, are illustrated in Figure 2–83. These patterns may be related to bedding patterns associated with specific envi-

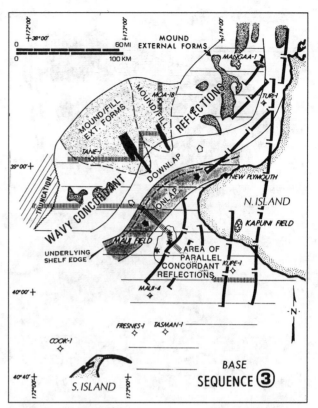

**Fig. 2-78.** Seismic facies map showing the different reflection configuration patterns for the base of sequence-3. From Bally, 1987, reprinted by permission.

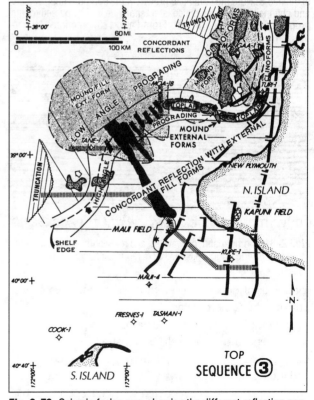

**Fig. 2-79.** Seismic facies map showing the different reflection configuration patterns for the top of sequence-3. From Bally, 1987, reprinted by permission.

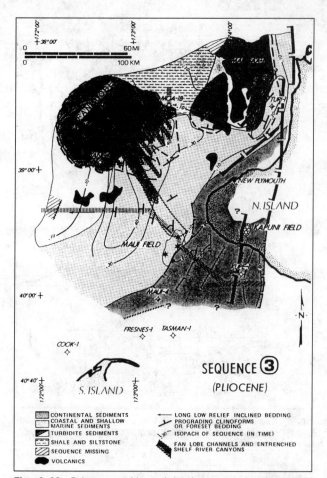

Fig. 2–80. Paleogeographic and isochron map for the top of sequence-3. Note the continued growth of the southern deep sea fan and channel complex as the coastal, shoreline and slope sediments prograde toward the northwest. From Bally, 1987, reprinted by permission.

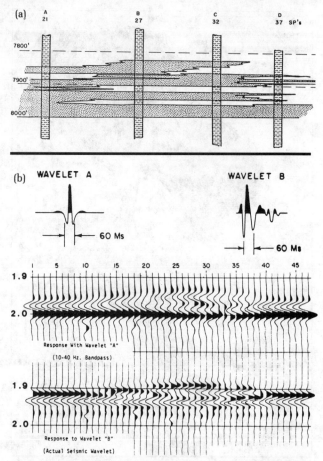

Fig. 2–82. A theoretical model of an interfingering sand-shale sequence developed by Geoquest. This shows both barrier island and channel sandstones. From Neidell and Poggiagliolmi, 1977, reprinted by permission.

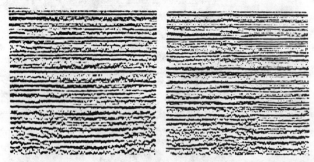

**CONVENTIONAL PROCESSING    WAVELET PROCESSING**

Fig. 2–81. Comparison of wavelet and conventional processing. From Neidell and Poggiagliolmi, 1977, reprinted by permission.

ronments of deposition. Note the enhanced resolution utilizing the 60 hertz signal.

Stratigraphic patterns associated with coastal sandstone deposition are well established. A regressive-transgressive pattern for shoreline deposition is common, and interval thicknesses from 4 to 3 meters may be developed (Fig. 2–84a). The modeled seismic response would show the proper polarity, impedance, and configuration of such units (Fig. 2–84b).

The geometry of fluvial channels has also been established, but in this case a more common pattern of development would be a valley-fill sequence with channel widths of from 8 to 5 kilometers (Fig. 2–85).

Many clastic sedimentary units reflect basal erosion or fill followed by abrupt cessation of clastic supply or erosion. These cylindrical log response patterns may have a wide variation in geometry, but their identification on seismic sections may lead to a better understanding of depositional processes (Fig. 2–86).

## Velocity Analysis

Seismic data can be analyzed to produce an estimate of interval velocities. This information can be crudely estimated by the use of stacking velocities with a velocity curve drawn through the time-depth velocity function. Reflections can be correlated and the velocity for a specific stratigraphic interval determined. Velocities can be calculated for specific travel time windows utilizing only the Root Mean Square (RMS) velocities.

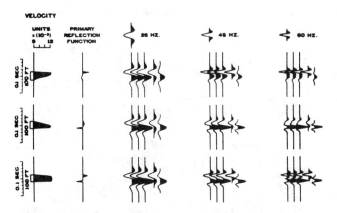

**Fig. 2-83.** Sandstone-shale interfingering model seismic response. From Harms and Tackenburg, 1972.

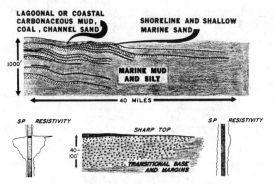

**Fig. 2-84a.** Stratigraphic patterns of coastal sand units. From Harms and Tackenburg, 1972.

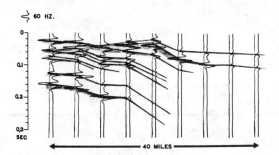

**Fig. 2-84b.** Seismic response model using a 60-Hz wavelet. From Harms and Tackenburg, 1972.

This procedure was suggested by Smith (1969) utilizing the Dix equations. The conceptual basis for stratigraphic interpretation was presented by Geophysical Services Inc. (Smith 1969) (Fig. 2-87). The interval velocity was developed by fitting a hypothetical velocity curve to arrival times and determining differences from expected arrival times (Fig. 2-88). Adjustments for lithologic variations were estimated by comparison to well logs (Fig. 2-89).

Application of this model to a Gulf Coast shelf area illustrates the usefulness of the technique. Anomalies are shown, and the data are contoured over a large area with minimum well control (Fig. 2-90). The interval velocity

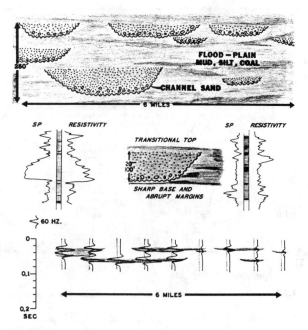

**Fig. 2-85.** Stratigraphic patterns of fluvial sand units. From Harms and Tackenburg, 1972.

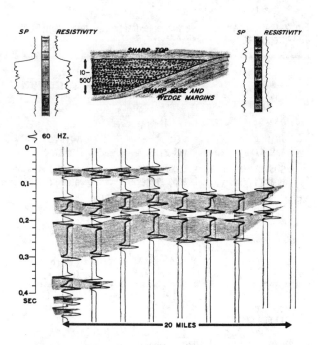

**Fig. 2-86.** Stratigraphic patterns of deep marine channels of the lateral pinchout of a channelized sandstone from a deltaic sequence. From Harms and Tackenburg, 1972.

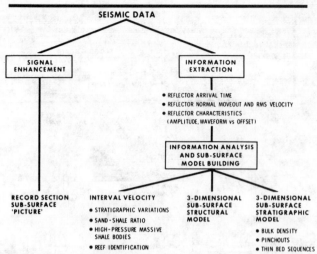

**Fig. 2–87.** Possible stratigraphic analysis flow chart. From Smith, 1969, reprinted by permission.

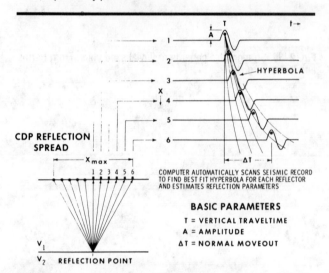

**Fig. 2–88.** Basic seismic information extraction system. From Smith, 1969, reprinted by permission.

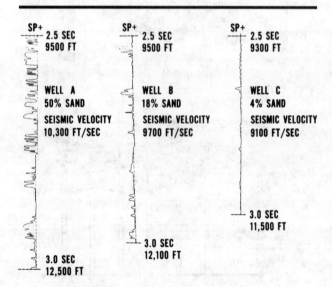

**Fig. 2–89.** From Smith, 1969, reprinted by permission.

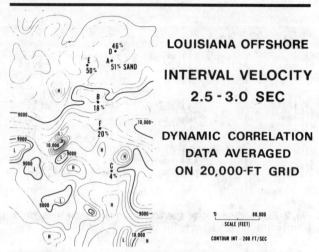

**Fig. 2–90.** Dynamic correlation data averaged on 20,000-ft grid. From Smith, 1969, reprinted by permission.

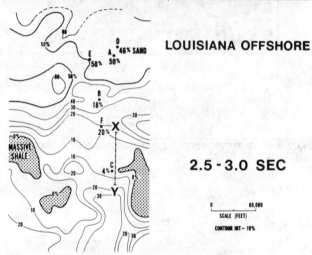

**Fig. 2–91.** Implied sand-shale ratio from internal velocity, 2.5–3.0 sec. From Smith, 1969, reprinted by permission.

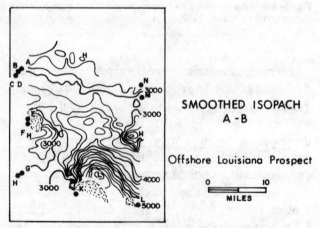

**Fig. 2–92.** Interval velocity A–B. From Smith, 1969, reprinted by permission.

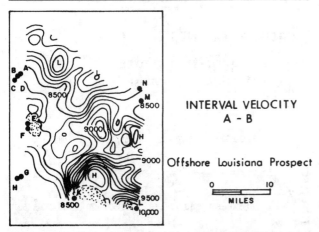

**Fig. 2-93.** Smoothed isopach A–B. From Smith, 1969, reprinted by permission.

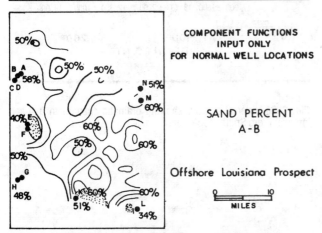

**Fig. 2-94.** Sand percentage A–B. From Smith, 1969, reprinted by permission.

map shows sandstone patterns, intrusive shale diapirs, shelf-edge and slope patterns, and most interestingly, the possible sand-filled canyon lying between shale diapirs (Fig. 2-91).

By utilizing reflectors to define stratigraphic intervals, it is possible to develop both a structural and a stratigraphic pattern within a time-rock interval. Figures 2-91, 92, 93, and 94 illustrate this approach.

Complex lithologic changes are present in many areas. Areas with little well control can be examined to determine if changes might indicate reservoir thickness and character, and also to aid in the interpretation of structural elevation of apparent reservoir intervals. There is an example from the Santa Barbara channel off the California coast. In this case both the structural closure and the changes in reservoir properties could be predicted by an anomalous pattern of interval velocities (Figs. 2-95 and 2-96).

Another example from the North Sea illustrates that, by knowing the interval velocity, it is possible to predict either lithology or porosity within an interval. In this case

the presence of a porous chalk reservoir was indicated (Figs. 2-97 and 2-98).

## Interval Velocity from Impedance

An integrated Syn-Log can be developed from wavelet processed data. Geoquest International Inc. has a patented software program for the extraction of velocity data from the amplitude changes in simple lithologic sequences (Figs. 2-99 and 2-100).

Interval velocities also can be calculated statistically from the moving average of the stacking velocities normalized against the digitized acoustic logs. This is the patented Seislog developed by Lindseth, which produces an acoustic velocity trace that can be scaled and presented in a color-coded display (Figs. 2-101 and 2-102).

## Stratigraphic Modeling

The interpretation of seismic record sections requires placing the data into a framework that can be interpreted from a geological point of view. This is accomplished by correlating field record sections to well data. Figures 2-103 and 2-104 indicate a disparate pattern of information. Little correlation can be seen between the seismic section and the correlated log section. By examining the velocity and bedding patterns, an impulse response is calculated and then a synthetic seismic section is produced (Fig. 2-104). The combination of these data can lead to development of a model stratigraphic section (Fig. 2-105). This section provides the basis for interpreting changes in reflection configurations and thickness of specific units. This information can be mapped utilizing available seismic and well control (Fig. 2-106). The maps are used for exploration for channel patterns in the pre-Belle City sandstone, the gravity-transported basinal fan (Medrano sandstone), and the channel patterns in the Lower Wade sandstone. It is the synthesis of the data that leads to predictable patterns and to profitable exploration.

## INTERPRETING SEDIMENTARY CHARACTERISTICS

### Developing a Data Base

The description of stratigraphic characteristics can produce an overwhelming array of observations; for example, there are aspects concerning bedding, sedimentary structures, textures, fabric, mineralogy, biota, chemistry of cements and matrix, and the areal configuration of all of these responses to the active processes within the depositional environment. The origin of these

## CALIFORNIA LINE 1

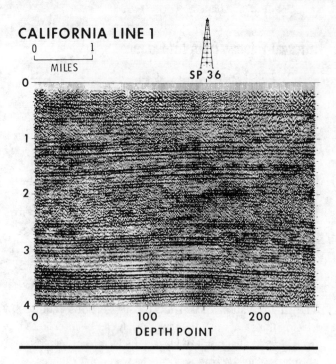

**Fig. 2-95.** Seismic line, Santa Barbara channel, showing change of reflector continuity near 3-sec interval. From Smith, 1969, reprinted by permission.

## CALIFORNIA LINE 1

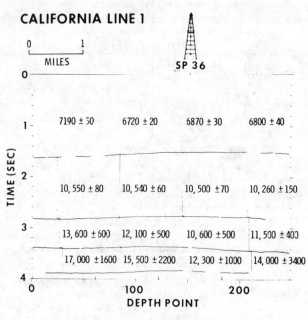

**Fig. 2-96.** Interval velocity showing dramatic changes in 3–4-sec interval. From Smith, 1969, reprinted by permission.

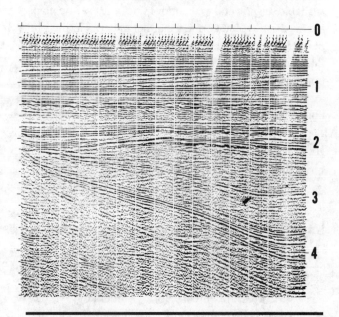

**Fig. 2-97.** North Sea prospect, stacked section. From Smith, 1969, reprinted by permission.

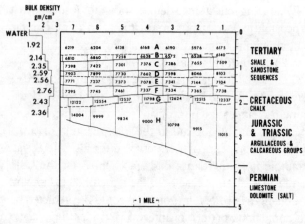

**Fig. 2-98.** North Sea prospect interpretation. From Smith, 1969, reprinted by permission.

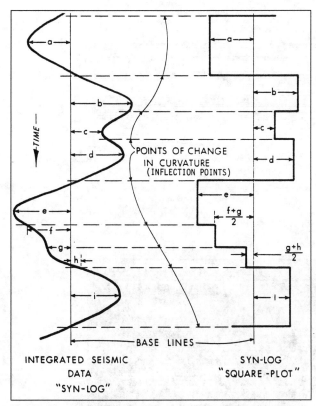

Fig. 2-99. Synlog square plot generation. Reprinted courtesy Geoquest.

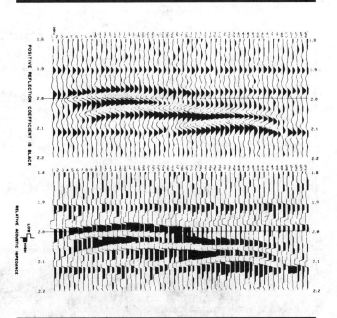

**Fig. 2-100.** Comparison of wavelet processed and synlog data. Reprinted courtesy Geoquest.

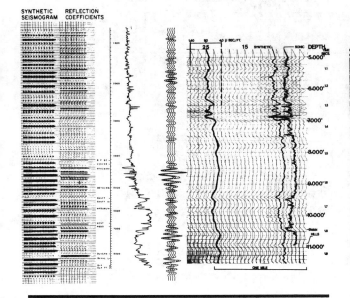

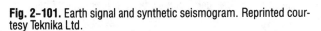

**Fig. 2-101.** Earth signal and synthetic seismogram. Reprinted courtesy Teknika Ltd.

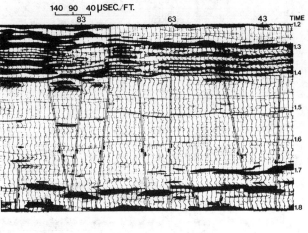

**Fig. 2-102.** Velocity changes within a faulted section. Reprinted courtesy Teknika Ltd.

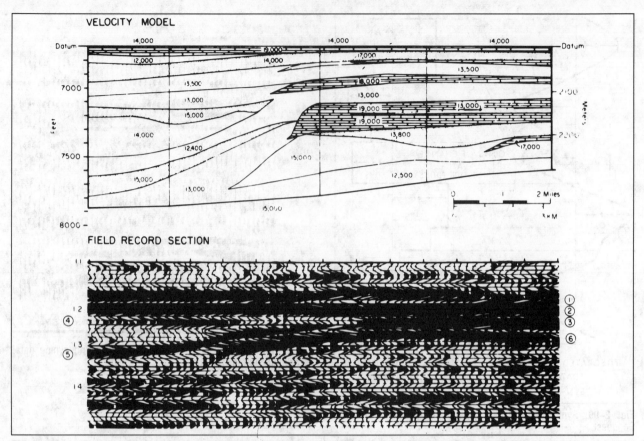

**Fig. 2–103.** (a) Geologic-model cross section and equivalent field-record section. From Galloway, Yancey, and Whipple, 1977, reprinted by permission.

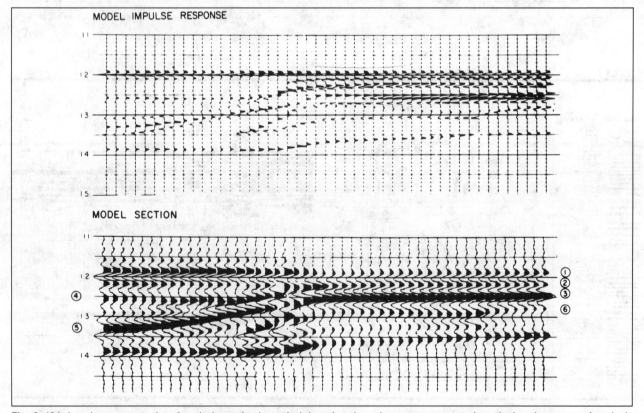

**Fig. 2–104.** Impulse-response plot of geologic section in vertical time domain and computer-generated synthetic seismogram of geologic model. From Galloway, Yancey, and Whipple, 1977, reprinted by permission.

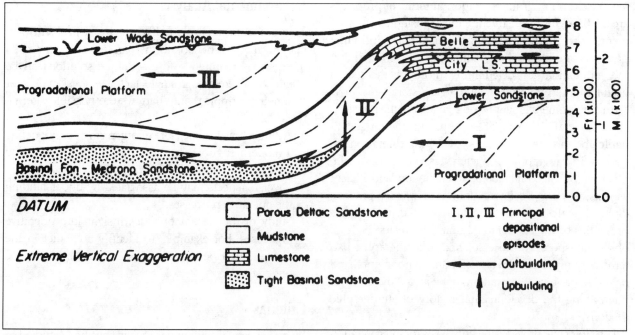

**Fig. 2-105.** Three depositional episodes producing upper Hoxbar shelf margin. No horizontal scale. From Galloway, Yancey, and Whipple, 1977, reprinted by permission.

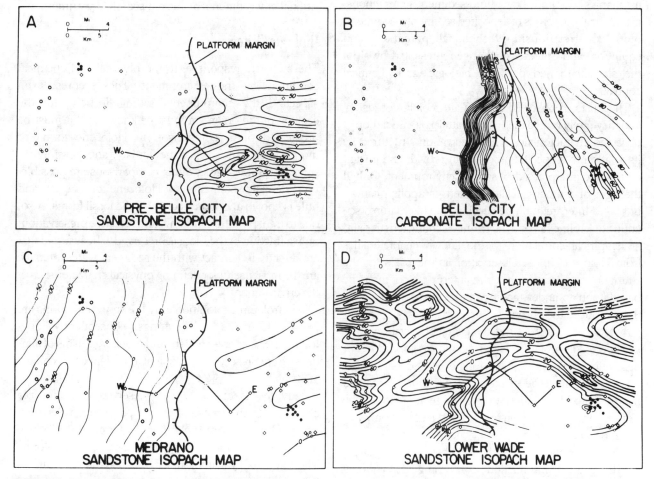

**Fig. 2-106.** Isopach maps of framework facies composing each depositional episode. Location of west-east log cross section is shown for reference. *A,* Dip-oriented delta-platform sandstones of progradational episode I. *B* and *C,* Shelf-carbonate unit and episode III. *D,* Dip-oriented distributary channel and delta-margin sandstones of progradational episode III, which buried Belle City shelf-edge carbonate complex and formed new topographic break west of map area. From Galloway, Yancey, and Whipple, 1977, reprinted by permission.

responses and their interpretation in terms of processes are the bases for reconstructing the depositional environment. The study and measurement of continuing processes are the goals of sedimentologic research and the foundation for determining those characteristics useful for interpreting sedimentologic history.

An environment is simply the sum total of the physical, chemical, and biologic processes operating within identifiable boundary conditions. The interpretation of observed stratigraphic characteristics cannot be a simplistic empirical correlation between modern and ancient aspects; it must be based upon a dynamic, process-directed analysis of the many interrelated, measurable properties of the stratigraphic unit. The relationship of observed aspects to processes and to stratigraphic patterns not only forms the basis for an environmental and paleogeographic interpretation, but it also is the basis for interpreting the developmental history of stratigraphic sections.

The Holocene provides the only basis for interpretation of process-response relationships. Sedimentology is a complex discipline that requires an understanding of the mechanics of fluid flow, fluid-sediment interactions, mechanical properties of deposited sediments, sediment-biotic interrelationships, chemical potentials and equilibria, and the rates of change represented by all of the aforementioned physical, biologic, and chemical controls.

Checklists of observed aspects are mostly confusing, and they rarely record the critical pattern or association of observed characteristics. The observation structured by its relation to a specific process can provide the insight required to interpret a depositional environment. Only if the rigor of a process-related pattern is applied can an unbiased interpretation be developed. This requires scientific discipline beyond the listing of characteristics and an empirical correlation to the known characteristics from Holocene units. Observation is the basis for the stratigraphic sciences, but the interpretation must be based upon an understanding of the causes or processes that produced the observed characteristic.

A data base includes those facts or observations that provide an insight into either boundary conditions or depositional processes. It has long been understood that observation must be directed by previously obtained insights into identifiable patterns and relationships. The significance of frosting on quartz sand surfaces only has meaning in terms of physical or chemical processes that produced the effect. Thus, the data base must be structured to yield information concerning depositional processes.

## Environmental Analysis

Environmental analysis is like solving a problem with many variables or unknowns, an equation for each unknown is required before a unique solution can be obtained. Therefore, a single factor, e.g., skewness or symmetrical ripples, only narrows the possible alternatives.

- A single physical measure or organism does not define a modern environment and, therefore, cannot define an ancient depositional environment.
- Diverse aspects may be associated by understanding the operative processes.
- Study of the Recent provides insight into operative processes. For example, to identify a chenier in the ancient, all aspects of the modern analogue must be understood.

## Criteria

Many aspects of sedimentary rocks have been studied in the Recent and are useful for determining processes. These measures form the basis for determining the depositional environments of ancient sedimentary units.

## Biologic Criteria

The biota are responsive to the physical and chemical conditions of the environment. These represent all aspects of the environmental setting. To be useful, the ecology of the organism must be known and it must be predicated on a comparison to Holocene forms or morphologic criteria. For the Tertiary, and possibly for some of the Mesozoic, direct comparisons are possible. For Paleozoic sediments few Holocene forms exist, and little is known about the ecology of the fossil forms. Also, due to transportation, mixing, and selective preservation, only a fraction of the biota is preserved. Those forms that are directly in contact with the sediment water interface are the most useful for defining physical characteristics of the environment.

In petroleum exploration only the very small forms are preserved in the well cuttings; consequently, only foraminifera, spores, pollen, and other organisms of similar size are preserved in sufficient abundance for ecologic analysis.

Body fossils often are not preserved in permeable clastic units that most often are petroleum reservoirs. Sediment-faunal interrelations however do produce traces of the faunas and, in some instances, the floras. The science of trace fossil analysis has matured during the past fifteen years, and now it is one of the more promising methods for determining physical, chemical, and biologic aspects of ancient environments.

Feeding mechanisms have not varied with time or with the changing biota. A primary depth control can be established related to the nutrient supply (Fig. 2–107). In addition the nature of the tracks, trails, burrows, and other traces provide information on sedimentation rates, bottom conditions, and physical aspects of sedimentary processes. Specific pattern types can be recognized and utilized as broad genetic groupings (Fig. 2–108).

Trace fossils can be used for environmental analysis in conjunction with both lithology and sedimentary structures (Figs. 2–109 and 2–110). This synthesis is based upon the distribution of Holocene forms and the recognition of similar trace patterns throughout the stratigraphic record. Studies of Holocene patterns has provided an effective basis for these types of comparisons. Very restricted environmental settings can be recognized (Fig. 2–111). The expansion of this types of work, however, has indicated that a specific form may not be restricted to a narrow ecologic niche. For example, the Crustacean, Callianassa, may occur from tidal estuaries to deeper marine environments, but it is most commonly developed in a shoreface environment (Figs. 2–112 and 2–113). This form was labeled Ophiomorpha and has been identified in Mesozoic to Holocene strata.

The amount and type of bioturbation are indicative of the chemical and physical hostility of the environment. Intervals may show a wide range of biological mixing (Fig. 2–114). In addition, other sediment faunal relationships, sedimentary structures, and bedding patterns can be used to infer environmental processes (Fig. 2–115). Examples from ancient stratigraphic units then can be more precisely interpreted (Fig. 2–116).

## Textural Criteria

The application of energy to the depositional interface is highly variable with respect to magnitude, duration, repetition, direction, competence, and form. Many environments have a unique energy distribution resulting in characteristic grain-size distributions. Individual aspects of a grain-size distribution may be used to reconstruct depositional processes.

Recent research has indicated that grain-size distributions are composed of one or more log-normal populations produced by mixing (Fig. 2–117). Each population is represented by a straight line on a log-probability plot of the grain-size distribution, and the straight line segments are joined together by abrupt changes in slopes (Fig. 2–118). These inflection points on the grain-size distribution-log-probability plot have been shown to have significance in terms of depositional processes (Sagoe and Visher 1977). Each curve segment may be related to a differing transportational and/or deposi-

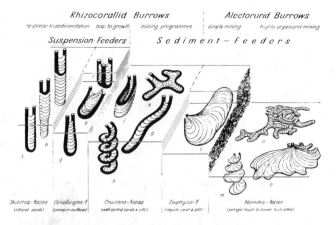

**Fig. 2–107.** Bathymetric zonation of fossil sprite burrows exemplifies the general rule that suspension feeders prevail in shallow and highly agitated waters. Elaborate sediment feeders appear in deeper and zoophycos facies in shallower positions due to local channeling or restriction, but nereites facies seem always to be restricted to the deepest zone. From Seilacher, 1967.

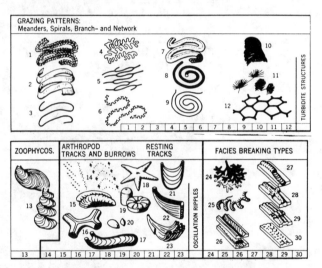

**Fig. 2–108.** All communites of true fossils regardless of geologic age can be assigned to one of three major types of ichnofacies, which show parallel differences in lithology and in organic sedimentary structures. From Seilacher, 1967.

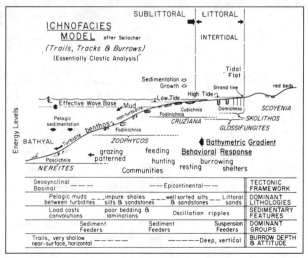

**Fig. 2–109.** Ichnofacies model. From Rodriguez and Gutschick, 1970.

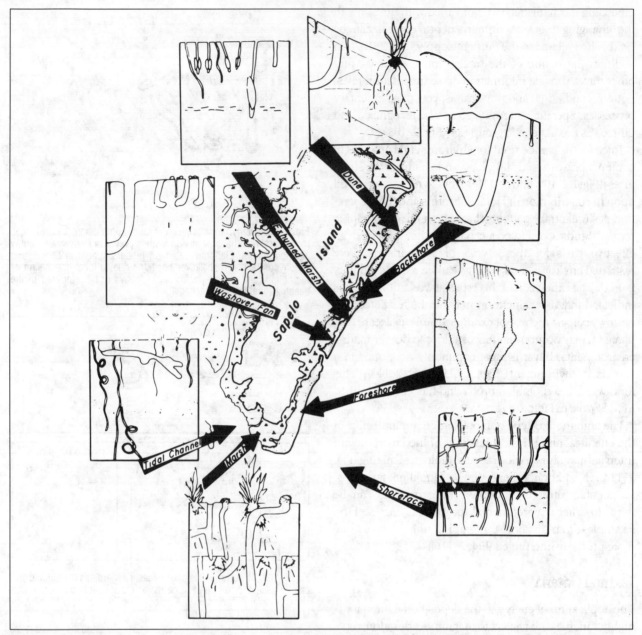

**Fig. 2–110.** Distribution of major assemblages of Recent lebensspuren in the vicinity of Sapelo Island, Georgia. From Frey, 1971.

tional mechanism. Suspension, saltation, and surface creep are separated by grain size, sorting, and relative proportion within the total distribution.

The precise mechanism of mixing log-normal populations is a continuing area for research. Experimental studies in the laboratory and actualistic sediment transport studies have indicated that the depositional process strongly controls the position of inflection points (Fig. 2–119). In this example sedimented sand from the depositional interface was sampled and the textural distribution determined. Specific wave and wind processes produced variations in the grain-size distribution. Similarly, in only a few days, a laboratory study of shoaling

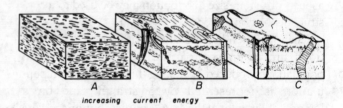

*increasing current energy* →

**Fig. 2–111.** Progressive changes in lebensspuren and substrates with increasing current energy in the depositional environment. *A,* Intensely burrowed siltstone containing few distinctive trace fossils, representing offshore deposition. *B,* Fine-grained sandstone containing larger but less abundant burrow mottles and many distinctive trace fossils, representing nearshore deposition. *C,* Fine-to-medium grained sandstone containing little burrow mottling but numerous distinctive trace fossils, representing nearshore and strandline deposition. From Frey, 1971.

**Fig. 2–112.** Ophiomorpha from a Holocene estuary

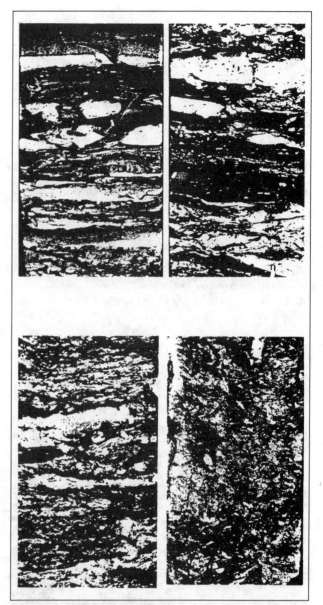

**Fig. 2–114.** Degrees of bioturbation: (3) moderate bioturbation (30–60%); (4) strong bioturbation (60–90%); (5) very strong bioturbation (90–99%) (churned); (6) completely bioturbated (100%) (churned). From Young and Rahmani, reprinted courtesy Canadian Society of Petroleum Geologists.

**Fig. 2–113.** Ophiomorpha, Cretaceous shoreface deposit, Texas

waves produced a modification of the source distribution comparable in detail to those from the Holocene (Fig. 2–120). Dimensional similitude was maintained, sampling was designed, and analytical techniques were utilized. From these comparisons it can be shown that textural analysis is a useful tool for determining specific depositional processes as an aid for environmental interpretation.

The scale of the textural changes in terms of the depositional event may also have significance. The specific process operating on a beach foreshore is swash and backwash developed by a run-up of each breaking wave.

**Fig. 2-115.** Pennsylvanian bay fill with pelecypods and worm burrows

**Fig. 2-116.** Arthropod (?) tracks Cambrian Tapeats sandstone, Arizona

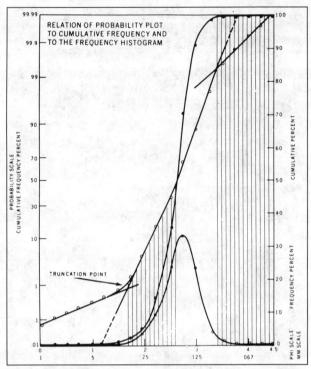

**Fig. 2-117.** Comparisons of grain-size distribution curves. The log-probability curve shows multiple curve segments and truncation points. From Visher, 1969.

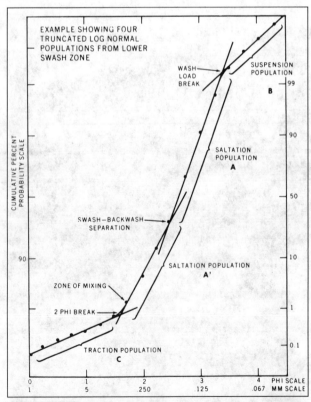

**Fig. 2-118.** Relation of sediment transport and dynamics to populations and truncation points in a grain-size distribution. From Visher, 1969.

By analyzing individual laminae produced by swash and backwash depositional events, it can be shown that swash events and backwash events produce differing grain-size distributions and that the mixing of laminae from these events produces a composite grain-size distribution (Fig. 2–121). The shape of this composite is similar to those developed on beaches both in the Holocene and in the laboratory. Thus, swash and backwash can be identified by the presence of an inflection point in the saltation population.

The depositional process can be related to the positions of curve inflection points. Sand samples from the Altamaha River Estuary indicate their importance for a variety of differing sedimentary conditions (Fig. 2–122). Understanding the casuality of these changes is subject of

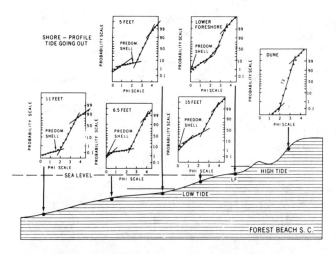

**Fig. 2-119.** Shore profile and sample distribution. Grain-size curves illustrate effects of environment on texture. From Visher, 1969.

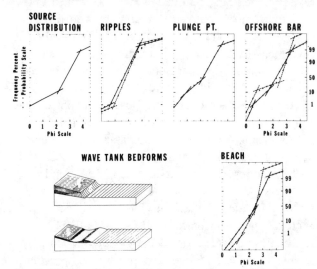

**Fig. 2-120.** Wave-tank modification by shoaling waves. From Visher, 1978a.

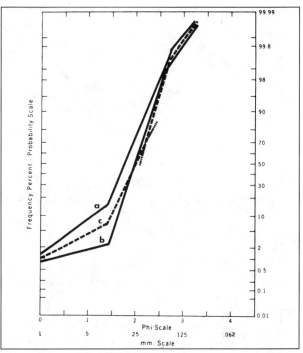

**Fig. 2-121.** Composite grain-size distribution of swash and back-wash samples taken from the wave tank. *a*, swash distribution; *b*, backwash distribution; and *c*, composite distribution of swash and backwash. From Kolmer, 1973.

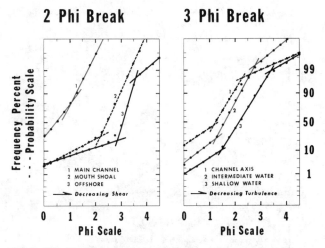

**Fig. 2-122.** Changing positions of inflection points in response to shear stress and turbulence. From Visher, 1978a.

continuing research. Sediment transport studies suggest it is possible to separate the effects of differing transport mechanisms. The 2-phi break shown in sand samples from the Altamaha River Estuary can be identified in bedload samples from a number of areas. The 3-phi break appears to be the result of the deposition at the sediment-water interface and reflects turbulent suspension.

Theoretical analysis of sediment transport produces a model to interpret the position of inflection points (Figs. 2-123 and 2-124). Application of this theoretical model is illustrated by comparison of modern distributary sands. The dimensionless power product (U×d/r) predicts the movement of the inflection point in a vertical direction. Other power products can be used to interpret horizontal movements. This approach is particularly important for interpreting sedimentary processes (Fig. 2-125).

In summary, it can be said that textural analysis is a useful tool to interpret sedimentary processes and, consequently, depositional environments. The application of

this tool to ancient stratigraphic units is also possible. Figure 2-126 is an example of grain-size distributions from modern beaches, and Figure 2-127 is an example from stratigraphic units interpreted to be beach deposits. Comparisons are easily made. However, differences are often developed, and these must be interpreted in the context of the developmental history of each stratigraphic unit.

## Sedimentary Structures

Sedimentary structures reflect depositional processes, which in turn are associated with specific sedimentary

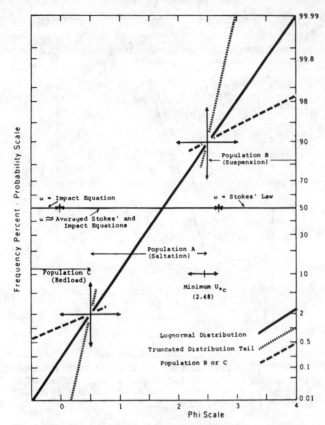

**Fig. 2-123.** An interpretation of the log normal grain-size distribution curve. From Sagoe and Visher, 1977.

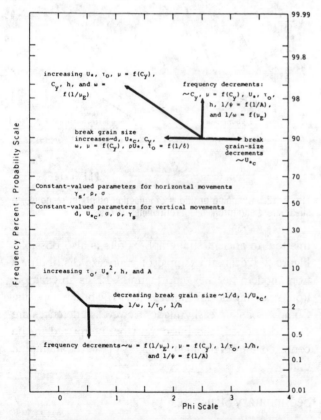

**Fig. 2-124.** Model developed indicating controlling parameters for grain-size distribution break positions. From Sagoe and Visher, 1977.

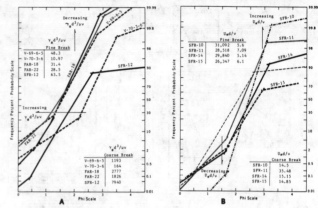

**Fig. 2-125.** Reynolds number and grain-size frequency. From Sagoe and Visher, 1977.

environments. Single structures may be formed under more than one environmental setting; consequently, similar sedimentary structures may occur in many different environments. Sedimentary structures are particularly valuable because direct analogues can be made between Holocene and ancient structures.

Crossbedding is probably the most widely noted sedimentary structure and may be formed in most environments. Crossbedding is related to migration of ripples, dunes, sand waves, bars, and other bedforms. These features are related to water depth and flow velocity, sand supply, shear stress, and grain size. A general hydraulic framework has been developed interrelating stream power and grain size (Fig. 2-128). This analysis does not consider water depth, viscosity, or turbulence scale and intensity, but it does provide a useful understanding of the interrelation of specific sedimentary structures.

In order to develop a genetic interpretation of sedimentary structures, it is necessary to define a differing basis for classification. Physical characteristics of the rheological properties of the transporting medium in relation to the operant physical process can provide a genetic basis for a classification of sedimentary features (Elliott 1968) (Table 2-11). This early attempt at interpreting the origin of sedimentary features has not been improved to date. It includes most features that have been identified in both Holocene and ancient stratigraphic units.

The interpretation of a simple bedform in relation to rheological flow is illustrated by the velocity distribution across a two-dimensional dune (Figs. 2-129 and 2-130). This has led to a better understanding of the origin of laminae and cosets. To expand this simple model, three-dimensional bedforms must be examined in relation to both stream power and grain size (Fig. 2-131). The nature of the internal bedding patterns for each type of bedform can be suggested. The form of these features as seen in a flume is shown in Figure 2-132. This experimentally

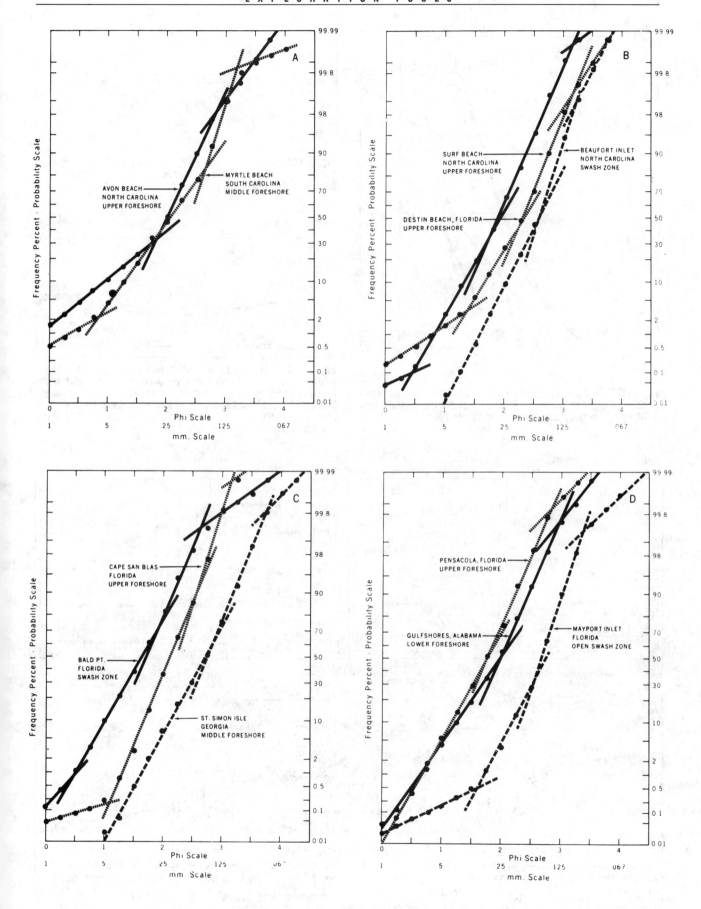

**Fig. 2-126.** Example of beach foreshore sands. From Visher, 1969.

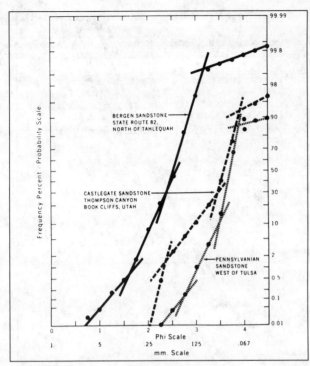

**Fig. 2–127.** Examples of sandstone with grain-size curve and shapes similar to modern beach sands. A beach origin for these sandstones is inferred from their curve shape. From Visher, 1969.

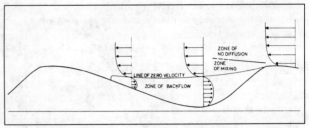

**Fig. 2–129.** Flow pattern over a leeface of a ripple. Velocity distribution, flow pattern separation, and three major zones on the lee sides are depicted. From Reineck and Singh, 1980.

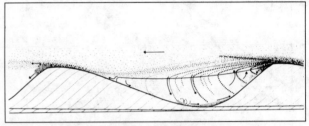

**Fig. 2–130.** Flow patterns and sedimentation processes on the lee side of a ripple. The heavy-fluid layer accumulates particles in the form of a proturberance at the crest, from where sediment is caught into backflow eddy and is deposited at the lee slope. Idealized path lines of sediment grains are shown modified after Jopling 1967. From Reineck and Singh, 1980.

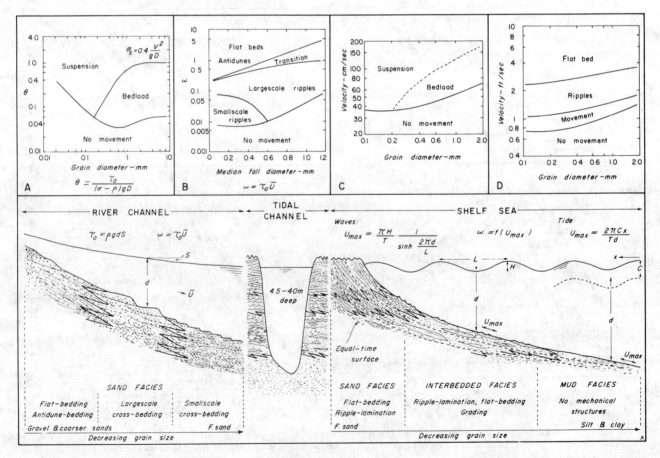

**Fig. 2–128.** Models for variation of grain size and primary structure with water depth for a freshwater and a shallow shelf-sea. From Allen, 1967.

**Table 2-11**
**Classification of Sedimentary Features**

| Nature of Operation | | Exokinematic | | | Endokinematic | | | Biokinematic |
|---|---|---|---|---|---|---|---|---|
| | | | Low | High | Translation | Transportation | | |
| Sediment Behavior | | Tranquil | F < 1 | F > 1 | Slumps | Horizontal | Vertical | |
| Liquid + detritus | Grain/grain | Even laminae | Crossbeds (forms drag) | Sheet structure | — | — | — | — |
| Quasiliquid detritus | Sheet flow washed out structure | — | — | Sand waves low bedding parting lineation | Graded bedding | Clastic intrusion | Sand volcanoes | No traces |
| Hydro-plastic detritus | Deformation flow of bed units | — | Erosion ripples diastems | Frondescent marks, flutes, convolute beds, tool marks, deformed crossbeds | Slurry, slump bedding | Crumpled bedding<br><br>Crumpled | Load structures, small diapir folds (air heave)<br><br>Mud lumps | Surface trace |
| Quasi-solid detritus | Pull aparts | — | — | Floating slabs | Slide, slump bedding | Crumpled bedding | Mud lumps | Bioturbated |
| Solid detritus | Discordant structures | — | — | Pot holes | Slide bedding | Block structures | | Good traces |

From Elliott; 1968.

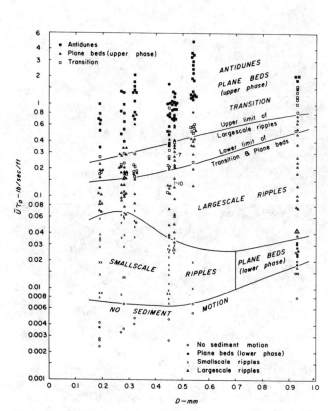

**Fig. 2-131.** Bedform in relation to stream power and grain size. After Allen, 1968. From Reineck and Singh, 1980.

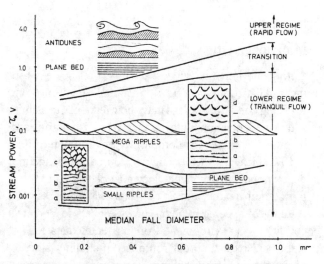

**Fig. 2-132.** Schematic representation of various bedforms and their relationship to grain size and stream power. a, straight-crested ripples; b, undulatory ripples; c, lingoid ripples; and d, lunate ripples. From Reineck and Singh, 1980.

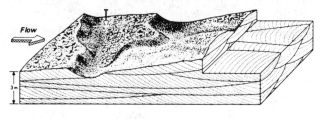

**Fig. 2-133.** Large-scale trough cross stratification formed by migrating dunes. From Potter, 1962.

developed feature can be recognized in Holocene sedimentary deposits (Fig. 2–133). Ripples and dunes are similar in form, and their internal bedding features can also be modified by the concentration of sediment in transport and changing flow velocities.

Another variable that must be considered is flow depth. Sedimentary features form in response to both currents and various types of waves. Velocity or stream power can be related to grain size, as we have previously seen, but the functional relationships also contain either implicitly or explicitly a statement of flow depth. This produces a model for interpreting the origin of specific sedimentary feature.

A complex pattern of flood and ebb, or changing flow velocities, produces a characteristic model of flaser bedding (Fig. 2–134). Similar patterns can be seen in the Holocene. An example of the formation of such units has been illustrated from the tidal environment (Fig. 2–135).

Similar sedimentary features related to high shear stress can be identified. Turbulent eddies as a result of high shear stress can occur in a number of environments. Only the process can be identified and not the environment of deposition (Figs. 2–136, 2–137, and 2–138).

## Mineralogic Criteria

Authigenic minerals suggest chemistry of the environment. Glauconite reflects normal marine salinity and reducing conditions. Clays may show response to salinity at time of deposition. Illite has been related to marine sediments, kaolinite to nonmarine environments (Fig. 2–139). Heavy minerals suggest source areas and not depositional environment. They may be related to transportational history and to depositional processes. The stability of specific minerals provide clues to chemical conditions of environment. Iron in a reduced state is not stable in oxidizing environment, including siderite, pyrite, magnetite, and chlorite. These minerals often indicate marsh environments.

Specific minerals suggest rates of erosion, transportation, and deposition. These indirectly can be related to depositional environments. Trace elements reflect changes in mineralogy and the amount and types of organic materials, primarily metals.

The equilibrium response between minerals and chemistry of the environment can be interpreted. For example, boron may be a salinity indicator. Sodium content of carbonates may reflect salinity. Also, many minerals are authigenic in the depositional environment and reflect its chemistry, which is one aspect of the environment.

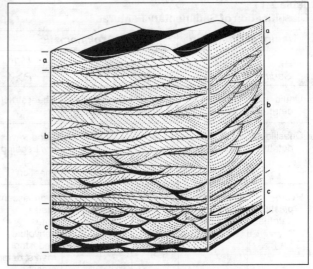

**Fig. 2-134.** Different types of bedding features formed as a result of wave action, depending on the intensity and type of wave action. From Reineck and Singh, 1980.

**Fig. 2-135.** Flaser bedding from Bartlesville sandstone

**Fig. 2-136.** Osage Layton formation

**Fig. 2–137.** Atokan formation, Arkansas

**Fig. 2–138.** Layton sandstone

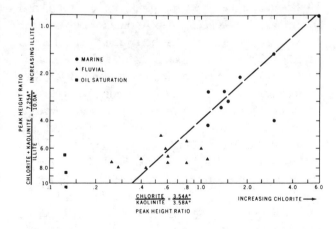

**Fig. 2–139.** Mineralogical data from fluvial and deltaic sandstone of the Bartlesville Bluejacket sandstone. Data suggest that illite and chlorite may be associated with marine sandstones and kaolinite with freshwater sandstones. From Visher, 1972.

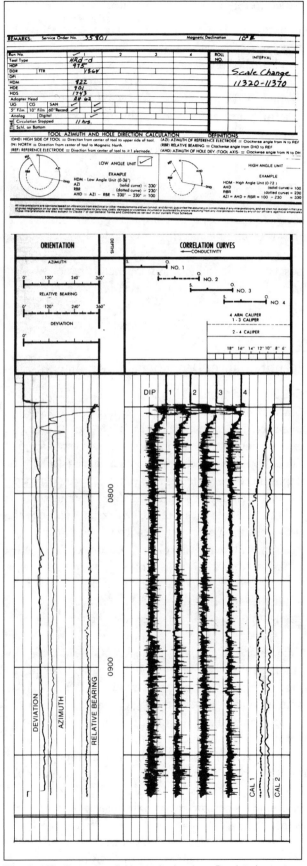

**Fig. 2–140.** Part of a field monitoring log. Reprinted courtesy Schlumberger Well Log Services.

115

## Morphology and Surface Texture

Impact, etch, and fracture surfaces of quartz grains have been shown to reflect sediment transport processes. U- and L-shaped marks are typical of beaches; smooth surfaces represent transport by suspension; and crescent shaped marks are produced by wind transport. High-relief fractures are due to crushing related to glaciers. These criteria are used by a number of researchers to identify transport processes and possible environmental history.

Shape of grains also suggests transport dynamics, discoidal grains are beach worn (coarse fraction), and elongate and equant grains are preferentially deposited in fluvial environments. Highly variable shapes and possibly platy and discoidal grains suggest transport by suspension. Roundness and sphericity are not environmentally sensitive, but they reflect transportation history, cycles of deposition, and cementation.

## Identification of Depositional Environments

Since a depositional environment is the sum total of all physical, chemical, and biological processes that operate in a specific physiographic area, the identification of an environment of deposition must include responses produced by the processes most characteristic of the depositional environment. There is no single process uniquely associated with a single environment. A multivariant data base must be analyzed. A unique solution to environmental identification is similar to solving a problem with many variables. A separate equation must be utilized for each response characteristic representing a differing process-response pattern.

In order to proceed toward a unique interpretation, stratigraphic response patterns must be compared to similar response patterns observed in the Holocene. Comparison to similar patterns on other ancient stratigraphic units has little scientific basis for interpretation. Processes cannot be observed by study of ancient stratigraphic units. What is the significance of any faunal association without the understanding of its ecology? What is the significance of a burrow or a biogenic trace if there is no understanding of morphological adaptation or the sediment-faunal interrelationship cannot be observed? What is the significance of a textural response if the physics of sediment transport has not been studied? Mineralogy, sedimentary structures, surface markings on grains, or fabric patterns all result from a range of environmental or transportational processes. The interpretation requires systematic comparison of Holocene patterns to stratigraphic responses.

It has been the author's observation that insufficient care has been taken in identifying specific responses resulting from characteristic processes associated with a specific depositional environment. For example, the beach process of swash and backwash is fundamental to this environment. Products of this process are specific patterns of sediment lamination, bioturbation, heavy mineral placers, and textural responses. Without direct observation and comparison to Holocene patterns, ancient beaches cannot be uniquely identified in the stratigraphic record. Similar analogies can be identified.

Since paleogeographic pattern is the goal of much stratigraphic analysis, a book on exploration stratigraphy must include a scientific basis for the identification of depositional environments. These patterns must be based upon a structured analysis of stratigraphic response patterns. This base can be used to predict reservoir patterns, geometry of stratigraphic units, and the three-dimensional pattern of permeability. These patterns interrelate source rock environments with permeability pathways allowing petroleum migration, and with associated permeability changes, the entrapment of hydrocarbons.

The environmental analysis flow chart (Table 2–12) is a scientific basis for the unique identification of depositional environments (Visher 1972). The data base is constructed, patterns identified, and comparison made to Holocene patterns. Developmental sequences are compared, and environmental associations are defined. The comparison of environmental sequences to paleogeographic patterns allows confirmation of the environmental interpretation. Such a rigorous analysis is the most useful basis for interpreting depositional environments and predicting their stratigraphic pattern.

## Summary

Holocene response patterns useful for environmental interpretation have been the focus of many books. Work on sedimentary structures has been recently summarized by Allen (1984), Reineck and Singh (1980), and Harms (1975). Work on textures reflects research by Visher (1978a), Sagoe and Visher (1977), and Visher and Howard (1974). Work on trace fossils have been summarized by McCall and Tenesz (1982), Frey (1975), and Reineck and Singh (1980). Books summarizing mineralogic studies include Pettijohn (1975), Fuchtbauer (1974), Pettijohn, Potter, and Siever (1973), and Garrels and Mackenzie (1971); those on surface textures include Krinsley and Doornkamp (1973). These summaries are based upon the process-response pattern so necessary for environmental interpretation. Synthesis of these observations into facies models has been the focus of books by Wilson (1975),

Schumm (1977), Reading (1986), Stanley and Kelling (1978), and Davis (1983). Countless other works have provided the stratigrapher with the scientific framework for predicting depositional patterns.

**Table 2-12**
**Environmental Analysis Flow Chart**

| *Start with most complete outcrop or core section of unit to be studied* |
| --- |
| • Make a detailed vertical section<br>Describe all structures—number, scale, and sequence<br>Determine current patterns if any are present<br>Describe the sequence of textural changes<br>Describe lithologic sequence<br>Describe character and pattern of bedding<br>• Determine sedimentary breaks, scour surfaces, unconformities, diastems, etc.<br>Look for unusual structures, burrows, etc.<br>Determine unusual mineral assemblages<br>• Determine presence or absence of fauna<br>Macro- or microfossils<br>Trace fossil types and sequences |
| *Determine and interpret possible genetic sequences* |
| • Relate sequences to a known depositional process<br>Correlate depositional sequence to a process of deposition<br>Determine possible boundary conditions—depth of water, rates of sedimentation, energy in depositional environment, and possible water chemistry<br>• Determine if repeating patterns are developed within the section<br>Simple pattern with one genetic sequence<br>Complex of many depositional periods, events, and depositional units<br>• Make a tentative assignment of depositional process<br>• Select samples for detailed analysis to identify depositional processes<br>Attempt a paleoecologic evaluation<br>Make detailed textural study of selected samples<br>Determine lithologic components of selected samples<br>Use other tests and studies deemed appropriate |
| *Assign a most probable depositional model for the genetic interval* |
| • Determine lateral variations of components analyzed<br>Contrast and compare vertical patterns with adjacent sections<br>Attempt a detailed bed-by-bed correlation<br>• Map variables that appear to be significant<br>Select a few variables that appear to vary systematically—grain size, bed thickness, crossbedding, and lithology<br>Map as many observed variables as can be conveniently seen or expressed on a single map<br>• Relate areal and vertical patterns to depositional processes<br>Compare Holocene and ancient depositional patterns<br>Select a depositional model most applicable |
| *Integrate and verify findings* |
| • Determine relation of section or sections to major tectonic features<br>Determine position in the basin<br>Determine overall position in relation to larger depositional cycles or patterns<br>• Obtain additional cores, outcrop sections, or other new data and use to confirm pattern obtained |

## DIPMETER INTERPRETATIONS

The dipmeter has been a useful tool for petroleum exploration for more than thirty years. Like most exploration tools the instrumentation and the analysis of the response patterns have dramatically changed. Where originally there were only three-arm dipmeters with correlation and dip calculation prepared by hand, now four-arm dipmeters, a wide range of computer correlations, and an array of dip calculations are available. There are low-density, high-density, CLUSTER, and GEODIP computer analysis methods of manipulating the data for easier analysis and interpretation. The sophistication of the instrumentation and the computer analysis, however, is not of value unless it is based upon sound stratigraphic principles. It has been amply demonstrated that the equipment does measure discontinuities across the borehole to a precision of 1/10th of an inch at the recording depth. This allows the calculations of dips of less than two degrees. Since the measurement does appear to be valid, the data generated should be interpretable in terms of the developmental history of the stratigraphic section.

Field record sections show the nature of the continuous physical measurements produced by the instrumentation (Fig. 2–140). To analyze such data, many decisions must be made by the petroleum explorationists. They must determine sampling interval, step interval, search angle, and the type of presentation to be developed. These decisions must be made in order to determine the type of information to be derived from the data. A fairly complete presentation of these choices is presented in the *Schlumberger Dipmeter Interpretation Manual* (1981). The purpose of this discussion is not to present the details of the technology but rather to provide a framework for interpreting the data.

### Data Presentation

The data are presented as dip magnitude and azimuth at some predetermined depth spacing (Fig. 2–141). The CLUSTER analysis produces dips that have a specifically defined variance limit (Fig. 2–142). The GEODIP processing expands the comparison of the resistivity curves to include pattern recognition. This is accomplished by examining segments of the curve and comparing each curve segment (Fig. 2–143). The resulting plot is illustrated in Figure 2–144. In addition, an azimuth-frequency plot is presented for predetermined intervals. The relating of either statistical confidence limits or curve shape provides a more coherent presentation of the data.

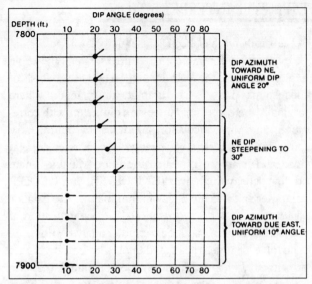

**Fig. 2-141.** A typical arrow (tadpole) plot. Reprinted courtesy Schlumberger Well Log Services.

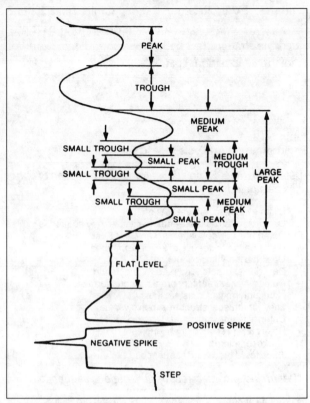

**Fig. 2-143.** GEODIP features defined. Reprinted courtesy Schlumberger Well Log Services.

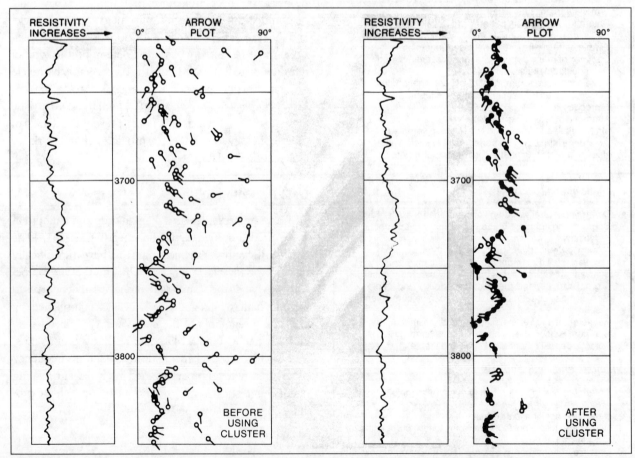

**Fig. 2-142.** Dipmeter results before and after using CLUSTER. Reprinted courtesy Schlumberger Well Log Services.

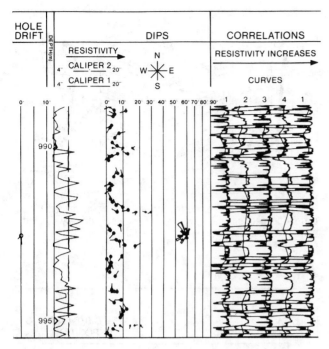

**Fig. 2-144.** Graphic GEODIP output. Reprinted courtesy Schlumberger Well Log Services.

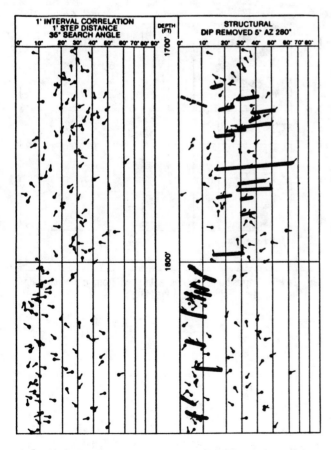

**Fig. 2-145.** Dipmeter arrow plots before (left) and after (right) DIPSUBTRACT. Reprinted courtesy Schlumberger Well Log Services.

## Stratigraphic Interpretation

As a guide to interpretation, it is useful to present the dipmeter data in conjunction with other logs. Dip patterns reflect sedimentary, structural, and depositional patterns, which must be included as a part of the interpretation (Fig. 2-145). The following are the most easily recognized patterns and their most probable causes (Fig. 2-146).

As an aid to recognizing these patterns, plotted log intervals using both the azimuth frequency plot and the Schmidt Plot are needed for additional interpretation (Fig. 2-147). These can aid in determining crossbedding dips and directions within zones with structural dip.

Crossbedding dips may be extracted from high-density dip information, and their form can be analyzed in relation to azimuth variation, inclination, and changes within individual cosets. This information can be usefully applied to the interpretation of stratigraphic intervals (Fig. 2-148).

The absolute range of dip variance also has depositional significance. As indicated by bedding patterns from seismic data, channel fill, mounded buildups, drape, and other patterns have depositional significance. By examining the vertical changes in magnitude of the dip spread (Fig. 2-149), a possible relationship to depositional energy can be shown. These data can then be correlated to biofacies information to produce an understanding of bathymetric cycles, patterns of onlap, and definition of shelf-slope transitions (Fig. 2-150).

Unconformity surfaces are easily identified from dipmeter logs. The absolute change in the dip direction and/or magnitude across a very narrow zone is highly suggestive of an unconformity surface (Fig. 2-151). In addition, crossbedding patterns in topographic lows above the unconformity surface and fracturing or weathering immediately below the surface produce systematic dip patterns.

## Structural Interpretations

Several applications have been suggested for interpreting faulting, folding, diapiric, and drape structures. Systematic changes of dip due to rollover, drag, and the position of fold axes all can be recognized.

A reef, or diapiric shale or salt pattern, is illustrated in Figure 2-152. A recent paper by Bengston (1982) describes the approach to structural interpretation. An example of this analysis illustrates its usefulness for interpreting complex structure (Fig. 2-153). The dipmeter is the exploration tool of choice in the prediction of oil occurrence in areas of complex structure. Patterns in overthrust belts, near salt domes, or complex diapiric shale

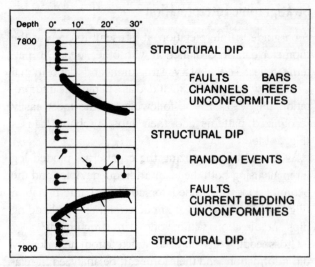

**Fig. 2-146.** Dip patterns and geologic anomalies commonly associated with them. Reprinted courtesy Schlumberger Well Log Services.

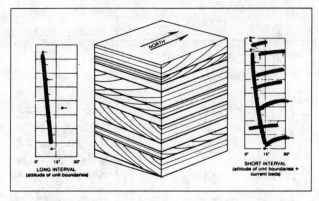

**Fig. 2-148.** Influence of correlation interval on computed dip in current-bedded channel sequence. Reprinted courtesy Schlumberger Well Log Services.

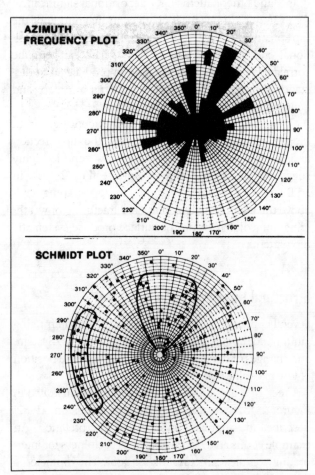

**Fig. 2-147.** AZR and POLAR plots of data from DIPSUBTRACT, Fig. 2-145. Reprinted courtesy Schlumberger Well Log Services.

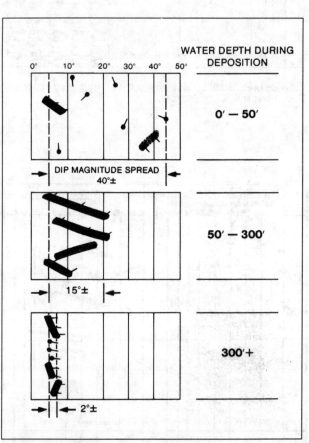

**Fig. 2-149.** Depositional depths estimated from dipmeter data. Reprinted courtesy Schlumberger Well Log Services.

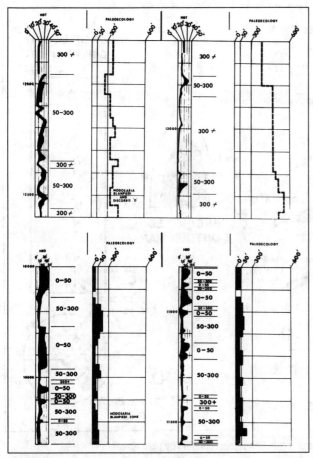

**Fig. 2-150.** Relation of dip spread to bathymetry as suggested by paleoecology. Reprinted courtesy Schlumberger Well Log Services.

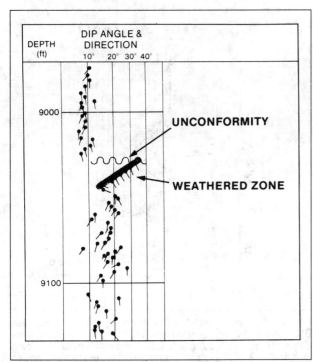

**Fig. 2-151.** Angular unconformity, Louisiana. Reprinted courtesy Schlumberger Well Log Services.

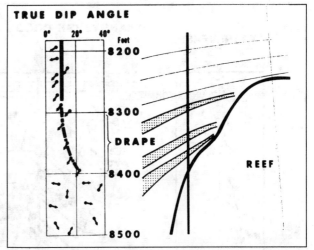

**Fig. 2-152.** Systematic increasing dip magnitude resulting from drape over a reef or diapiric features. Reprinted courtesy Schlumberger Well Log Services.

anticlines can be interpreted by this type of data.

Fracture analysis is another area of structural interpretation that recent advances in dipmeter analysis can provide. Since the dipmeter produces four mutually perpendicular microresistivity traces, the tool will cross vertical mud-filled fractures as it rotates. By plotting pairs of traces, it is possible to detect simultaneous differences in absolute resistivity values (Fig. 2-154). Fracture detection is one of the more important tools available to the explorationist in complex structural areas and low permeability formations. Downward increasing dip magnitude for three cycles A, B, and C is probably due to crossbedding. The interval below possibly is related to growth faulting (Fig. 2-155).

## Summary

Coherent patterns of dip change in a well section are information that reflects stratigraphic or structural patterns. It would be foolhardy to throw away useful information, and an effort must be made at interpretation. Figure 2-120 shows a series of increasing patterns of dip. Which patterns are sedimentary, which may be structural, and which may reflect an unconformity surface? These patterns must be placed into a larger context of stratigraphic data before an interpretation can be developed.

## REFERENCES FOR FURTHER STUDY – CHAPTER 2

Allen, J.R.L., 1984 Sedimentary Structures. Their Character and Physical Basis: Developments in Sedimentology, no. 30: New York, Elsevier, 1256 p.

Bally, A.W., ed., 1987, Atlas of Seismic Stratigraphy: Am. Assoc. Petrol. Geol., Studies in Geology, no. 27.

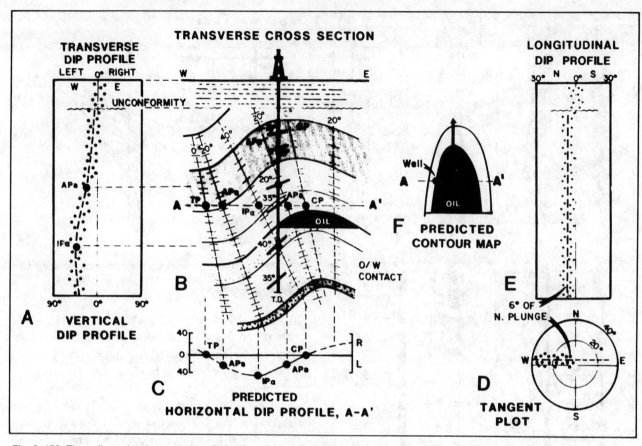

**Fig. 2–153.** Three-dimensional structural prediction from limited dip control. From Bengston, reprinted by permission.

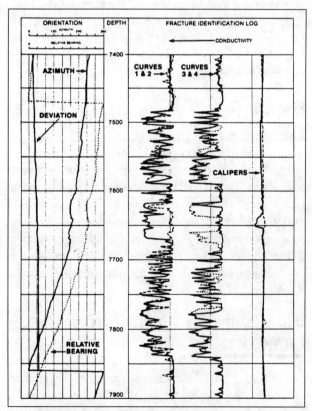

**Fig. 2–154.** Fracture identification (FIL). Reprinted courtesy Schlumberger Well Log Services.

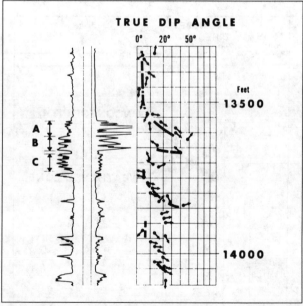

**Fig. 2–155.** Interpretation of dip patterns requires additional data. Downward increasing dip magnitude for three cycles, A, B & C probably is due to cross-bedding. Interval below possible related to growth faulting (Fig. 2–155). Reprinted courtesy Schlumberger Well Log Services.

Bengston, C.A., 1982, Structural and stratigraphic uses of the Dip profiles in Petroleum Exploration, *in* Halbouty, M.T., ed., The Deliberate Search for the Subtle Trap: Am. Assoc. Petrol. Geol. Mem., no. 32, 351 p.

Berg, O.R., and D.G. Woolverton, eds., 1985, Seismic Stratigraphy II An Intergrated Approach to Hydrocarbon Exploration: Am. Assoc. Petrol. Geol. Mem., no. 39, p. 79–92.

Collins, A.G., 1975, Geochemistry of Oilfield Waters Developments in Petroleum Science, v. 1: New York, Elsevier, 496 p.

Demaison, G.J., and G.T. Moore, 1980, Anoxic environments and oil source bed genesis: Am. Assoc. Petrol. Geol. Bull., v. 64, p. 1179–1209.

Frey, R.W., 1975, The Study of Trace Fossils. A Study of Principles, Problems, and Procedures in Ichnology: New York, Springer-Verlag, 562 p.

Leythaeuser, D., R.G. Shaefer, and M. Radke, 1987, On the primary migration of petroleum: 12th World Petroleum Congress Proceedings, v. 2, Exploration: New York, John Wiley & Sons Ltd., p. 227–236.

Reineck, H.E., and I.B. Singh, 1980, Depositional Sedimentary Environments, 2nd ed.: New York, Springer-Verlag, 549 p.

Tissot, B.P., and D.H. Welte, 1984, Petroleum Formation and Occurrence: New York, Springer-Verlag, 699 p.

Roberts, W.H., III, and R.J. Cordell, eds., 1980, Problems of Petroleum Migration: Am. Assoc. Petrol. Geol., Studies in Geology, no. 10, 273 p.

Waples, D.W., 1980, Time and temperature in petroleum formation Application of Lopatin's method to petroleum exploration: Am. Assoc. Petrol. Geol. Bull., v. 64, p. 916–926.

# THE MODEL APPROACH TO EXPLORATION STRATIGRAPHY

No substantial part of the universe is so simple that it can be grasped and controlled without

abstraction. Abstraction consists in replacing the part of the universe under consideration by a

model or similar but simpler structure. Models, formal or intellectual on the one hand, or

material on the other, are thus a central necessity of scientific procedure. **A. Rosenblueth and N. Wiener, 1944–1945.**

## INTERRELATION OF GEOLOGIC OBSERVATIONS

The synthesis of scientific theories and the search for a basis of interrelation of all physical laws was a goal of science in the first part of the twentieth century. Mill (1865) recommended a "Deductive Method" be employed in the investigation of composite causation by:
- the formulation of a set of laws;
- the deduction of a statement of the resultant effect from a particular combination of these laws; and
- verification.

The hypothetico-deductive method thus could form a basis of scientific verification.

G.K. Gilbert was the principal proponent in the Americas for the application of causality to the historical sciences (Gilbert 1886). Hypotheses need to be tested, and "...one seeks diligently for the facts which may overthrow his tentative theory...". His approach embodied four essential elements:
- (1) observing the phenomenon to be studied and systematically arranging the observational data,
- (2) inventing hypotheses regarding the antecedents of

the phenomena,
- (3) deducing expectable consequences of the hypothesis, and
- (4) testing these consequences against new observations (Gilluly 1963, 221).

It was Chamberlin (1897) that brought to a culmination the ideas developing during the period of "Rationalism" in geologic thought. His principle of the "multiple working hypothesis" brought credence to the method of deduction from observation as the basis for the construction of hypotheses. An hypothesis could only be validated by testing against further observation.

In the first part of the twentieth century geological scientists were busy applying the physical, chemical, and biological laws developed in the nineteenth century. The use of analogy led to the statement of geological "laws" or principles, such as Walther's "Principle of Facies" (Walther 1893–94). Gilbert (1914) formulated a number of "laws", or hypotheses, related to sediment transport. Laws of physical chemistry were used by N.L. Bowen in his important synthesis, the *Evolution of Igneous Rocks* (Bowen 1928). Bucher (1936) went beyond simple formulations and suggested the concept of "natural law" in geology.

Application of physics and mechanics to the understanding of structural systems was usefully based upon dimensional analysis and scale model theory by Hubbert (1937). Experimental studies can lead to understanding of large systems, such as mountain chains. His suggestion was that tectonic processes can be easily understood in terms of uniformitarianism, and not as catastrophic events outside of accepted mechanical and physical processes, within a time-space continuum. In addition, experimental methods can be used to verify hypotheses relating to large scale earth systems.

The first part of the twentieth century has been the period of search for quantitative measures, and the escape from the geological approach, stated as the "art of the geological inference." Few quantitative descriptions were possible for most geological systems, and statistical descriptions, using normative or probabilistic statements resulted.

As an outgrowth of the need for testing of hypotheses, new bases of geologic observation were required. Many of these were based upon historical approaches to causality, experimental studies, and comparisons of the Holocene to the rock record. Expansion of the trend towards quantification of observations continued using mechanistic, physical, chemical, statistical, and mathematical descriptions.

## EXPERIMENTAL METHODS IN GEOLOGICAL SYNTHESIS

Empirical inferences based upon incomplete observation must of necessity be compared to rational deductions from systems of previous observation. Most geologists based their syntheses upon empirical evidence without regard to theory and conceptual formulations based upon broader observational bases. The development of geological thought necessitated a methodology that could include both empirical evidence, and a conceptual synthesis.

The most useful methodology which could address these problems was the application of experimental methods to the understanding of geologic patterns. This approach led to new formulations of systems, uses of analogy, and causality. During the 1950s and 1960s, these approaches broadened the observational basis of the geological sciences.

Actualistic studies of fluvial hydrology (Mackin 1948), oceanographic processes and responses (Shepard 1948, and Shepard and Moore 1955), research programs in sedimentology and sedimentary facies (Lowman 1947, 1949), and reefs (Ladd 1944, 1950) provided an actualistic basis for interpreting responses to measurable processes. A landmark contribution was the publication of "Recent Sediments, Northwest Gulf of Mexico" (1960), summarizing work commissioned by the American Petroleum Institute and carried out between 1951 and 1958 under the direction of Francis Shepard. In addition, research in geomorphology (Wolman and Miller 1960) and a summary textbook by Scheidegger (1961) suggested the need for the use of causality in the development of geological inferences. The result of this approach essentially changed the nature of the development of geologic hypotheses. Without actualistic studies based upon processes quantitatively measured, a hypothesis could not be verified.

The use of scale-model theory allowed the verification of physical processes. Geological systems beyond time or scale frames of reference required experimental verification. Laboratory experimentation could be expanded and made useful to examine space and time in geological dimensions.

Experimentation led to a broader understanding of causality. What are the causes of paleogeographic changes? What causes evolution? What are the reasons for observable stratigraphic patterns? Can we understand what produces observable geologic patterns? The expansion of the use of cause and effect brought new syntheses of geologic observation. Most geologic synthesis published during the first half of the twentieth century usefully applied these developmental concepts.

Studies in physics and mathematics have cast doubt upon the possibility of a unique synthesis of physical laws and experimental procedures so usefully applied in the development of geological hypotheses. First, with Heisenberg's "Uncertanty Principle" (1927), a hypothesis interrelating quantum mechanics and wave theory, but the mathematical proof suggested that the synthesis could not be verified by observation. Secondly, even more crushing, was the mathematical theorem of Gödel (1931), which proved that a thesis could not be demonstrated beyond its physical-mathematical limits.

## THE SYSTEMS APPROACH TO SYNTHESIS

Reading the papers of historians and philosophers of science suggests the need for new definitions of science, new understandings of universal "laws," the need for new bases for syntheses of observation, and new syntheses of theory and observation. Gödel and Heisenberg's insights loom large over all scientific theories. Is it possible to synthesize all observation and theory using concepts of causality, verifiable by experiment, experience, and predictive schemes?

In the geological sciences, statistical analysis, factor, principle component, and other methods of analysis of variance have provided more quantitative description, but rarely insights into causality (Krumbein and Graybill 1965). Synthesizing concepts such as the depositional environment, based upon process-response patterns (Krumbein and Sloss 1963), have provided an actualistic base for synthesizing geologic observation. These concepts were the basis for the synthesis of observations in geomorphology (Pittendrigh 1958; Hack 1960; Scheidegger 1961): and in the biological sciences by Mayr (1951) and Simpson (1963).

The world is complex, heterogeneous, and sensually overwhelming. The necessity to filter or in some manner structure what we perceive is constantly a problem.

A new approach to synthesis that is increasingly prevalent in the philosophy of science is the concept of the conceptual model introduced by Rosenblueth and Wiener (1944–45) and described by Visher (1984). The model concept utilizes the observational base, but goes beyond empiricism and deductive hypotheses to examine a system or paradigm for causal patterns. The *schema* or theme that can "explain" the interrelation of all observations is useful for suggesting new or additional observations, and also provides the basis for verification by prediction of new observations. This approach includes concepts of inference, ideas, and projection beyond the experiential base of the scientist. It truly is a creation by the scientist.

## What Is a Model?

A model has been defined by Kendall and Buckland (1960) as "a formalized expression of a theory or the causal situation which is regarded as having generated observed data." If the role of the scientist is to order observations and to construct hypotheses and laws to aid in prediction, then he is faced with the task of organizing the complex world into simple systems and determining "an overview of the essential characteristics of a domain" (Apostel 1961, 15). This simplification requires both sensual and intellectual creativity (Kuipers 1961, 132). Reality then, as perceived by the scientist (stratigrapher), is a "patterned and bounded connexity which has been explored by the use of simplified patterns or symbols, rules and processes" (Meadows 1957, 3–4). A simplified way of expressing the above is in context of the model concept. Deutsch (1948–1949) suggests that the model format has greater implication than the sum of its individual parts, and models can be used as a basis for prediction in the real world. Rosenblueth and Wiener (1944–1945, 321) suggest:

"... models, imperfect as they may be, are the only means developed by science for understanding the universe. This statement does not imply an attitude of defeatism but the recognition that the main tool of science is the human mind and that the human mind is finite."

## Types of Models

Models can be used to organize observations based upon a scaled representation of the real world or material models. Real world properties are represented by the same properties with only a transformation in scale (Ackoff, Gupta, and Minas 1962, 109). Scale models were formalized by Hubbert (1937) and have been used extensively in the development of practical solutions to engineering problems. Also, they may be analogous where, for example, elevation is represented by a color or line rather than a three-dimensional model.

Another group of models using symbolic or formal assertions, and using mathematical or verbal forms to construct the logical basis, has been classified into abstract or theoretical and symbolic or formal models (Rosenblueth and Wiener 1944–1945, 317). Abstract models deal with the scientific search for a structure equivalent to that of a given experience. Scientific knowledge consists of a sequence of abstract models that form the basis for organizing observations from the universe. A formal model is a symbolic assertion in logical terms of an idealized and relatively simple system sharing its structural properties with the original factual system. Theoretical and symbolic models are approaches to the development of a construct that is useful in predicting patterns or connexity within observations.

For a model to have validity, its construct must have a logical basis, and its application to the real world must provide a basis for prediction. The concepts involved in deterministic and stochastic models suggest their logical basis. If the thread that holds the model together is deterministic, then possibilities of cause and effect can be entertained. Construction of formalized mathematical statements is possible, and within this context the data used in the model provide the basis both to interpret and predict the association of natural events or processes. However, if a random event or a nonsystematic pattern is present, then probability statements must be included and statistical, for example, stochastic, models must be constructed (Krumbein and Graybill 1965, 17). These models can be explicitly stated in statistical terms, or they may have a conceptual basis.

No one type of model is better than another. The value of a model is in suggesting new observations that need to be made, organizing previously unstructured observations, and most importantly, predicting natural associations or events that occur in the observational data base of the science.

## Applications of Models

The model approach goes beyond synthesis of observations, it is the tool that can lead to new syntheses of observation. The geological sciences rarely could define their systems in relation to mechanics, or to mathematical systems that could include all variables within the system. In keeping with Godel and Heisenberg, they required new approaches to synthesis of observation. The model *schema* has become the basis for advances in geological thought.

Conceptual models were first applied to geological data by Miller (1954). The concept of facies models was introduced by Potter (1959), process-response models (Krumbein and Sloss 1963), and conceptual models by Visher (1963; 1965). New syntheses utilized the conceptual model approach, for example, the plate tectonic model as developed and presented by Isacks, Oliver, and Sykes (1968).

The model hypothesis, as illustrated by the development of plate tectonics, shows its usefulness in understanding and predicting natural patterns. When Vine (1963) and Vine and Mathews (1966), introduced their hypothesis that if polarity reversals of the magnetic field do exist, there should be strips of sea-floor material having reversed polarity spreading out symmetrically and parallel to midocean ridges. Thus, if spreading of the ocean floor occurs, blocks of alternately normal and reversely magnetized material would drift away from the centre of the ridge and parallel the crest.

Vine and Mathews hypothesized that the whole ocean crust is made of striped material parallel to the axes of midocean ridges, and having alternately, reversed and normal magnetization. This was consistent with Dietz's (1961) "spreading floor" hypothesis. A model is based upon its predictability, and when the analysis of the Eltanin 19 cruise data (Pitman and Heirtzler 1966), which measured the magnetic anomalies across the Pacific Antarctic and from the Reykjanes Ridges, near Iceland, produced striking corroboration of the Vine-Mathews hypothesis, a whole decade of research on crustal thicknesses; first motion of crustal movement as a result of earthquakes, fault patterns, the topography of the ocean floor, the distribution and configuration of continental margins, patterns of seismicity, and focal mechanisms to support the concept of transform faulting; and heat flow data could be understood in the framework of the sea floor spreading hypothesis. The concept of plate tectonics was firmly developed, and corroborating evidence from seismology, paleontology, age dating, and countless other lines of evidence produced a synthesizing model that has been a basis for continuing research and understanding. But even now, this concept cannot be used to interpret observations concerning patterns of mountain building on continental margins and within cratonic plates. The model framework has been demonstrated as a basis for synthesis, but it is only a useful device that must be superseded by new observations and a new synthesis.

Much of the more recent literature in geology refers to models as the basis for synthesis. The model approach is a multivariant analysis of independent types of observation. New approaches in linear programming, information processing, and topologic map comparisons may become the bases useful for the synthesis of geologic data. Geologic thought is again in transition to new methods and approaches. New approaches are beyond inference, "look alike" comparisons, and statistical statements of association: they are creative syntheses verifiable by experience, experiment, and new observation. The use of computers for information processing, and possibly the development of logical systems through artifical intelligence or programs of pattern recognition, may provide new approaches to scientific synthesis.

Models are tools or devices for aiding in organizing observations, but their application is not simple. What types of observations are codified into a model? What levels of integration of data are needed or possible? In the model approach to scientific analysis, it is necessary to find an integrating principle or structure that can be used to interrelate all observations. In stratigraphy this rational thread can be paleogeography and the controls or processes that affect changes in paleogeographic patterns. The stratigraphic goal is history of the earth. If a basis can be found for developing historic models, then it may be possible to use stratigraphic data in a dynamic manner to predict patterns.

## CLASSIFICATION OF MODELS

A sophisticated structuring of models is required (Table 3–1). This table indicates a possible interrelationship of some of the more commonly used stratigraphic models and is based upon both dimensional and process controls.

Placing a type of model within a particular block is arbitrary, but the usual application is at the level indicated. Conceptual models also may contain elements of developmental history or stochastic variables; three-dimensional and possibly four-dimensional variations can be illustrated by Fourier or other derivative models. Basin models may simply be descriptive and have little relation to history or to dynamics of change. Certainly, the goal of most stratigraphic analysis must be along the four-dimensional line and result in a developmental analysis. From this a simulation model may have the most power in interpreting the stratigraphy of a particular area. Many of the other models would be tools or steps that

**Table 3-1**
**A Stratigraphic Paradigm**

|  | Static Models | Dynamic Models | Developmental Models |
|---|---|---|---|
| Descriptive | Classification<br>Linear regression<br>Taxanomic | Experimental<br>Mechanistic<br>Probability | Theoretical<br>Diagrammatic |
| Two Dimensional | Topographic<br>Structural<br>Paleogeographic<br>Paleocurrent | Derivative<br>Fourier<br>Harmonic<br>Trend Surface | Measured sections<br>Cross sections |
| Three Dimensional | Geologic maps<br>Facies<br>Scale<br>Isolith | Sedimentary<br>Tectono facies<br>Sequence<br>Ecologic | Stochastic<br>Actualistic<br>Basin |
| Four Dimensional | Evolutionary<br>Time stratigraphic<br>Unconformity | Conceptual<br>Symbolic<br>Deterministic<br>Process response | Simulation |

lead toward more complete statements regarding the relationships between observations.

As in many other sciences, the ability to go from the specific to the more general is often severely limited by the observational data base. The transition from evolutionary biology to population dynamics and ecology required many cycles back to more careful observations. Descriptions of population diversity, population abundances, interrelationships between the environment and the faunal and floral populations, and even the diversity within a gene pool of a particular taxon provide bases for dynamic models. Countless other examples can be cited, including sedimentary models. Models relating to the pattern of a stream through time, physical measurements in modern rivers, scale models, historic events, dimensional analysis, deterministic relationships, and even conceptual insights were necessary before a process-response model could be suggested. Stochastic and simulation models have been developed. An example of the elements of the process-response model for fluvial sedimentation is illustrated by Table 2-2. Only the measurable aspects useful in identifying fluvial deposits are indicated.

## Types of Data

After observation of interrelated data, some large body of information or measurement needed to construct a model almost always is missing. The information necessary to provide a basis for the model may require a whole new program of research and observation. In many cases a new method of measurement or another method of analysis must be developed before a model can be suggested.

Of all the aspects of stratigraphic analysis, the most critical and the least understood is what constitutes useful data. How many attempts were made to understand the pattern of successive depositional units within the stratigraphic section before the stratigrapher found this to be a powerful tool for predicting the areal pattern of depositional units? Johannes Walther (1893–1894) suggested the observation was important, but he lacked the understanding of the developmental mechanism.

## Organization of a Data Base

The problem in organizing a data base to show the logical basis for a model is the discovery of the theme or basis for the organization. For example, the literature on sedimentary structures is extensive, and much of the collated information has either an actualistic or a strongly based process control, but still little can be generalized from a list of sedimentary structures. Few generalizations concerning the depth of water, strength of the current or waves, fluid content of the sediment at the time of deposition, sequence of events necessary for the formation of complex compactional or synsedimentary structures, rates of sedimentation, or many of the process-related aspects of a depositional environment are suggested. What is the problem? Simply, if there is no classification or ordering scheme meaningful in terms of process of formation, all the observations and description of sedimentary structures goes for naught. However, if some basis can be established for observation, a classification scheme can be developed. The relation of fluid content and the energy applied to the sedimentation unit has been a basis for such a classification (Elliott 1968). Table 2-11 illustrates the method used in organizing observations. Without the concepts of fluidity and rheology there would be no basis for such a classification. These are not directly observable aspects, but they can be measured, experiments can be constructed, and a synthesis can be applied to determine if the model is useful in an actualistic sense.

The unrelated observation has little meaning in stratigraphy, but the relationship may be on an objective basis and still have meaning in terms of less quantitative models or at a lower level of integration. Data simply are not available to develop quantitative, deterministic models of basin evolution or the history of a particular stratigraphic succession in both space and time.

## Verification of a Model

Models are simply useful constructs, and they do not represent truth. If models are dynamic, they are useful in developing larger scales or more quantitative levels of data integration. If a model is valid, it has built into it the seeds of its own destruction. Therefore, verification of a model comes to mean only synthesis of data useful in predicting a pattern or patterns that are found within the data upon which it is based. Unifying constructs of stratigraphy, such as unconformities, time-stratigraphic subdivision of the stratigraphic record, stratigraphic sequences, and sedimentologic models, all have proven to be highly useful concepts. These have been verified by bringing together different lines of evidence, diverse types of data, and their ability to predict stratigraphic patterns.

The key to verification of a model is in its power to interrelate diverse types of observations. For example, plate tectonics not only had an impact upon understanding the history of the oceans, but also upon tectonics at plate margins, basin evolution, sequences of related stratigraphic units, population diversity, and periods of explosive evolution, volcanic activity, and many apparently unrelated physical, biological, and stratigraphic observations. Plate tectonics is an accepted model because of this diversity of supporting evidence, not because of magnetic stripes, the fit of continental plates, or age dates from oceanic areas. In fact, many of these observations do not fit very well into the plate tectonic model. However, this provides the basis for expanding the model, adding new processes to the overall theme, and pointing to areas where more research, additional observations, and new levels of synthesis are possible.

If a model can be used only to predict patterns in a particular formation or a restricted area, it truly is not a model. The integration of observation is simply a description of that formation, albeit a good description, if it allows detailed comparisons of one section to another. If the model does not have a thread or theme to fit it together, it is not verifiable against any standard except more observation. If one observation fails to confirm a pattern, the entire system loses validity. If there is a unifying principle, such as that for plate tectonics, it cannot be destroyed on the basis of individual observations, since the model is greater than the sum of its parts. Consequently, the challenge in petroleum exploration is to find those unifying principles that can be used as a basis for interpretation and prediction.

## MODELS APPLIED TO STRATIGRAPHIC DATA

Stratigraphic data represent such a diversity of observations that it has been difficult to find a basis for synthesis, and even more difficult to quantify observations so that they can be repeated by subsequent workers. Each stratigrapher relates observations to experience and places differing emphasis on the same observation. It has been true that the explorationist with the widest range of experience was best able to make comparisons and then by analogy draw conclusions concerning the significance of his observations. Such an approach has led to subjective interpretations, disagreement among experienced explorationists, and a lack of understanding of the significance of individual observations.

### Choosing a Model

Paramount in scientific analysis is the statement of the study's purpose or the reason for making systematic observations. The purpose not only suggests the types of observations needed, but also the form of the unifying principle or model to be applied to the data.

If the problem is one of identifying an unconformity, some basis must be found to indicate time. The methods of determining time require some application of paleontology or a basis for recognizing the absence of a particular event or events. Many papers on stratigraphy attempt to show the presence of an unconformity by using descriptive or a two- or three-dimensional model; the results of these efforts must of necessity simply be nonsense if no measurement of the time dimension is used.

Other examples can be cited the problem of determining the depositional environment of a stratigraphic unit. Depositional environments are dynamic interrelationships of physical, chemical, and biologic processes operating in an areally restricted site. A model that contains only descriptive aspects cannot be used for environmental recognition. Processes operate through time, and some indication of the time relationships of stratigraphic units must be demonstrated. Also, since an environment represents events taking place in an area, a pattern needs to be established. The unrelated observation of cross bedding, textural characteristic, or the presence of a particular taxon cannot be sufficient to define the depositional theme or model.

## Types of Models Used in Stratigraphy

The value of the definition of types of models is that each model proscribes the form of the observation, the basis for integration of the data, and the method or methods of verification. This rigor makes it more difficult to make interpretations, construct hypotheses, and draw conclusions beyond that warranted by the data.

***Four-Dimensional Static Models.*** Of particular importance in stratigraphic analysis is the understanding of areal variations through time. In a static sense data can be integrated, which is useful in organizing stratigraphic observations. Of specific importance was the great contribution made by Charles Darwin in his observational synthesis that led to his book *The Origin of Species* (1859). The concept or model of evolution was the foundation for evolutionary biology and paleontology. Also, the evolutionary concept was a landmark in the philosophy of science and provided an approach to the organization of data that could not be presented in a rigid symbolic form. In those instances where data are closely interrelated in a deterministic manner, rigid symbolic models were possible and laws constructed were useful in predicting events. However, this was impossible with the data used by Darwin.

The concept of a time-stratigraphic framework of stratigraphy not only required the use of evolutionary biology, but it also required the same model principles. This approach allowed the reconstruction of areal variations of time-rock units and their development and change through the history of the earth. The application of this model to understanding earth history in a developmental sense has only been utilized during the past few years. The model requires both a careful subdivision of the stratigraphic record and a three-dimensional integration of lithologic and biologic information from individual time-rock units. Data of this type are only presently becoming available, and their application to carefully defined time-stratigraphic subdivisions has been utilized in only a few areas. The ability to interpret and predict patterns derived from this type of data requires that carefully collected quantitative lithologic and biologic observations be integrated over short time intervals. Chapter 1 discussed the methods of obtaining the necessary time-stratigraphic subdivision and presentation of lithologic and biologic information.

An unconformity is another four-dimensional static type of construct. An unconformity is represented by a three-dimensional surface related to time by the absence of time-rock units. Patterns of nondeposition and erosion form the basis for identifying an unconformity. The analysis of unconformities often does not include one or more elements in the construct or model. Few studies show either the areal pattern of missing time-rock units or the amount of time missing in the stratigraphic section. If no data relate to one or both of these measurements, the analysis lacks rigor and little basis is available for predicting the cause or the pattern of the unconformity.

***Three-Dimensional Dynamic Models.*** A group of models that are dynamic and show three-dimensional patterns have particular importance in the interpretation of stratigraphic observations. Sedimentary models reflect the vertical or time-dependent pattern of lithologic units. Models useful in predicting these patterns can be a basis for interpreting dynamic processes that produce specific associations. These models provide a level of integration that goes far beyond description of areal and vertical sedimentary patterns. The dynamic nature of the model indicates what is to be expected in comparable sections produced by similar processes. Comparison between Holocene and ancient stratigraphic units is possible, thus allowing process observations from the Holocene to be used directly in interpreting similar ancient associations. The theme or unifying principle of the model allows the structuring of observations from Holocene and other ancient examples, plus it provides the basis for the interrelation of observations from all examples of stratigraphic associations produced by similar processes.

An example of a dynamic three-dimensional model is related to the association of sedimentary patterns produced by changes in the shoreline position. Tectonics or change in sea level can be related to a regressive-transgressive model of sedimentation, which can provide the theme that interrelates vertical and areal patterns of depositional environments and sedimentary facies. The regressive-transgressive model was shown by Curray (1964) to be one of the most important integrating principles in stratigraphy (Fig. 3–1). The dynamic interrelationship between sea-level changes and tectonism is related in this model to rates of sedimentation, erosion, subsidence, or uplift and to the position of the shoreline.

Two additional models related to the regressive-transgressive theme are tectono-facies and sequence models. Tectonic and eustatic controls of sedimentation are separated into different models. Broader implications of tectonism and eustatic changes are used to interpret stratigraphic patterns. Tectono-facies models relate to rates of uplift, elevation of source areas, climatic changes and topography, distance of transport, and to comparable aspects of the depositional site. The range of these data provides the basis for developing models that are valuable in interpreting and predicting many stratigraphic associations. In a similar fashion the sequence model based upon systematic changes in sea level can be related to unconformity patterns, faunal and floral associations,

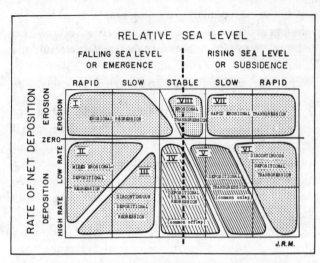

**Fig. 3–1.** Regressive-transgressive model. From Curray, 1964.

patterns of changing water depths, vertical changes in lithology, thickness of sedimentary units, and patterns of sedimentary facies.

***Four-Dimensional Dynamic Models.*** Conceptual models using quantitative and nonquantitative data, such as process-response models, should be included in this category. Process-response models have been used in geology for more than 25 years (Krumbein and Sloss 1963); their value in geology was indicated by Whitten (1964, 455) in the following statement:

> Most geological processes are very complex and involve: (1) interaction of a large number of variables; and, (2) simultaneous variation of all or most of the variables. . . However, despite certain restrictions imposed by the nature of the geological problems, most are susceptible to quantitative analysis.

If processes are considered factors affecting the cause of variation within a set of observations and responses are measurable variations or effects of the processes, then models interrelating these two aspects are useful constructs in complex systems. Stratigraphy is one of these systems, and application of this type of model appears to be particularly useful.

Process-response models need not be conceptual constructs. They may be stated by precise definition of both the processes and responses they produce. Usually there are insufficient data, and the interrelationships among the processes are sufficiently complex that little chance exists for developing deterministic models. They can provide the framework for understanding the cause of variation. Experimental, descriptive, and other types of dynamic models may be used as part of the structure of a process-response model. The rigor implied in the definition of this model, the structuring of observations, and its

application to specific problems can be the basis for interpreting complex stratigraphic associations. Validation of a process-response model requires comparison to other sets of data or additional observations to determine if it can be used to predict the pattern of variation observed. This pragmatic approach to scientific analysis may not be as rigorous as the testing of a mathematical or symbolic statement, but in simpler models, for example, taxonomic classification, the periodic table, or organic evolution, they can be powerful in their application.

***Three-Dimensional Developmental Models.*** One of the major unifying principles of stratigraphy is the understanding that the stratigraphic record represents events and processes that are operating today. From this, a three-dimensional historic reconstruction can be accomplished by using Holocene sedimentary patterns as actualistic models. These models require understanding process-response models and deterministic models to integrate data from the Holocene. With this base, it is possible to relate observations to the stratigraphic record. However, the observations from the stratigraphic record must be modeled so that direct comparisons of objective data are possible. A key to this type of analysis is recognition of the process relationship common to Holocene areal patterns and ancient stratigraphic sections. Selected cores from the Holocene represent historical development, but they also show a direct relationship to areal patterns at the present depositional surface. Thus, by careful study of both processes and patterns of sedimentation, actualistic models can be developed and these applied to the interpretation of ancient stratigraphic units.

***Four-Dimensional Developmental Models.*** Simulation studies have been applied to a wide range of problems. They are widely used in integrating reservoir data from oil fields and are useful in predicting reserves, production history, and the ultimate performance of a reservoir. Many types of data are used in these simulation studies, and static, dynamic, and developmental aspects are required in the model construction.

Other examples of simulation studies are related to sedimentary process-response models. Processes of deltation, wave and current simulation on shorelines, and growth of reef communities are all subjects of simulation studies (Harbaugh and Merriam 1968). These smaller scale studies using actualistic data can provide the basis for more insight into processes and lead to larger scales of integration of stratigraphic data.

***Depositional Models.*** The development of sedimentologic models requires synthesis of diverse types of data. Sedimentology is a complex discipline that requires an understanding of the mechanics of fluid flow, fluid-sediment interactions, mechanical properties of deposited sediments, sediment-biotic interrelationships, chemical potentials and equilibria, and the rates of change represented by all of the aforementioned physical, biologic, and chemical controls. The interpretation of observed stratigraphic characteristics cannot be a simplistic empirical correlation between modern and ancient aspects: it must be based upon a dynamic process directed analysis of the many interrelated measurable properties of the stratigraphic unit.

This approach requires development of a comprehensive data base, an understanding of the process-related vertical and areal association of stratigraphic units, and an understanding of the unifying model that interrelates all of the identifiable sedimentologic and stratigraphic aspects.

Observation of stratigraphic relationships is as important as the detailed observation of internal characteristics of stratigraphic units. Mapping stratigraphic aspects is the basis for comparison to modern depositional patterns.

Vertical and areal patterns of deposition have become the basis for interpreting sedimentary responses in the Holocene, and these patterns then can be used to interpret depositional responses in ancient stratigraphic units. The formalization of the patterns of change with regard to both depositional characteristics and sequences is now the basis for sedimentologic and stratigraphic research. This framework provides the structure into which observations can be interrelated. Without such a structure, or if there is a break in deposition or nonrecurring events, then the significance of the relationship is not interpretable. A larger or more inclusive sequence or pattern must be obtained that can provide insight into the continuing processes.

## Summary

Rarely are there sufficient data to state unequivocally all the processes that are utilized in the formalization of models. The model serves as a goal with additional insights providing new bases for observation and for interpretation. Confirmation of a model is always based upon its ability to interrelate all observed characteristics. This requires evaluation, alteration, and expansion of the model as additional data are obtained. Application of the model to stratigraphic observations is a part of this process, and the model is useful only so far as it is able to interrelate and predict stratigraphic patterns. Modifications may be necessary based upon stratigraphic relationships as well as theoretical and actualistic considerations.

Sedimentologic models are used as the basis for stratigraphic interpretation. Each environmental realm is analyzed in terms of ongoing processes, boundary conditions, and those responses preserved in the stratigraphic record. A suggested model may represent only the initial stages of synthesis, or sufficient research may be available to provide a rigorous structure for interpreting observations. The approach is consistent with processes forming the foundation for synthesis. Process-related responses are described both in terms of internal characteristics and the organization of stratigraphic units. Responses are evaluated in terms of boundary conditions, sedimentologic processes, and preserved stratigraphic characteristics. This approach provides the conceptual basis for a stratigraphic association and the observational criteria useful in environmental and paleogeographic interpretation.

## SEDIMENTARY MODELS

In order to interpret the stratigraphic record, it is necessary to determine those continuing themes or association of processes that produce specific depositional patterns. Analysis of Holocene facies patterns suggest that only a few continuing depositional themes are developed. Examination of environmental realms, such as shorelines, regressive depositional sequences, shelf sedimentary processes, and channeling suggests that only a few generalized models have been responsible for most depositional histories. Subenvironments with differing paleogeographic and tectonic boundary conditions or nonrecurring processes may produce variations, but these can also be explained within the model framework.

## Shoreline Models

Within the coastal environmental framework, the physical, chemical, and biologic processes operate through extended periods of time. These reflect either tidal or wave processes with a biochemical or a terrestrial sedimentary source. Low-energy shorelines are dominated by biologic or suspension processes, leading to two differing shoreline models. Figure 3–2 illustrates these patterns for stratigraphic units with terrestrial clastic as the source materials. Distinctions require careful observation, and most responses can occur in more than one model. An illustration from an outcrop section from the Dakota formation near Morrisson, Colorado, is not easily interpreted (Fig. 3–3). It can be formed as a tidal flat, a beach, or a bay-levee-estuary. Some specific process-related observations are required to make the correct bay-levee-estuary interpretation.

SHORELINE SAND BODIES

| CHARACTERISTIC | TIDAL FLAT | BEACHES | COASTAL MARSH | BAY-LEVEE-ESTUARY |
|---|---|---|---|---|
| Texture | FINE ↑<br>COARSE ↓ | COARSE ↑<br>FINE | FINE | FINE ↑<br>MEDIUM TO FINE |
| Sorting | FAIR | GOOD | POOR | POOR TO FAIR |
| Lithology | INTERBEDDED<br>SAND-CLAY | DOMINANT<br>SAND | CLAY-SAND<br>PEAT | CLAY-FINE SAND |
| Trace Fossils | VERTICAL<br>BURROWS | VERTICAL<br>BURROWS | RARE<br>BURROWS | TRAILS ON BEDDING PLANE<br>PLANE |
| Body Fossils | SHELL LENSES | SHELL LAYERS | RANDOM SHELLS | SHELL BANKS |
| Structures | WAVE & CURRENT<br>RIPPLES | LAMINATED | CURRENT RIPPLES<br>ROOT MOLDS | GRADED BEDS<br>CURRENT & WAVE RIPPLES |
| Sand Thickness | 5-20 FT. | 5-10 FT. | 5-50 FT. | 10-100 FT. |
| Area-Sq Mi | 10-100 | 10-100 | 500-5000 | 500-5000 |

**Fig. 3-2.** Comparison of physical characteristics of shoreline sand bodies. From Visher, 1970.

**Fig. 3-3.** Dakota sandstone, Alameda section, Morrison, Colorado

## Regressive Marine Sand Bodies

The process of shoreline progradation requires a surplus of sediment supply in relation to the available energy. This limits the processes of sedimentation that can be utilized in constructing models (Fig. 3-4). A stratigraphic unit dominated by waves can only be developed for the time when either longshore or shoreward transport of clastics is higher than sediment dispersal. The sequence of units, therefore, must have specific characteristics (Fig.

3-5). Other patterns also require a surplus of supply for the available energy, and both the areal and vertical patterns will reflect the dispersal rate of supply balance. In these relationships the shape of the depositional site also will produce changes in the geometry of the stratigraphic unit (Fig. 3-6).

## Marine Shelf Sand Bodies

Models are based upon the processes associated with bottom currents, slumping, turbulence, biologic activity, and sediment supply (Fig. 3-7). Distinctions among these environments have led to controversy in the interpretation of these units in the ancient stratigraphic record (Figs. 3-8 and 3-9).

## Channel Sand Bodies

Relatively few response characteristics are useful for interpreting which environmental pattern of channeling is developed (Fig. 3-10). Changes in processes of channel fill produce textural sequences, and water depth controls the fauna and sedimentary structures. These are minor differences, but exploration will be quite different for each type of reservoir.

To illustrate this, a more detailed examination of fluvial channels suggests differing controls for hydrocarbon

REGRESSIVE MARINE SAND BODIES

| CHARACTERISTIC | BARRIER ISLAND | DISTRIBUTARY MOUTH BAR | LOBATE DELTA | TIDAL DELTA |
|---|---|---|---|---|
| Texture | COARSE ↑<br>FINE | MEDIUM ↑<br>FINE | MEDIUM ↑<br>FINE | FINE ↑<br>MEDIUM TO FINE |
| Sorting | POOR TO GOOD | FAIR TO GOOD | FAIR TO GOOD | GOOD |
| Lithology | SAND ↑<br>CLAY | SAND ↑<br>CLAY | PEAT-SAND ↑<br>CLAY | PEAT-SAND ↑<br>CLAY |
| Trace Fossils | COMMON BURROWS | RARE BURROWS | RARE BURROWS | RARE BURROWS |
| Structures | LAMINATED ↑<br>WAVE RIPPLES | LAMINATED ↑<br>CURRENT RIPPLES | CURRENT RIPPLES ↑<br>CROSS BEDDING | CURRENT RIPPLES ↑<br>THICK X-SETS |
| Sand Thickness | TO 35 FT. | TO 60 FT. | TO 200 FT. | TO 100 FT. |
| Area-Sq Mi | 50-500 | 100-500 | 1000-10,000 | 500-1000 |

**Fig. 3–4.** Comparison of physical characteristics of regressive marine bodies. From Visher, 1970.

**Fig. 3–5.** Atoka formation, Mountainburg, Arkansas

**Fig. 3–6.** Almond formation, Rock Springs, Wyoming

accumulations. Geomorphic models have been suggested for fluvial sedimentology (Allen 1965a) (Fig. 3–11). These four models are based upon study of Holocene fluvial patterns. A fifth model, associated with the depositional history, can also be suggested. The five models would include: valley fill, coastal plain, marginal basin fill, alluvial fans, and strike-valley fill. Each of these models produces differing patterns of reservoir geometry, permeability, relation to hydrocarbon source rocks, and seals and require differing exploration strategies.

The meander-belt-valley fill model has been recognized since the early 1960s (Visher 1965b) (Fig. 3–12). The reservoir geometry is blanket deposition with a characteristic textural and sedimentary structure pattern. Exploration has indicated that these reservoirs are not commonly filled with hydrocarbons.

By comparison, the coastal plain has a great many narrow channels, and a closer proximity to potential source rocks (Fig. 3–13). The single channel tends to be straight, not more than one kilometer wide, and often associated with coastal plain faulting, anticlines, and salt and shale pillows. Compactional drape structure may also be developed. Seals may be flood-plain shales or marsh deposits.

The marginal basin reflects a tectonic and depositional history that leads to very large accumulations of hydrocarbons. The basin is the site for stacking of straight fluvial channels to thicknesses of hundreds of meters (Fig. 3–14). These basins are filled due to rising base level

| CHARACTERISTIC | MARINE SAND BODIES | | | |
|---|---|---|---|---|
| | SHELF SANDS | PRO-DELTA | DEEP WATER | NEARSHORE |
| Texture | COARSE-FINE | MEDIUM-FINE | COARSE-FINE | FINE-VERY FINE |
| Sorting | FAIR-GOOD | POOR | POOR-FAIR | VERY POOR |
| Lithology | SAND | FINE SAND-CLAY | INTERBEDDED SAND & CLAY | MIXED SAND & CLAY |
| Trace Fossils | VERTICAL BURROWS | BURROWS ON BED SURFACES | TRACKS & TRAILS | MIXED |
| Structures | PLANAR X-BEDS | CURRENT RIPPLES | CURRENT RIPPLES LAMINATED | WAVE RIPPLES |
| Sand Thickness | 5-100 FT. | 50-300 FT. | 500-5000 FT. | 10-100 FT. |
| Area-Sq Mi | 1000's | 100-500 | 500-2000 | 50-200 |

**Fig. 3–7.** Comparison of physical characteristics of marine sand bodies. From Visher, 1970.

**Fig. 3–8.** Weber sandstone, Dinosaur National Monument, Colorado

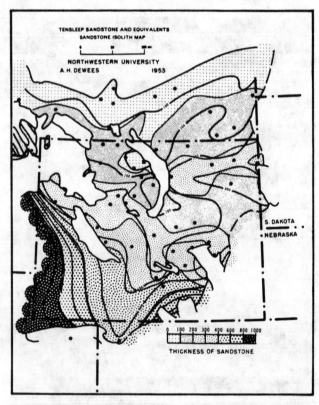

**Fig. 3–9.** Tensleep sandstone and equivalents, sandstone isolith map. From Sloss et al., 1960.

| CHARACTERISTIC | CHANNEL SAND BODIES | | | |
| | FLUVIAL CHANNEL | TIDAL CHANNEL | SUBMARINE CHANNEL | DELTAIC DISTRIBUTARY |
| --- | --- | --- | --- | --- |
| Texture | FINE / COARSE ↑ | FINE / COARSE-MEDIUM ↑ | NO PATTERN | FINE-MEDIUM / LITTLE PATTERN |
| Sorting | GOOD | GOOD | FAIR-GOOD | GOOD |
| Lithology | SAND / CLAY DRAPE | SAND / CLAY LAYERS | SAND / CLAY LENSES | FINE-MEDIUM SAND / SILT LAYERS |
| Trace Fossils | VERY RARE BURROWS | VERTICAL BURROWS | COMMON BURROWS | RARE BURROWS |
| Structures | CURRENT RIPPLES / LAMINATED / FESTOON X-BEDS | PLANAR X-BEDS FACING UPSTREAM / CURRENT RIPPLES | CURRENT RIPPLES / LAMINATED | PLANAR X-BEDS FACING DOWNSTREAM / LAMINATED |
| Lag Deposits | WOOD / CLAY CLASTS | WOOD, SHELLS / CLAY CLASTS | SOME / CLAY CLASTS | WOOD / CLAY CLASTS |
| Sand Thickness | 10-60 FT. | 50-200 FT. | 100-500 FT. | 50-200 FT. |
| Length | 10-100 MI | 1-5 MI | 1-20 MI | 20-200 MI |
| Width | 1-30 MI | 1-30 MI | 1-5 MI | 1-3 MI |

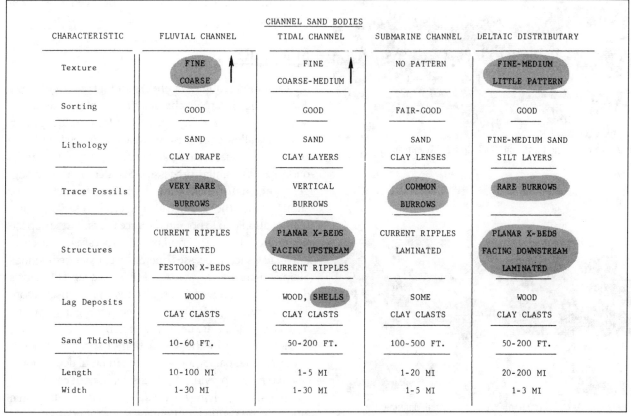

**Fig. 3-10.** Comparison of physical characteristics of channel sand bodies. From Visher, 1970.

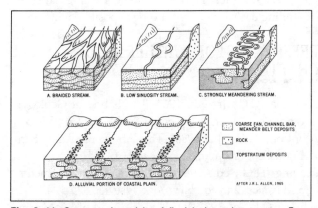

**Fig. 3-11.** Conceptual models of fluvial channel geometry. From Allen, 1965.

## VERTICAL SIZE CHANGES

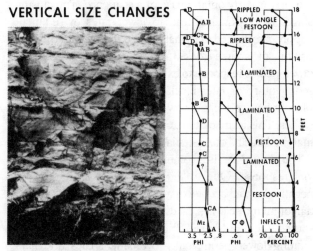

**Fig. 3-12.** Relation of textural changes to outcrop features. Vertical size data include mean size, sorting, and percent deposited by traction (Inflect, %). From Visher, 1965b.

**Fig. 3–13.** Almond formation, Rock Springs, Wyoming

**Fig. 3–14.** Dakota formation, Colorado National Monument

during the early onlap phase of a sequence. The closing stages of a sequence lead to lower rates of sediment supply, a restricted marginal basinal setting, and the deposition of source rocks. Many marginal basins are developed in response to extensional tectonics, and synsedimentary or later block faulting is developed, thus placing source rocks below thick fluvial reservoir sandstones. This is the pattern for the Sirte basin of Libya and marginal basins of Australia and Brazil.

The alluvial fan is a common post- or synorogenic fluvial response. Due to the presence of a tectonically negative area adjacent to many uplifts, there is the potential for inland seaways, lakes, or other environments with source rock potential. Trapping must be related to faulting lower permeability overbank shales and silts against more porous and permeable reservoir units.

The most likely site for major hydrocarbon accumulations is in unconformity sandstones. Fluvial deposition on unconformity surfaces is commonplace, and the paleotopographic pattern of the surface provides sites for channel deposition and the permeability pathway from onlapping source rocks on the shelf margin. A large proportion of hydrocarbon accumulations occur within 100 meters of an onlap unconformity surface. The geometry of channel sandstones deposited on this surface leads to both stratigraphic and structural traps. Seals are related to onlapping finer-grained sedimentary units. Source rocks may immediately underlie or overlie reservoir sandstones.

## SUMMARY

Exploration requires an understanding of the geometry and the internal characteristics of stratigraphic sequences. Sedimentary models reflecting equilibrium processes allow comparisons to be made between Holocene depositional patterns and ancient stratigraphic units deposited by similar processes. Such comparisons suggest the nature of traps, whether structural or stratigraphic, the geometry of specific reservoirs, and their relationship to potential source rocks. These depositional models reflect equilibrium processes, and since only a few such processes operate over extended periods of time, there are only a few possible models. This allows their recognition with limited stratigraphic observation. The observer is guided by the understanding that specific response patterns are unique to each model. Each model is characterized by the association of specific response patterns. Both paleogeographic and internal depositional characteristics provide the basis for identification and comparison. Applying the model to specific stratigraphic units can be verified and enlarged, or refined, by additional observation. The model approach to stratigraphic interpretation thus is both a self-correcting and an expanding basis for understanding depositional history.

## REFERENCES FOR FURTHER STUDY — CHAPTER 3

Albritton, C.C., Jr., ed., 1963, The Fabric of Geology: Stanford, California, Freeman, Cooper, & Co., 374 p.

Albritton, C.C., Jr., 1980, The Abyss of Time: San Francisco, Freeman, Cooper and Co., 251 p.

Reading, H.G., ed., 1986, Sedimentary Environments and Facies, 2nd ed.: New York, Elsevier, 625 p.

Schneer, C.J., ed., 1979, Two Hundred Years of Geology in America: Univ. New Hampshire Press, 383 p.

# 4 FLUVIAL SYSTEMS

In the testing of hypotheses lies the prime difference between the investigator

and the theorist. The one seeks diligently for the facts which may overthrow his

tentative theory, the other closes his eyes to these and searches only for those

which will sustain it. **G.K. Gilbert,**
*Inculcation of the Scientific Method by Example, 1886*

## INTRODUCTION

The fluvial milieu is a complex product of atmospheric elements including chemical, biologic, and physical processes. The effects of the physical forces developed by the rotation of the earth, the effects of solar insulation producing the hydrologic cycle, and the complex equilibrium between biochemical and inorganic chemical reactions produce the materials transported by aeolian, glacial and fluvial processes. Preserved sediments in the stratigraphic record may be associated with pedogenic regimes, surficial dunes, and glacial deposits, but more importantly, they are principally deposited in stratigraphic units associated with streams and rivers.

Depositional areas for streams and rivers include alluvial fans, alluvial valleys, coastal plains, and a complex of non-marine stratigraphic units associated with marginal basins. Other non-marine deposits, including pedogenic, aeolian, playa, and glacial deposits, are only rarely preserved.

Extensive geomorphic research on surficial deposits carried out during the past forty years has provided a detailed process-response understanding of most depositional environments. This research changed from a statistical description of response patterns, to a renewed process directed research, leading to a synthesis of depositional systems. The summary of this research by Scheidegger (1961) provided the basis for this synthesis, as illustrated by the 1964 Society of Economic Paleontologists and Mineralogists symposium on the hydrodynamic interpretation of sedimentary structures (Middleton 1965). A review of the literature on process geomorphology illustrates the breadth and depth of the reexamination of surficial processes which was conducted during the 1950s and 1960s. The identification of a limited number of fluvial depositional systems was the result of this intensive period of research. In a summary article, Allen (1965a) developed conceptual models of fluvial channel geometry (Fig. 4-1). This framework is still useful as a basis for describing the origin and preservation of fluvial stratigraphic units.

## BOUNDARY CONDITIONS

In any attempt to make some sense out of the complexity of non-marine processes and responses it is

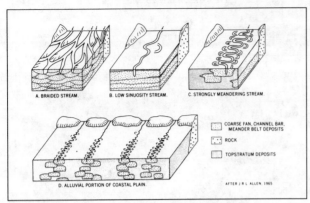

**Fig. 4–1.** Conceptual models of fluvial channel geometry. From Allen, 1965a.

necessary to abstract a limited number of depositional themes. The work of Allen is seminal in this process. Since the thrust of this book is to interpret stratigraphic depositional patterns, it is necessary to understand those non-marine themes that have contributed to significant thicknesses in the preserved stratigraphic record. Few themes lead to thick stratigraphic sequences. The interactive comparison of the many Holocene stratigraphic patterns with those preserved in older sequences is essential to this process. Due to the temporal proximity of Holocene glacial events, surficial deposits commonly developed include aeolian dune fields and playas, together with a broad areal distribution of pedogenic sequences. This has resulted in extensive research in the formation of these regimes, with research on the processes and patterns for the formation of these deposits. Their stratigraphic importance, however, is quite different from their present areal distribution or research emphasis. Unconformities leave little tangible depositional record. Non-marine coastal sedimentary sequences may be preserved by onlap or rise in base level, and the history of these events recorded in the stratigraphic record. Preservation of glacial, aeolian, playa, or soil horizons requires unique eustatic, tectonic, or depositional events. An understanding of the boundary conditions is required to determine the timing and magnitude of these events.

Rates of sea-level rise, tectonic subsidence, coastal geometry, and energy associated with coastal processes all need to be understood prior to interpreting the origin of a stratigraphic sequence. Other parameters, including climate, rates of sediment supply, and the topographic pattern of marginal shelves and depocenters, are important to the understanding of the preservation of these stratigraphic sequences.

Since a limited number of preserved response patterns have been identified in the stratigraphic record, it suggests that the considerations outlined in the preceeding paragraph are important to the genetic interpretation of the stratigraphic record. Only a few themes are important

for understanding the stratigraphic record. Since the principle preserved stratigraphic intervals are those associated with fluvial-alluvial depositional systems (Table 4–1), these will be emphasized. Each will be discussed individually, and their stratigraphic importance documented.

# FLUVIAL DEPOSITIONAL THEMES

Desert environments are associated with specific boundary conditions, and can usefully be discussed separately. Preserved deposits are usually related to alluvial fan and marginal basin depositional themes. Specific pedogenic, aeolian, playa, and saline lake themes may be described as a part of this larger depositional framework.

## Aeolian and Playa Depositional Models

Two principle themes have been identified: (1) fault bounded topographic lows (often below sea level), with alluvial fans, wadis, dunes, and ephemeral playas and saline lakes or (2) stable cratonic interior and marginal basins, fringed by alluvial fan or pedogenic platforms, and containing, in flood basins, ephemeral streams, playa deposits, and dune fields (Collinson 1986). Examples of fault bounded basins would be Death Valley, California, rift valleys of Ethiopia and the Sudan, and the Dead Sea rift between Israel and Jordan. These settings have been extensively described by various authors (Glennie 1970). Cratonic interior basins are exemplified by the interior dune fields of Saudi Arabia, the Sahara, the Tarim-Junggar basins of northwest China, and central Australia. Sedimentology in these basins has been described by Wilson (1973).

Each of these depositional themes are characterized by internal drainage, with evaporation in excess of precipitation. The principle supply of sediment into the basin is by infrequent storms, with high runoff through braided stream and ephemeral wadi channels. The basin floor can be the site of playa, aeolian dune, or saline lake deposition. Preservation of these basin floor deposits requires either progradation of wadi or alluvial fan units across the basin floor, or a catastrophic flooding event.

Preservation of aeolian and playa deposits in the stratigraphic record has been widely reported, but most descriptions are not based on careful comparison to Holocene aeolian deposits. The mechanism for preservation often has not been reported, causal processes to interpret mineralogy have not been addressed, detailed textural comparisons are lacking, and patterns of trace and body fossils have not been compared to Holocene depositional sequences. Comparison of response patterns from Holocene to ancient stratigraphic units, is

**Table 4-1**
**Fluvial Response Characteristics**

|  | Alluvial Fan | Valley Fill | Coastal Plain | Marginal Basin |
|---|---|---|---|---|
| Geomorphic Elements | Braided Streams with Flood Deposits | Point Bar Sequences | Straight Single Channels: Meander Belts: Braided Channels. | Low sinuosity Channel Complex |
| Slope Gradient | High - tens of Feet/Mile | Moderate | Low | Very Low |
| Geometry | Lensoid up to 2000 kms² | 20m Thick: 5000 kms² | 1–5 kms Wide, Up to 40 m Thick | Ovoid, Greater than 1000 kms² |
| Migration Patterns | Vertical Stacking of Channel Fill Sequences | Lateral Scour | Vertical Stacking | Cut and Fill of Channel Sequences |
| Vertical Sequences: Interchannel Deposits | Levee and Fine Grained Overbank Deposits | Levee and Flood Plains | Alluvial Valleys with Splays and Flood Deposits | FLood Plains and Marsh Deposits |
| Channel Interval | Braid Bars, 1–5 m Thick, Fining or Coarsening Upward | Fining Upward, Sequence of Sed. Structures | Channel Fill Sequence | Channel Fill Sequences, 1–5 km 5–10 m thick |
| Basal Unit | Channel Scours with Coarse Lag Deposits | Lag Deposits of Clay Chips and Coarse Detritus | Large Scale Sand Waves 2–10 m High | Cut and Fill with local Lag Deposits |

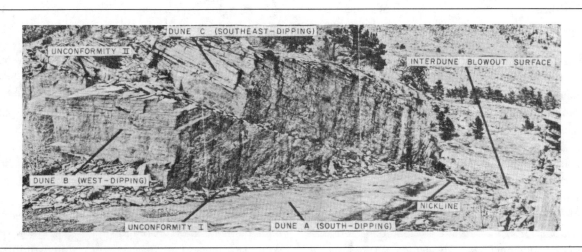

**Fig. 4-2.** From Walker and Harms, 1972.

typically based upon "look-a-like comparisons, and detailed physical, chemical, and biologic processes comparisons are universally lacking. One example where many of these sedimentologic and stratigraphic considerations have been discussed is the Lyons sandstone from the "Lyons Quarry," in Boulder County, Colorado (Walker and Harms 1972). Comparisons to Holocene dunes have been made, and sedimentologic responses shown in Figures 4-2 and 4-3, are comparble to Holocene process-response patterns. More recent papers in *Desert Sediments Ancient and Modern,* edited by Frostick and Reid (1987), show detailed comparisons of Holocene to ancient depositional response patterns.

Soil horizons are easily identified in all stratigraphic sequences. They show patterns related to topography, and illustrate the vertical sequence of soil horizons, including the development of calcrete, silcrete, and ferricrete regoliths (Ollier 1986). The presence of root zones or coastal marsh deposits, sabkha evaporites, and specific pedogenic chemical response patterns are recognized in many stratigraphic intervals.

These surfaces are important to the interpretation of the depositional history. Regoliths have been identified in the stratigraphic record from Precambrian to Holocene and are significant to the interpretation of depositional, tectonic, and eustatic histories.

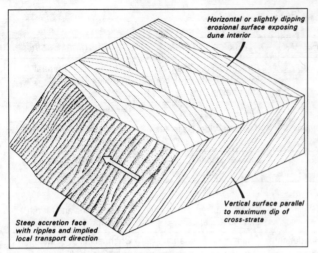

**Fig. 4-3.** From Walker and Harms, 1972.

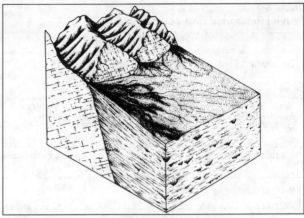

**Fig. 4-4.** Recognition of alluvial fan deposits in the stratigraphic record. From Bull, 1972.

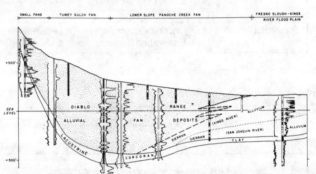

**Fig. 4-5.** Cross section of Diablo alluvial fan. From Bull 1972.

## Alluvial Fan Model

Sedimentology of alluvial fans is complex due to strong climatic, provenance, and slope effects. These deposits are often preserved flanking tectonically active fault scarps associated with wrench faulted, extensional, and foreland basins. The interaction of tectonics and sedimentology is a modifying influence on fan geometry, slope, and facies patterns. With this heterogeneity of possible variation it would appear that a simple unifying theme could not be expressed. Based upon Allen's model for braided stream deposition, and extensive work on both arid and humid molasse deposits preserved in the stratigraphic record, a generalized theme may be suggested (Fig. 4-1). The controlling theme for deposition is the braided river, resulting from an overload of coarse bedload, and a high stream gradient.

The surface of the fan is not constrained, and consequently the fan form is the natural product of sedimentary and tectonic processes. Boundary conditions influence grain size and mineralogy of the alluvial detritus, the fan size and slope, the nature of interchannel sheet flood deposits, and locally, debris flows that may cross the fan.

The sedimentology of modern fans has been extensively studied, but few genetic sequences have been described, and little understanding of the interaction of boundary conditions to the development of stratigraphic sequences. For comparison purposes a summary article by Bull (1972) describes alluvial fan deposition related to a faulted uplift, and presents a depositional sequence that may be usefully applied to ancient alluvial fans (Figs. 4-4 and 4-5).

Modern descriptions of arid fans from Death Valley, California (Denney 1967) and from southern Nevada (Bluck 1964) are important contributions to our understanding. These sedimentologic studies describe braided channels, debris flows, and "sieve deposits," but little is presented about their vertical sequence or their architecture. Braid channels are the principle mechanism of sediment transport, and these channels are confined by natural levees, producing braid channel sequences 3–10 meters thick. Important sandstone reservoir sequences are thus developed in California, Columbia, Venezuela, and southern Oklahoma. The Cañon Limon field, on the border of Columbia and Venezuela, is a giant field with high permeability. Exploration for hydrocarbon reservoirs on these fans is difficult, and requires an understanding of down-fan changes in permeability, with faulting an important control for defining patterns of fluid flow. The reservoir geometry patterns are highly variable (Table 4-2), but by mapping presssure patterns, related to down fan artesian flow, and mapping of salinity patterns, it may be possible to define traps showing higher salinity and lower pressures. Sealing characteristics would be enhanced by changes in grain size, and pressure gradients.

The study of a humid fan from India, by Wells and Dorr (1987), is one study where comparisons can be developed to the extensive braided stream deposits associated with carbonaceous molasse stratigraphic intervals (Figs. 4-6 and 4-7). Another example by Kochel and Johnson (1984) describes humid temperate alluvial fans in Vir-

**Table 4-2**
**Alluvial Fan Reservoirs**

Interval is lensoid with no preferred elongation

Area of fan interval greater than 2000 kms²

Reservoir units, 3–10 m thick, up to 200 m wide, and greater than 5 times in length

Flow units, either coarsening or fining upward, from 3–10 m thick, 10–30 m wide, and length 5 times width

Fine-grained flow boundaries are irregular in thickness and distribution

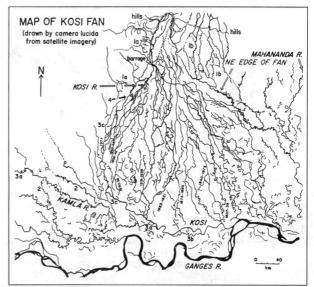

**Fig. 4–7.** New map of Kosi drainage system, constructed from 19 March 1977 false color satellite image. Numbered areas are as follows. *1a:* Interfan area west of Kosi fan, where dry watercourses feed into small distributary fans and commonly fail to reach the Kosi River. *1b:* Interfan east of the fan. Channels contain water near the foothills, but many rivers drain underground while crossing the bhaber gravel zone. *2:* Long discontinuous sections of meandering channels show the mobility of drainage patterns on adjacent fans. *3a:* Kamla (Kamala) River on Kosi-Gandak interfan. *3b:* Former part of Kamla? (adjacent reaches overwhelmed by the Kosi). *3c:* Possibly, the next major shift of the Kosi River will be into these channels. *4:* Meandering secondary channel in the modern Kosi. From Wells and Dorr, 1987.

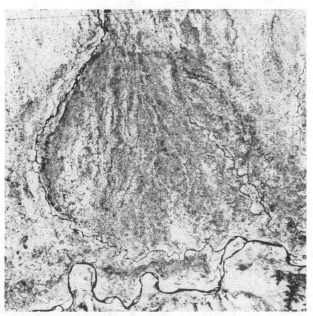

**Fig. 4–6.** Late dry season Landsat image of the Kosi alluvial fan (19 March 1977). From Wells and Dorr, 1987.

ginia. From these limited studies it is difficult to determine depositional themes of importance to the interpretation of humid molasse deposits in the stratigraphic record. However, ancient examples of similar stratigraphic sequences are abundant in the Oligocene and Miocene Siwalik sequences south of the Himalayas in India and Pakistan, the Pennsylvanian, Pottsville formation of west Virginia, and the carbonaceous Cretaceous-Tertiary fans of the Raton Basin (Flores and Pillmore 1987) (Figs. 4-8, 4-9, 4-10, and 4-11).

A detailed analysis of braided stream fan deposition has been developed for giant oil fields of northeast China. The paleogeographic setting is from a marginal lacustrine basin (Qiu, Xue, and Xiao 1987). A characteristic vertical sequence of a short coursed braided stream sandstone body, more than 10 meters thick, from the Shengli oil province illustrates the complexity of vertical change in grain size and permeability (Fig. 4-12). The rate of lateral change can be seen from a cross section and a map section (Figs. 4-13 and 4-14), where the the sand-body

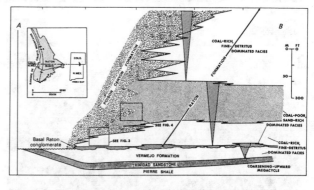

**Fig. 4–8.** Diagrammatic cross section of the western part of the basin, displaying coarsening-upward megacycles (1-3) that contain clastic wedges (coal-poor, sand-dominated facies) thickening to the west. Location shown on the inset map. From Flores and Pillmore, 1987.

density ranges from 25.9 to 33.3% in the fan interval. From these data the alluvial fan architecture appears similar to that described for semi-arid fans, and the predictive geometric model may be applicable as a general scheme for alluvial fan depositional systems (Table 4–2). Many more detailed studies of Holocene and preserved ancient fans are needed to substantiate detailed sandstone depositional patterns useful for reservoir prediction.

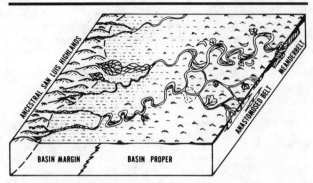

**Fig. 4-9.** Depositional model illustrating basin alluvial paleoarchitectu and evolution of fluvial systems during deposition of coal-rich fine detritus-dominated facies. From Flores and Pillmore, 1987.

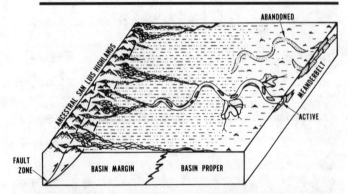

**Fig. 4-10.** Depositional model illustrating basin alluvial paleoarchitecture and evolution of fluvial systems during deposition of coal-poor, sand-dominated facies. From Flores and Pillmore, 1987.

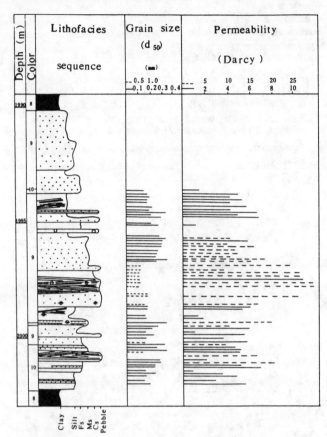

**Fig. 4-12.** Lithofacies sequence of short-coursed braided sandstone body. From Qiu et al., 1987.

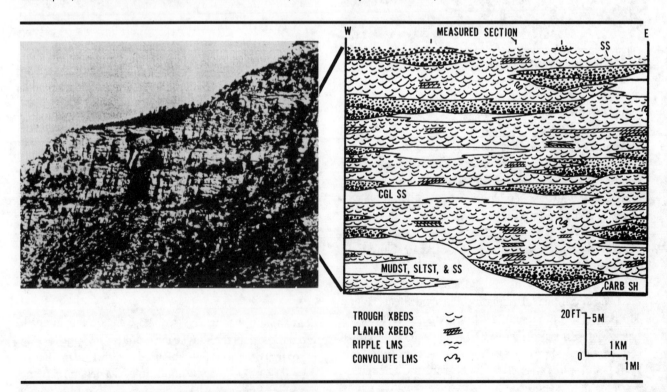

**Fig. 4-11.** Photograph and outcrop diagram of the Raton and Poison Canyon Formations along the basin margin, comprising braided stream deposits. The diagram was constructed from the measured section shown. From Flores, 1987.

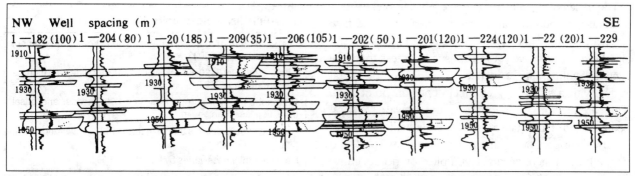

**Fig. 4-13.** Correlation section of subzones Sall$_2$, Sall$_3$, the short-coursed braided sandstones bodies, Sheng-li field. From Qiu et al., 1987.

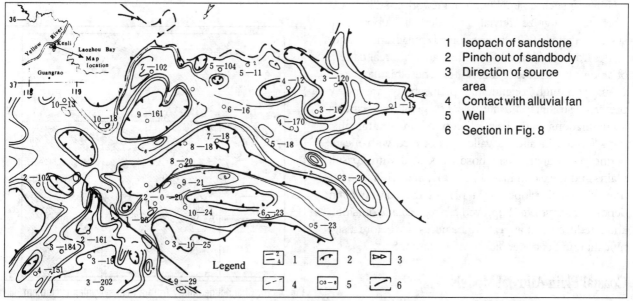

1   Isopach of sandstone
2   Pinch out of sandbody
3   Direction of source area
4   Contact with alluvial fan
5   Well
6   Section in Fig. 8

**Fig. 4-14.** Geometry of short-coursed braided sandstone bodies. (Subzone Sall$_2^4$, Shengtuo field). From Qiu et al., 1987.

## Valley Fill Depositional Model

The valleyfill-meander belt depositional system has been recognized since the early 1960s by Bernard et al. (1962), by Allen (1964) (Fig. 4-15), and by Visher (1965). It is one of the four depositional systems found useful by Allen (1965a) (Fig. 4-1).

Research on Holocene depositional history and patterns has indicated that there is a great deal of heterogeneity between meandering rivers, and even within a single river reach. Schumm, in a summary book *The Fluvial System* (1977), has outlined the great degree of variability in depositional patterns as related to the amount and nature of the sediment load, flood history, gradients, and other boundary conditions. From the many studies he reports, it is unlikely that a single characteristic fluvial point bar sequence can be identified, making it impossible to recognize a single "point bar model." Most of the variability described from Holocene meandering streams relates to types to vertical sequences of sedimentary structures, and to interbedding of con-

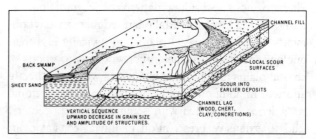

**Fig. 4-15.** Geomorphology of the fluvial depositional model. It suggests that each depositional unit may be related to specific depositional processes and positions within a fluvial channel complex. From Visher, 1972, after Allen, 1965.

trasting textural or lithologic units. Depositional systems, however, also interrelate depositional geometry, mineralogy, faunal and floral elements, and areal patterns of textural and sedimentary structure vertical sequences. This broader perspective allows the identification of only a few fluvial themes, each with a process-response framework (Visher 1970) (Table 2-3). The geometric aspects of the meander belt depositional theme can be suggested for comparison to ancient depositional patterns (Table 4-3).

The reservoir geometry is relatively simple, with a blanket distribution (Table 4-3). Permeability, and flow units are strongly parallel to valley elongation, and lateral pinchouts are relatively abrupt, usually in less than 200 meters. Sedimentary sequences are generated by lateral scour, with a shallowing channel depth on point bars. This architecture produces a fining upward textural sequence, and a sequence of large scale bedforms, overlain by an upper flow regime, current laminated unit, and capped by a heterogenous complex of ripple cross-bedded channel fill, levee, and crevasse splay units (Visher 1972) (Fig. 4-15).

Utilizing such a simplifying framework it still is possible to recognize fluvial meanderbelt depositional systems. Potter (1962), in describing Pennsylvanian sandstones from the Illinois basin, illustrates the areal pattern of a point bar sequence (Fig. 4-16). These blanket reservoirs, with high longitudinal permeability are rarely preserved stratigraphic intervals. Few giant hydrocarbon accumulations have been discovered in stratigraphic traps formed in alluvial valleys associated with meandering rivers apart from those associated with coastal plains and wave-dominated delta systems. This association will be developed as a part of the coastal plain depositional system. Improved seismic data acquisition and a reduced cost appears to be necessary before a significant number of new fields will be discovered.

## Coastal Plain Alluvial Models

Evidence has been presented for periodic eustatic sea-level changes of many tens of meters throughout the Phanerozoic. The resulting patterns of progradation and transgression involve coastal plain sedimentary environments. Slopes on coastal plains are typically less than one/half meter per kilometer: consequently, portions of coastal plains many tens of kilometers wide, are directly affected. In addition, the rise in base level and tectonic subsidence, associated with sedimentary depocenters, lead to extensive preservation of fluvial coastal plain sequences. Coastal plains are incised by riverine channels, and alluvial depocenters include both channels and associated riverine valleys. Coastal plains are of varying widths, slopes, and climatic regimes, and the streams and rivers that cross them contain differing types of clastic detritus. These variables produce differing types of stream valley and alluvial plain sequences. Of particular importance to the development of a dynamic equilibrium amongst these variables is the nature of the coastline and coastal processes. Surplus sediment supply relative to rise in base level and tectonic subsidence leads to progradation of the coastline, and if there is a surplus of wave or tidal energy along the coastline, the response is not of

**Table 4-3**
**Valley-Fill Fluvial Reservoirs**

| |
| --- |
| Interval pattern irregular, greater than 3000 kms², with channel width varying up to 50% |
| Channel interval 5–20 m thick, from 2–50 kms wide, and elongation greater than 5:1 |
| Channel pinchouts abrupt, less than 200 m wide, and basal transition less than one m |
| Reservoir flow units less than 6 m thick, 100 m wide, and 300 m long |
| Reservoir thickness varies by less than 25% |

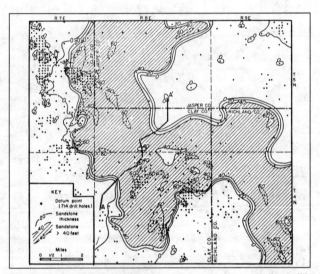

**Fig. 4-16.** Channel geometry from a Pennsylvanian cyclothem, Illinois basin. From Potter, 1962.

progradation, but of transgression. Study of modern coastal plains indicates that the delta process may be the controlling theme for alluvial coastal plain response patterns.

Fan deltas are related to high rates of sediment supply, coarse sediment, tectonically active coastal margins, narrow continental shelves, and to relatively high coastal energy. These boundary conditions produce high slopes across the coastal plain, braided river channels, and extensive sheet flood deposits across alluvial fan flood plains (Table 4-4). A modern example may be from the western coastal plain of southern Alaska. The Copper River formed a complex braided channel plain more than 15 kilometers wide, composed of coarse detritus as a response to glacial outwash (Galloway 1975) (Fig. 8-66).

An ancient example may be from the ancestral Yellow river, near where it discharges into the Gulf of Bohai (Qiu et al. 1987) (Figs. 4-17, 4-18). In this example, from a producing oil field, the long, braided-river channel deposits show composite channels up to 30 meters thick and several kilometers wide. The alluvial valley is up to 120

**Table 4-4**
**Coastal Plain Alluvial Plain Reservoirs**
*Coastal Plains Associated With Fan Deltas*

| |
|---|
| Alluvial valley patterns controlled by the slope from the sediment source to the depocenter |
| Reservoir channels are 5–20 m thick, with both fining and coarsening upward reservoir sequences |
| Flow units are 2–5 m thick, less than 100 m wide, and 100 m wide |
| Areal distribution of interval greater than 1000 kms², with width to length elongation greater than 1:4 |
| Interval thickness varies by more than 50% |

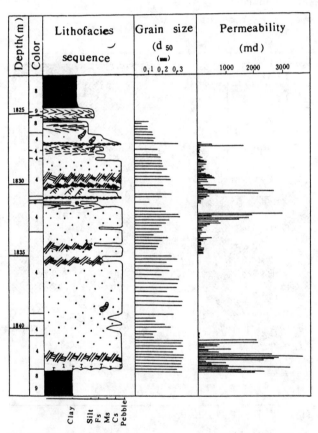

**Fig. 4-17.** Lithofacies sequence of long-coursed braided sandstone body (Subzone Ngl, Well G212, Gongdong field). From Qiu et al, 1987.

kilometers long. The stratigraphic interval contains polymictic conglomerate units, sandstones with limited clay flasers, and typically contain abundant silt and carbonaceous interbeds separating flow units. In composite sequences there are both coarsening and fining upward units.

Similarly, wave and tidal dominated deltas are developed on narrow continental shelves, with the depocenter near the shore-line, or down the continental slope in submarine fans. The coastal plain has a relatively steep

gradient, due to restricted progradation, and due to subsidence associated with a coastal or slope depocenters. The gradient produces river meandering across both coastal and delta plains. Coastal plain-meander belt fluvial sequences show both lateral scour and vertical stacking in response to rise in the base level, or tectonic subsidence (Table 4-5). The Niger River coastal plain illustrates these characteristics (Fig. 4-19). Another example, from a high wave dominated coastal plain from the Texas Gulf Coast, are the Brazos-Colorado alluvial valley systems illustrated by Kanes (1970) (Fig. 4-20). Both show the high degree of meandering typical of a steeper coastal plain gradient.

An ancient example, the Upper Cretaceous Tuscaloosa sandstone reservoirs of southwestern Mississippi, was described by Hamilton and Cameron (1986). Coastal units are high energy deltaic strandplain deposits, and the alluvial valley contains highly irregular fining upward point bar deposits. Another ancient depositional interval is represented by the Morrowan Foster sandstone. Meanderbelt sequences are deposited on the coastal plain, marginal to the Arkoma basin in southeatern Oklahoma (Lumsden, Pittman, and Buchanan 1971) (Fig. 8-65). The associated coastal and shelf sandstone deposits are also wave-dominated deltaic strandplain deposits, with little coastal progradation.

Riverine channels, dominated by suspension load and related to riverine deltas, show high rates of progradation, and are characterized by lobe switching, which results in a low coastal plain gradient. Slopes are less than .5 meter/kilometer, and rivers with such a low gradient do not meander. Single channels have a half-pipe cross section, and channel position changes by avulsion. Individual channels are leveed up to kilometers in width, and are tens of meters deep (Table 4-6). Marginal to channels are flood basins and swamps, partially filled by crevasse splay deposits. Much of the lower portion of the Mississippi River alluvial valley contains straight river channels, more than 2 kilometers wide (Gould 1970) (Fig. 4-21). These river channels are nearly filled by vertical channel aggradation during the avulsion process, thus producing in channel-fill sequences a few tens of meters in thickness.

Ancient examples of hydrocarbon-productive, lenticular channel-fill sequences are common in the stratigraphic record. These have been described by Visher (1971), for the Pennsylvanian, Bartlesville sandstone in Oklahoma (Fig. 4-22); by Busch (1974), for the Redfork sandstone (Fig. 8-47); and by Root (1980), for the Cretaceous Muddy sandstone (Figs. 8-49 and 8-50).

These coastal plain–delta plain process-response themes are consistent with studies from Holocene coastal depositional areas throughout the world, and provide a

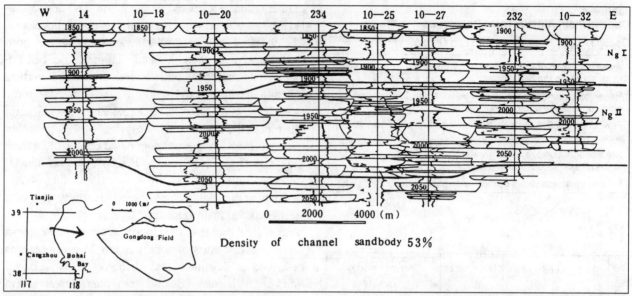

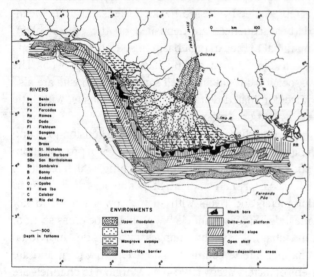

**Fig. 4–18.** Correlation section of Ng, the long-coursed braided sandstone bodies, Gongdong field. From Qiu et al., 1987.

**Fig. 4–19.** From Allen, 1970.

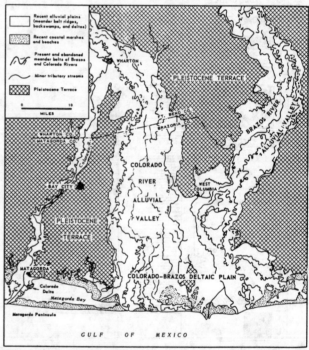

**Fig. 4–20.** From Kanes, 1970.

**Table 4-5**
**Coastal Plain Alluvial Valley Reservoirs**

*Wave-Dominated Coastlines: Sequences 10-30 m Thick*

| |
|---|
| Individual fining upward reservoirs, 10–20 m thick, 5–15 kms wide, and irregular in areal pattern |
| Flow units less than 50 m wide and 200 m long |
| Alluvial valleys controlled by topography and structure |
| Intervals with multiple stacked fining upward reservoirs, areal distribution greater than 2000 kms² |
| Intervals are lensoid with thickness changes less than 50% |
| Intervals between reservoir channels are irregular in both thickness, with greater than 50% change, and in areal pattern |

**Table 4-6**
**Coastal Plain Alluvial Valley Reservoirs**

*Low-Energy Coastlines - Riverine Channels 20-40 m Thick*

| |
|---|
| Alluvial valleys reflect structural and topographic controls |
| Less than 25% change in sequence thickness |
| Each sequence is from 2–6 kms wide, with elongation greater than 5 times, with flow units less than 200 m wide and 500 m long |
| Channels are straight, with avulsion angles greater than 20 degrees |
| Stacked multiple sequences are distributed over an area greater than 3000 kms², with width to length ratio 1:3 |
| Intervals between channel sequences highly irregular, with greater than 25% change in thickness |

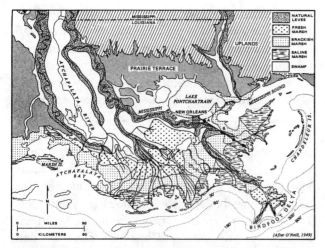

**Fig. 4–21.** Distribution of swamp and marsh environments on Mississippi deltaic plain. From Gould, 1970.

basis for organizing observational data useful for prediction of stratigraphic patterns.

# MARGINAL BASINS—LOW SINUOSITY STREAMS AND FLOODBASINS, LACUSTRINE COMPLEXES

**M**arginal coastal basins are important depositional sites, often with kilometers of sedimentary fill. These basins typically are structurally controlled, contain composite fluvial or lacustrine intervals, or where restricted, evaporites or fine grained biogenic sediments. Where these basins are associated with fluvial drainage systems they often are filled by low sinuosity stream deposits, but in other areas and times, they may be the site of lacustrine

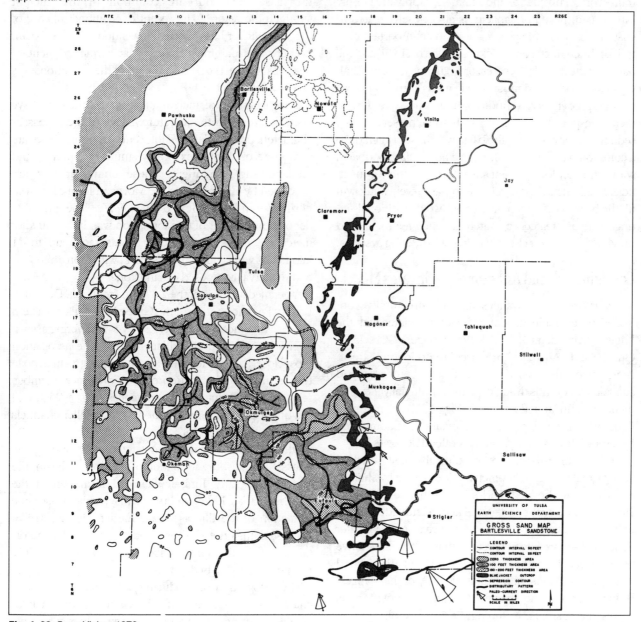

**Fig. 4–22.** From Visher, 1972.

depositional complexes, including fan deltas, density current, mass flow, and slump deposits. At low latitude, and in areas of low rainfall, these basins may become the site of either biogenic or evaporite deposition. The complex of boundary conditions leads to many possible response patterns, and consequently many types of stratigraphic intervals. Those basins that contain evaporite depositional intervals are most often associated with carbonate depositional systems, and will be discussed in conjunction with these depositional systems.

Tectonic boundary conditions are also complex, with fore-arc, back-arc, synsedimentary wrench-fault, and post orogenic extensional basins, all leading to the possible formation of a topographic basin on the continental margin. In most all of the above described structural frameworks, subsidence and sediment supply are closely interrelated. Receptor capacity is also closely related to eustatic sea-level changes, resulting in an important control for the character and thickness of the basin fill and the nature of the fluvial processes, or lacustrine history. Analysis of modern depositional areas, and ancient stratigraphic intervals, indicate that only two fundamental response patterns are likely to be preserved: 1) if sediment supply is high, the basin will fill, and aggrade to a common base level, resulting in a nearly flat floodbasin with extremely low gradient streams; or 2) if there is insufficient sediment supply relative to base level, water will fill the basin to its spill point. These two depositional settings can be treated separately as end members, and provide the basis for characterizing two differing systems.

## Low Sinuosity and Anastomosing Streams Model

Many rivers occupy valleys that are structurally controlled; for example, the Benue-Niger, associated with the Niger River delta (Fig. 4-19). Cooper's Creek in Queensland, Australia, produces a continuous stream profile across the coastal plain, and deltaic processes and subsidence controls the delta plain and alluvial valley gradients (Legun and Rust 1982) (Fig. 4-23). A marginal flood basin is developed in the alluvial valley, dominated by levees, crevasse splay, and other flood deposits.

In the Ganges-Brahmaputra alluvial valley, a broad valley, greater than 20 kilometers wide is filled by shallow anastomosing, braided, and straight channels (Bristow 1987). Channels are up to several kilometers wide, and up to 20 meters deep. The flood basin covers an area greater than 5000 square kilometers (Coleman 1969). Illustrative of the depositional process, recently (in 1988) flood waters covered most of the alluvial valley of the Ganges-Brahmaputra flood basin. Channel deposition is represented as bars, including diagonal, scroll, medial, tributary, and lateral types. Channel cross sections show

a high degree of channel asymmetry, and extremely complex patterns related to bar development (Bristow 1987). The resultant is a multistory and multilateral sand body 20 kilometers wide and up to 200 kilometers long. Internal depositional geometry is marked by a hierarchy of internal erosion and depositional surfaces, scaling with the size of bars and channels. Due to the high width/depth ratios, lateral accretion surfaces are extremely low angle. Stream gradients possibly affect channel geometry, and data suggests that gradients as low as .05–.09 meter/kilometer may produce a range of geometries from anastomosing, braided, to "meandering." Other flood basins may be more restricted, resulting in internal drainage patterns (Legun and Rust 1983) (Fig. 4-24). Such patterns have not often been described in modern depositional settings, but are common in the ancient stratigraphic record. The recent depositional history of the Yellow River has reflected, low-sinuosity flood basin; deeper water marginal basin; and lacustrine sedimentary histories (Chien 1961). A summary of these relationships is presented in Table 4-7.

Many examples of ancient fluvial sheet sandstone have been described in the literature. Many of these possess characteristics similar to Holocene examples, and are composed of a complex of low sinuosity, braided, and "meandering" channels. These characteristics are described by Campbell for the Morrison formation from northwestern New Mexico (Fig. 4-25). The probable depositional extent of this formation was 10 × 160 kilometers. Composite channel intervals range up to 11 kilometers wide and 15 meters thick, with more than 90% of the interval composed of sandstone.

A detailed analysis of the 2.3 kilometer thick, Devonian Battery Point formation of the Gaspe Peninsula of eastern Quebec Province is characterized by four differing associations (Lawrence and Williams 1987). The association most closely associated with low sinuosity stream deposition is in the lower portion of the Fort Prevel member, represented by channel fill sequences (Fig. 4-26a), and the upper portion (Fig. 4-26b) reflects smaller channels and sheet flood deposits.

Other examples have been suggested including the Triassic Hawkesbury formation of the Sydney Basin, the late Triassic–early Jurassic Statfjord formation of the North Sea (Falt 1986), and the Pennsylvanian Caseyville formation of the Illinois basin. Earlier, Hobday (1978) suggested that the Carboniferous, Ecca, and Beaufort Groups, from the Karoo basin of South Africa, possibly represent flood basin deposits.

The recognition of producing subsurface flood-basin channel sequences is difficult, owing to the lack of detailed mapping of channel geometry. Two examples, however, have been reported in the literature. Qiu et al.

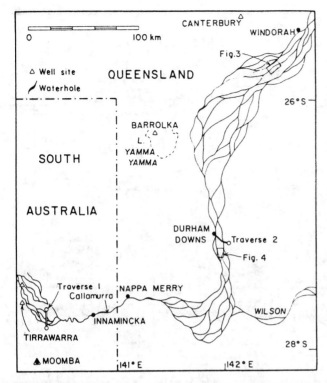

**Fig. 4-23.** Principal features of Cooper's Creek between Windorah, Queensland and Tirrawarra, South Australia. From Legun and Rust, 1982.

**Table 4-7**
**Geometry of Low Sinuosity Fluvial Reservoirs**

Intervals of stacked multi-story channels, area greater than 2000 kms², and interval is ovoid with less than 3:1 elongation

Channel sequences show basal and lateral scour surfaces 5–10 m deep, .5–3 kms wide, and from 5–10 kms long

Reservoir 5–10 m thick, .5–3 kms wide, 5–10 kms long, and reservoir flow units less than 5 m thick, 300 m wide, and 500 m long

In single reservoir channels, abrupt lateral change in thickness

Fine-grained and carbonaceous units define flow boundaries, and intervals from 1–20 m thick, over areas of up to 50 kms²

(1987) suggest that reservoir intervals from the giant Daqing oilfield are deposited as low sinuosity, or "meandering" channels (Figs. 4-27 and 4-28). Both fining upward and channel sequences with little vertical change in grain size are reflected. The sandstone body density is reported to be 60%, characteristic of a flood-basin depositional setting. The second example, the basal Cretaceous Mannville formation, has been attributed to a variety of fluvial depositional settings. The areal channel pattern suggests an anastomosing channel depositional system overlying a lower Cretaceous unconformity surface (Putnam 1983) (Fig. 4-29). The paleogeographic setting and depositional sequence suggests deposition across a low-gradient coastal plain (Fig. 4-30). Comparison to the Columbia River flood basin of southeastern British Columbia to the Mannville fluvial system is suggestive of a similar process-response depositional setting (Table 4-8). This extensive valley system is more than 150 kilometers long and 50 kilometers wide. Channel sequences also show both fining upward and sequences with little vertical change in texture (Table 4-8).

Seismic lines illustate the continuity of reflections over a single channel system (Fig. 4-30), and the development of a reflection doublet associated with the base and top of the channel sequence (Putnam 1982) (Fig. 4-31).

## Siliciclastic Lacustrine Depositional Models

The sedimentology of Holocene lakes has been a common subject of study. Lakes from Eurasia, western and central United States, and Africa have been extensively studied (Matter and Tucker 1978; Picard and High 1981). Many of these lakes are in arid or semiarid environments and are dominated by chemical or biogenic sediments; others, however, are related to rifts or to glacial processes. These lakes contain extensive siliciclastic depositional intervals, and are comparable to many ancient lake sequences. A balance between water depth, clastic imput, and chemical sedimentary sequences is the

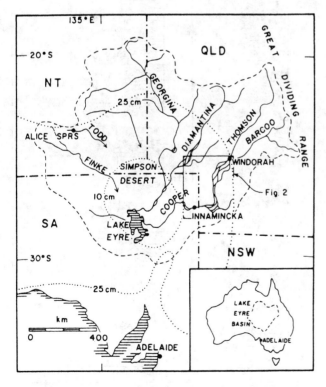

**Fig. 4-24.** Location map of Lake Eyre Basin. Dotted lines are contours of mean annual rainfall. From Legun and Rust, 1982.

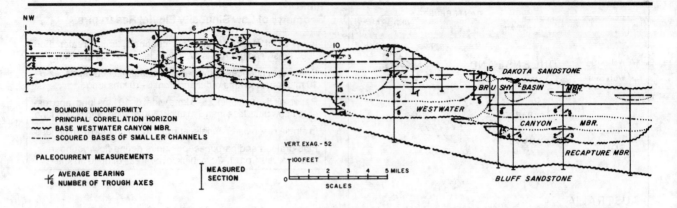

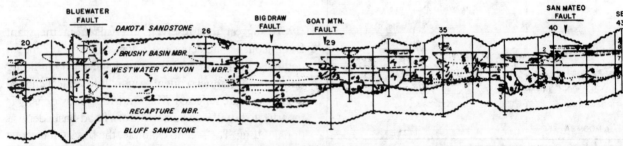

BOUNDING UNCONFORMITY
PRINCIPAL CORRELATION HORIZON
BASE WESTWATER CANYON MBR.
SCOURED BASES OF SMALLER CHANNELS

PALEOCURRENT MEASUREMENTS

AVERAGE BEARING
NUMBER OF TROUGH AXES

MEASURED SECTION

VERT. EXAG.- 52

100 FEET

0   1   2   3   4   5 MILES
SCALES

**Fig. 4–25.** Stratigraphic cross section showing relations of Westwater Canyon Member to enclosing rocks. This section is approximately transverse to paleocurrent directions and shows cross-sectional dimensions of channel systems that compose continuous sheet of Westwater Canyon sandstone. From Campbell, 1976, reprinted by permission.

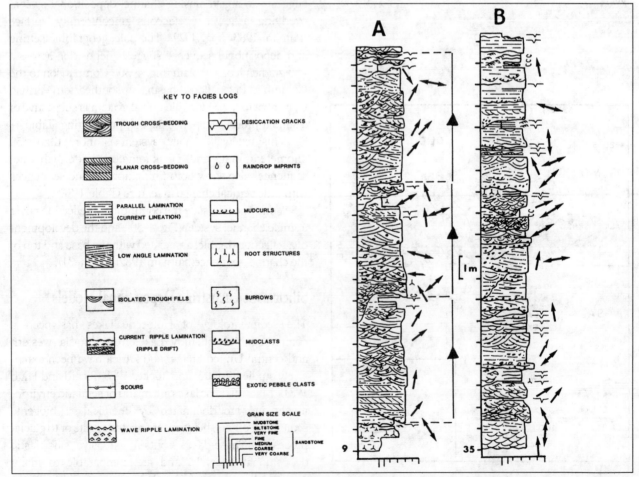

KEY TO FACIES LOGS

TROUGH CROSS-BEDDING
PLANAR CROSS-BEDDING
PARALLEL LAMINATION (CURRENT LINEATION)
LOW ANGLE LAMINATION
ISOLATED TROUGH FILLS
CURRENT RIPPLE LAMINATION (RIPPLE DRIFT)
SCOURS
WAVE RIPPLE LAMINATION

DESICCATION CRACKS
RAINDROP IMPRINTS
MUDCURLS
ROOT STRUCTURES
BURROWS
MUDCLASTS
EXOTIC PEBBLE CLASTS

GRAIN SIZE SCALE

MUDSTONE
SILTSTONE
VERY FINE
FINE
MEDIUM
COARSE
VERY COARSE

SANDSTONE

**Figure 4–26.** Association 1. *A,* Fining-upward sequences from the lower part. *B,* Sheet sandstones from the upper part. Numbers refer to meters above the base of the association. From Lawrence and Williams, 1987.

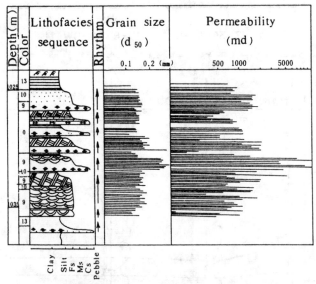

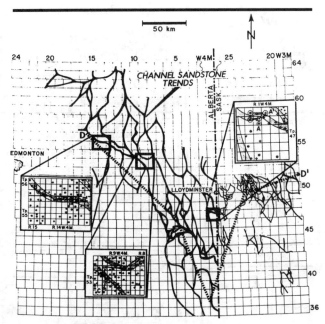

**Fig. 4-27.** Lithofacies sequence of low sinuousity meandering sandstone body (Subzone Pl2+3, Well L7-28, Daqing). From Qiu et al., 1987.

**Fig. 4-29.** Areal distribution of the thickest channel sandstones (16 m minimum). From Putnam, 1983.

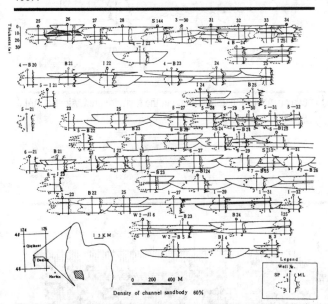

**Fig. 4-28.** Geometry of low sinuousity meandering sandstone body, (Subzone Pl2=3, Daqing). From Qiu et al., 1987.

principle response pattern seen in most Holocene lacustrine deposits. The predominant facies patterns in the deep basin may include fans, varved clays and silts, or varves with varying proportions of chemical or biochemical laminae. Shoreline intervals may also include chemical or biochemical deposits, but also may be dominated by deltaic or coastal clastic sequences (Picard and High 1981). Characterization of depositional themes is difficult due to the diversity of climatic, provenance, and eustatic conditions; length of time for deposition; and tectonic boundary conditions. Preservation of lacustrine intervals in the stratigraphic record requires special boundary conditions. Typically they are associated with eustatic sea-level rise, or disruption of drainage patterns

by local tectonism. Often they reflect a short time interval or a complex history of subsidence and fill. Summary sections representing basin fill sequences for a number of Holocene lakes are detailed by Picard and High (Fig. 4-32). These schematic sections do not present a pattern that can be usefully compared to many ancient lake deposits. Premodern glacial and tectonic events may have produced sequences difficult to compare to many ancient lake deposits.

A dynamic depositional theme has been proposed for Lake Brienz in Switzerland by Sturm and Matter (1978) (Fig. 4-33). In this example the importance of slumping and gravity flows is illustrated. Many ancient lakes appear to reflect these processes in depositional sequences, or in the history of lacustrine basin fill.

The Pliocene fill of the Ridge basin reflects elements of both clastic and chemical lacustrine basin sedimentation (Link and Osborne 1978) (Fig. 4-34). As illustrated by Figures 4-35 and 4-36, more than 1 kilometer of deep basin clastic gravity flow and submarine fan deposits are interbedded with laminated basin mudstone intervals. Offshore lacustrine mudrock-chemical facies are reflected in basinal areas distal to the clastic facies (Fig. 4-36). Marginal to the basin center are stromatolitic carbonate mudflats with gypsum nodules, fluvial-deltaic lobes, and, adjacent to a bounding fault on the western margin, alluvial fans. The overall interval reflects a transition at the base from marine (more than 600 meters thick), overlain by deltaic, followed by lacustrine, and capped by alluvial sequences (Crowell 1975). The entire interval is more than 9000 meters thick, with the central basin lacustrine interval the dominant stratigraphic interval.

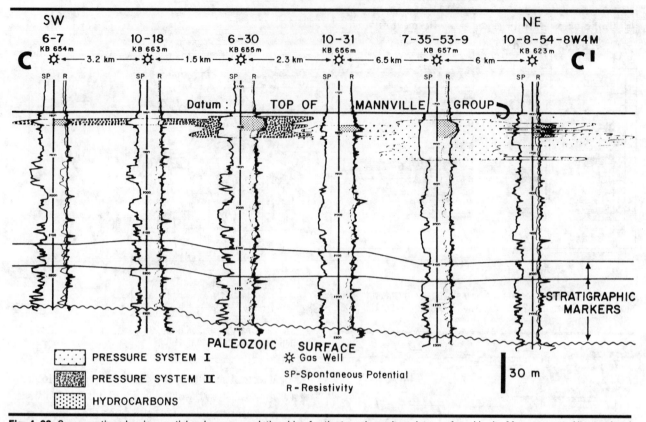

**Fig. 4-30.** Cross-section showing spatial and pressure relationships for the two channel sandstones found in the Myrnam area, Alberta. Areal distribution of the thickest channel sandstone (16 m. minimum). From Putnam, 1983.

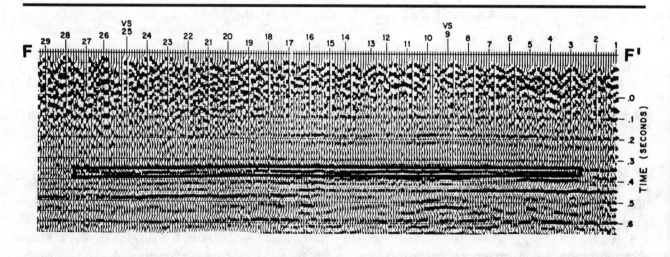

**Fig. 4-31.** Seismic section showing bright spot oriented along channel sandstone in east-central Alberta. Modified from Putnam, 1982, reprinted by permission.

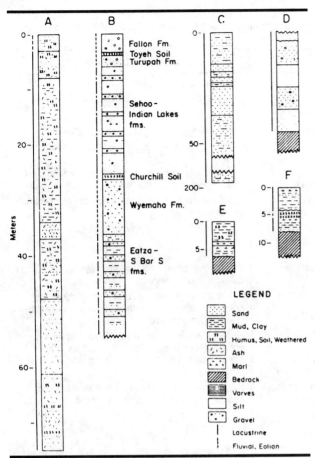

Fig. 4-32. From Picard and High, 1981.

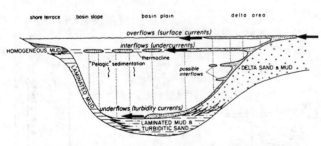

**Fig. 4-33.** Distribution mechanisms and resulting sediment types proposed for clastic sedimentation in oligotrophic lakes with annual thermal stratification. Note that hypothetical shore terrace is situated higher than depth of thermocline. Width of basin and sediment thickness are not to scale. From Sturm and Matter, 1978.

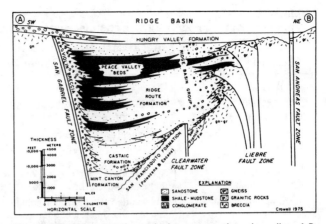

**Fig. 4-34.** Diagrammatic geologic cross-section along line *A-B* showing major stratigraphic and structural relationships in the Ridge basin. From Link and Osborne after Crowell, 1975, 1978.

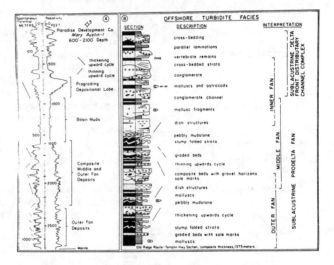

**Fig. 4-35.** Electric log (A) and columnar section (B) of the offshore lacustrine (?) turbidite facies. From Link and Osborne, 1978.

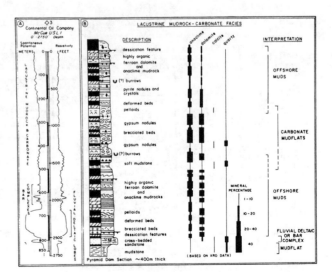

**Fig. 4-36.** From Electric log (A) and columnar section (B) of the offshore lacustrine mudrock-carbonate facies. From Link and Osborne, 1978.

**Table 4-8**
**Comparison of the upper Mannville channel system to the anastomosing upper Columbia River of southeastern British Columbia. Putnam, 1983**

| Channel Fills | Upper Mannville | Columbia River |
|---|---|---|
| Lithology | Dominantly sandstone, some shale fills | Dominantly sand, some mud fill |
| Thickness (max.) | approx. 35 m | 10 m |
| Width (min.) | approx. 300 m | 40 m |
| Sedimentary Structures/ Bedforms | High-angle (30°) cross-beds with laminae up to 2 cm thick. | Dominant bedforms are sand waves with main sedimentary structure being tabular planar cross-beds. |
| Nature of Contacts | Scoured lower contact commonly overlying coal, upper contact can be sharp to gradational. | Scoured lower contact commonly overlying peat or mud, upper contact can be sharp to gradational. |
| Textural Trends | Multi-storied with individual fining-upward stories 2–6 m in thickness. Channel sandstones commonly exhibit an overall upwards-fining appearance. | Multi-storied with individual fining-upwards stories about 1 m thick. |
| *INTERCHANNEL SAND DEPOSITS* | | |
| Thickness (average) and Area | 6-9 m, thicker near major channel sandstones. Several km² up to tens of km². | 2–3 m near active channel crevasse and thinning to 0.3 m. Larger splays up to 1 km². |
| Sedimentary Structures/ Bedforms | Current ripples, climbing current ripples, wave ripples. | Current ripples, parallel laminations, rare cross-beds, bioturbation by rooting. Bedforms are ripples, dunes, transverse bars. |
| Nature of Contacts | Scoured lower contact proximal to major channel sandstones, gradational lower contact away from major channel sandstones. Upper contacts sharp to gradational. | Scoured lower contact proximal to channel crevasse, gradational lower contact away from crevasse. Upper contacts sharp to gradational. |
| Vertical Textural Trends | Fining-upwards proximal to major channel sandstones, coarsening-upwards away from major channel sandstones. | Fining-upwards proximal to channel crevasse, coarsening-upwards away from crevasse. |
| *INTERCHANNEL FINE-GRAINED DEPOSITS* | | |
| Lithology | Mudstone, shale, siltstone, coal. | Clay, silt, peat. |
| Sedimentary Structures | Thin parallel laminae (scale of mm), bioturbation in varying degrees, coarser siltstones commonly current and wave - rippled, soft sediment deformation features. | Thin parallel laminae (scale of mm), rare bioturbation. |

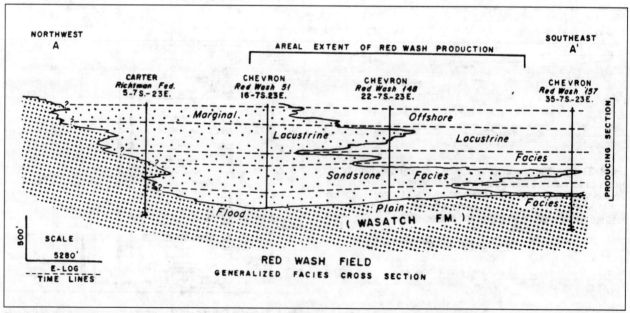

**Fig. 4–37.** Generalized stratigraphic section across Red Wash area showing distribution of facies changes. From Chatfield, 1972, reprinted by permission.

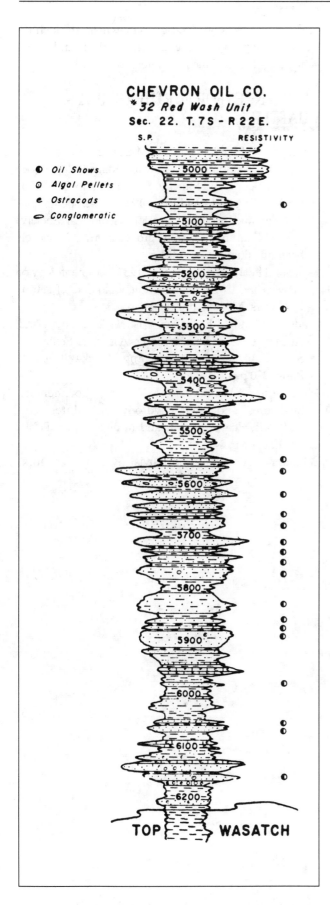

**Fig. 4-38.** Typical electric log and lithology of producing section at Red Wash field. From Chatfield, 1972, reprinted by permission.

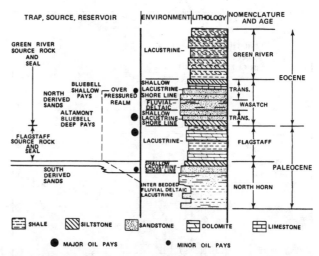

**Fig. 4-39.** Stratigraphy, nomenclature, and environmental relations of northern Uinta basin. From Lucas and Drexler, 1976, reprinted by permission.

Similar ancient stratigraphic intervals have been found to contain giant hydrocarbon accumulations. Included are lacustrine intervals from the Reconcavo basin of Brazil, a west African marginal basin in Cabinda, sedimentary intervals in the central Sumatra, Minas basin, an interval in the Paleocene Fort Union formation of the Wind River basin in Wyoming.

Clastic and chemical intervals from the Eocene Green River basin of Utah are the site of giant hydrocarbon accumulations. Marginal clastic deposition is illustrated by the Red Wash formation (Chatfield 1972) (Fig. 4-37), with the interbedding of reservoir and source rock intervals (Fig. 4-38). In the central basin there is a complex interbedding of coastal, fluvial-deltaic, and biogenic-chemical facies (Fig. 4-39) (Lucas and Drexler 1976). A typical well log from the giant Altamont-Bluebell oil field is illustrated (Fig. 4-40). These examples and the Daqing oilfield of China (Qiu et al. 1987) illustrate that of principle importance to hydrocarbon exploration is the presence of biogenic sediments which can be a major source rock. Type I kerogen, predominatly of algal origin, may produce normal parafins, and organic detritus from continental plants may add high molecular weight waxes, resulting in a low API gravity crude oil.

Recognition of ancient lake deposits is dependent upon the recognition of the boundary conditions, the presence of varved or laminated fine grained units, and unusual mineral assemblages. Also, the absence of marine fauna, the paucity of bioturbation, and the presence of gravity-dominated sedimentary response patterns are also indicative of restricted lacustrine basin fill intervals. Facies patterns may be complexly interrelated with patterns of subsidence and sedimentary fill. Climate cycles may produce changes in rates of sedimentation, miner-

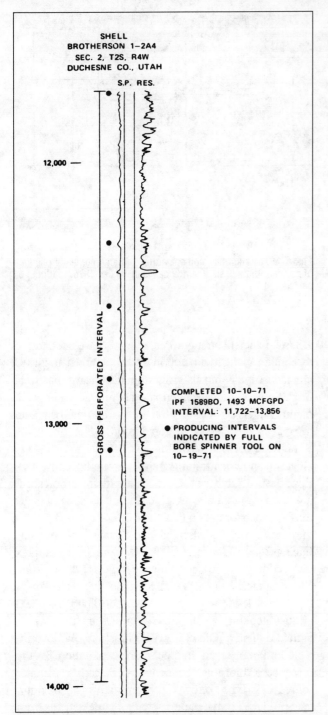

SHELL
BROTHERSON 1–2A4
SEC. 2, T2S, R4W
DUCHESNE CO., UTAH

S.P. RES.

12,000 —

GROSS PERFORATED INTERVAL

COMPLETED 10–10–71
IPF 1589BO, 1493 MCFGPD
INTERVAL: 11,722–13,856

● PRODUCING INTERVALS
INDICATED BY FULL
BORE SPINNER TOOL ON
10–19–71

13,000 —

14,000 —

**Fig. 4–40.** Typical electric log of Altamont-Bluebell producing well.
From Lucas and Drexler, 1986, reprinted by permission.

alogy, and water depth. These variables make the identification of specific characteristic vertical and areal sequence patterns difficult to interpret or predict.

# REFERENCES FOR FURTHER STUDY — CHAPTER 4

Ethridge, F.G., R.M. Flores, and M.D. Harvey, eds., 1987, Recent Developments in Fluvial Sedimentology: Soc. Econ. Paleont. and Mineral. Spec. Publ. no. 39, 389 p.

Frostick, L.E., and I. Reid, eds., 1987, Desert Sediments Ancient and Modern: London Geological Society, Special Publ. no. 35, Palo Alto, California, Blackwell Scientific Publications, 401 p.

Koster, E.H., and R.J. Steel, eds., 1984, Sedimentology of Gravels and Conglomerates Canadian Soc. of Petrol. Engineers, Memoir, no. 10, 441 p.

Matter, A., and M.E. Tucker, eds., 1978, Modern and Ancient Lake Sediments: Inter. Assoc. of Sediment. Spec. Publ., no. 2, Oxford, Eng., Blackwell Scient. Publ., 290 p.

Middleton, G.V., ed., 1965, Primary Sedimentary Structures and their Hydrodynamic Interpretation: Soc. Econ. Paleont. and Mineral. Spec. Publ., no. 12., 265 p.

Schumm, S.A., 1977, The Fluvial System: New York, John Wiley & sons, 338 p.

# 5 COASTAL SEDIMENTOLOGY

Science. . . is not merely a discovery of pre-existent fact: it is also, and more importantly, a creation of

something new. It is just as creative as art, though in a different way. Scientific laws are not something

existing from eternity in their own right or in the mind of God, waiting to be discovered by man: they did

not exist before men of science formulated them. **J. Bronowski and B. Mazlish**
*The Western Intellectual Tradition,* 1960

## INTRODUCTION

The coastal zone reflects an equilibrium between the sediment supply, wave and tidal forces, geometry of the coastal area, and relative change in sea level with respect to the depositional surface. This is a complex association of factors that would appear to require a complex of depositional patterns. However, due to the development of equilibrium, a limited number of specific depositional profiles is produced and these profiles reflect an integration of the sedimentologic controls. A shoreline has three possible responses:

▸ Progradation resulting from a surplus supply of sediment for the available dispersive forces, a geometric receptor capacity, or a lowering in the relative position of the depositional interface;

▸ A stable shoreline representing a balance between sediment supply, dispersive forces, and the position of the depositional interface, or

▸ Erosion and transgression of the shoreline across the continent reflecting a relative rise in the position of the depositional interface with respect to sediment supply and dispersive forces.

Of these possibilities only the first two result in the accumulation of sedimentary units, and the second possibility is simply a special case of the first. For extensive shoreline sedimentary units to be preserved, the first case is significant.

Shoreline progradation results from sediment accretion on an equilibrium shoreface surface in response to deposition from a sediment transport system either parallel to the shoreline or from offshore. The overwhelming sediment source parallel to the shoreline is from deltas, and two types of coastal depositional patterns are developed:

1. Strandplains, barrier islands, and inlets
2. Bays, coastal marshes, and lagoons

Sediment transport from offshore results in two types of coastal depositional patterns:

1. Siliciclastic tidal flats, estuaries, beaches, and barrier islands
2. Calcareous tidal flats, estuaries, bays, and beaches

Stratigraphic units resulting from these four model types may not always be easily separated. Boundary conditions and process controls can be analyzed to evaluate their relative importance along coastlines.

# BOUNDARY CONDITIONS

The shoreface geometry, tectonic and sea-level responses, and climate must be evaluated as a basis for interpreting coastal sedimentology. These boundary conditions along Holocene coastlines must be understood as the framework for interpreting sedimentary responses. With this base the four sedimentologic models can be developed from Holocene actualistic studies and can be applied to the recognition and interpretation of shoreline stratigraphic units. (Table 5–1).

## Shoreface Geometry

The distribution of energy reaching a shoreline depends upon the offshore slope and the curvature of the shoreline. Wave fronts are refracted as the wave progression is retarded by frictional forces. Wave energy per unit length of a wave must remain equal; the dissipation of energy by bottom friction reduces wave height and speed as energy is distributed across the shoreface profile. Thus, embayed coastlines and coastlines with a low slope in the shoreface zone dissipate energy in the offshore shoal areas resulting in reduced energy release at the shoreline. The opposite effect may be seen by tidal waves. Embayed coastlines concentrate the wave by increasing its amplitude at shallow depths near the shoreline. Strongly embayed coasts or funnel-shaped estuaries develop the highest tidal amplitude. In deep water the progression of the waves toward the shoreline results in bottom friction and sediment transport. The equilibrium between the shoreface geometry and the progressive tidal wave or the surface wave depends upon the relative importance of these processes.

The offshore slope reflects a dynamic equilibrium between deposition and sediment dispersion. The profile is adjusted to the prevailing wave and current processes and undergoes seasonal changes as wave direction, height, and period change. Similarly, changes in water density and tidal amplitude produce systematic seasonal changes. Deposition along coastlines is a response to nonequilibrium events, such as unusual tides, storms, and wave conditions, resulting in a periodicity of deposition and development of many discontinuous surfaces or bedding planes.

Modern coastlines dominated by bays and estuaries reflect a low slope, and water interchange from the offshore is primarily by changes in the relative elevation of water surfaces. These changes may be produced by semidiurnal or diurnal tidal cycles, onshore or offshore winds, barometric pressure patterns, surface water drainage, and evaporation. Deposition in these areas modifies the coastal geometry, and stratigraphic continuity reflects either unusual tectonic conditions or an equilibrium between surface water discharge patterns and offshore wave, tidal, and current forces. An equilibrium coastline reflects the continuing forces and processes operating through time, and topographic or geometrical irregularities reflect nonsystematic events or continuing climatic or tectonic events operating at a sufficient rate to prevent equilibrium. Specifically, lagoons, bays, and estuaries reflect nonequilibrium conditions and often are not stratigraphically continuous either areally or vertically.

## Tectonic and Sea-Level Changes

Systematic changes in either sea level or uplift and subsidence are reflected in the shoreline position and geometry. The shoreline is the stratigraphic indicator most easily identified, and changes in position result in changes in both areal and vertical facies patterns. To determine the cause of changes requires an understanding of rates of change in boundary conditions. Shoreline deposition is rarely at the same rate described for alluvial and deltaic sedimentary patterns. Systematic changes in relative sea level with respect to the depositional interface has a stronger influence on stratigraphic patterns.

A relative rise in the depositional interface produces landward migration with consequent destruction of an equilibrium shoreface. Preserved stratigraphic patterns reflect both the rates of deposition and the rise of the depositional interface. Similarly, a subsiding depositional interface results in seaward migration of the shoreface, and the continuity of stratigraphic units reflects both rates of deposition and fall of the depositional interface. Shoreface processes are dominant, and the resulting facies pattern reflects the relative rate of deposition and sea-level change.

Rapid changes in relative sea level produce a discontinuous pattern of sedimentary units, and a stacking of depositional units may show successive landward or seaward jumps in the shoreline positions (Fig. 5–1).

This pattern results in preservation of only portions of the depositional sequence, and the systematic changes in shoreline position are reflected in a discontinuous pattern in thickness and the depositional environment of preserved stratigraphic units.

Relative sea-level changes interact with the geometry of the depositional site and result in local bars, bays, estuaries, and lagoons. Sediment accumulation and energy are concentrated in specific areas. For example, headlands receive more wave energy, and barrier islands develop across bays and estuaries in response to onshore and longshore sediment transport and topographic irregularities. Thus, many sedimentologic responses must be related to systematic changes in sea level and topographic boundary conditions.

**Table 5-1**
**Coastal Response Characteristics**

| | Wave | Biologic | Tidal | Biochemical |
|---|---|---|---|---|
| Geomorphic elements | Beaches, barrier islands, offshore bars | Bay, marsh, lagoon | Tidal flats, channels | Mud and tidal flats, channels |
| Shoreline geometry | Straight | Irregular | Irregular or embayed | Straight or embayed |
| Geometry | Linear-blanket 5–10 m thick | Lensoid | Blanket-Lensoid 2–6 m thick | Blanket 2–6 m thick |
| Migration pattern | Horizontal offlap | Vertical or shoreward onlap | Horizontal offlap | Horizontal offlap |
| Vertical sequences | | | | |
| Subaerial zone | Berm-backshore dunes | Alluvial channels | Marsh, clay-dunes | Sabkha aeolian dunes |
| Littoral zone (intertidal) | Low-angle truncation surfaces | Swamp with local shell concentration | Burrowed flasers fining upward channeled | Algal mats and pinnacles channeled |
| Upper shoreface | Coarsening upward current ripples and planar cross sets | Burrowed silt and clay shell banks | Current and wave bedded sands burrowing | Pelleted mud, skeletal grains burrowed |
| Lower shoreface | Irregular bedding bioturbation wave ripples | Burrowed muds | Interbedded and burrowed sands and muds | Mud and green algal skeletal pelleted sand |
| Marine platform | Shell and relict sand lag | Phosphatic, shell and glauconitic, biogenic shale | Relict sand or sand waves | Skeletal and oolite pelleted algal muds |

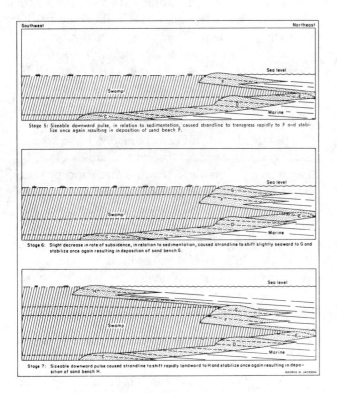

**Fig. 5-1.** Diagrammatical cross sections illustrating irregularity of sand deposition along transgressive Cliff House shoreline. From Hollenshead and Pritchard, 1961, reprinted by permission.

Relative sea-level changes are mostly in response to eustatic controls, not to tectonic uplift and subsidence. Sediment load is insufficient to cause subsidence, as suggested for the deltaic framework. Tectonic coastlines, although common, reflect rates of change sufficient to limit preservation of stratigraphic units. Coastlines where uplift is dominant result in subaerial erosion, and coastlines of submergence produce marine scour and erosion (Curray 1964) (Fig. 3–1).

## Sediment Supply

The interrelation of strandline progression and sediment supply is important in interpreting the continuity of sedimentary units. The shape of the depositional site, the transport or dispersion of sediment in the marine environment, and the relative change in sea level modify the progradation rate. Separating the effect of sediment supply from these boundary conditions can only be suggested by examining a larger area of sedimentation and, likewise, a larger stratigraphic area and interval.

The following independent sedimentologic and stratigraphic observations suggest the possible interrelationship of sedimentation rate with boundary conditions. A study must include:

▸ Determination of water depths from the biota,

▸ Shelf depositional patterns from bedforms and textural studies,

▸ Wind and wave patterns from paleocurrents indicated by preserved bedforms,

▸ Proximity to sediment sources from facies studies, and

▸ Rates of sedimentation from study of bioturbation.

## Climate

Rarely is the nature of the climate on the adjacent continental area considered in stratigraphic interpretation. The nature of the land vegetation not only controls the mechanics of sediment trapping and erosion, but it reflects the climate of the coastal area. Carbonate sedimentation on coastal areas is strongly affected by evaporation, rainfall distribution, and temperature. Carbonate deposition is a complex of chemical, biochemical, and biologic processes, and the sedimentology of coastal tidal flats, shoals, bays, and estuaries must be based in part upon an understanding of climatic factors. Similarly, the deposition of siliciclastics is strongly affected by the nature of vegetation on tidal flats, the biota in the shoreface, the rainfall distribution, and the temperature. Arctic beaches and coastlines are demonstrably different from those on humid-temperate coastlines or even wet-dry, arid, or tropical coastlines. Sedimentologic responses can be studied, and, in relation to other stratigraphic units, a paleogeographic framework can be useful in interpreting the sedimentology.

Climate indicators include the nature of land vegetation, preserved marine organisms, biochemically or chemically precipitated minerals, diagenetic mineral associations, and sediment-faunal traces. The interpretation of climatic boundary conditions can be based upon many types of observations. Each of these sedimentologic characteristics has been the focus of study in Holocene depositional sites. These data are useful as the basis for stratigraphic interpretation of climatic boundary conditions.

## Summary

As has been suggested, many boundary conditions are important in coastal sedimentology. Identification of boundary conditions and the interpretation of specific stratigraphic responses are not easily obtained. A synthesis of Holocene data is required before interpretation of ancient stratigraphic units is possible. The development of actualistic models from the Holocene provided the necessary basis, and these aspects are included in the description of the four types of coastal sedimentologic models (Table 5–1).

## SEDIMENTARY PROCESSES AND RESPONSES

Sedimentary processes produce and shape the geomorphic elements developed under equilibrium conditions. Waves shape the beach shoreface in response to the pattern of energy release. Changes in water level result in formation of inlets if the coastal geometry includes lagoons, bays, and estuaries. Tides produce broad flats and channels, and progradation is the response to onshore transport of sedimentary detritus. If the predominant source is offshore, marine sands, silt, and mud constitute the detritus, and if the offshore marine clastic materials are principally biogenic, it constitutes the source material. Wind processes are important as a control for longshore currents, the development of beach ridges, and the transport of clay, silt and sand resulting in coastal dunes, deposition on sand flats, and the fill of lagoons, marshes, and bays. These four processes (wave, tides, biologic, and wind) are interrelated, but their response may be uniquely determined in the framework of coastal sedimentology (Table 5–1).

## WAVE DOMINATED THEMES

Waves are an important mechanism of energy release on coastlines. Waves are described in terms of height, length, and period; shoaling waves also change in steepness. These aspects are fundamental to the nature of their effects on coastal profiles. The depth of wave activity is directly proportional to wavelength, and most wind-produced waves have an effective depth of less than 10 meters (Curray 1969). Thus, most waves produced by wind stress on the water surface are important in sediment transport and deposition near shorelines. Waves of this type are instrumental in forming the coastal profile. Transport of clastic detritus is primarily in the zone of shoaling waves. Shallow water waves are modified by interaction with the depositional surface and produce bedforms reflecting the character of the wave and the form of the depositional surface (Clifton, Hunter, and Phillips 1971) (Fig. 5–2), (Clifton 1976) (Fig. 5–3).

The high-energy, nonbarred coast described by Clifton, Hunter, and Phillips (1971) and by Ingle (1966) illustrates the interaction of waves and deposition. The wave period reflects the number of waves reaching a coast in a measured length of time, wave height reflects wave energy, and wavelength reflects the depth that the wave expends frictional turbulent energy in sediment transport and erosion.

Under conditions of low wave height and length, the sediment responses are different. For example, the southeast coast of Georgia has a low-wave energy regime with bedforms strongly modified by biogenic activity and

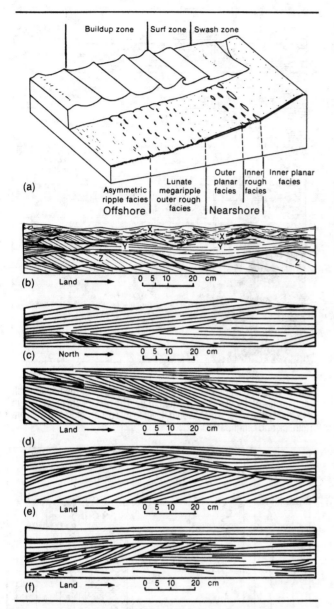

**Fig. 5-2.** Zonation of wave activity and facies of sedimentary structures within and adjacent to the high energy nearshore. (b) Internal structure of sand just seaward from the offshore-nearshore boundary on an average day. (c) Internal structure produced by lunate megaripples. (d) Outer portion outer planar facies. (e) Internal structure of the inner rough facies on a relatively gently inclined beach near shore profile. (f) Internal structures from the seaward edge of the inner planar facies (swash zones). From Clifton, Hunter, and Phillips, 1971.

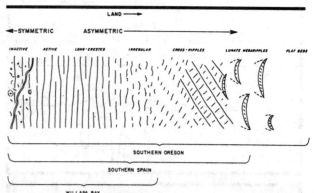

**Fig. 5-3.** Sequence of structures commonly observed off the coast of southern Oregon, southeastern Spain, and Willapa Bay, Washington. Brackets below diagram indicate the range of structures observed in each environment. From Clifton, 1976.

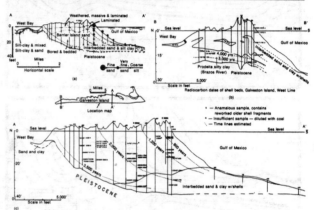

**Fig. 5-4.** Radiocarbon dates of shell beds, Galveston Island eight-mile load. From Bernard, LeBlanc, and Major, 1962.

sand transport restricted to shallow depths (Howard 1972).

Waves with long wavelengths and unusually long periods affect coastlines in a different manner. These waves or swells also are present along most coastlines, and their propagation toward the shoreline results in sediment transport at depths well below those for the shorter wavelength waves resulting from wind stress. The energy in these waves is carried by internal waves within the water column, and turbulence dissipates much of the

energy. The pattern of internal waves is reflected in water-density distributions, and their action on the sedimentary interface depends upon the geometry of density layers, the shape of the shelf, and the current patterns within the water column. Bedforms developed in deeper water indicate sediment transport, but the effectiveness of these waves in forming an equilibrium profile has been argued. (Swift 1967; Otvos 1970; Southard and Cacchione 1972).

## Wave Process Model

Process models useful for stratigraphic purposes require vertical profiles or sections that reflect history. Only a few cored sections of coastal areas dominated by wave processes are available for analysis. However, these few sections can be used as the basis for interpreting depositional and, therefore, stratigraphic response patterns.

To interpret the dynamic history of beach or barrier inlet coastlines requires either careful dating of depositional surfaces (Bernard et al. 1962) (Fig. 5-4) or tracing

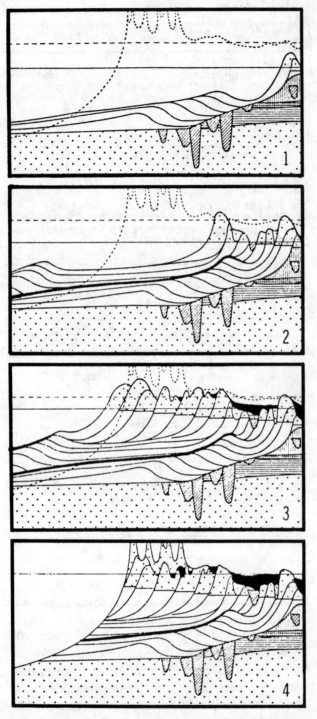

**Fig. 5-5.** Part of the complex sequential development in a barrier island along the North Sea. From van Straaten, 1965.

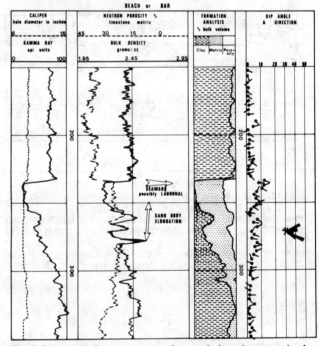

**Fig. 5-6.** Compilation of the sedimentary features of the beach-shelf mud profile off Licola, Gulf of Gaeta, Italy. The results are presented in a vertical sequence, which would develop in a prograding coast. From Reineck and Singh, 1980.

**Fig. 5-7.** Log and dipmeter patterns from an Indonesian example of a beach-bar sequence. From Goetz, Prins, and Logar, 1977.

bedding planes in closely spaced core holes (van Straaten 1965) (Fig. 5-5). These two examples are carefully documented, and the historic interpretation is consistent with observed sedimentary characteristics. Questions concerning the origin of a barrier island and its subsequent history have received a great deal of attention. The two cited examples and summary works by Hoyt (1967), Fisher (1968), Otvos (1970), and Schwartz (1971) suggest

that their formation relates to topographic irregularities along a drowned coastline, and the form of the beach-barrier system is in response to wave and longshore current processes. The stratigraphic implication is that the equilibrium between sediment supply and wave energy suggests both the internal and areal patterns of the

**Fig. 5–8.** A Pleistocene shoreline sequence from bluff of St. Mary's River, Georgia.

**Table 5-2**
**Wave-Dominated Shoreface Sandstone Reservoirs**

| |
|---|
| Interval thickness from 3–13, width up to 10 kms and elongation greater than 5 times width |
| Flow units up to 200 m wide and 2 kms long |
| Thickness variation from event marker below sandstone to top of sandstone ranges from 50–100% |
| Thickness variation from an overlying marker to the top of the sandstone less than 50% |
| Top sandstone-shale transition abrupt, less than 1 m, basal transition greater than 2m |

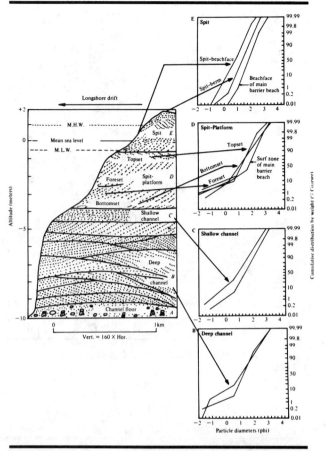

**Fig. 5–9.** Profile and section through east bank of Fire Island inlet showing sequence of strata deposited as inlet shifts westward. View is north along axis of inlet. *AE* designate units in inlet sequence. Cross strata, not drawn to scale, are shown in views revealing their distinctive aspects. Except spit-platform foresets, most cross strata in all units show maximum dips in a north-south line parallel to the axis of the inlet and at right angles to the plane of this figure. Characteristic grain-size curves of sediments from each unit and subunit of the inlet sequence have been drawn on the right. The scale on the y-axis for grain-size sequences is a probability scale. From Kumar and Sanders, 1975.

sedimentologic responses (Figs. 5–6, 5–7, and 5–8).

High rates of sediment supply result in rapid progradation of the wave-formed shoreface and produce a blanket stratigraphic unit. In this case sand thickness reflects the depth of the equilibrium wave base; inlets are rapidly obliterated; and a regressive shoreface sand sheet is formed (Table 5-2). Under conditions of lower sediment supply, extensive wave reworking and longshore extension of the shoreface may result in replacing the wave dominated shoreface by inlet depositional patterns (Figs. 5–9 and 5–10). Progradational units are modified by transgression as suggested for the Delaware coast by Kraft and Allen (1975) (Fig. 5–11), and only in areas of high subsidence or relative lowering of the depositional interface can these upper zones be preserved. The resulting pattern is destructive, and the preserved stratigraphic unit is only a few tens of centimeters thick and produces deposition on the marine platform.

The rarity of well-documented examples of interdeltaic, wave-dominated, shoreface stratigraphic units suggest that preservation of these deposits requires either an unusual lowering of the depositional interface or an unusual rate of supply of clastic detritus. Both of these possibilities are common to deltaic depocenters, and often preserved shoreface depositional patterns are associated with deltas. This association was recognized by Barrell (1906).

## Exploration Strategies

Barrier island sandstone bodies are not important reservoirs outside depocenter areas. Barrell (1906) suggested that only within the deltaic framework is there sufficient subsidence to ensure preservation of shoreface sandstone units. The many examples cited in the literature are universally within this depositional framework.

One of the first documented subsurface examples of a barrier island system being the site of a major accumulation of hydrocarbons is the Bisti field of the San Juan

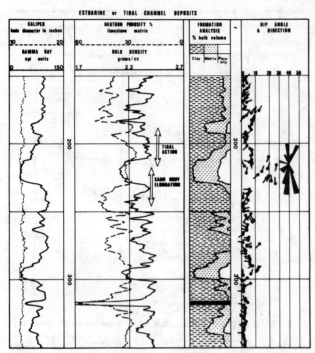

**Fig. 5–10.** Dipmeter and log data from a tidal inlet. From Goetz, Prins and Logar, 1977.

basin of New Mexico (Sabins 1961) (Fig. 5–12). This example also relates to deltaic strandplain sedimentation. The presence of a depocenter is supported by the thickness of the sandstone interval, and the continuing supply of clastic in a longshore drift system is indicated by the straight shoreface pattern. The presence of a reworked transgressive interval at the top of the shoreface is documented by the presence of an authigenic marine shelf mineralogy.

An exploration model can be developed from the stratigraphic patterns and depositional history of a barrier bar system. The Saber bar of Cretaceous age in the Denver-Julesburg basin is an important example of the application of subsurface data to the prediction of the bar geometry and the distribution of hydrocarbons (Griffith 1966).

Most depositional surfaces have been warped by basement and synsedimentary tectonic movement. Consequently, any lenticular or laterally discontinuous depositional pattern has a potential for stratigraphic entrapment of hydrocarbons. Mapping depositional patterns on such a structural base is useful for defining the position of traps (Fig. 5–13). Drilling of structural noses is a useful exploration strategy (Fig. 5–14). The use of strati-

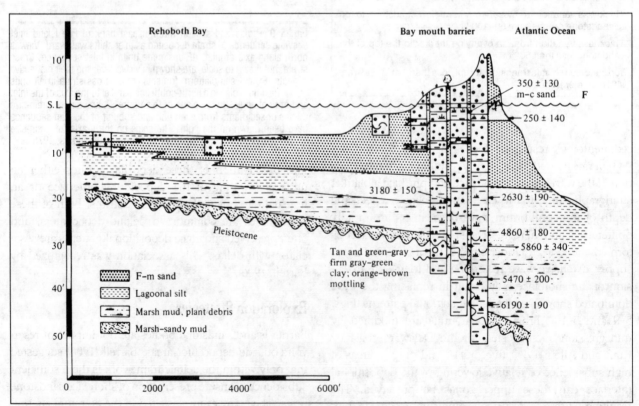

**Fig. 5–11.** An interpretive cross section of a bay-mouth barrier south of Dewey Beach showing the vertical sequence of Holocene sediments and associated radiocarbon dates. The date of 2360 years BP at 18 feet below sea level establishes the rapid transgression of Atlantic coastal wash over barriers across the low-lying adjacent shoreline. The somewhat lenticular cross section of the bay-mouth barrier should be considered to be undergoing erosion in the emerged and submerged beachface and then washed over landward and upward in space and time. Part of this stratigraphic record may be preserved in the nearshore area under shallow marine sands and gravels. The marsh surface overridden by the barrier sands may crop out at low tide, undergo erosion, and be redeposited as reworked material. From Kraft and Allen, 1971.

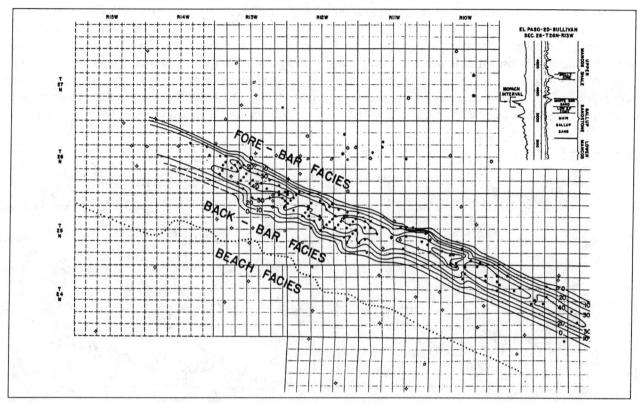

**Fig. 5–12.** Bisti oil field in a barrier island sandstone from the Gallup sandstone in Utah. From Sabins, 1961, reprinted by permission.

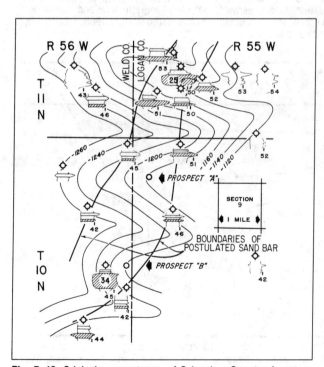

**Fig. 5–13.** Original prospect map of Saber bar. Structural contour datum in top of D sandstone. Contours in feet. Note locations of Prospects A and B. From Griffith, 1966, reprinted by permission.

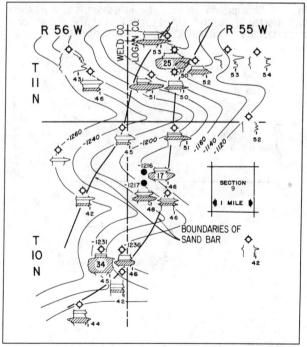

**Fig. 5–14.** Original prospect map of Saber bar, uncorrected, after drilling two wells on Prospect A and one well on Prospect B. Substantial differences between original structural interpretation and actual D sandstone tops in wells drilled are caused by variations in bar thickness, a factor not taken into account in original map. Contours in feet. Locations of Prospects A and B are shown in Fig. 5–13. From Griffith, 1966, reprinted by permission.

graphic patterns can enhance the chances for exploration success. If structure is based on a datum below the sandstone unit, it can reflect a structural-topographic pattern that may influence deposition of shoreface and shelf sedimentary units (Fig. 5–15). The model of a barrier island sequence is well understood, and an isopach map of sand distribution can then be related to the topographic-

structural base (Fig. 5–16). The combination of these two maps leads to a combination of paleotopography and structural warping and compactional drape. A structure map on the top of the sandstone results in a facies and structural pattern of the reservoir (Fig. 5–17). This combination is an exploration map and can significantly reduce the number of wells necessary to define the hydrocarbon

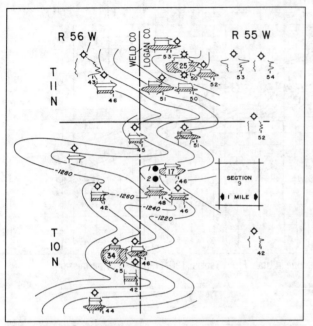

**Fig. 5–15.** Structural contour map. Structural datum is base of Saber bar (corrected to new control) and correlatives at base of bar. Contours in feet. From Griffith, 1966, reprinted by permission.

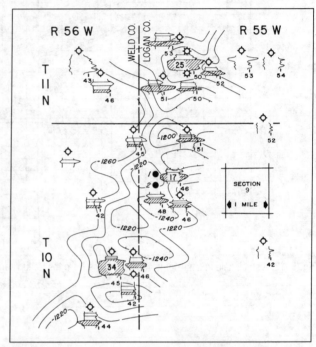

**Fig. 5–17.** Composite structural map using top of Saber bar as datum. Map is derived from combining structural and isopachous contour values. Contours in feet. From Griffith, 1966, reprinted by permission.

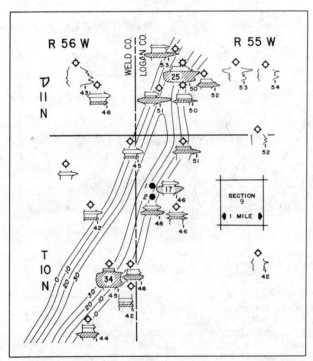

**Fig. 5–16.** Isopachous map of Saber bar. Contours in feet. From Griffith, 1966, reprinted by permission.

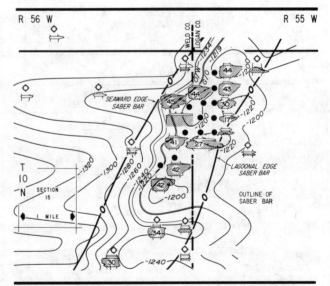

**Fig. 5–18.** Composite structural map, using top of Saber bar as datum. Map completed to September 1965 after thirteen wells had been drilled. Note similarity with Fig. 5–16. Contours in feet. Heavy zero contours are eastern and western edges of bar (see Fig. 5–15). From Griffith, 1966, reprinted by permission.

accumulation (Fig. 5–18). Exploration thus utilizes all of the available data to enhance probabilities for finding commercial hydrocarbons in each well.

## Summary of Wave Process Model

The most commonly observed sedimentologic criteria are the coarsening upward textural pattern, destruction of bedding and sedimentary structures by bioturbation, and the absence of current structures. Wave processes are indicated by textures and sedimentary structures, and the marine environment is shown by trace and body fossils, mineralogy, and bedding patterns. Caution should be exercised in attributing every marine, wave-deposited sand to beach or barrier island paleogeographic patterns. Additional information from many stratigraphic sections is required to suggest such an interpretation.

## TIDE DOMINATED THEMES

Similar to surface waves, tides are described in terms of period, amplitude or height, wavelength, and steepness. These measurable aspects are modified by the coastal boundary conditions. The simple rise and fall of the sea level in response to moon-sun tidal relationships are modified into both internal and surface progressive waves; changes in steepness are a response to friction; and increased wave height is due to bottom and shoreline configuration. These modifications are highly variable along a single coastline, and the sedimentologic responses produced by tidal action are complex.

The principal effect of tides along coastlines is the alternate raising and lowering of the head or relative elevation of the water surface on land and in the ocean. This change in elevation produces alternate shoreward and seaward flows of water. Channels, inlets, or simply the coastal margin are characterized by this pattern, and the resulting currents modify the coastal profile and transport sediments. The pattern of currents reflects the rate of sea-level rise and fall, the rate and volume of drainage from the continent, the density of the waters, and the Coriolis force. Sediment transport is controlled by current patterns. Consequently, the nature of the bedforms, the pattern of geomorphic elements, and the depositional profiles reflect the complex of tidal forces. Study of Holocene coastlines that are dominated by tidal forces

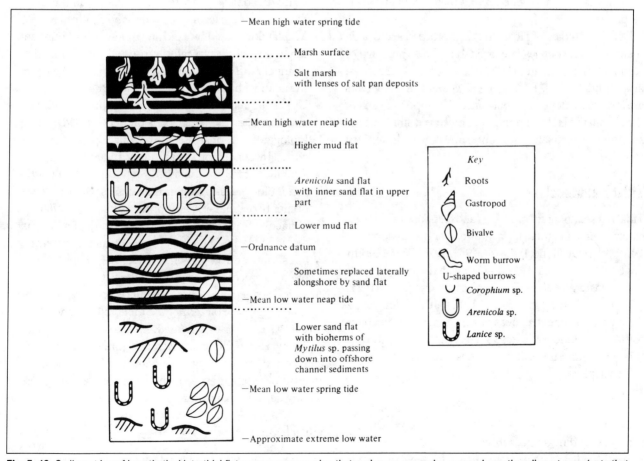

**Fig. 5–19.** Sediment log of hypothetical intertidal flat sequence, assuming that each zone progrades seaward over the adjacent zone (note that lower mudlfat is not always present. This sequence is cut by the deposits of the creeks, which may in places entirely replace it. From Evans, 1975.

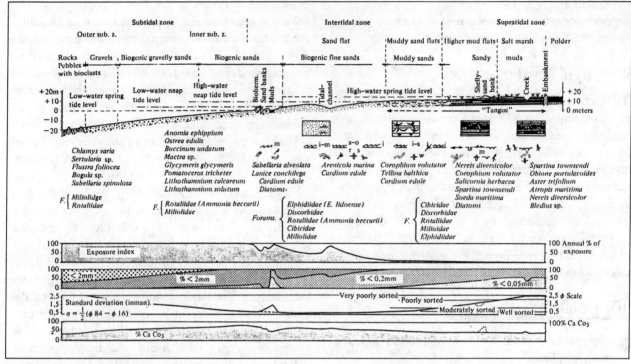

**Fig. 5–20.** Typical sedimentary sequence. After Larsonneur, 1975.

provides the basis for interpreting these sedimentologic responses.

Detailed studies of Holocene tidal processes and sedimentologic responses are available. The three major response elements are: (1) tidal flats, (2) tidal estuaries, and (3) tidal inlets. Each of these geomorphic elements are formed or altered by tidal effects. Examples of each of these from the Holocene and from preserved stratigraphic sections are useful for interpreting sedimentary responses.

## Tidal Flat Model

Tidal flat sedimentary processes and responses have been carefully documented for a number of areas (Terwindt 1988). Work by Reineck (1967a, 1972, 1975) on the North Sea tidal flats, Evans (1965, 1975) (Fig. 5–19) on The Wash on the eastern coast of England, and Larsonneur (1975) (Fig. 5–20) for the Mont Saint-Michel area of France are three well-documented areas of siliciclastic tidal flat deposition. Sedimentary responses can be arbitrarily subdivided into five differing zones that reflect process-related aspects:

▸ Shallow subtidal zone dominated by well-sorted sands, wave and megaripples (planar and tangential cross beds), burrows formed by suspension feeders, and depending upon energy, body fossils and/or algae.

▸ Low tidal flat, ranging from rarely exposed to approximately 50% exposure dominated by both wave and current ripples, abundant tracks, trails and vertical

burrows, some admixture of clay and silt in the sand, and local mud flasers separating sand ripples.

▸ Midflat dominated by sand and mud flasers, abundant fauna and flora, typically transected by meandering channels with fining upward cycles and coarse lag at the channel base.

▸ Upper flat represented by deposition from slack tide and characterized by suspension deposits. Reflecting biologic (flora) processes or chemical processes suggested by deposition of secondary mineral phases.

▸ Supraflat, rarely submerged, reflecting dominantly wind, biologic, and chemical processes with the development of marsh, evaporite pans, or deflation surfaces. Sediment alteration after deposition is common, and new mineral phases, sedimentary structures, and burrow patterns reflect the chemical and biologic processes operating on the flat.

A vertical sequence of these five units is produced by progradation, and the pattern of the units reflects shoreface geometry, rates of sedimentation, nature of the clastic sediment source, climate, tidal range, time-velocity asymmetry, and offshore current patterns. Thus, detailed analysis of sedimentary sequences may be useful in interpreting depositional history.

## Siliciclastic Tidal Process Model

Criteria for identifying tidal flat sedimentation have been developed for both Holocene and ancient examples (Ginsburg 1975). The association of fining upward

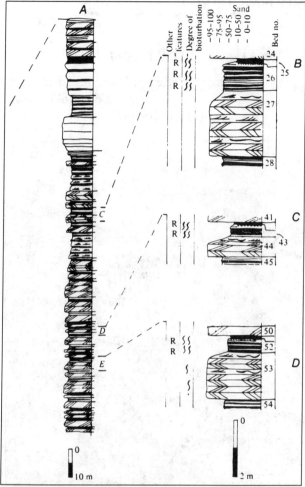

**Fig. 5–21.** *A,* Sedimentary log, Salt Spring Hills, California. *B, C,* and *D,* Detailed logs for selected intervals showing complete paleotidal range sequence. *R* = runzel marks. From Klein, 1975a.

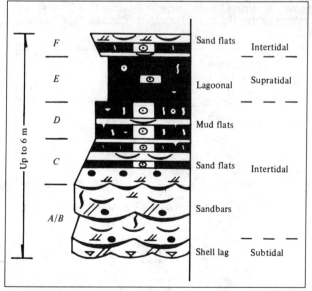

**Fig. 5–22.** Idealized tidal cycle of Uan Caza formation. From Rizzini, 1975.

### Table 5-3
### Tide-Dominated Shoreface Sandstone Reservoirs

Individual fining upward sandstone reservoirs from 3–12 m thick, 2 to 10 kms wide, with width to length ratio of 3:1

Area, of a single reservoir, with inclined bedding surfaces dipping less than 20 degrees, less than 10 kms²

Flow units less than 3 m thick, 50 m wide, and 200 m long

Between event markers, interval containing multiple stacked sandstone reservoirs, lensoid with thickness variations greater than 50%

Area of interval of stacked sandstone reservoirs less than 2000 kms²

channel sequences with preserved evidences for exposure and of bioturbation by an indigenous invertebrate fauna is nearly diagnostic of a tidal-dominated shoreface. A dynamic model, however, must suggest the areal and vertical arrangement of local tidal flat sequences. Several stratigraphic examples illustrate repetitive tidal flat depositional cycles, and these suggest that a dynamic regressive-transgressive pattern is a common response to changes in provenance or in climatic, subsidence, and topographic boundary conditions. Particularly good examples are the 160-meter Late Precambrian Wood Canyon formation of eastern California and western Nevada (Fig. 5–21) and the lower Devonian van Caza formation of southern Tunisia (Fig. 5–22).

A recent literature review by Terwindt (1988), citing carefully documented sequences of possible tidal origin, suggests that extreme care is required before such an interpretation can be postulated. Terwindt (1988, 235) reviews criteria for subtidal, intertidal, and supratidal deposits based upon both Holocene and ancient analogues. Utilizing these criteria, it is evident that one or more complete sequences and a paleogeographic framework is required for an unambiguous identification (Table 5–3). Vilas (1988) suggests a vertical sequence for a sedimentary sequence associated with tidal flat channels (Fig. 5–23).

Petroleum accumulations in coastal tidal sequences are preserved in deltaic areas where tidal processes can be demonstrated. In context of this association, tidal reservoirs are developed in the ancestral Mahakam Delta (Allen et al. 1979), in Tertiary intervals from the Niger Delta (Weber 1971), the Orinoco Delta, and Upper Carboniferous Morrowan-Desmoinesian stratigraphic sequences from southern Oklahoma.

A well-documented outcrop and subsurface analysis of Upper Cretaceous depositional sequences from Alberta, Canada, of probable nearshore tidal origin, has been summarized by Rahmani (1988). Sequences are compared to the mesotidal Georgia coastal estuaries (Fig. 5–24). Regional paleogeographic patterns are illustrated and also are comparable as to scale and areal distribution of depositional environments (Figs. 5–25 and 5–26). Preservation of such a complex only a few tens of meters thick requires an overall rising sea level, with a third order eustatic sea-level rise.

Demonstration of intertidal deposition is often only possible from preserved vertical sequences of lithologic units. Tidal depositional processes are an important adjunct to delta, beach-barrier, and low-energy coastal sedimentary models. The separation of processes is useful to characterize response aspects, but stratigraphic examples must be interpreted from the broader sedimentologic framework.

Physical processes are dominant in the formation of most siliciclastic tidal flat deposits. Carbonate tidal flats, however, combine physical with biologic and chemical processes. The important characteristics of these tidal deposits are discussed within the framework of biologic processes and compared to the physical processes.

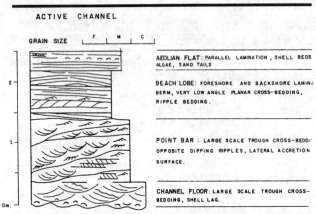

**Fig. 5–23.** Theoretical sequence in the active channel of the tidal inlet. From Vilas et al., 1988.

## BIOGENIC CARBONATE THEMES

### Holocene Depositional Processes

Biogenic productivity in areas of low rate of detrital siliciclastic supply are particularly important along many coastlines in subtropical and tropical climatic regimes. In

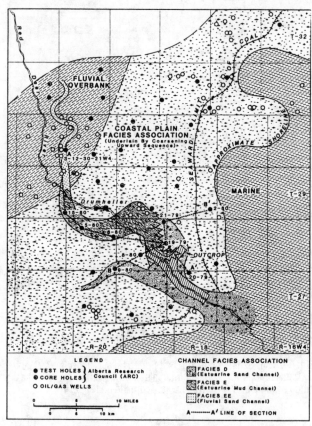

**Fig. 5–25.** Facies map of the estuarine channel and associated environments. The following tripartite up-the-estuary facies transition is clearly evident: (1) Lower Estuary (Facies D: Estuarine Sand Channel); (2) Middle Estuary (Facies E: Estuarine Mud Channel); and (3) Upper Estuary (Facies EE: Fluvial Sand Channel). From Rahmani, 1988.

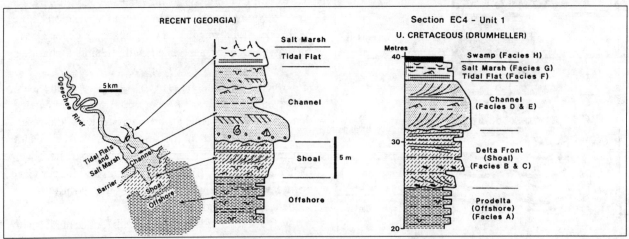

**Fig. 5–24.** Comparison between a prograding estuary-shelf sequence of Ossabaw Sound, Georgia and the delta-estuarine sequence of the Bearpaw-Horseshoe Canyon Formations transition, Drumheller. From Rahmani, 1988.

these cases organic productivity is reflected in the growth of algae. Blue-green and green algae mantle the littoral zone and produce laminations, stromatolitic structures, coated grains, and algal biscuits and pinnacles. In the associated shallow-water areas, calcareous algae, *Halimeda, Penicillus,* and *Rhiphocephalus,* are abundant and produce detrital grains of carbonate in the sand to submicron-size range. Turtle grass or *Thalassia* and other grasses are abundant in shallow offshore areas. In addition to the flora, a heterogeneous population of invertebrates produces detrital clastic carbonate particles and modifies the texture of carbonate muds by pelletization. The sedimentology of these shorelines reflects both a complex of boundary, provenance, and energy conditions, and of physical, chemical, and biochemical processes. The possible sedimentologic responses are many, but conditions required for equilibrium reduce the possibilities, and characteristic patterns reflect primarily climatic and energy aspects. Three specific examples include the range of conditions: (1) the Persian-Arabian Gulf, (2) Shark Bay of western Australia, and (3) the Bahama Island chain off the southeast coast of Florida. Shark Bay of western Australia represents a subtropical (average minimum temperature of 15 °C and average maximum temperature of 29 °C), low rainfall (23 centimeters),

and intermediate tidal range (from coastline (Hagan and Logan 1975). These conditions are reflected in a progradational sequence of units as illustrated for the outer portion of the bay area (Fig. 5–27). Biologic activity is represented by the littoral *Fragum* community, smooth algal mats in the lower intertidal zone, pustular and tufted to gelatinous algal mats in the middle and upper intertidal zones, and blister and film (dry) mats in the supratidal zone (Hagan and Logan 1975, 217).

Contrasting climatic and energy conditions of the Abu Dhabi area of the Persian-Arabian Gulf are reflected in a differing areal sedimentary pattern. Tidal range is from 0.75 to 2.1 meters, average maximum temperature is 41 °C with a minimum of 13 °C, and annual rainfall is less than 4 centimeters. The biotic zonation is subtidal coraline and skeletal sands stabilized by grass, lower intertidal pelletized algae muds, invertebrate burrows, mangrove roots and organic clutter, upper intertidal algal mats, and lower supratidal cerithid shell coquinas. The sequence is capped by deflation surfaces and by local accumulations of aeolian sands (Schneider 1975, 214). Post-depositional chemical alteration modifies the structures, textures, and mineralogy of these deposits resulting in an association of calcite, dolomite, gypsum, anhydrite, and other mineral phases.

Higher rainfall (yearly average of 114 centimeters) and lower tidal range (from 17 to 41 centimeters) result in the development of differing geomorphic features and sedimentary patterns in the Bahama Islands (Ginsburg and Hardie 1975). The areal pattern is of a highly irregular coast cut by tidal channels, and the vertical sequence is developed on the sides of the channels (Figs. 5–28 and 5–29). Biogenic carbonate detritus is the product of calcareous algae and the disintegration of invertebrate shells. This detritus is trapped by blue-green algae, *Thalassia,* and mangrove roots and is pelletized by the indigenous invertebrate fauna. Progradation of the

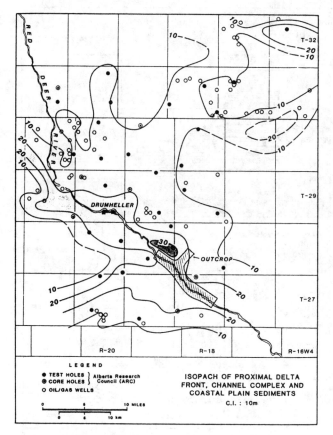

**Fig. 5–26.** Isopach map of the combined thickness of the proximal delta front. From Rahmani, 1988.

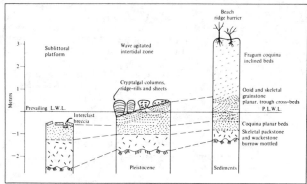

**Fig. 5–27.** Cross section (C–D) showing the upper part of the Holocene-Recent sequence on the outer margin of Hutchison embayment. From Hagan and Logan, 1975.

channel shoreface results in a stratigraphic sequence similar to that deposited on channel banks with a lower subtidal zone of strongly bioturbated pelletized muds, a sequence of algae mats becoming more finely laminated upward with increasing intraclasts and mud cracks, and capped by a beach ridge composed of coquinas, or possibly oolites.

The three biogenic carbonate depositional patterns that include most of the features recognized in ancient stratigraphic examples were summarized by Lucia (1972) (Table 5-4). Most of the response characteristics similar to Abu Dhabi are identifiable in these units from the Upper Clearfork Formation of the Flanagan Field, Texas (Lucia 1972, 169).

Most Holocene coastlines dominated by biochemical formation of carbonate have a low offshore gradient, and shelf areas are restricted by reefs or shoals with water depths rarely exceeding 10 meters. This topographic and environmental pattern is controlled by the production of carbonate by algae, which requires light for photosynthesis and, consequently, water depths less than 15 meters. Holocene examples of three differing carbonate tidal flat models that are formed in response to variations in climate and tidal range are presented. In each of these examples, a single depositional surface is described. The migration of Holocene shoreface surfaces results in depositional units with a high preservation potential, but these units do not reflect the dynamic regressive-transgressive pattern preserved in ancient stratigraphic sections. Thick carbonate sequences with preserved characteristics indicating supratidal, intertidal, and shallow low subtidal

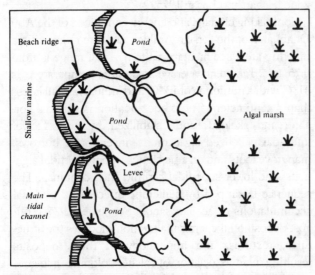

**Fig. 5-28.** Physiographic-hydrographic subdivisions. From Ginsburg and Hardie, 1975.

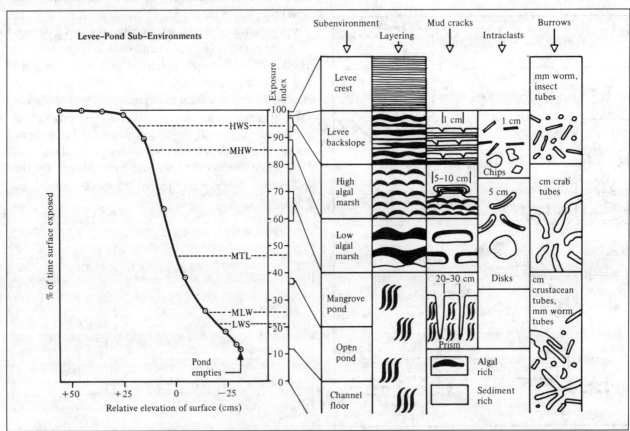

**Fig. 5-29.** At the left, the curve shows the exposure index of levels within the channeled belt; at the right, the columns show zonal distribution of the major sedimentary features. From Ginsburg and Hardie, 1975.

depositional responses are common in the stratigraphic record.

## Biochemical Tidal Response Model

The biochemical formation of carbonate and the deposition of detrital carbonate grains at coastlines is an important sedimentologic response model. Tens of thousands of kilometers of Holocene shoreline are dominated by this process. The energy available may be wave, tidal, or wind, and deposition may reflect any or all of these processes.

A particularly good example illustrating the biogenic communities from the Devonian Manlius Formation of New York and the relation to the tidal facies similar to the Bahamas has been developed (Fig. 5–30).

The thick (1,000 to 1,500 meter) late Triassic Dachstein Limestone section in northern Switzerland and western Austria reflects a tidal process model (Fig. 5–31). The dynamic equilibrium among sediment production, shoreline regression, and rising sea level is suggested as the framework for producing repeated coastal tidal response cycles. Rising sea level is consistent with the pre-Rhaetian sequence boundary.

The origin of stacked sequences of tidal units has been the focus of considerable study by modern sedimentologists. The classical model suggested by Ginsburg (1971) is an autocyclic pattern produced by progradation of a very flat depositional ramp across the shallow carbonate generating shelf. Water depths over areas of thousands of square kilometers rarely exceed 15 meters, and are the site of intense production of carbonate muds and sands by algal and invertebrate faunas and floras. Progradation may be rapid, and can fill all shallow shelf areas with a subtidal to supratidal depostional sequence. Only intrashelf basinal areas would not fill with detrital bioclastic carbonates (Table 5–5). Progradation and fill would reduce the carbonate productivity of the shelf. Continued relative sea-level rise would produce a rhythmic stacking of shelf sequences marginal to areas of pelagic carbonate deposition associated with intrashelf basins.

The alternate concept is that in some fashion allocyclic sequences may be developed. Fischer (1975), suggests eustatic "glacial" cycles for the "lofer" cycles developed in the Triassic in Austria (Fig. 5–31). Milankovitch cycles may produce periodic climatic changes, resulting in either changes in sea level or organic productivity, and with continuing relative subsidence, the stacking of coastal sequences can result. Such a pattern has been described for the Middle Triassic of the Italian Dolomites (Goldhammer 1988). He describes fourth, fifth, and sixth order cycles, with the latter approximating 21,000 years.

Read (1973, 1975) described stacked tidal sequences for the Devonian Pillara formation of the Canning basin of western Australia. Approximately 7 cycles from 2 to 10 meters thick are indicated, reflecting response characteristics from supratidal exposure to subtidal (Fig. 5–32).

**Table 5-4**
**Sequence of Sedimentary Features in the Upper Clearfork, Flanagan Field, Texas, From Lucia, 1972.**

| Interpreted Sedimentary Environment | Sedimentary Structure | Fossils | Particle Size |
|---|---|---|---|
| Supratidal | Irregular laminations<br>Lithoclasts<br>Desiccation features<br>Quartz silt beds | *Rare*<br>Thin-shelled small forams, ostracods, molluscans | Lithoclasts to lime mud |
| Intertidal | Distinct burrows<br>Churned to wispy-mottled structures<br>Quartz silt beds<br>Algal stromatolites<br>Discontinuous fractures | *Very few*<br>Thin-shelled small forams, ostracods, molluscans<br>Filamentous algae | Fine sand-size pellets to lime mud |
| | Current-laminated rocks<br>Crossbedding | *Very few*<br>Echinoids<br>Small molluscans | Fine sand-size pellets to mud with some lithoclasts |
| Marine | Churned rocks<br>Burrowed rocks | *Locally abundant*<br>Echinoids, bryozoans, large fusulinids, molluscans, algal-formas | Coarse sand-size pellets to lime mud |

**Table 5-5**
**Tide-Dominated Carbonate Shoreface Reservoirs**

Fining upward sequences 2–10 m thick, distributed over areas greater than 2000 kms², interval thickness from 30 to more than 500 m thick

Biohermal and/or algal mounds, .3-3 m high, irregular in distribution, elongated, greater than 5:1 parallel to shoreline over areas of less than 10 kms²

Bioclastic and/or biostromal reservoir units, 3-10 m thick, and distributed over an area greater than 5000 kms²

Flow units at top of individual sequences, from 1-5 m thick and distributed over areas of more than 1000 kms²

Nodular and laminated anhydrite units are lens shaped, 10 to more than 100 m thick, elongated 3:1 parallel to shoreline, with areas of greater than 1000 kms²

Other rhythmic cycles have been observed for Cambro-Ordovician sequences from eastern and southern shelves proximal to the North American craton.

Associated with arid coastlines, carbonate tidal sequences often are associated with subaerial evaporite lagoons, diagenetic gypsum and/or anhydrite fill of desiccation cracks, or the formation of nodular gypsum or anhydrite units. These units may thicken, and in some instances (for example, the Hith evaporite of Upper Jurassic age in Saudia Arabia, the Buckner formation of the United States Gulf Coast, and the San Andres of the west Texas Permian basin), become substantial stratigraphic units up to a few hundreds of meters thick, occurring over areas greater than 1000 square kilometers (Table 5-5). These intevals are important in providing both lateral and vertical seals to hydrocarbon migration.

Major petroleum accumulations are associated with coastal carbonate sequences. They span stratigraphic intervals from Cambro-Ordovician to Tertiary, and are the focus of continued petroleum exploration. The Smackover formation of the Jay Field in Florida is a typical example of a coastal, a probable microtidal system with two or more stacked sequences of leached pelletal dolomite up to 30 meters thick (Fig. 5-32) (Ottmann et al.

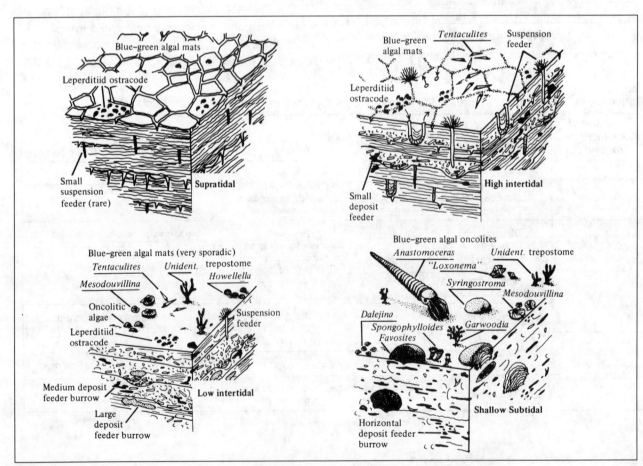

**Fig. 5–30.** Inferred reconstructions of Manlius facies. From Walker and LaPorte, 1970.

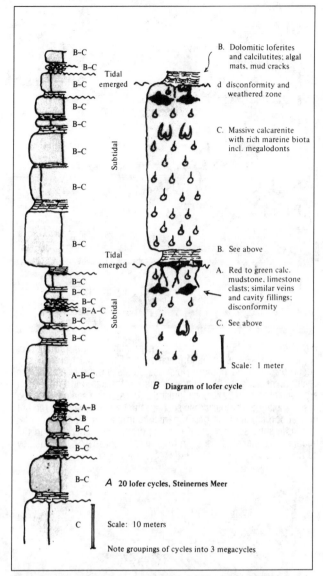

B. Dolomitic loferites and calcilutites; algal mats, mud cracks

d disconformity and weathered zone

C. Massive calcarenite with rich mareine biota incl. megalodonts

B. See above

A. Red to green calc. mudstone, limestone clasts; similar veins and cavity fillings; disconformity

C. See above

Scale: 1 meter

B Diagram of lofer cycle

A 20 lofer cycles, Steinernes Meer

Scale: 10 meters

Note groupings of cycles into 3 megacycles

**Fig. 5–31.** The Lofer cycle: *A*, segments of 20 cycles and *B*, the cycle as typically developed. From Fischer, 1975.

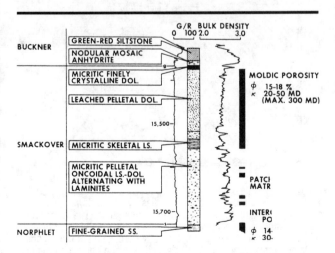

**Fig. 5–32.** Log of sedimentary sequence in Exxon-LL&E No. 1 Jones McDavid well, Jay field. From Ottman et al., 1976, reprinted by permission.

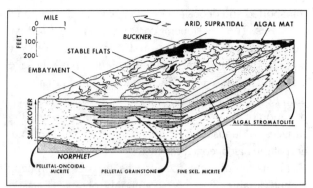

**Fig. 5–33a.** Smackover Facies, Jay field area. From Ottman et al., 1976, reprinted by permission.

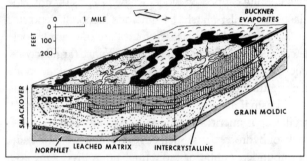

**Fig. 5–33b.** Smackover porosity, Jay field area. Present porosity distribution is indicated by vertical hachers. From Ottman, et al., 1976, reprinted by permission.

1976). The reservoir interval is underlain by pelleted micrite, and overlain by algal stromatolite flats and nodular anhydrite (Figs. 5–33a and 5–33b). Modern analogies are microtidal flats of the Bahamian platform, but the evaporite association is similar to flats from Abu Dhabi.

# TRANSGRESSIVE BIOLOGIC SHOREFACE THEMES

Without the continuing addition of clastic or biochemical detritus to the shoreface, the stratigraphic response pattern is of necessity one of transgression. Environments associated with such coastal intervals are dominated by marsh, bays or lagoons with associated wave, tidal, and subaerial processes. Four principal responses are possible, high and low energy clastic depositional sequences, and high and low energy bioclastic and/or evaporite sequences. All of these are commonly developed along Holocene coastlines. The process/response characteristics include: (1) low energy clastic processes, dominated by marsh and swamp deposits; (2) high energy wave and tidal reworking of regoliths or relict clastic detritus; (3) low energy bioclastic lagoonal and marsh deposition, with or without evaporite mineral associations; and (4) high energy redeposited bioclastic detritus. These units are overlain by nearshore sediments transported and deposited by shallow water depositional processes. The low slope of coastal plains, the relatively slow rates of sedimentation, and if there is a relatively rapid landward movement of the shoreline in response to a eustatic rise in sea level, the result is a thin stratigraphic unit. Recognition of these units is important to the interpretation of depositional history, definition of bounding surfaces to depositional sequences, and the reconstruction of the paleogeographic history of basins.

The distribution of these surfaces in space and time is highly variable. The most commonly preserved clastic patterns are those associated with coastal marsh deposits, overlain by lagoonal and thin bioturbated clastic units capped by biogenic shelf deposits. For bioclastic intervals, marsh, fossiliferous lagoonal deposits, or laminated and/or algal laminates are overlain by pelleted micrite and capped by fine grained argillaceous micrites with biogenic phosphorite and glauconite. During the deposition of these coastal sequences, intrashelf basinal areas may be the site for the deposition of argillaceous micrites and marls, providing a major source rock for hydrocarbons. Shelf edge depositional units may be mounded oolitic or biohermal units producing the restriction for shelf sedimentary intervals.

The two major categories of clastic and bioclastic transgressive coastlines can be conveniently separated and Holocene and ancient sedimentary patterns compared.

## Clastic Transgressive Model

A well-documented Holocene depositional history has been developed for the high energy Delaware coastline by Kraft and coworkers (Kraft and John 1979; Kraft et al. 1987). Core holes together with age dates provide the necessary time-stratigraphic framework for the depositional history (Fig. 5–34). The depositional response to a eustatic rise in sea level produces a sequence that may be usefully compared to ancient stratigraphic units (Figs. 5–35 and 5–36).

A similar, but lower energy, sequence is illustrated by the coastal plain adjacent to the Mississippi River delta (Penland, Boyd, and Suter 1988) (Figs. 5–37 and 5–38). The abandoned delta plain is dominated by marsh, lagoonal, and bay deposits. Shoals, however, are the product of transgressive reworking of deltaic and associated shelf faunal and clastic detritus. A lower rate of sediment supply results in destruction of the barrier ridges, and the result is a sequence dominated by coastal lagoon and marsh sedimentation, overlain by foreshore Gulf sands, silts and clays.

## Ancient Examples

Ancient examples of these high and low energy clastic transgressive shorelines are abundant in the stratigraphic

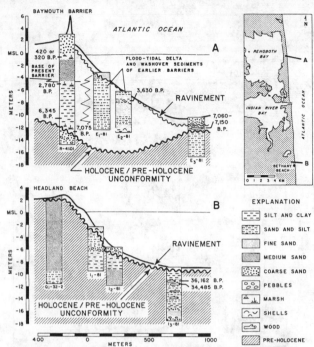

**Fig. 5–34.** Two shoreface cross sections showing interpretation of cores drilled across the shoreface zone. The basal transgressive unconformity (holocene/pre-Holocene) is shown and related to the revanement surface. The two surfaces are the same in the case of erosion against a preexisting highland of Pleistocene sediment. In the case of a lagoon-barrier cross section, the two surfaces diverge but may at some points merge. Of particular importance is the fact that very little sand-size sediment is being eroded or transported in the middle and upper shoreface. From Kraft and John, 1979, reprinted by permission.

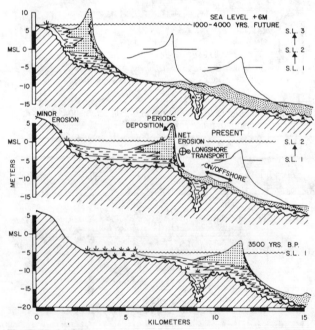

**Fig. 5–35.** A profile projection of morphology and sedimentary environmental lithosomes, based on studies on the Delaware coastal zone. The middle profile is the present configuration. The lower profile is adjusted to a local relative sea level approximately 5 m below present about 3.5 ka. The upper profile is a projection of the coastal profile and stratigraphic sequence anticipated at a peak sea level of +6 m from 1,000-4,000 yrs in the future. The upper profile is very similar to many laterally adjacent Pleistocene stratigraphic sequences in the Delaware coastal plain. From Kraft and John, 1979, reprinted by permission.

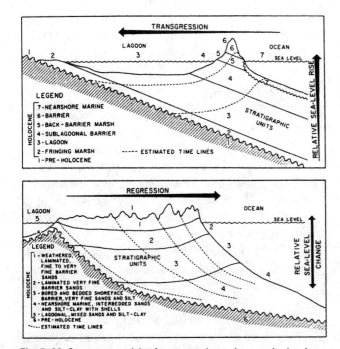

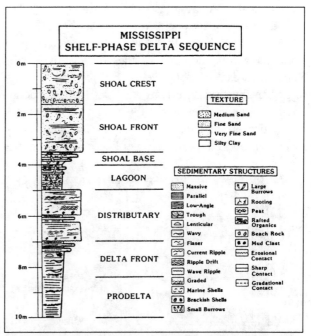

**Fig. 5–36.** Sequence models of transgressive and regressive barriers showing variants of depositional environmental sequences. Lithostratigraphic units cross time lines based on radiocarbon dates. From Kraft and John, 1979, reprinted by permission.

**Fig. 5–38.** A generalized stratigraphic model for an abandoned shelf-phase Mississippi River delta complex illustrates the significance of the transgressive component. In this new stratigraphic sequence, shelf-phase delta complexes, which differ considerably from the traditional deep-water Mississippi River delta complex model, are seen as the primary depositional constituents of the Holocene Mississippi River delta plain. From Penland, Boyd, and Suter, 1988.

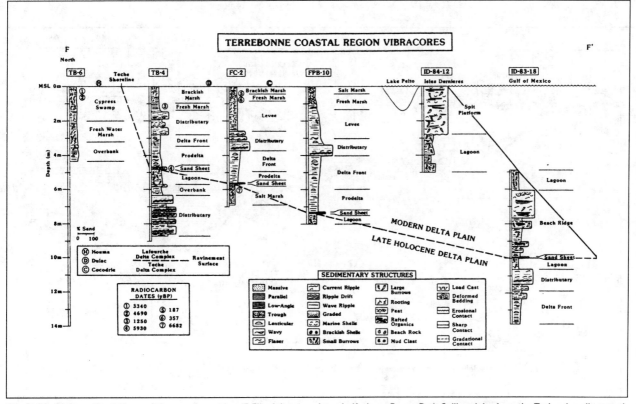

**Fig. 5–37.** Diagram illustrates a vibracore dip section (F-F') of the complete shelf-phase Bayou Petit Caillou delta from the Teche shoreline south to the Isles Dernieres barrier island arc. This shallow-water delta lies on a revanement surface 7-8 m in the subsurface near the Isles Dernieres that merges up-dip to a relict transgressive barrier shoreline. Note the total thickness of this deltaic sequence and the significance of the transgressive sequence component that becomes thicker toward the coast. From Penland, Boyd, and Suter, 1988.

record. Most coal deposits associated with Pennsylvanian cyclothems are of this origin (Wanless et al. 1970) (Fig. 5–39) (Cavoroc and Ferm 1968) (Fig. 5–40). Other examples from Cretaceous deposits of the Rock Springs and Kemerer areas of southwestern Wyoming exhibit response characteristics consistent with a low energy coastal sedimentology (Hale 1962). The possible areal and vertical extent of these coastal units is well illustrated by the Latrobe Valley formation of Eocene age of south-

eastern Australia (Gloe 1960). In this instance more than 200 meters of lignite is preserved along a belt more than 200 kilometers in length.

## Bioclastic Transgressive Coastal Model

Both high and low energy sequences are present along Modern coastlines. The basal transgressive unit may consist of a reworked regolith, a coastal marsh, often either fresh water or mangrove swamp, or a lagoonal evaporitic or laminated algal unit. These units may be overlain by a nearshore pelleted molluscan blanket carbonate unit. Coastal units on mid-latitude marginal shelves on east sides of ocean basins are overlain by fine-grained intervals containing biogenic phosphate, glauconite, chert, and micrite. These units are often capped by glauconitic, phosphatic, and argillaceous micrite (Table 5–6).

A good example of the low energy coastal pattern has been described by Scholl and Stuiver (1967) (Figs. 5–41 and 5–42) from the Florida Bay. This is not an evaporitic sequence, but a similar pattern is observed with transgressive lagoonal and lake sediments containing fine grained dolomite on the southeastern coast of Australia (Thom et al. 1986, 81–87). In this instance the coast is high energy and transgressive units are capped by a regressive carbonate sand sequence.

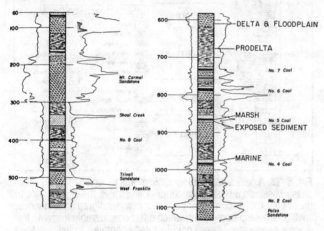

**Fig. 5–39.** Environmental interpretation of Middle and Upper Pennsylvanian rocks as displayed in part of an electric log in White County, Southern Illinois (interpretation by Wanless based on regional environmental maps of units present at this locality). From Wanless et al., 1970.

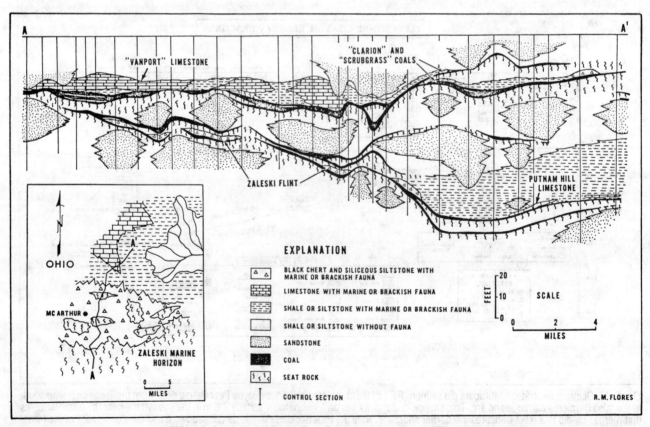

**Fig. 5–40.** Cross section and paleogeographic map of the Zaleski flint and associated strata near McArthur, Ohio. Data used in preparing this cross section are from Flores (1966) and the open file of stratigraphic sections of the Ohio Geological Survey. From Cavoroc and Ferm, 1968.

**Table 5-6**
**Biogenic Transgressive Shoreface and Shelf Interval**

Basal carbonaceous and fine-grained, 50% of grains less than 44 um in diameter, interval, from .1–10 m thick, with an area greater than 10,000 kms²

Overlain by a pelleted molluscan sand, and/or siliciclastic sand unit, 5-20 m thick, over an area greater than 5000 kms²

In some intervals locally developed, algal or biogenic mounds, 1-10 m thick, .2-3 kms in diameter, with reservoir up to 8 m thick

Overlain by a fine-grained argillaceous and biogenic unit, containing one or more of the following: glauconite, phosphate, chert, micrite, and kerogen, with more than 50% of the grains less than .44 um in diameter, unit from 5-20 m thick, with an area greater than 5000 kms²

Interval may be capped by prograding clastic or carbonate units

**Fig. 5–41.** See graph p. 150. Paleogeographic map of Florida Bay area. From Scholl and Stuiver, 1967.

A clearly transgressive evaporitic lagoonal interval has been described by Phleger (1969) (Figs. 7–22a and 7–22b). The ancient analogue for this setting is illustrated by the Jurassic Todilto formation of northwest New Mexico (Figs. 7–24 and 7–25). This occurrence is discussed in the context of evaporitic depositional systems in Chapter 7.

Sequences described represent disconformities in the stratigraphic record, and have been described by Bennison (1972) for Missourian limestone sequences from Oklahoma. These intervals include lagoonal, swamp, thin algal limestones, capped by phosphatic shales. These intervals are related to the Checkerboard, Hogshooter, Dewey, and Paola limestones. Some of these intervals reflect high energy shoal deposition overlain by biogenic black phosphatic and glauconitic shale. The transgressive-regressive shelf depositional systems have been described by Heckel (1986). He suggests a depth related facies framework for the Kansas "cyclothem" (Fig. 5–43). In this example the condensed interval is the offshore micrite, overlain by overlapping outershelf phosphatic shale. The nearshore algal and bioclastic carbonate units are overlain by prograding prodelta clastic shales or coastal sandstone units. Due to shallowing water, the algal carbonate units do not form ovoid or elongate biohermal reservoir units in the near shore zone, and are not often productive of hydrocarbons.

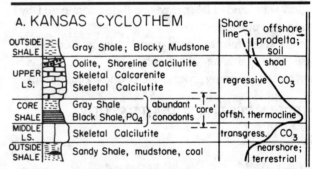

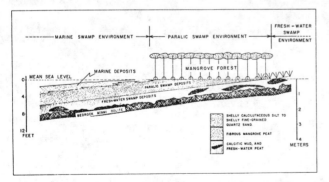

**Fig. 5–42.** Idealized cross section of coastal swamps representing sediments underlying edges of bays and waterways. Stratigraphic sequence is clearly transgressive as freshwater deposits overlying bedrock in turn are overlain by paralic swamp and marine deposits. Cross section shows that mangrove peat begins to form over freshwater calcitic mud essentially at sea level. From Scholl and Stuiver, 1967.

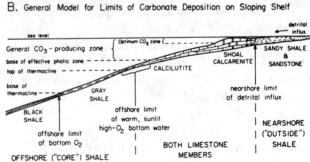

**Fig. 5–43.** A: Basic four-member, transgresive-regressive, major ("Kansas") cyclothem in Midcontinent Pennsylvanian. B: Model for deposition on sloping tropical shelf, showing positions of rock types that became superposed with transgression and regression to produce Kansas cyclothem. From Heckel, 1986.

## SUMMARY

Coastal sedimentology was shown to be influenced by a complex of boundary conditions and processes, but in terms of preserved stratigraphic sequences, equilibrium is developed and maintained as only a few depositional patterns. This insight is the foundation for interpreting ancient shoreface sequences. Models based upon continuing processes reflect the balance of forces operating along coastlines. Their usefulness is based upon the dynamic interrelationship of sedimentologic conditions with developmental history. Local tectonic, provenance, or topographic variations along a shoreline may result in differing depositional response patterns, but these rarely are widely distributed areally or temporally. For a stratigraphic model to be useful, characteristic patterns must be generally applicable to a range of Holocene and ancient sedimentary units.

Four types of coastlines represent equilibrium conditions. Continuing deposition results in the formation of areally or vertically extensive stratigraphic units:

- High-energy wave dominated regressive coastlines,
- Tidal-dominated coastlines with longshore or offshore supply of siliciclastic detritus,
- Coastal areas dominated by biochemical formation of carbonate detritus with low to intermediate tidal energy, and
- Low-wave and tidal-energy coastlines dominated by biologic processes.

Each of these four response models can be used as the basis for organizing stratigraphic and sedimentologic data and determining process and boundary conditions.

## REFERENCES FOR FURTHER STUDY — CHAPTER 5

Davis, R.A., Jr., and R.L. Ethington, eds., 1976, Beach and Nearshore Sedimentation: Soc. Econ. Paleont. and Mineral., Spec. Pub. no. 24, 669 p.

de Boer, P.L., A. van Gelder, and S.D. Nio, eds., 1988, Tide-influenced Sedimentary Environments and Facies: Dordrecht, Holland, D. Reidel Pub. Co., 530 p.

Ginsburg, R.N., ed., 1975, Tidal Deposits A Casebook of Recent Examples and Fossil Counterparts: New York; Springer-Verlag, 428 p.

Nummedal, D., O.H. Pilkey, and J.D. Howard, eds., 1987, Sea-level Fluctuation and Coastal Evolution: Soc. Econ. Paleont. and Mineral., Spec. Pub. no. 41, 267 p.

Reineck, H.E., and I.B. Singh, 1980, Depositional Sedimentary Environments, 2nd ed.: New York, Springer-Verlag, 549 p.

# 6 SHELF SEDIMENTOLOGY

It is precisely the fundamental problem of scientific method to state the principles of scientific methodology that are to

be used to answer these questions of measurement, of goodness of fit, of parameter estimation, or identifiability, and

the like. The principles needed are entirely formal in character in the sense that they have as their subject matter

set—theoretical models and their comparison. **E. Nagel, P. Suppes, and A. Tarski**
*Logic, Methodology and Philosophy of Science,* **1962**

## INTRODUCTION

The geometrical, physical, and historical development of continental shelves has been the focus of recent research. Problems of equilibrium, the significance of the post-glacial rise in sea level and Holocene sedimentologic responses, and climatic changes through the Quaternary have made interpretation of shelf sediments difficult. Consequently, depositional models useful for stratigraphic analysis have only recently been developed.

Shelf sampling by remote sensors and physical devices has been discontinous, so there is a poor understanding of areal and vertical variation of sedimentary attributes. Also, shelf geomorphology was poorly understood. Use of submersibles, remotely operated television, side-looking sonar, and deep-tow instrument packages allows direct observation of topographic, sediment, and faunal patterns. The shelf may represent a dynamic equilibrium in response to long-period waves, tides, and currents. Physical, chemical, and biological processes are reflected in both areal and vertical sedimentologic responses. Dynamic process-response models may be suggested, and properties of the sedimentary models that reflect

boundary conditions and the depositional history can be determined.

## BOUNDARY CONDITIONS

The boundary conditions significant to development of depositional models include:
- Climatic conditions,
- Relation of shelf-ocean basin patterns,
- Topography of the shelf,
- Systematic changes in sea level, and
- Patterns of provenance.

Processes are related to these aspects, and the sediment responses are controlled by the interrelation of processes with these boundary conditions. Preserved shelf stratigraphic units reflect these boundary conditions. Their interpretation provides the basis for determining paleogeography, depositional history, and stratigraphic patterns.

### Climate

Biologic processes are directly controlled by climate and

patterns in the oceanic circulation. In the absence of strong physical process controls, biologic aspects may be dominant. Shelf areas dominated by biologic processes also reflect internal and surface waves, tidal and oceanic currents, and the topography of a shelf area. Most carbonate shelves are adjacent to ocean basins and occur in climatic belts where the minimum water temperature is greater than 15°C. Temperature is the most critical climatic factor, but precipitation, evaporation, surface runoff, and storms are important aspects in carbonate sedimentology. Ecologic studies in the Holocene suggest that a tropical to subtropical climate related to areas of moderate addition of fresh water from rains or surface discharge are optimum for biologic development. If biologic productivity is high, wave, tidal, and storm processes modify the sediment distribution but they are not controlling factors in their development.

## Shelf-Ocean Basin Relationships

Physical aspects of the shelf-ocean basin interrelationship affect changes in patterns of internal waves, tides, amplitude and period of surface waves. In conjunction with topography, they produce changes in patterns of sedimentary units on shelves. The area of the ocean basin, its depth, its geometry in relation to continents and prevailing wind patterns, and the shelf-basin margin geometry have been shown to be important in tide development. Similarly, waves may be affected by these boundary conditions and, consequently, shelf sedimentology (Fig. 6-1).

In biologic systems, productivity is related to nutrient supply, and this is a critical aspect of shelf sedimentology. In the absence of an overwhelming supply of clastic detritus and in favorable climatic belts, the sedimentology reflects biologic productivity. Nutrients are derived from ocean currents, areas of upwelling, and from continental runoff.

Few studies of Holocene depositional patterns on shelves have been reported with regard to basin-shelf boundary controls. Patterns of carbonate sedimentology are related to ocean basin-shelf geometry. Carbonate platforms of the Bahamas, the northwest Australian shelf, and the Persian-Arabian Gulf reflect these controls.

## Shelf Topography

The shelf topography is in part a self-actualizing stratigraphic response. Subsidence in response to sediment and/or water load, biologic productivity, and an equilibrium profile related to physical processes are indicated in the shelf topography. However, the initial shelf geometry and gradient provide the framework for development and

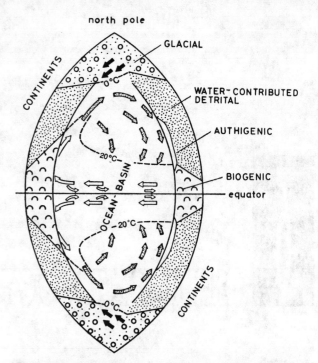

**Fig. 6-1.** Idealized distribution of chief classes of sediment on continental shelves bordering an ocean when sediments are in equilibrium with their environment. Note exaggerated width of shelves. From Emery, 1969, reprinted by permission.

preservation of sedimentary units. Specifically, the development and fill of channels, canyons, estuaries, bays, and lagoons reflect shelf topography. Also, the formation of beaches, barrier islands, shoals, and banks are a demonstration of the initial shelf topographic expression prior to transgression.

The pattern and type of energy distributed over a shelf not only depend upon conditions in the ocean basin, but also upon the shape and geometry of the shelf. An equilibrium between these two aspects is rarely observed on Holocene shelves (Emery 1969) (Fig. 6-1). With a longer period of time, or under conditions of lower rates of change than those developed in the Quaternary, such a condition may be developed (Curray 1965; Swift 1970) (Fig. 6-2).

The width/depth ratio of a continental shelf or an epeiric sea is a particularly important factor in terms of energy distribution for long-period waves. Bottom currents or waves may be developed at the sediment-water interface under favorable geometric conditions and may produce characteristic sediment responses. Also, density stratification and internal waves and currents may be developed to transport suspended sediments; to produce or to maintain submarine channels and canyons; and to distribute sediment across the shelf. These patterns for clastic sediments were studied in the Holocene (Curray 1965) (Fig. 6-2); additional studies may yield an understanding of the interrelation of shelf topography to the sedimentology of shelf units.

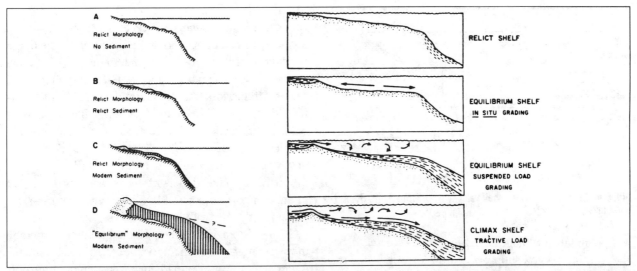

**Fig. 6–2.** *A,* Diagrammatic representation of relict (inherited from previous environmental conditions) vs equilibrium sediment and shelf morphology. *B, C,* and *D,* Hypothetical sequence with relict topography and basal transgressive sediments being covered with sediments in equilibrium with the new environment. Eustatic ability and gradual regional subsidence assumed. From Curray, 1965.

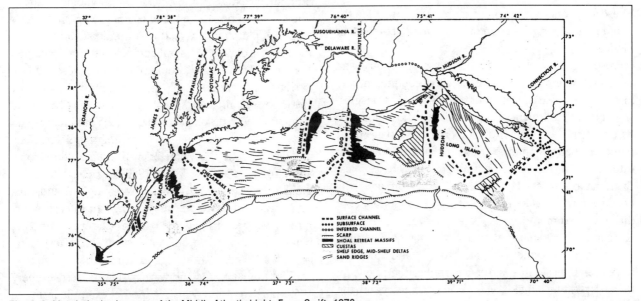

**Fig. 6–3.** Morphologic elements of the Middle Atlantic bight. From Swift, 1976.

## Eustatic Changes

Systematic changes in sea level related to stratigraphic sequences is a particularly important boundary condition on continental shelves. Onlap across the continental margin provides the mechanism to preserve thick continental shelf sedimentary units. Progressive rise in sea level is reflected by blanket stratigraphic units of uniform lithology, thick carbonate platform and reef units, regressive-transgressive deltaic sequences, and broad epeiric sea and shelf units of varying thickness and aspect.

The continental shelves are particularly sensitive to changes in sea level; the migration of shorelines, continuity of sedimentary units, and the pattern of lithologies are responses to these changes (Fig. 6–3). The patterns of regional unconformities are directly a result of changes in sea level. These can be interpreted to show directions of regression and transgression, source areas, and paleogeographic patterns. Depositional history in response to these systematic changes can be interpreted, and if the pattern of sea-level changes is known, the specific stratigraphic response can be placed into the historic framework.

## Provenance Patterns

Shelves may reflect an equilibrium condition with sea level, tectonism, and physical processes, resulting in:

▸ Repeated cycles of regression and transgression of delta-related stratigraphic units,

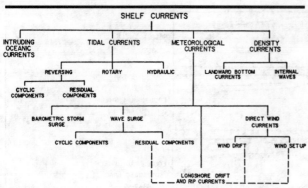

**Fig. 6-4.** Components of the shelf velocity field. From Swift, Stanley, and Curray, 1979.

- ▸ A dominance of biogenic detritus,
- ▸ A continuing slow supply of detritus for long periods of time, or
- ▸ A nonequilibrium condition where sediment primarily of a chemical nature is accumulated.

These four conditions of provenance are exhibited in the four types of process-boundary condition patterns or models.

Provenance is related to continental geometry and drainage patterns, climate both in the continental and marine source areas, history of regression and transgression on the continental plate and the shelf area, the nature of the continental plate margin, and patterns of sea-level change. Thus, provenance is the aspect that integrates most shelf sedimentologic responses. Examination of shelf sedimentary units provides the observational basis for interpreting other boundary conditions. Sedimentary aspects of these units are particularly useful in this type of interpretation. Textural, mineralogic, sedimentary structures, and biologic responses indicate the nature and application of energy to the depositional interface, the water chemistry, and the paleogeography, both on the adjacent continent and the shelf. From the areal and vertical patterns of stratigraphic units, an understanding of equilibrium conditions, tectonic and sea-level history, and the sedimentary controls can be determined. Thus, the depositional history of shelf stratigraphic units can be interpreted.

## Depositional Processes

Study of depositional processes on Holocene shelves has been hindered because of lack of continuous observation of shelf current patterns, including responses to oceanic, geostrophic, and climatic patterns and events. Changes are related to latitude, coastal and shelf geometries, eustatic changes in sea level, and tectonic patterns related to the distribution and history of movement of cratonic plates, and subsidence patterns related to water and sediment loading. To establish equilibrium process models

for understanding autocyclic and allocyclic sedimentary sequences is a challenge. Swift et al. (1979) suggested some of the components of a the shelf velocity field as a product of various types of shelf currents (Fig. 6-4). In a recent personal communication with Swift, he indicated that this framework was far too simple as a causative basis for understanding shelf current patterns. If this indeed is true, data collection will be required over large areas and over long periods of time before predictive models can be developed. As an example of the complexity of the problem, computer models are currently being developed in an attempt to predict "El Nino" events (Kerr 1987b), but currently only with limited success.

Geomorphologists have additional methods whereby they can suggest sedimentary processes from response patterns. For example, mathematical sediment transport models and actualistic comparison to bedform response patterns are useful to predict patterns of sedimentary structures, textures, lithology, distribution of fauna, flora and bioturbation, and mineral coatings on grains. These response patterns may lead to an understanding of the history of sediment transport and depositional processes. For example, bedform geometry has been described for various flow conditions and sediment supply by Belderson et al. (1982) (Fig. 6-5). Also, a theoretical model has been developed utilizing a .25 millimeter grain size, 4.25 meter wave height, wave length of 210 meters, and water depth of 24.5 meters with both unidirectional and bidirectional flows (Allen 1980) (Fig. 6-6). Increased grain size, changes in velocity, water depth, and tidal asymmetry: all can appreciably change the internal structures and bedform geometry. An example from the Nantucket Shoal area off Long Island, New York, suggests that a tidal sand ridge area may be in dynamic equilibrium with tidal and wave current patterns (Emery 1969) (Figs. 6-7 and 6-8). Yet another example from the North Sea by Houbolt (1968) suggests the dynamic nature of sediment transport on shoals and banks due to the presence of prograding, bedding patterns, reflected by internal seismic reflection patterns (Fig. 6-9).

Texture also has recently been studied across sand waves on the Belgium coast (de Maeyer and Wartel 1988) (Figs. 6-10 and 6-11). These data can be usefully compared to textural patterns from sand waves associated with the Altamaha River estuary (Visher and Howard 1974).

The distribution of grain-size sediment modes across shelves suggests sedimentary processes not directly related to equilibrium sand-wave transport. Density stratification, related to salinity and temperature, appears to be an important control for the distribution of finer-grained sediments. Fine sediment eroded from the bottom, or derived from deltaic sources, has been shown

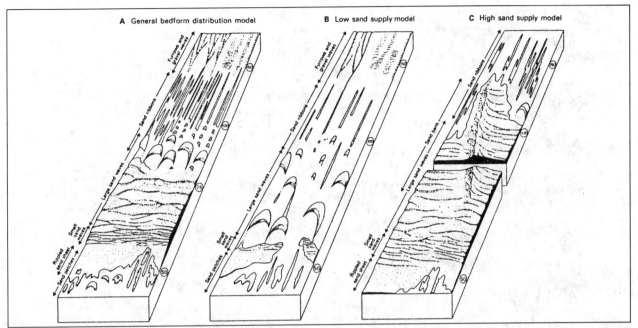

**Fig. 6-5.** *A,* general distribution model of bedform zones along tidal current transport paths, *B,* with variations resulting from low rates of sand supply; and *C,* high rates of sand supply. The bedform zones are aligned parallel with mean spring peak near-surface tidal current velocities (shown in cm/s). From Belderson, Johnson and Kenyon, 1982).

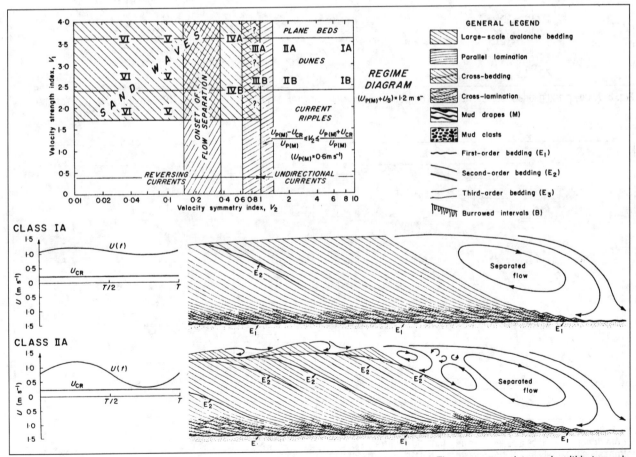

**Fig. 6-6.** Regime diagram and predicted categories of sand-wave and dune internal structure. These structures intergrade within two sub-classes separated by values for *B* when large-scale flow separation becomes inevitable: sub-class 1 (no separation, Classes *V* and *VI*), sub-class 2 (separation; Classes *I-IV*). From Allen, 1980.

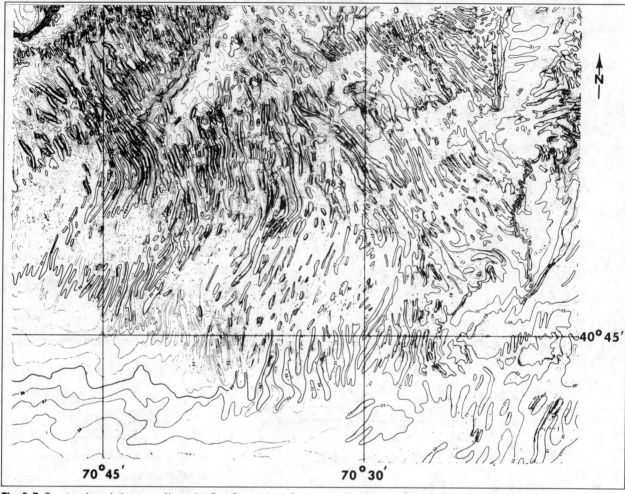

**Fig. 6-7.** Coast and geodetic survey, Nantucket Bay, Connecticut. Contours are in fathoms. From Emery, 1969, reprinted by permission.

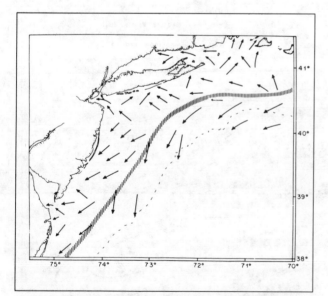

**Fig. 6-8.** Diagrammatic representation of paths followed by plastic bottom drifters carried by a general residual (non-tidal) current along bottom of continental shelf off northwestern U.S. Dashed line is 200-m contour near edge of shelf, and vertically lined stripe represents general zone of divergence of bottom currents as indicated by bottom drifters. From Emery, 1969, reprinted by permission.

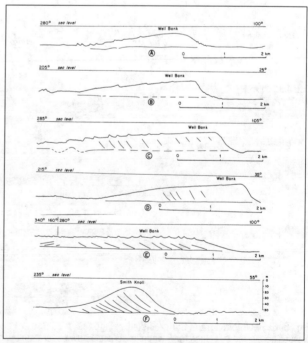

**Fig. 6-9.** Interpretive sketch of continuous seismic profile records, Well Bank and Smith Knoll, North Sea. From Houbolt, 1968.

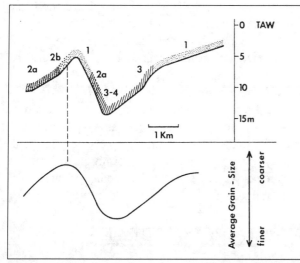

**Fig. 6-10.** Distribution of average grain size and grain size types along a transect perpendicular to the coast and crossing the Nieuwpoort Bank. (*TAW*=Belgian ordnance level). From de Maeyer, 1988.

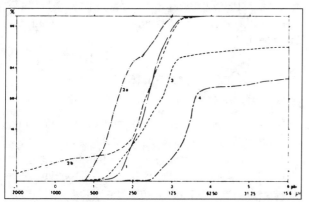

**Fig. 6-11.** Some typical grain size curves on log-probability paper (see text). From de Maeyer, 1988.

to be primarily distributed by advection (net horizontal water movement), not by diffusion (McCave 1972, 242). Advective currents relate to surface winds, tidal circulation, and areas of upwelling and intrusion of denser waters. These forces may concentrate turbid water at any place on the shelf.

The tracing of sediment modes across the United States Gulf Coast shelf has been described by Curray (1960). A sand mode may be a product of winnowing of finergrained sediment by internal waves associated with the mid-shelf, 30–50 fathom (55–91 meters), depth interval (Fig. 6–12). The fine sand mode is much more broadly distributed (Fig. 6–13), and appears both in shallower water to a depth of 20 fathoms (36 meters), and marginal to the coarser sand mode (Fig. 6–12). It may be the product of storm surge and long period waves developed during hurricane events. The silt-sized mode is even more broadly distributed in relation to water depth and potential source areas (Fig. 6–14). This distribution may be

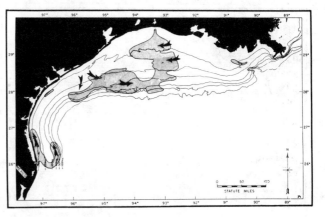

**Fig. 6-12.** Distribution of grain-size modes of 02.9 φ (1.00-0.134 mm) of surface sediments. Most of these modes are between 2.0 and 2.9. From Curray, 1960, reprinted by permission.

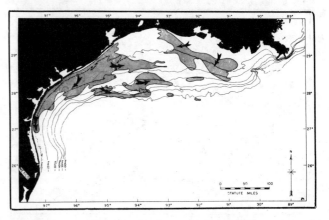

**Fig. 6-13.** Distribution of grain-size modes of 3.0–4.2 φ (0.125–0.054 mm) of surface sediments. From Curray, 1960, reprinted by permission.

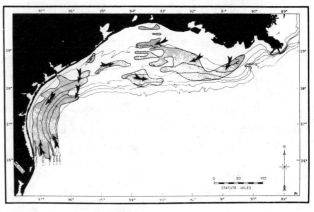

**Fig. 6-14.** Distribution of grain-size modes of 4.3–7.1 φ (0.05-.φφ68 mm) of surface sediments. From Curray, 1960, reprinted by permission.

related to advective currents. Patterns of a clay-sized mode of sedimentation (Fig. 6–15) is possibly the product of pelletization of muds by nektonic and pelagic organisms. This interpretation is supported by the presence of a higher portion of organic carbon and the lower content of oxygen near the 100 fathom (180 meter) contour

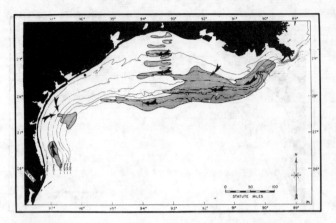

**Fig. 6-15.** Distribution of grain-size modes of finer than 7.2 $\phi$ (0.0068 mm) of surface sediments. From Curray, 1960, reprinted by permission.

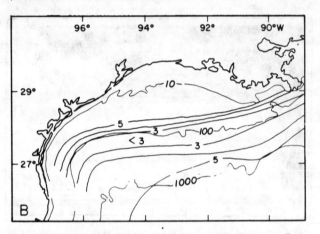

**Fig. 6-16.** Oxygen content (ml/l.) of waters next to bottom. From Friedman and Sanders, 1978. Modified from Trask, 1953.

(Friedman and Sanders 1978) (Fig. 6-16). Removal of sediment from suspension is a complex process. Pelletized or flocculated clay may settle through a denser water layer in contact with the shelf bottom. This mechanism may lead to the accumulation of fine-grained organic rich sediment at the shelf margin, and possibly to the transport, by low density fluidized flow of organic-rich, fine-grained sediments down the continental slope or into submarine canyons cut into the edge of the continental shelf.

## SEDIMENTOLOGIC THEMES

Shelf and interior sea sedimentology are characterized by the same processes. Wave and current patterns are common to all deeper water environments. Tidal forces, although most important on continental shelves, also occur in marginal and interior basins. Wind and storm tides and other processes related to the biological, biochemical, and chemical aspects are reflected in the sedimentology. These processes within the framework of water depths that are typical of shelves and interior basins result in only a few dynamic equilibrium sedimentologic

patterns. Sedimentologic and stratigraphic models that exhibit these patterns can be characterized by:

▸ Density current and progradational responses,
▸ Low-energy transgressive responses,
▸ Tidal-wave responses, and
▸ Biochemical responses.

Each of these responses result from a combination of physical and developmental processes and produce characteristic stratigraphic units. Other processes that result in sedimentary patterns may be locally important, but they are not broadly distributed or do not produce developmental sequences.

## SILICICLASTIC SHELF THEMES

Three depositional systems may be suggested based upon process themes. Storm, tidal, and biogenic dominated sedimentologic themes appear to be useful for organizing Holocene sedimentary patterns, and they also appear to be important stratigraphic response patterns (Table 6-1).

### Storm and Current Dominated Clastic Shelf Model

Equilibrium shelves, as suggested by Curray (1965) (Fig. 6-2), indicates that tractive load grading may be a possible response pattern to depositional processes, or events. This geometry may be in response to regressive-transgressive coastal progradational patterns, storm and current dominated shelf processes, or to glaciomarine suspension and mass flow depositional processes. Little information is available on these mechanisms of sediment transport, but local studies of reworking of relict sediments may provide information on the origin of depositional systems, and the mechanism for their preservation in the stratigraphic record.

A Holocene system of furrows, sand ribbons, sand waves, sand patches, and a rippled sand sheet has been described for the Celtic sea (Johnson and Kenyon et al. 1982) (Fig. 6-17). The conceptual model is that this system may produce coarsening upward depositional sequences in response to periodic storm events (Johnson 1978, 162) (Fig. 6-18). Such a model could usefully be employed to describe ancient stratigraphic intervals from the Precambrian (Johnson 1978, 163) (Fig. 6-19) to the Jurassic (Anderton 1976). Belderson (1986) contrasts and compares patterns of storm generated sand ridges with tidal active sand banks (Table 6-2). A sequence of sedimentary structures from a marine lower Carboniferous sedimentary sequence from Cork, southern Ireland, may be a useful model for the interpretation of internal bedding patterns of sand waves (Raaf et al. 1977) (Fig. 6-20).

**Table 6-1**
**Shelf Models**

| | Siliciclastic Shelves | | | Biochemical Shelves | | | |
| --- | --- | --- | --- | --- | --- | --- | --- |
| | *Storm & Current Dominated* | *Tidal Dominated* | *Low-Energy Biogenic Transgressive* | *Epeiric Sea Carbonate Platforms* | *Regressive-Transgressive Ramps* | *Progradational Shelf Basins* | *Shelf Edge Bioherms* |
| Geometry | Thickening across shelf edge | Thickening wedge to shelf edge | Thin blanket | Blanket with mounds around basins | Thickest at shelf edge | Abrupt thickness changes | Elongate barriers |
| Boundary Conditions | Eastern continental margins, mid- to high latitude | Western continental margins, mid-latitude | Eastern continental margin, mid-high latitude | Cratonic platform, low latitude | Eastern continental margin, low-mid latitude | Eastern continental margin, low latitude | Continental margin, low latitude |
| Dominant Depositional Area | Mid shelf | Outer shelf | Outer shelf | Cratonic basin margins | Outer shelf | Landward of shelf basins | Shelf edge |
| Depositional Sequences Depositional Surface | Disconformity | Subaerial unconformity | Disconformity | Subaerial unconformity | Disconformity | Hardground | Subaerial karst |
| Basal Unit | Onlapping shales and sand | Coarse clastics | Marsh | Pelleted sand | Clastic interal or marsh | Salt or argillaceous micrite | Skeletal Calcareous sands |
| Dominant Lithology | Coarse sand & shale sequences | Medium, grained quartzose sand | Pelleted argilloceous micrite & argillaceous | Molluscan pelleted sand | Pelleted and skeletal sands oolite or biohermal mounds | Calcareous sand and Bioherms | Bioherms or reep |
| Topmost Unit | Shoreline clastics or marsh | Lagoonal carbonate | Phosphatic and glauconitic marl, or siliceous shale | Local mounds | Sabkha | Nodular anhydrite | Carbonate breccia |

**Table 6-2**
**Comparison Between Offshore Tidal Sand Banks and Storm-generated Sand Ridges. From Belderson, 1986.**

| | Tidal Sand Banks (Active) | Storm-generated Sand Ridges |
|---|---|---|
| Angle with coastline | Related primarily to peak tidal current direction | Up to a 60°, with primary mode 35°-40° |
| Sand waves | Abundant and semipermanent | Rare to absent |
| Obliquity to main flow | 0°-20°, but generally 7°-15° | Up to 60°, primary mode 35°-40°. (With coast-parallel main flow). |
| Height | Up to 43 m (55 m for moribund banks). | 3-12 m (average 7 m) |
| Crests | Frequently sharp (except where crests are near sea surface) | Smooth-crested |
| Slope angles | 6° or less | 2° or less (mean slope 0.5° for offshore ridges) |
| Spacing | 2-30 km | 0.5-7 km |
| Length | Up to 70 km (200 km for moribund banks) | up to 20 km |
| Internal structure | Pervasive "normal" cross-stratification | Hummocky cross-stratification? |

**Table 6-3**
**Storm Deposited Shelf Reservoirs**

Interval containing reservoir sandstones is lenticular, with up to 25% change in thickness, elongation more than 5 times width, bounding surfaces of interval continuous over areas greater than 5000 kms²

Linear sandstone reservoirs 2–20 m thick, and 1–10 kms long, and up to 5 kms wide, elongation parallel to shoreline or shelf edge, flow units up to 500 m wide and one km long

Low angle truncation surfaces, less than one degree, at base and top of sandstone reservoir units

Bounding surfaces of shelf interval continuous over areas greater than 5000 kms²

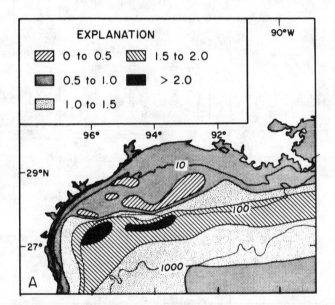

**Fig. 6–17.** Coincidence between higher-than-usual values of organic matter in bottom sediments and lower-than-usual proportions of dissolved oxygen in water next to bottom. Organic matter–based on 1.8 times measured content of organic carbon—in bottom sediments. After Richards 1957, from Friedman and Sanders, 1978.

These bedding patterns are similar to those theoretically developed by Allen (1980) (Fig. 6-6).

The occurrence of sand waves as patches of relict sediment of Pleistocene age (Figs. 6-3 and 6-8) may reflect stranded coastal progradational shoreface deposits. These units may be reworked and preserved by transgressive eustatic events. In a number of areas these mid-shelf sand waves show evidence of active transport (Fig. 6-9), and may reflect equilibrium deposition by Modern current and storm patterns. A table of geometric aspects for these Modern depositional systems may be useful for interpreting bedding patterns and depositional history (Table 6-3). Seismic reflection geometries from possibly a similar pattern may be illustrated by the Cretaceous off the Alaskan North Slope (Fig. 6-21).

Hydrocarbon accumulations have been associated with similar patterns from the Cretaceous of Wyoming. Regressive-transgressive patterns associated with Frontier, Shannon, Sussex, Teapot and Parkman formations have produced complex patterns of offshore sandstone depositional sequences. These units are related to eustatic changes associated with sequence unconformities.

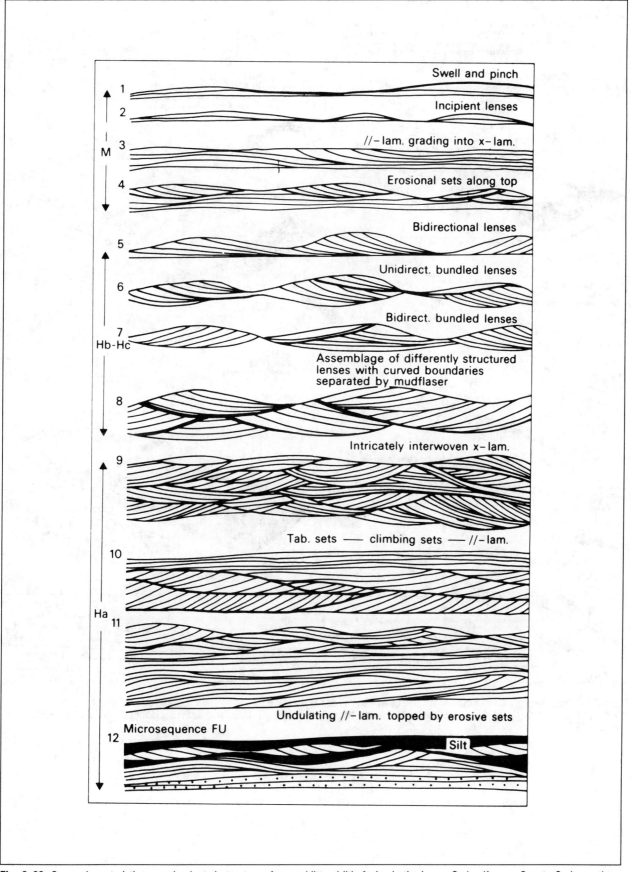

**Fig. 6-20.** Some characteristic wave-dominated structures from sublittoral lithofacies in the Lower Carboniferous, County Cork, southern Ireland. From Raaf et al., 1977.

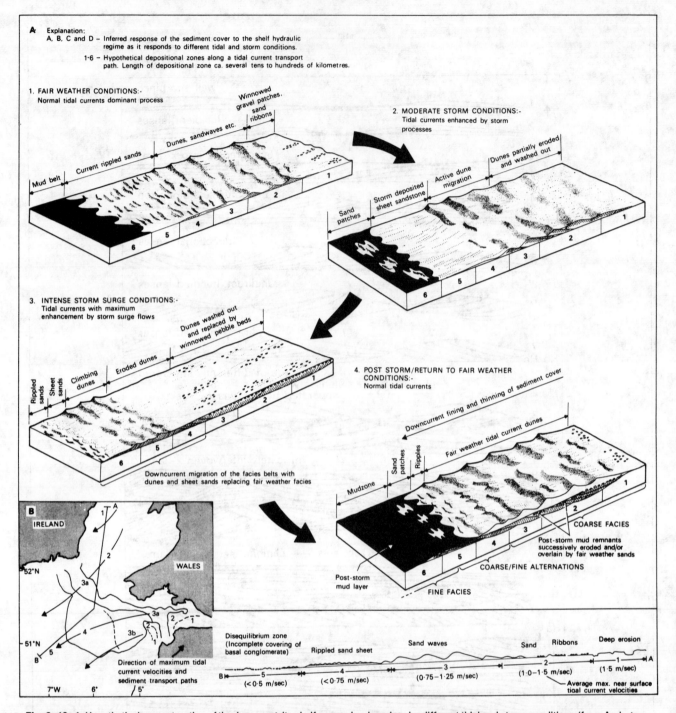

**A** Explanation:
A, B, C and D – Inferred response of the sediment cover to the shelf hydraulic regime as it responds to different tidal and storm conditions.

1–6 – Hypothetical depositional zones along a tidal current transport path. Length of depositional zone ca. several tens to hundreds of kilometres.

1. FAIR WEATHER CONDITIONS:-
Normal tidal currents dominant process

2. MODERATE STORM CONDITIONS:-
Tidal currents enhanced by storm processes

3. INTENSE STORM SURGE CONDITIONS:-
Tidal currents with maximum enhancement by storm surge flows

4. POST STORM/RETURN TO FAIR WEATHER CONDITIONS:-
Normal tidal currents

**Fig. 6-18.** *A,* Hypothetical reconstruction of the Jura quartzite shelf sea as developed under different tidal and storm conditions (from Anderton, 1976). *B,* Possible modern analogue as exemplified by the tidal current transport path in the Celtic Sea (after Belderson and Stride, 1966). From Johnson, 1978.

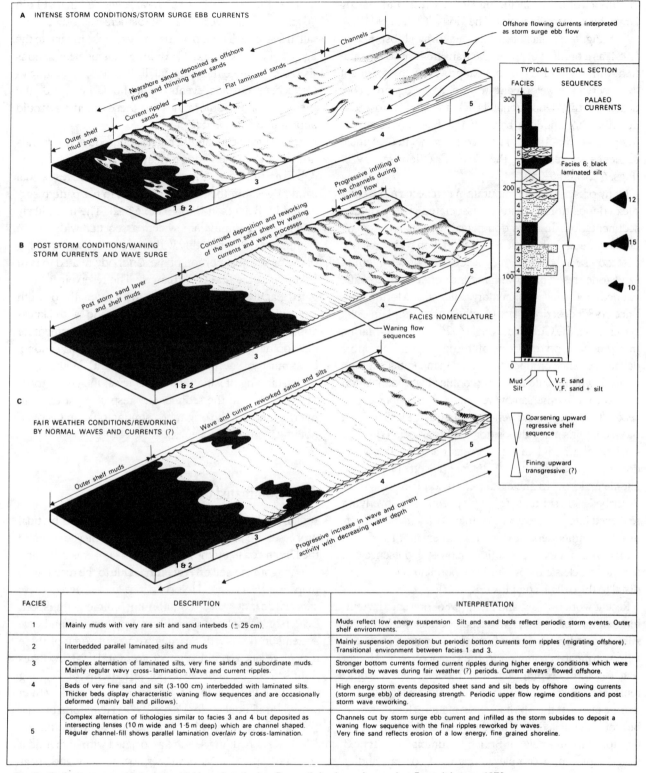

| FACIES | DESCRIPTION | INTERPRETATION |
|---|---|---|
| 1 | Mainly muds with very rare silt and sand interbeds (± 25 cm). | Muds reflect low energy suspension  Silt and sand beds reflect periodic storm events. Outer shelf environments. |
| 2 | Interbedded parallel laminated silts and muds | Mainly suspension deposition but periodic bottom currents form ripples (migrating offshore). Transitional environment between facies 1 and 3. |
| 3 | Complex alternation of laminated silts, very fine sands and subordinate muds. Mainly regular wavy cross-lamination. Wave and current ripples. | Stronger bottom currents formed current ripples during higher energy conditions which were reworked by waves during fair weather (?) periods. Current always flowed offshore. |
| 4 | Beds of very fine sand and silt (3-100 cm) interbedded with laminated silts. Thicker beds display characteristic waning flow sequences and are occasionally deformed (mainly ball and pillows). | High energy storm events deposited sheet sand and silt beds by offshore  owing currents (storm surge ebb) of decreasing strength. Periodic upper flow regime conditions and post storm wave reworking. |
| 5 | Complex alternation of lithologies similar to facies 3 and 4 but deposited as intersecting lenses (10 m wide and 1·5 m deep) which are channel shaped. Regular channel-fill shows parallel lamination *overlain* by cross-lamination. | Channels cut by storm surge ebb current and infilled as the storm subsides to deposit a waning flow sequence with the final ripples reworked by waves. Very fine sand reflects erosion of a low energy, fine grained shoreline. |

**Fig. 6–19.** Storm surge-ebb facies model based on the late Precambrian Innerelv member. From Johnson, 1978.

A number of fields, including House Creek and Hatzog Draw, have been described. The House Creek field (Berg 1975) (Fig. 6–22) has been interpreted by Hobson et al. (1984) as a product of regressive-transgressive sedimentation, with a stranded offshore bar providing the sediment for reservoir development (Fig. 6–23). Log and core details of the depositional facies are similar to those described for modern bars off the Belgium coast (de Maeyer et al. 1988), and storm generated vertical sequenes suggested for the Celtic Sea (Johnson 1978) (Figs. 6–24 and 6–25).

Many preserved reservoir units may reflect a more complex depositional history associated with deltaic sedimentation. In wave dominated strandplain depositional systems, subsidence, in conjunction with a relative eustatic rise in sea level, may lead to the preservation of sandstone reservoirs overlain and underlain by fine grained mid-shelf sedimentary intervals. Morton and Price (1987) describe a late Quaternary strandplain delta system from the Gulf Coast (Figs. 6–26 and 6–27). Marine transgression and reworking of the upper portions of this deltaic sequence produced a coarser, glauconitic, capping interval, which could easily be confused with deposition related to a marine sandwave origin. The Upper Cretaceous (Campanian) La Ventana tongue of the San Juan basin contains many depositional units associated with a stacking of strandplain sequences associated with a wave-dominated deltaic system (Palmer and Scott, 1984). A 50 kilometer long, 18 meter thick strand-plain system is described for subunit H (Fig. 6–28). The overlying reworked interval, locally more than 6 meters thick, overlies the earlier-deposited deltaic interval (Fig. 6–29). Other units, for example, unit E, shows the presence of a 6 meter thick sandstone unit of possible shelf origin, distal to the deltaic complex (Fig. 6–30).

Recent work by Cant and Hein (1986) presented a reinterpretation of many Cretaceous sandstone units of Alberta and British Columbian. They suggest three differing sequences common to the Cretaceous of Alberta (Fig. 6–31). In keeping with their models, Type 1 would represent a strandplain system characteristic of the marine facies of the Spirit River formation; Type 2 would be storm deposited shelf sandstones, similar to the Gog Group, Spirit Tunnels, of British Columbia; and Type 3 would represent a complex stacking of coastal, and/or marine shelf sands, overlain by an unconformity with valley-fill conglomerates and a high energy transgressive reworking of the topmost unit of the sequence, typical of the Viking and Cardium formations of Alberta (Figs. 6–32

and 6–33)). Haq et al. (1987) have suggested that at both 90 and 94 million years (Table 1–5) there was a significant fall in sea level. These periods are very close in time to the age of the Viking and Cardium Sandstone formations (Fig. 6–34). Since there are small differences in dating of Cretaceous eustatic events throughout the world, the presence of unconformities at these intervals would appear to be reasonable.

In Saskatchewan, another series of productive oil fields is developed in the Viking formation (Evans 1970) (Fig. 6–35). Evans interprets these linear sandstone ridges as tidal sand ridges, but they do not conform to the pattern suggested by Belderson (1986) (Table 6–2). The more likely origin is as a strandplain system associated with a wave dominated delta, possibly with a transgressive reworking of the uppermost portion of the sandbody related to post Viking eustatic sea-level changes. The reservoir is elongated parallel to the shelf edge, with a length to width elongation greater than five to one, and a thickness approximating 10 meters (Fig. 6–36). The requirement for developing genetic interpretations for reservoir sandbodies must of necessity be multivariant. Without an understanding of variations in sea level, the geometry of the shelf, and in the sedimentary response patterns, an interpretation of the origin of a stratigraphic unit is often not much more than a guess.

## Tidal Dominated Clastic Shelf Model

Broad, shallow, high energy shelves dominated by tidal processes are present today in the North Sea adjacent to the Dutch coast, off the northwest coast of Australia, and the Patagonian shelf off Argentina. Due to the recent (less than 700 years ago) Holocene rise in sea level, these shelves do not reflect equilibrium sediment transport. Belderson describes an equilibrium pattern for sandwave systems in the North Sea (Fig. 6–5), but some of these may represent a combination of storm and tidal processes (Johnson 1978). Process-response patterns of bedding patterns reflect sediment transport associated with a semi-diurnal tidal cycle (Visser 1980). He reports bundle cyclicity associated with spring-neap tidal cycles (Fig. 6–37). Thicker bedding sets are associated with higher tidal amplitudes, indicative of higher flow velocity and shear stress. Internal structures reflect either unidirectional or bipolar cross bedding sets, with cross bedding surfaces rarely exceeding 22° (Belderson 1986). Textural and mineralogic response patterns reflect the higher energy

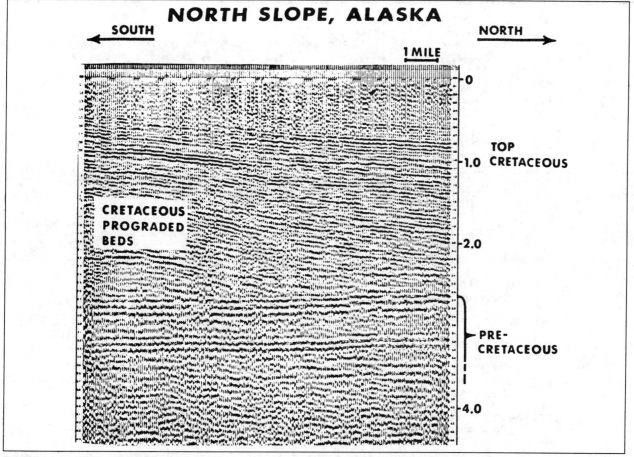

**NORTH SLOPE, ALASKA**

← SOUTH          NORTH →

CRETACEOUS PROGRADED BEDS

TOP CRETACEOUS

PRE-CRETACEOUS

**Fig. 6–21.** Seismic reflections may be produced by thin sand and silt units formed by deepwater waves, density current, and slump patterns. Or they may result from regressive-transgressive depositional cycles. In either case, the depositional history of a shelf sequence is indicated. Reprinted courtesy Petty-Ray Geophysical.

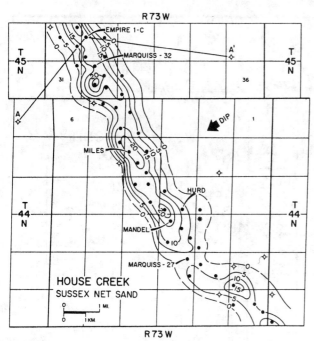

**Fig. 6–22.** Net sandstone isolith of Sussex sandstone (Cretaceous), House Creek Field, WY. Contour interval is 5 ft (1.5 m). Isolith pattern suggests linear sand-body alignment parrallel to depositional strike with an asymmetrical cross-section. From Berg, 1975, reprinted by permission.

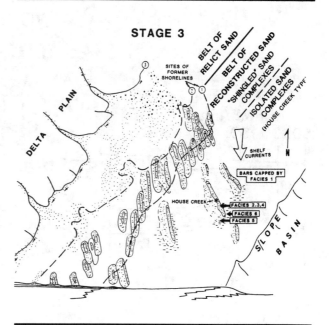

**Fig. 6–23.** Continued rapid transgression as in Stage IIB; reconstruction and migration of relict sands as offshore bars; final burial of bars by marine muds. From Hobson et al., 1984, reprinted by permission.

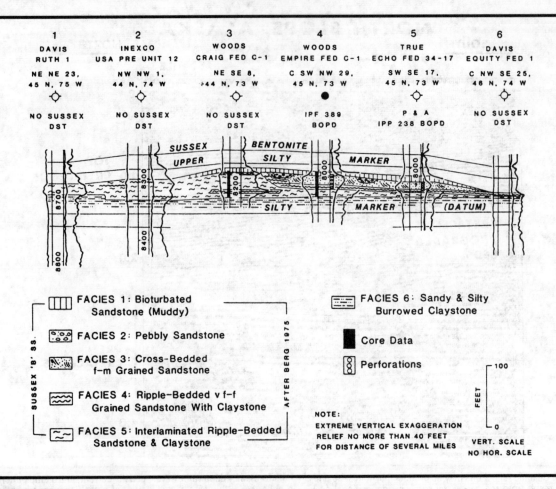

Fig. 6-24. Interpretive dip cross section showing probably distribution of facies in Sussex B sandstone complex. Core interpretation in well 3 is from Core Laboratories' and scout ticket data. From Hobson et al., 1984, reprinted by permission.

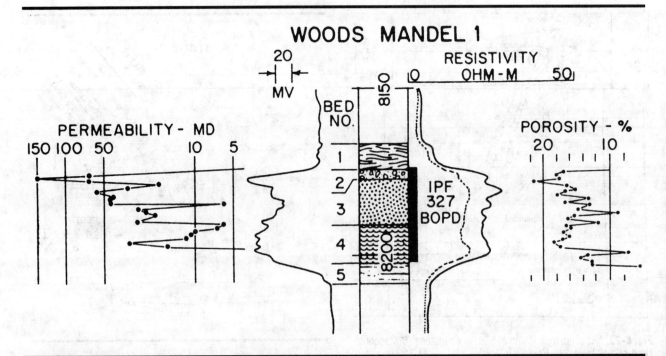

Fig. 6-25. Permeability and porosity of the Sussex sandstone in Woods Petroleum Mandel Federal #1, House Creek Field, Wyoming, showing also blunt-base, blunt-top SP and resistivity log pattern. From Klein, 1985.

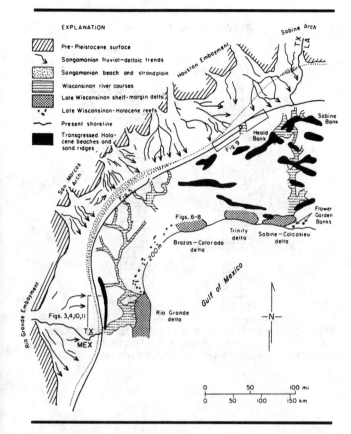

**Fig. 6-26.** Tectonic elements and regional distribution of late Pleistocene and Holocene depositional systems. Texas coastal plain and continental shelf. From Morton and Price, 1987.

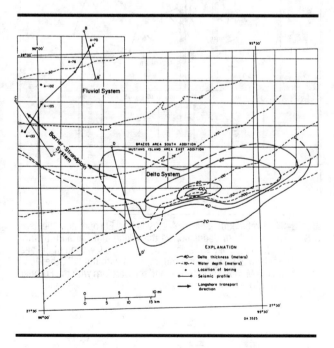

**Fig. 6-27.** Locations of sediment borings and seismic profiles in relation to bathymetry and shelf-margin deltaic deposits. From Morton and Price, 1987.

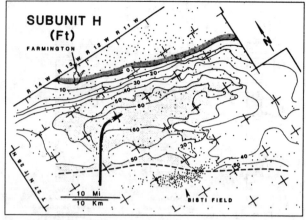

**Fig. 6-28.** (A) Net-sandstone map of subunit G. (B) SP log and facies map of subunit G. From Palmer and Scott, 1984.

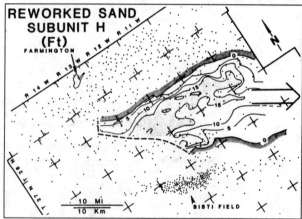

**Fig. 6-29.** Net-sandstone map of subunit H. From Palmer and Scott, 1984.

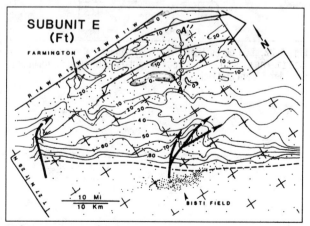

**Fig. 6-30.** Net-sandstone map of subunit E. From Palmer and Scott, 1984.

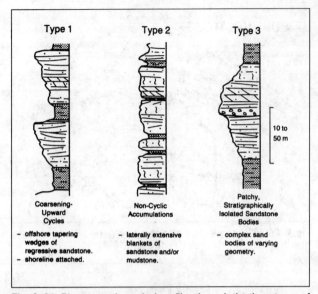

**Fig. 6–31.** Diagrammatic vertical profiles through the three types of shallow marine deposits discussed in the text. Coarser zones project farther to the left. From Cant and Hein, 1986.

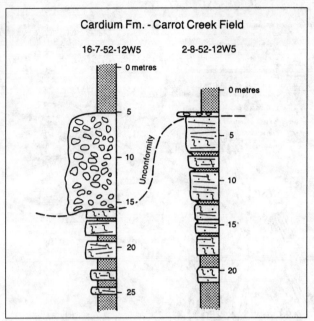

**Fig. 6–33.** Two cores typical of the Cardium formation in many oil fields. The lower regressive sequence is cut out by the unconformity on which the conglomerate rests. From Cant and Hein, 1986.

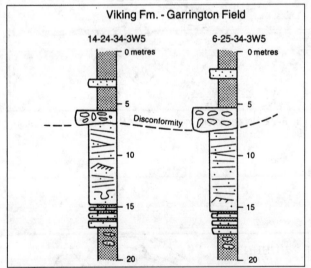

**Fig. 6–32.** Two cores from the Viking sandstone in south-central Alberta. The erosion surface does not cut down through strata in this case. From Cant and Hein, 1986.

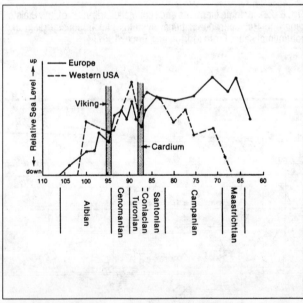

**Fig. 6–34.** Cretaceous sea-level curves for Europe and western U.S.A. The periods of deposition of the Viking and Cardium formations are indicated. They were deposited during a lowering of the sea level. The Viking is almost contemporaneous with the "J", the Newcastle, and the Muddy sandstones, and the Cardium with the Frontier in the western U.S.A. From Cant and Hein, 1986.

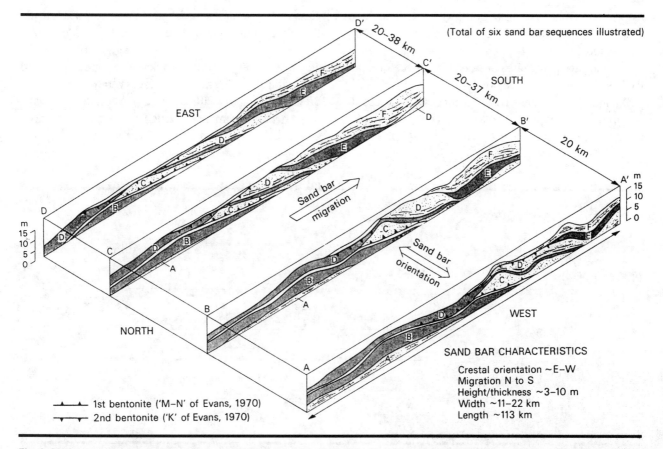

(Total of six sand bar sequences illustrated)

SAND BAR CHARACTERISTICS

Crestal orientation ~E–W
Migration N to S
Height/thickness ~3–10 m
Width ~11–22 km
Length ~113 km

⊢—▲—⊣  1st bentonite ('M–N' of Evans, 1970)
⊢—▼—⊣  2nd bentonite ('K' of Evans, 1970)

**Fig. 6–35.** Sections through the Lower Cretaceous Viking formation illustrating the preservation of six sand bar sequences (*A-F*) interpreted as representing E-W oriented linear tidal sand ridges which gradually migrated southwards. From Evans, 1970, reprinted by permission.

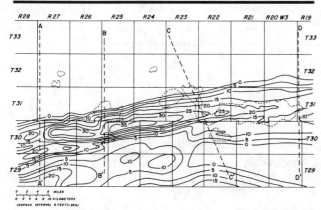

**Fig. 6–36.** Isopachous map of Cretaceous Lower "L" Member of Viking formation, Doddsland Hoosier area, southwestern Saskatchewan. From Evans, 1970, reprinted by permission.

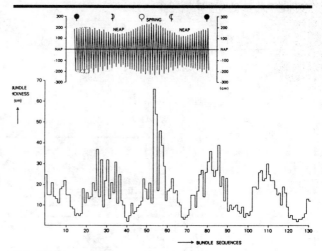

**Fig. 6–37.** Measurement of 131 adjacent bundles in one set reveals cyclic sinusoidal pattern. Graph of general tidal pattern at Vissingen is included for comparison. From Visser, 1980.

associated with semi-diurnal tides, with the presence of well sorted, orthoquartzitic sandstone containing abundant glauconite and phosphate. In areas of lower shear stress and sediment transport, patterns of bioturbation (Wilson 1986) may be developed.

Stratigraphic response patterns suggest the necessity for a process/response model to produce marine blanket quartzose sandstone intervals. These stratigraphic units contain well sorted, glauconitic sandstones, with greater than 95% detrital quartz. Internal structures reflect either unidirectional or bipolar cross bedding sets, often dominated by fourteen day tidal bundles (Allen and Homewood 1984) (Figs. 6–38 and 6–39), locally bioturbated, and often associated with marine carbonate intervals. Sequences of sedimentary structures are similar to those theoretically described by Allen (1980) (Fig. 6–6), and by

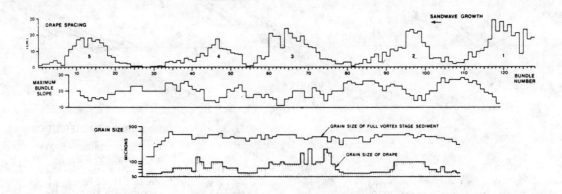

**Fig. 6–38.** Record through 128 bundles of drape spacing, maximum bundle slope and grain size. Drape spacings show a marked periodicity of about 27, indicating a semi-diurnal tidal regime. Thick-thin alterations, as in bundle sequence 1, are due to the diurnal inequality of the tides. From Allen and Homewood, 1984.

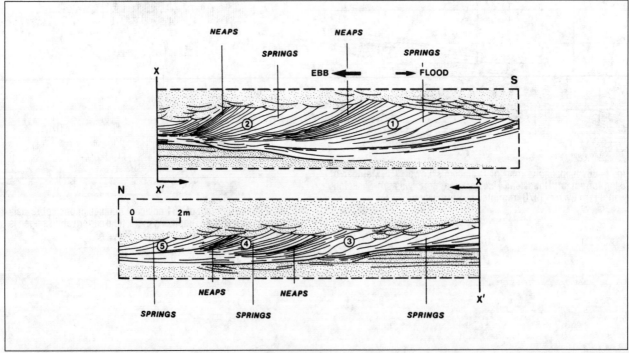

**Fig. 6–39.** Field sketch of the cliff exposure at Bois du Devin, near Fribourg. Bundle sequences are numbered 1–5. True dip of foresets is approximately 30° to strike of exposure. From Allen and Homewood, 1984.

Raaf et al. (1977) from the Lower Carboniferous, Cork Co., Ireland (Fig. 6–20). Tidal bundles also have been recognized from the Jurassic Curtis formation from Utah (Kreisa and Moiola 1986). Since the recognition of the theme by Visser (1980) many additional examples will undoubtedly be described from the stratigraphic record.

The interaction of the shelf geometry with incident tidal wave energy has long been a focus of research (Belderson 1986) (Table 6–3). In addition to this summary paper, a number of detailed studies concerning tidal shelf process and response patterns, and also ancient stratigraphic examples, have been collected into a single volume edited by Knight and McLean (1986).

In an attempt to place this research into a process-response framework it appears to be useful to examine the energy distribution across a tidally dominated shelf (Fleming 1938) (Fig. 6–40). Redfield (1958) discussed the possibility of tidal resonance on very broad and low gradient shelves marginal to large ocean basins. Visher and Hyne (1973) suggested that resonance could be an equilibrium response, and could produce the requisite tidal amplitudes and frequencies that would result in the generation of blanket orthoquartzitic sandstone intervals that are so common in the stratigraphic record (Figs. 6–41, 6–42 and 6–43).

Webb (1976) developed a resonance model for the Patagonian shelf based on the analytic properties of the Green's function. The analysis illustrated the requisite interrelation of the shelf geometry to the frequency of open oceanic tidal periods. He found that at the frequency of the dominant shelf resonance, more than 95% of the incident energy is absorbed. The dominant resonance was near 14 radians/day, and resonance produced a geographically localized increase in tidal amplitude. He also suggested that one would expect to find resonance on shelves approximating a quarter wavelength of the semi-diurnal tide (Webb 1976).

Energy absorption is distributed across the entire shelf, resulting in the focus of energy on winnowing and transport of sediment. Semi-diurnal tidal periods would result in the maintenance of a dynamic shelf surface of the requisite slope and width to efficiently dissipate all incident energy. Can such a model be responsible for the generation of blanket sandstones in the stratigraphic record? Models are simply useful tools or devices that allow us to

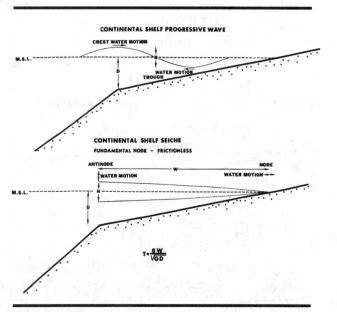

**Fig. 6–41.** Progressive tidal wave model with antinode at or near shelf edge. From Visher and Hyne, 1973.

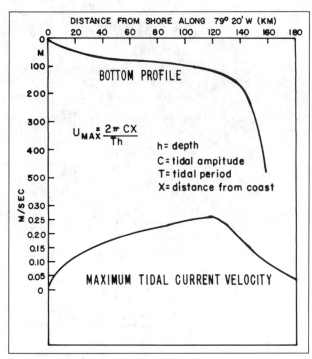

**Fig. 6–40.** From Fleming, 1938.

| UNCONFORMITY | FORMATION | AREA | AVERAGE SHELF WIDTH | ADJUSTED SHELF SLOPE | ADJUSTED DEPTH AT SHELF BREAK | SANDSTONE THICKNESS |
|---|---|---|---|---|---|---|
| SAUK BASAL CAMBRIAN | TAPEATS | ARIZONA | 260 km | 45 cm/km | 118 m | 300 m |
| SAUK LOWER CAMBRIAN | BRIGHAM | IDAHO | 160 | 80 | 130 | 380 |
| SAUK MIDDLE CAMBRIAN | FLATHEAD | WYOMING | 290 | 20 | 56 | 150 |
| TIPPECANOE ORDOVICIAN | EUREKA | UTAH | 160 | 53 | 85 | 240 |
| TIPPECANOE ORDOVICIAN | ST. PETER | OKLAHOMA | 320 | 10 | 31 | 180 |
| ABSAROKA PENNSYLVANIA | WEBER | COLORADO | 480 | 32.5 | 156 | 450 |
| ABSAROKA PENNSYLVANIA | TENSLEEP | WYOMING | 160 | 32.5 | 52 | 150 |
| PRE-LEONARD PERMIAN | COCONINO | ARIZONA | 390 | 45 | 175 | 450 |
| PRE-LEONARD PERMIAN | GLORIETA | NEW MEXICO | 195 | 36 | 71 | 275 |
| ZUNI UPPER TRIASSIC | NUGGET | WYOMING | 320 | 50 | 156 | 450 |
| ZUNI UPPER TRIASSIC | NAVAJO | UTAH | 400 | 52.5 | 210 | 600 |
| AVERAGE | * | | 285 km | 44 cm/km | 113 m | 330 m |

**Fig. 6–42.** Geometry of blanket sandstones. From Visher and Hyne, 1973.

interrelate our observations, and predict results, and if the response patterns are consistent with the model it may be useful for interpretion. Such is the possibility for a large number of stratigraphic examples.

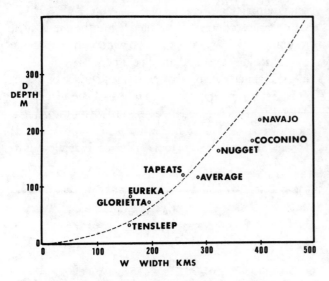

**PALEOZOIC CONTINENTAL SHELF AND RISE SEICHE**

$$T=\frac{5.2W}{\sqrt{GD}} = 40,000 \text{ SEC (11 HRS)}$$

**Fig. 6-43.** Resonance model. From Visher and Hyne, 1973.

*Paleozoic Examples.* The stratigraphic response pattern for shelf sedimentary units is indicated by facies patterns for a number of ancient units. Sedimentary aspects related to mineralogy, bioturbation, and vertical sequences of deposited units from ancient stratigraphic units are available.

Internal sedimentary characteristics for each of these sandstone units suggest a marine origin and were interpreted as formed by tidal processes: the Cambrian Zabriske Quartzite and Wood Canyon Sandstone (Klein 1976), with the paleogeographic pattern suggested by Mallory (1972) (Figs. 6–44 and 6–45), the Ordovician Eureka Quartzite (Klein 1975), and the St. Peter Sandstone (Pryor and Amaral 1971), with the paleogeography pattern suggested by Sloss et al. (1960) (Fig. 6–46). Each unit overlies a sequence unconformity and was deposited during a major period of sea-level rise. The preserved geometry of the sandstones is a wedge-shaped blanket with a thickness of 100 to more than 600 meters at the shelf edge and thinning to a few tens of meters near the shoreline. The width-depth shelf geometry has been obtained by adjusting the preserved thickness to reflect onlap (Fig. 6–42). The adjusted shelf geometry is favor-

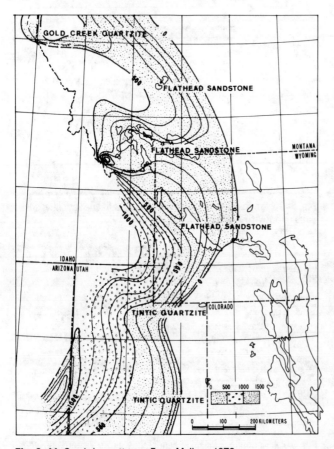

**Fig. 6-44.** Cambrian patterns. From Mallory, 1972.

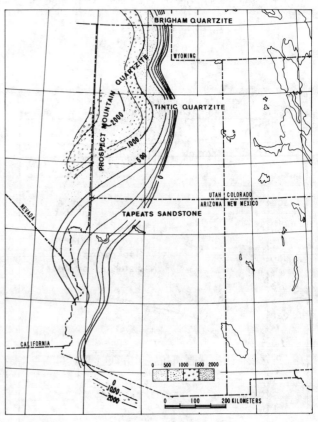

**Fig. 6-45.** Ordovician patterns. From Mallory, 1972.

able for the development of high wave and tidal energy (Redfield 1958, Webb 1976). The length of time represented by these units (tens of millions of years) suggests a dynamic equilibrium between deposition and sea-level rise.

The complex of Permian formations, Coconino, DeChelly, and Glorieta, are quartzose sandstones thickening from more than 150 meters to more than 600 meters toward a shelf edge (Fig. 6–47). Continental equivalents include the Organ Rock shale, Cutler, Abo, Yeso, and Rancho Rojo formations (Kreisa 1986). These formations are arkosic and were deposited in fluvial-deltaic and tidal flat sedimentary environments. The geometry and mineralogy of the quartzose sandstones strongly suggest a

change in depositional environment. Evidence from internal characteristics is useful for determining both the processes of sedimentation and the depositional environment.

Photographs illustrate the bidirectional nature of cross-bedding, the presence of parting lineation on depositional surfaces, and the presence of amphibian, molluscan, and other faunas (Figs. 6–48, 6–49, and 6–50). In addition, textural studies show a similarity in depositional processes to marine shoals and sand waves (Fig. 6–51). This combination of geometry and internal sedimentary features indicates that these units formed on a marine shelf.

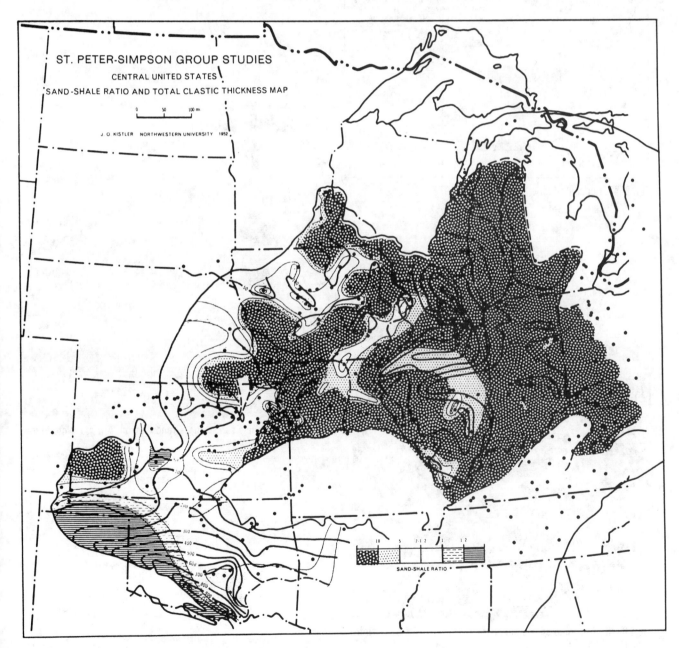

**Fig. 6–46.** St. Peter-Simpson Group studies; sand-shale ratio and total clastic thickness map. From Sloss et al., 1960.

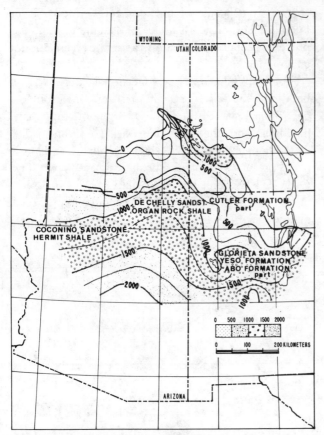

**Fig. 6–47.** Coconino-Glorieta. From Mallory, 1972.

**Fig. 6–49.** Coconino sandstone, Grand Canyon, Arizona. Note parting lineation resulting from unidirectional currents.

**Fig. 6–50.** Coconino sandstone, Grand Canyon, Arizona. Note four-toed amphibian tracks.

**Fig. 6–48.** Coconino sandstone, Grand Canyon, Arizona. Note bimodal crossbedding.

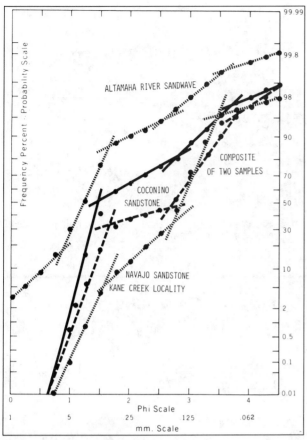

**Fig. 6-51.** Textural responses are similar to Holocene patterns. From Visher and Freeman, 1976.

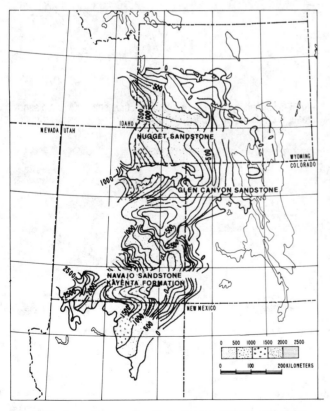

**Fig. 6-52.** Isopach map, Navajo sandstone and its equivalents, western U.S. From Mallory, 1972.

***Mesozoic Examples.*** Another stratigraphic unit with similar characteristics is the Nugget, Glen Canyon, and Navajo sandstones (Fig. 6-52). This interval has facies equivalents of lacustrine, fluvial-deltaic, and marine limestones (Fig. 1-17) and represents a blanket of quartzose sandstone thickening from only a few meters to more than 700 meters toward the shelf edge. Facies patterns indicate that deep-water biogenic shales occur on the western margin of the shelf (Fig. 6-41). This paleogeographic framework indicates the source of the clastics to be arkosic sediment from the continental interior with an array of depositional environments reflecting marine shelf sedimentology. Sedimentary response patterns preserved in these sandstone units support this interpretation.

These patterns were studied and reported in detail by Freeman and Visher (1975) and Visher and Freeman (1976). Crossbedding patterns illustrate the similarity between reported patterns from Holocene sand waves (Fig. 6-18) and the outcropping Navajo sandstone (Fig. 6-54). The presence of fossiliferous limestone lenses, an abundant trace fossil assemblage (Fig. 6-55), and sedimentary structures reflecting viscous and inertial fluid flow indicates these units have a subaqueous origin. This is supported by the textural response patterns, which are directly comparable to those generated by high-velocity confined flow in channels and tidal estuaries (Fig. 6-56).

The physical process model developed by Redfield and by Webb suggests that high-velocity flow, as illustrated in Figure 6-40, can be used to explain the response patterns described. The resonance geometry of these shelf deposits is illustrated by Figure 6-43. From this association of facies frameworks, geometries, internal sedimentary features, mineralogies, and faunal associations, the clastic tidal shelf model of sedimentation can be recognized and usefully applied to the interpretation of ancient stratigraphic sequences.

***Exploration Strategy.*** A well-log cross section shows the stratigraphic patterns for a sequence of units commencing with an unconformity (Fig. 6-57). This section illustrates the depositional mounding of the Entrada Sandstone and infilling by the overlying Todilto formation. The geometry of the ridges was mapped with seismic and available well control (Fig. 6-58). An isopach, thick with high impedence contrast at both the top and the base of the formation, is indicated.

A structural section illustrates that oil distribution is partly controlled by depositional topography at the top of the reservoir unit, and also is partly a result of structural tilt. The structural section also indicates a regional north dip with differential entrapment related to a spill point (Fig. 6-59). Seismic lines across two oil fields show the

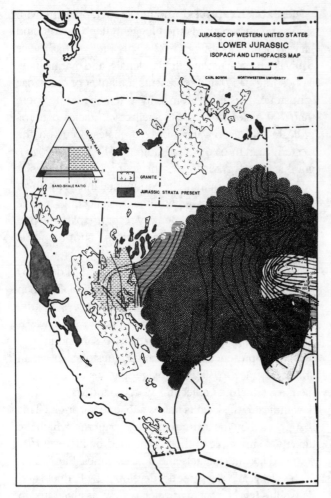

**Fig. 6-53.** Isopach and lithofacies map, Lower Jurassic of the western U.S. From Sloss, Dapples, and Krumbein, 1960.

**Fig. 6-55.** Navajo sandstone, Lake Powell, Utah. Note presence of limestone lens in center of large dunes.

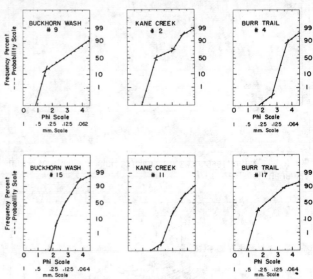

**Fig. 6-56.** Basic types of log-probability curve shapes, Navajo sandstone. From Freeman and Visher, 1975.

**Fig. 6-54.** Navajo sandstone, Lake Powell, Utah. Note interwoven crossbed sets.

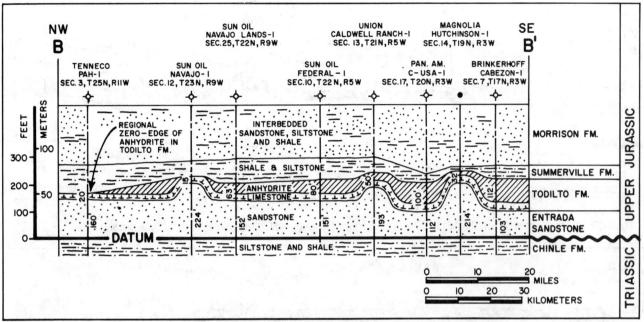

**Fig. 6–57.** Regional stratigraphic cross section *B-B¹* illustrating thickness changes in Entrada sandstone and overlying Todilto formation. Base of Entrada sandstone is used as horizontal stratigraphic datum. From Vincellette and Chittum, 1981, reprinted by permission.

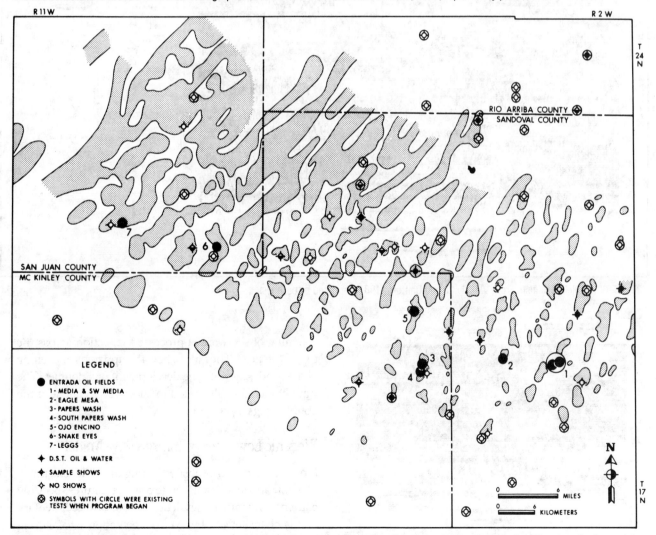

**Fig. 6–58.** Map showing general location and configuration of Entrada seismic anomalies along southwestern flank of San Juan basin; areas shown by dotted pattern outline Entrada sand thicks or topographic highs. Several anomalies had more than one test, but only the initial wildcat location is shown. Drilling activity shown is as of January 1980. From Vincellette and Chittum, 1981, reprinted by permission.

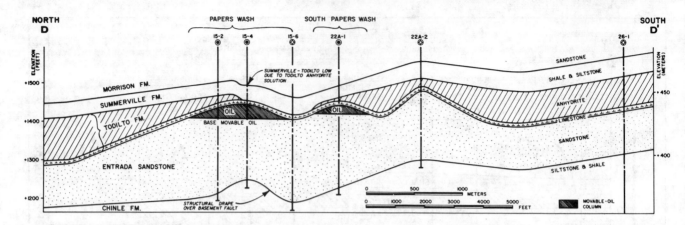

**Fig. 6–59.** Structural cross section *D-D¹*, Papers Walsh field. From Vincellette and Chittum, 1981, reprinted by permission.

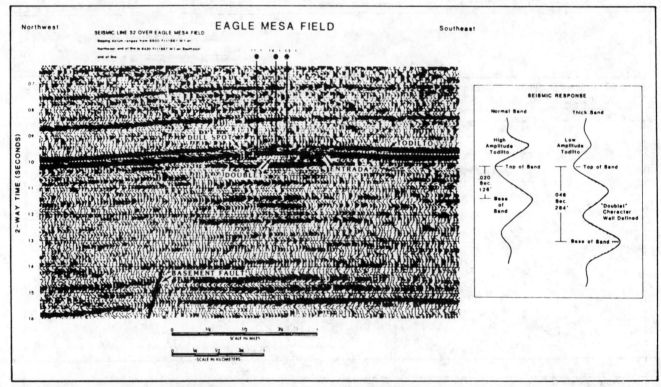

**Fig. 6–60.** Eagle Mesa field. From Vincellette and Chittum, 1981, reprinted by permission.

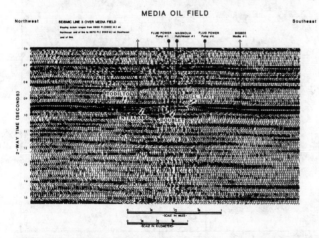

**Fig. 6-61.** Media oil field. From Vincellette and Chittum, 1981, reprinted by permission.

softening of the wavelet-processed reflection across the top of the field. They also show the apparent thickening of the sandstone with the development of a doublet. The wavelet model is illustrated to show development of the doublet (Figs. 6–60 and 6–61).

## Biogenic Low-Energy Transgressive Model

Modern sediments deposited during the Holocene transgression are often similar in texture, structures, bedding patterns, and mineralogy. The transgressive sedimentary unit is characterized by low sedimentation rates, strong biologic modifications, and a mineralogy reflecting chemical equilibrium. The deposited units consist of relict sediments but with phosphate pellets and clastic debris, glauconite, nodules of metal oxides, and shell fragments of various forms and composition. In periods and areas of slow sediment accumulation, sedimentary materials reflect the productivity of the overlying column of water. The provenance is the biological and chemical constituents of the overlying water column acting upon and accumulating on a pre-existing depositional surface. The surface may be formed by wave, tidal, subaerial, or chemical erosion, but the sedimentary response reflects transgressive low-energy deposition. The model, typical of Holocene continental shelves, is common throughout the stratigraphic record (Table 6–1).

The occurrence at unconformity surfaces of thin stratigraphic units composed of biogenic sediments of the type observed on the continental shelves indicates the importance of the low-energy transgressive shelf stratigraphic model. Thick phosphate and glauconite units occur in the Danian (lower Paleocene) in New Jersey overlying the sequence unconformity. A similar association occurs in the Eocene of the Gulf Coast (Triplehorn 1966). These same stratigraphic associations commonly occur at sequence or parasequence boundaries in shelf stratigraphic sections. Similar stratigraphic units can be recognized in interior basins, for example, mid-Cenomanian unconformity surfaces in Wyoming and in Alberta reflect transgressive sedimentary patterns followed by deepening water.

The association of glauconite with phosphate suggests that phosphate minerals also may reflect marine transgressive conditions and may accumulate in significant amounts under conditions of low energy and low rates of sedimentation. The marine Phosphoria formation deposited during the Upper Carboniferous (Absaroka sequence) transgression in the western Cordillera is consistent with this interpretation (Sheldon 1964).

This depositional model suggests profound changes in water depths, patterns of oceanic circulation, and regional changes in provenance and depositional patterns. Similar textural, mineralogic, and biologic responses may occur locally in response to changes in water depth or sediment supply. Identification of the model requires recognition of regional patterns of change and an understanding of the necessary boundary conditions and patterns that reflect shelf or basin sedimentologic processes. A larger stratigraphic context is the only basis for interpreting the shelf sedimentologic framework from specific sedimentary responses.

## BIOCLASTIC SHELF THEMES

Shelf sedimentology is strongly controlled by the development of biochemically produced carbonates. The base of the food chain is the algae developed in warm, shallow marine waters. Three groups of algae are present, and each group fills a differing ecologic niche. Blue-green algae are tolerant to changes in salinity, wet-dry periods, and temperature changes. Green calcareous algae require more equitable conditions. Red algae encrust the surface of deposited sediments. All three forms require sunlight for photosynthesis and, therefore, concentrate in waters less than 10 meters deep. The products of biogenic processes are concentrated in this same shallow water depth or are transported by density currents into deeper water. Algae not only form the base of the food chain, but they also are important producers of carbonate detritus. The nature of clastic particles, the rates of carbonate detritus accumulation, and its distribution reflect biochemical processes. To develop an equilibrium carbonate shelf, the rate of detritus accumulation must be sufficient to maintain a shallow water depth. The development of stratigraphic carbonate units reflects slower rates of sea-level rise and tectonic subsidence.

Physical transport processes are controlled by the geometry of the biogenically constructed and maintained shelf. The interrelation of tidal, wave, density, and wind-driven currents with biogenic processes provides the basis for interpreting sedimentologic patterns. Reefs and shallow shoals develop in areas of high nutrient supply and high wave and current energy, and they absorb much of the energy from the ocean basin to produce a fringing or barrier pattern on the seaward margin of the shelf.

Lower energy areas on the shelf are sites of high carbonate productivity from algae and the disintegration of molluscan shells. Areas of higher wave or tidal energy on the shelf develop as shoals, mounds, and local biogenic communities. These patterns can be related to the shelf geometry and to inlets or breaks in the marginal shoal or reef area. Topographic irregularities in the shelf are submerged in the interrelated physical and biogenic sediment patterns. Local channels may develop in response to tidal currents; shoals may be in response to wave and current patterns on the shelf; biologic communities may reflect local changes in nutrient supply, salinity, turbidity, or water depth.

The shape of the depositional unit is produced by changing boundary conditions, such as increased water depth. Reefs on the shelf margin may show progradation, onlap, or simply form a vertical face producing a steep forereef escarpment. These responses are identifiable, and the pattern of other elements on the shelf and in the ocean basin reflects this type of systematic change. Similar changes in boundary conditions (such as a restriction of the shelf), and changes in climate, terrigenous clastic provenance, and the geometry of the ocean basin are reflected in sedimentary patterns.

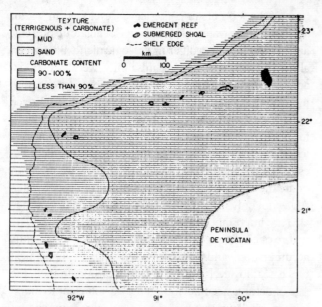

**Fig. 6-62.** Texture and carbonate content of surface sediments of the Yucatan shelf, Mexico. From Ginsburg and James, 1974.

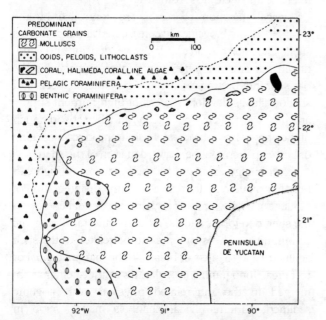

**Fig. 6-63.** Predominant carbonate grains in surface sediments of the Yucatan shelf, Mexico. From Ginsburg and James, 1974.

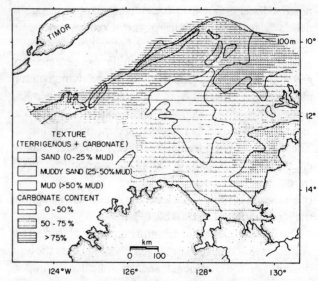

**Fig. 6-64.** Texture and carbonate content of surface sediments on the Sahul shelf. From Ginsburg and James, 1974.

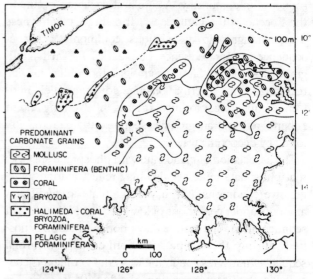

**Fig. 6-65.** Predominant carbonate grains in surface sediments of the Sahul shelf. From Ginsburg and James, 1974.

Within the framework of these boundary conditions, stratigraphic units are interpretable. Many examples of carbonate banks, transgressive sheet carbonate units, patch and pinnacle reefs, and biostromal and biohermal mounds are present in the stratigraphic record. These shelf units are typically associated with ocean and basin margins, and form extensive blanket deposits. Their presence indicates a very characteristic and specialized response to specific boundary conditions and sedimentary processes. Because of the sensitivity of these responses to small environmental changes, stratigraphic models are easily developed and sedimentologic aspects interpreted (Table 6-1).

## Epeiric Sea Carbonate Platform Model

This model occurs on open shelves and in interior epeiric seas and results in thick sedimentary units during periods of slowly increasing sea level. The Yucatan and the Sahul shelves are characteristic of this type of deposition (Ginsburg and James 1974) (Figs. 6-62, 6-63, 6-64, and 6-65). These shelf areas contain a depth-controlled pattern of patch reefs and biostromal mounds. Nutrients are supplied by wind- and tidally-produced currents that cross the shelf area. These currents distribute and winnow the

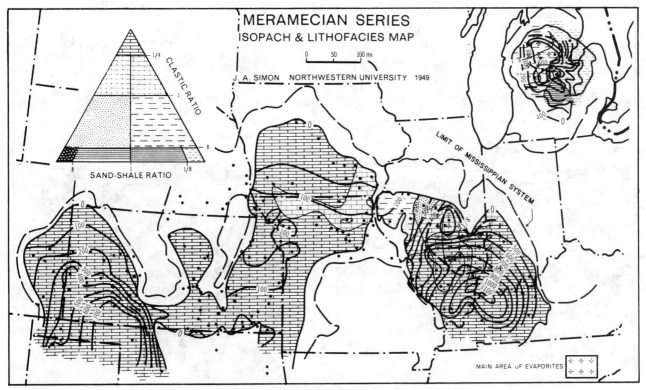

**Fig. 6-66.** Isopach and lithofacies map, Meramecian Series. From Sloss, Dapples, and Krumbein, 1960.

bioclastic detritus and produce sand waves, ribbons, and ripples. Sedimentary structures produced by these currents were studied in the Bahamas by Imbrie and Buchanan (1965); their presence indicates wave and current depositional responses. The sorting and deposition of fine and coarse clastic debris reflect the patterns of currents and the shelf topography. Areas with less energy supply are filled with poorly sorted bioclastic debris. The shelf surface adjusts to a dynamic equilibrium among sediment supply, waves and currents, and biogenetically produced bioherms and biostromes. Shelf depositional patterns reflect an equilibrium sedimentologic model that can be usefully applied to interpretation of stratigraphic units (Table 6-4).

Stratigraphic units from interior seas are characteristic of this type. Mississippian (Lower Carboniferous) units from Illinois are predominantly bioclastic. Their pattern and internal textures and structures have been fully documented (Sloss et al. 1960) (Fig. 6-66).

A similar pattern is suggested for the Madison Limestone of Montana and Wyoming (Sloss et al. 1960) (Figs. 6-67, 6-68, 6-69, and 6-70). Other units from the Ordovician of midcontinent United States and the Silurian (Fig. 6-71) and Devonian of Ohio suggest the stratigraphic framework for bioclastic carbonate units. In each case the Ordovician, Silurian, Devonian, and Mississippian carbonate sections occur in the upper portion of a sequence near the limit of a transgression in a time and an area of relative continental stability. The units reflect repeated local unconformities or paraconformities, large-scale crossbedded units, winnowed fossil fragmental detritus, and many zones reflecting wave and storm depositional processes.

## Regressive-Transgressive Ramp Model

Regressive-transgressive shelf depositional patterns are common in the stratigraphic record. The Holocene history of transgression has left a relict pattern of shelf sediments that can be used to construct a facies framework for comparison to ancient stratigraphic sequences. The west Florida shelf illustrates the depth, textural, and facies patterns similar to those described for the Yucatan and Sahul platforms. On the west Florida shelf, nearshore detrital siliciclastic sands grade seaward into bioclastic and pelleted molluscan sands, and nearer to the shelf edge, vermitid and coralline algal mounds, oolitic sands, and deeper water, foraminiferal carbonate muds (Ginsburg and James 1974) (Figs. 6-72 and 6-73). On the south Florida shelf, a low energy sequence is developed, but also molluscan and pelleted sands and muds occur nearshore. The shelf edge is dominated by calcareous and coralline algae (Ginsburg and James 1974) (Figs. 6-74 and 6-75). The depositional pattern is mostly a ramp,

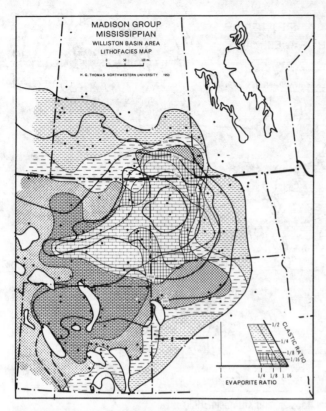

**Fig. 6–67.** Lithofacies map, Madison Group, Mississippian, Williston basin area. From Sloss, Dapples, and Krumbein, 1960.

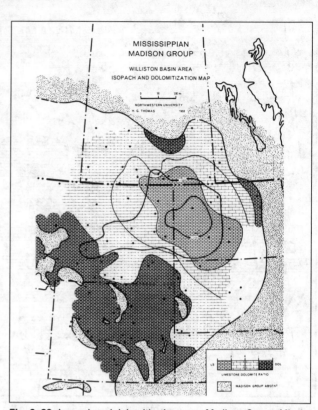

**Fig. 6–69.** Isopach and dolomitization map, Madison Group, Mississippian, Williston basin area. From Sloss, Dapples, and Krumbein, 1960.

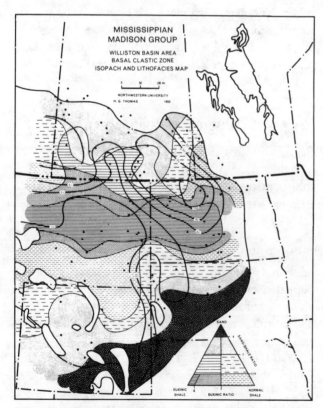

**Fig. 6–68.** Isopach and lithofacies map, Madison Group, Mississippian, Williston basin area basal clastic zone. From Sloss, Dapples, and Krumbein, 1960.

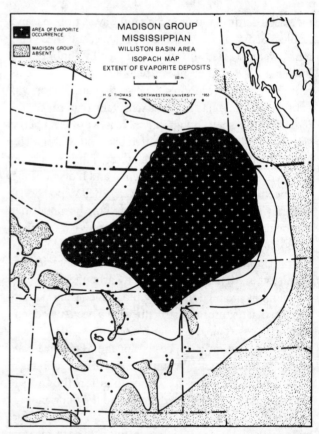

**Fig. 6–70.** Isopach map showing extent of evaporite deposits, Madison Group, Mississippian, Williston basin area. From Sloss, Dapples, and Krumbein, 1960.

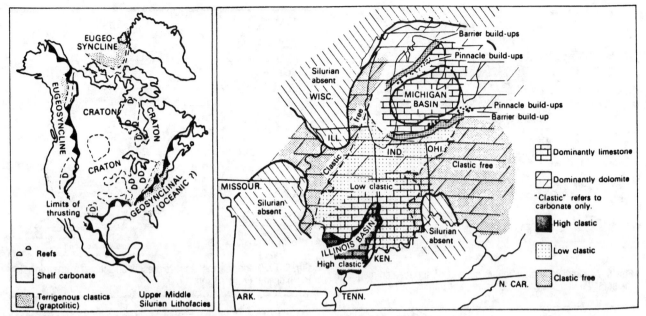

**Fig. 6–71.** Silurian patterns. From Wilson, 1975.

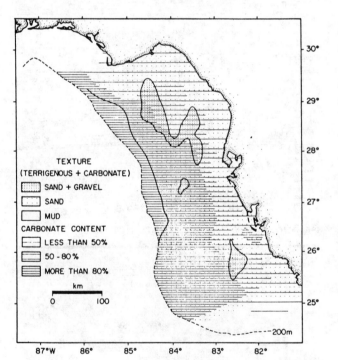

**Fig. 6–72.** Texture and carbonate content of surface sediments on the eastern Gulf of Mexico shelf. From Ginsburg and James, 1974.

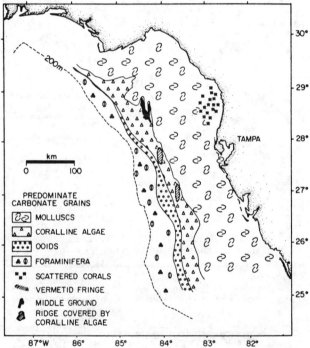

**Fig. 6–73.** Predominant carbonate grains in surface sediments and location of bathymetric highs on the eastern Gulf of Mexico shelf. From Ginsburg and James, 1974.

**Table 6-4**
**Epeiric Sea Carbonate Reservoirs**

Interval of Bioclastic carbonate reservoirs, 5–40 m thick, distributed over areas greater than 20,000 kms²

Oolitic grainstone reservoirs, 5–20 m thick, width to length ratio of 1:5

Ovoid areas, greater than 5000 kms², intervals 10–40 m thick, with greater than 50% of grains and fragments less than 44 um

Biohermal and biostromal sequences from 5–30 m thick, 2–6 kms in diameter, marginal to ovoid areas with grain size less than 44 um

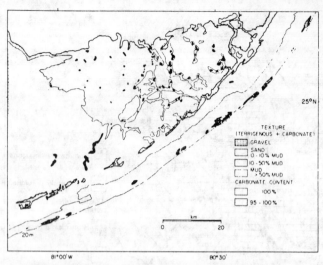

**Fig. 6-74.** Texture and carbonate content of surface sediments on the south Florida shelf. From Ginsburg and James, 1974.

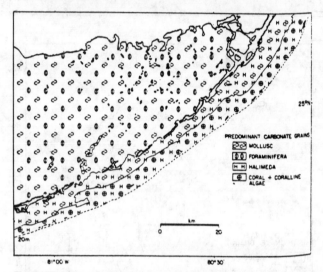

**Fig. 6-75.** Predominant carbonate grains in surface sediments on the south Florida shelf. From Ginsburg and James, 1974.

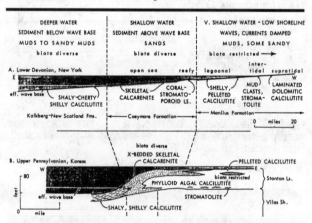

**Fig. 6-76.** Recent marine sedimentation model applied to environmental reconstruction of carbonate units in stratigraphic record on two different scales. *A,* Lower Devonian Helderberg Group in central and eastern New York. *B,* Upper Pennsylvanian Stanton Limestone (Lansing Group, Missourian stage) in central Wilson County near Benedict, Kansas. Cross section shows present topography on top of Stanton corrected for 20 feet per mile regional westward dip computed along algal calcilutite facies for seven miles to east. From Heckel, 1972.

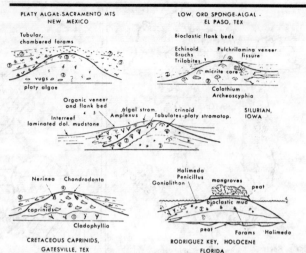

**Fig. 6-77.** Biohermal mounds are present on shelves from Ordovician to Holocene. From Wilson, 1975.

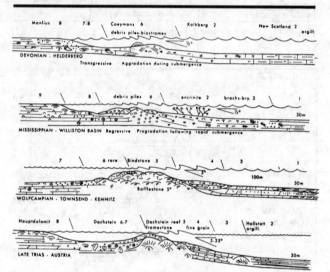

**Fig. 6-78.** Bank marginal facies patterns. From Wilson, 1975.

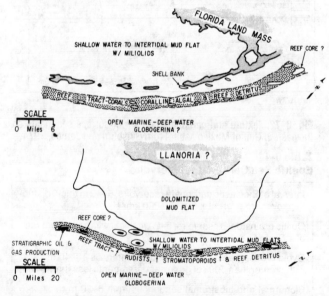

**Fig. 6-79.** Analogy of Lower Cretaceous and Recent reef complexes. From Griffith, Pitcher, and Rice, 1969.

with little change in slope gradient (Heckel 1972) (Irwin 1965) (Fig. 6–76). The distribution of the facies trend containing mounds is marginal to the molluscan sand facies. During transgression, resulting from an eustatic rise in sea level, mounds may grow to 3-20 meters in thickness (Wilson 1975) (Fig. 6–77). Closer to the shelf edge, the mounded trend may become more continuous, and produce a marginal bank (Wilson 1975) (Fig. 6–78). With continued rise in sea level, transgression may result in the development of a parasequence boundary with a deeper water outer shelf, low energy argillaceous micrite, and/or biogenic shale overlying the mounded interval (Table 6–5).

***Ancient Analogues.*** The similarity of the Holocene Florida reef tract to Lower Cretaceous patterns is shown by Figure 6–79. The regressive-transgressive pattern of the Gulf Coast Lower Cretaceous shelf is illustrated by Lozo and Strickland (1956) (Fig. 6–80). Hydrocarbon production is developed from a discontinuous series of mounds in the Edwards formation (Griffith et al. 1969) (Fig. 6–81). Both seismic sections and paleogeographic maps are useful to define the pattern of the bank (Fig. 6–82). Other mounded trends also occur in the Sligo, and Cow Creek formations, and are capped, respectively, by Pine Island and Hosston black, outer shelf shales. These mounds and banks build to near sea level, and small eustatic changes periodically expose the mounds to subaerial leaching. The transgressive framework of deposition provides the seal capping the mounded interval.

Another example from the Midcontinent Pennsylvanian has been described by Heckel (1972) (Fig. 6–83). A similar pattern is developed, including the trend of mounds occurring marginal to the pelleted molluscan sand facies. The Lansing-Kansas City Groups, of Missourian age, are productive of hydrocarbons from the developed mounds. In this example, the overlying Heebner shale is a phosphatic deeper water shale unit (Watney 1984) (Fig. 6–84).

In some instances, if the shelf slope is low, and the relative rise in sea level is rapid, the lower transgressive interval is thin, dominated by laminated or argillaceous micrite. In this case, the interval is overlain by a prograding ramp, with the facies trends vertically stacked from nearshore to deeper water. A pattern similar to the west coast of Florida is developed, but the facies pattern is the reverse of that described for the United States Gulf Coast Lower Cretaceous shelf. This sequence may be associated with a high stand parasequence as described by Vail (1987) (Fig. 2–55).

**Table 6-5**
**Regressive-Transgressive Ramp Reservoirs**

| |
|---|
| Interval from 30–200 m thick, distributed over an area greater than 5000 kms² |
| Basal interval, possibly carbonaceous or siliciclastic, or fine grained argillaceous micrite, with more than 50% of grains less than 44 um |
| Overlain by a bioclastic and oolitic grainstone unit, 5–20 m thick, with a width to length elongation of 1:10 |
| Overlain by ovoid, 2–5 kms in length, biohermal-biostromal porous reservoir sequence, 3–20 m thick |
| Overlain by a molluscan sand sequence, 10–20 m thick, with an area greater than 3000 kms² |
| Many intervals are capped by a lens-shaped nodular or laminated anhydrite sequence form 10 to more than 100 m thick |

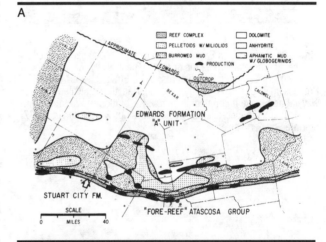

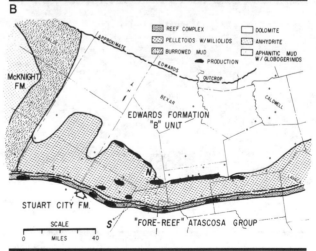

**Fig. 6–80.** (a) Instant-in-time map of the *A* unit of Edwards formation. (b) Instant-in-time map of the *B* unit of Edwards formation. From Griffith, Pitcher, and Rice, 1969.

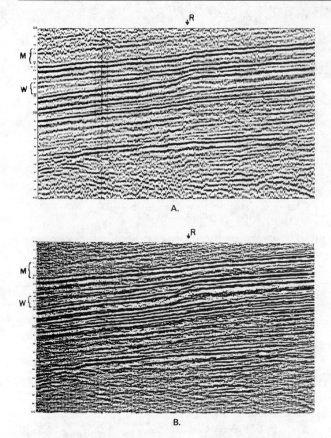

**Fig. 6–82.** A dip line in East Texas. Major reflections generally show how section breaks into its major units. The Midway (*M*) and Woodbine (*W*) are prograding sediments. Note Edwards Reef edge (*R*). Processing (PULSE) is designed to shorten the equivalent source wavelet and make it constant in shape. Resolution of Woodbine pinchouts has been increased by this processing. From Sherriff, 1977, reprinted by permission.

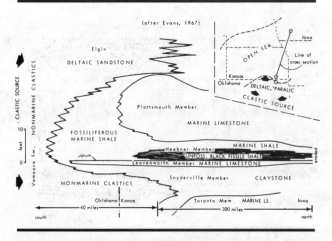

**Fig. 6–83.** Stratigraphic framework of an Upper Pennsylvanian black shale in midcontinent North America, mostly after Evans (1967). Plattsmouth, Heebner, Leavenworth, Snyderville, and Toronto members constitute Oread formation (Shawnee Group, Virgilian stage). From Heckel, 1972.

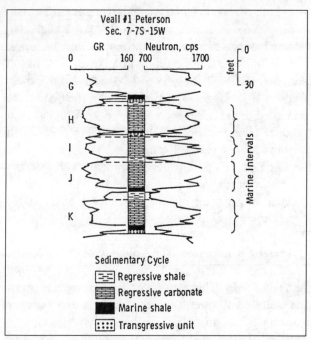

**Fig. 6–84.** Definitions on gamma ray–neutron logs of intervals chosen for mapping intervals are defined and labeled on the appropriate log curve. From Watney, 1984.

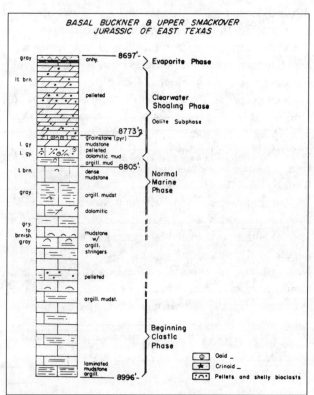

**Fig. 6–85.** Transgressive Jurassic of the Gulf Coast shelf illustrates the depositional sequence. From Wilson, 1976.

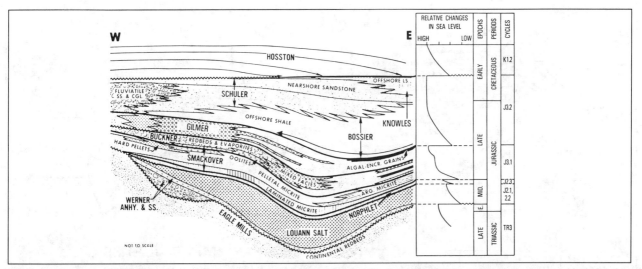

**Fig. 6-86.** Generalized west-to-east cross section across East Texas basin, showing formational units within each sequence (designated by *heavy black lines*). A cycle chart summarizes regional relative changes of sea level for each sequence. Cycles are not plotted to a linear time scale. From Todd and Mitchum, 1977, reprinted by permission.

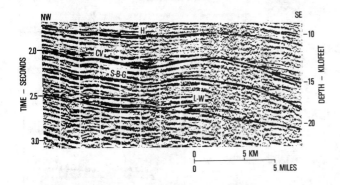

**Fig. 6-87.** Seismic line A, East Texas basin, showing Triassic and Jurassic sequence boundaries. Arrows mark cycle terminations of onlap, downlap, and truncation, which provide criteria for recognizing sequence boundaries. Formations present: Hoston (*H*), Cotton Valley (*CV*), Gilmer (*G*), Buckner (*B*), Smackover (*S*), Louann (*L*), Werner (*W*), and Eagle Mills (*EM*). From Todd and Mitchum, 1977, reprinted by permission.

A sequence of this type is illustrated by the Smackover formation of the United States Gulf Coast (Wilson 1975) (Fig. 6-85). The transgressive-regressive pattern is illustrated by Figure 6-86, and shows the relationships between facies patterns and relative changes in sea level. Two major cycles are present in the Smackover-Gilmer interval (Todd and Mitchum 1977). The seismic record section (Fig. 6-87) shows a thin interval of onlap at the base of the Smackover formation, overlain by continuous reflectors sloping towards the basin center to the southeast. Patterns shown by the stratigraphic and seismic sections suggest both vertical stacking and lateral facies changes to finer grained, deeper water units. An upper Jurassic facies map, which includes the dominantly clastic Cotton Valley formation, supports a paleogeographic interpretation similar to the Holocene west Florida shelf (Sloss et al. 1960) (Fig. 6-88). The eastern Texas basin pattern shows a shelf sequence from coastal

detrital clastics to offshore calcareous pelleted sands and muds, to a central basin filled with argillaceous micrite to carbonate shoals or mounds at the shelf edge (Fig. 6-85). Between the Smackover and Gilmer formations, a nodular anhydrite unit is present, reflecting a slowed rate of relative rise in sea level (Fig. 6-86). This unit may represent lagoonal deposition at a parasequence or a second order sequence boundary. Hydrocarbon production associated with the Smackover formation is related to both oolite and mounded buildups marginal to the shelf pelleted molluscan sand facies. The deeper water intrashelf basinal argillaceous micrite has been indicated to be the hydrocarbon source rock.

The regressive-transgressive carbonate ramp model relates to the largest hydrocarbon accumulations in the world. The Jurassic intervals of the Persian-Arabian Gulf, particularly the Arab zones of the Ghawar field, are similar to those described for the Jurassic of the United States Gulf Coast.

## Intrashelf Basin Biohermal Model

On many shelves, during a relative rise in sea level, the interior shelf basin continues to subside, but does not fill with sediment. This combination leads to a different facies framework than that found on shelves of low slope. The flank of the biohermal bank may steepen, due to the development of a of a basin margin flexure, which results in bioherms up to hundreds of meters in thickness. The potential for such a development may be seen on the Sahul shelf, where the central area of the shelf is topographically depressed (Fig. 6-64). Reefs in this area are shown to develop on the oceanward side of the shelf basin (Fig. 6-65). The relative sea-level rise, sedimentation patterns, and the proximity of a shelf edge: all are impor-

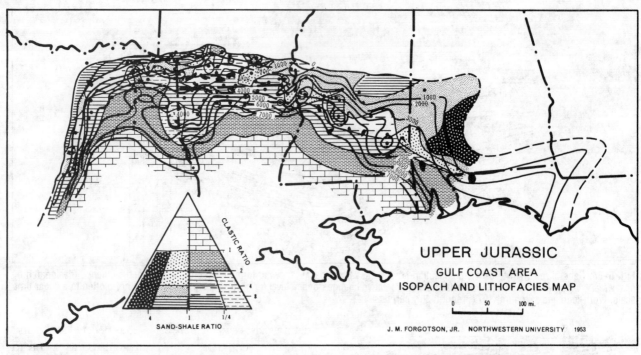

**Fig. 6–88.** Isopach and lithofacies map of Upper Jurassic, Gulf Coast area. From Sloss, Dapples, and Krumbein, 1960.

### Table 6-6
### Intra-shelf Basin Biohermal Reservoirs

Wedge shaped interval of molluscan carbonate sand thinning from shelf edge to the shoreline, 20–200 m thick, with an area greater than 3000 kms²

Discontinuous mounded biohermal and biostromal reservoirs, 20–200 m thick with individual mounds up to 25 kms², on the seaward side of an interval of fine-grained micrite, with greater than 50% of grains less than 44 um

A basinal ovoid interval, from 2000 – 5000 kms², with greater than 50% of grains less than 44 um in diameter

Within the ovoid interval, possibly mounded debris flows, from 5–25 m thick, over areas of 5–50 kms²

Sequences, starting at the base of the interval, may stack and prograde towards the ovoid fine-grained interval

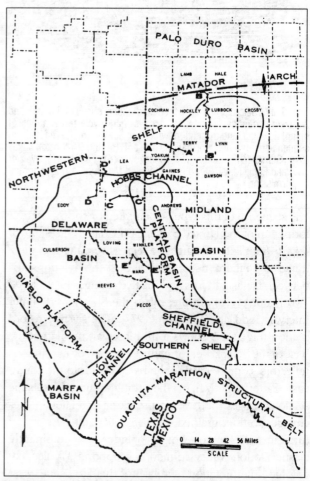

**Fig. 6–89.** Major Permian geologic features, including location of cross sections, West Texas and southeastern New Mexico. From Silver and Todd, 1969, reprinted by permission.

tant in the development of these patterns. Facies sequences and patterns derived from both Holocene examples and ancient stratigraphic patterns are summarized in Table 6–6.

Well-documented examples are developed in Pennsylvanian and Permian stratigraphic units from the Delaware and Permian basins of West Texas (Silver and Todd 1969) (Fig. 6–89). Stratigraphic sequences are developed during periods of rapid rise in relative sea level. Section D-D′ is a stratigraphic section across the northern flank of the Delaware basin, illustrating a portion of the Permian depositional history (Fig. 6–90). This section illustrates the history of basin subsidence, biohermal development, and shelf sequences. The stratigraphic sequence illustrated by the seismic section on the eastern side of the basin does not show the high angle reef flank slope, and well control indicates this area is dominated by shelf edge bank sedimentary patterns (Bubb and Hatlelid 1977) (Fig. 6–91). The Permian Leonardian Series lithofacies map also illustrates the carbonate builups on the north and western sides of both the Delaware and Permian Basins (Sloss et al. 1960) (Fig. 6–92). Similar facies patterns are suggested for the upper Pennsylvanian Horeshoe Atoll of the Permian basin (Vest 1970) (Fig. 6–93). The seismic section shows continuous shelf reflections, mounded buildup at the shelf edge, and an abrupt bank margin slope into a deeper water central basin (Sheriff 1977) (Fig. 6–94). Later depositional events show onlap on both flanks of the shelf edge carbonate buildup.

Similar patterns have been described for many other marginal basins, including the Devonian of western Alberta, Canada, a basinal margin reef development from north Africa (Bubb and Hatlelid 1977) (Fig. 6–95).

The shelf basin reef association is particularly important for major hydrocarbon accumulations. The closed basin allows the preservation of fine grained, organic rich source rocks. Basin margins are often flanked by lagoonal or coastal evaporite units which can provide seals for hydrocarbon traps. In addition the sequence history may produce onlapping patterns after subaerial exposure, leaching, and dolomitization of reefal reservoir sequences.

## Shelf Edge Reef Model

Extensive research has been carried out on Holocene fringing or barrier reefs on the shelf margin and surrounding marine carbonate platforms. The Great Barrier Reef of northeastern Australia has been studied and reported on in detail by Davies and co-workers (e.g., Davies et al. 1985). In addition, the fringing reef on the margin of the Belize shelf off British Hondurus has been studied in detail (Purdy 1974). Maps of sediment facies, and shelf topography have been prepared for both of these examples by Ginsburg and James (1974) (Figs. 6–96, 6–97, 6–98, and 6–99). Reefs grow rapidly in areas of upwelling of marine waters adjacent to narrow shelves. Organic productivity is controlled by water temperature, nutrient supply from both continental and marine sources, and the presence of relatively high tidal or wave

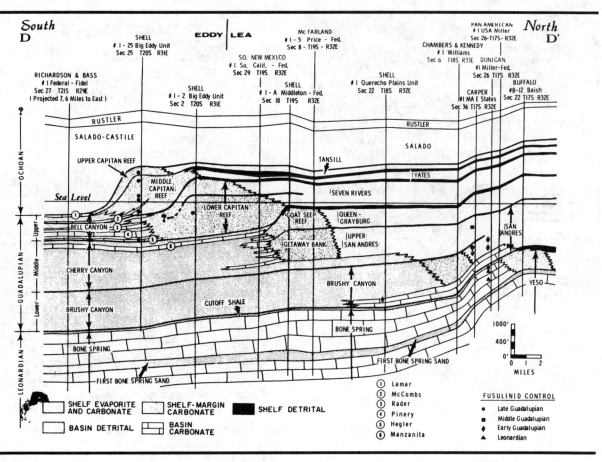

**Fig. 6–90.** South-north cross section D-D¹, Guadalupian physical stratigraphic framework, northern Permian basin, Eddy and Lea Counties, New Mexico. Vertical scale in feet. From Silver and Todd, 1969, reprinted by permission.

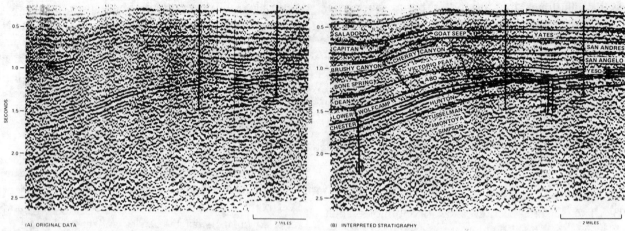

(A) ORIGINAL DATA     2 MILES     (B) INTERPRETED STRATIGRAPHY     2 MILES

**Fig. 6–91.** Central basin platform, Lea County, New Mexico (12-fold CDP vibrosis® data). Shelf-margin carbonate-bank buildup on this line is indicated by (1) abrupt change in dip at shelf edge and (2) seismic facies change from high-amplitude, continuous reflections to low-amplitude, nearly reflection-free zone at shelf edge. Leonardian and Guadalupian shelf-margin banks, composed mainly of dolomitized skeletal limestones of the Abo, Victorio Peak, Goat Seep, Getaway, and Capitan formations, are documented by wells in this part of Permian basin. Basinward of shallow-water banks are siltstones, shales, and micritic limestones of Dean, Bone Spring, Brushy Canyon, and Cherry Canyon formations; shelfward of banks are thin-bedded, dolomitized micritic and dolomitized algal-laminated limestones and sandstones of the Yates, Seven Rivers, Queen, Grayburg, San Andres, San Angelo, and Yeso formations. From Bubb and Hatlelid, 1977, reprinted by permission.

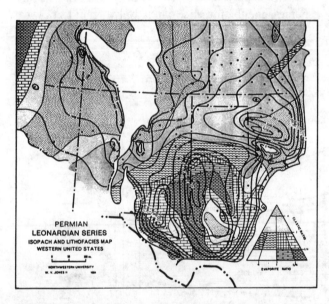

**Fig. 6–92.** Isopach and lithofacies map, Permian Leonardian Series, western U.S. From Sloss, Dapples, and Krumbein, 1960.

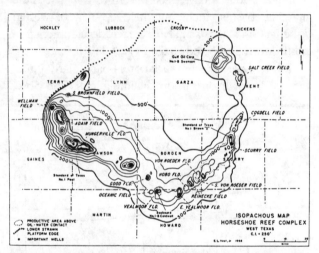

**Fig. 6–93.** Isopachous map of the Horseshoe reef complex in west Texas showing thickness of reef limestone and location of significant fields producing along the crest of the atoll. From Vest, 1970, reprinted by permission.

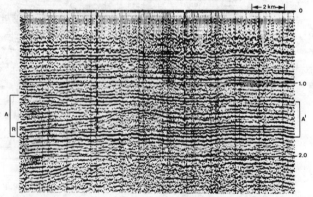

**Fig. 6–94.** Section across Horseshoe Atoll in West Texas; R denotes the portion of the section that contains the reef (just left of center). The backreef area of flat-lying, strong, continuous reflections is to the right. The forereef showing an entirely different progradational reflection pattern is to the left. Note the deterioration of data quality below the reef. From Sheriff, 1977, reprinted by permission.

energy. The thickness of the reef interval is a product of subsidence and eustatic sea-level rise, with intervals more than 1000 meters thick preserved in the stratigraphic record. The cause of subsidence is little understood, but appears to be related to carbonate productivity (Menard 1983) (Fig. 6–100). His study of atolls, guyots, and volcanic islands in the South Pacific suggests a direct relationship between the presence of thick carbonate intervals and subsidence. In the northern area of the Great Barrier Reef of Australia, more than 3000 meters of Neogene reefal carbonate is preserved (Davies et al. 1985). Thicknesses of these magnitudes suggest flexuring of the crust at the shelf edge, or in the case of the South Pacific atolls, possibly crustal thinning.

Stratigraphic patterns illustrate the potential for thick reefal buildups kilometers in thickness. Seismic profiles of a Jurasssic reefal buildup offshore west Africa (Fig. 6–101), and a Miocene builup off Papua, New Guinea (Fig. 6–102), illustrate these geometric patterns (Bubb and Hatlelid 1977).

Of particular interest to the explorationist is the low productivity of many shelf edge carbonate buildups. Without later progradation of the shelf edge, the reef tract is not overlain by low energy sealing shales. Due to the narrowness of reefed shelves, there is often an insufficient volume of source rocks preserved in depth restricted basins on the shelf to produce commercial accumulations of hydrocarbons. The forereef, however, may be the site of isolated patch reefs, such as the Golden Lane reef in the Gulf of Mexico, off Mexico. Also many forereef areas are the site of deposition of debris flows derived from the reef

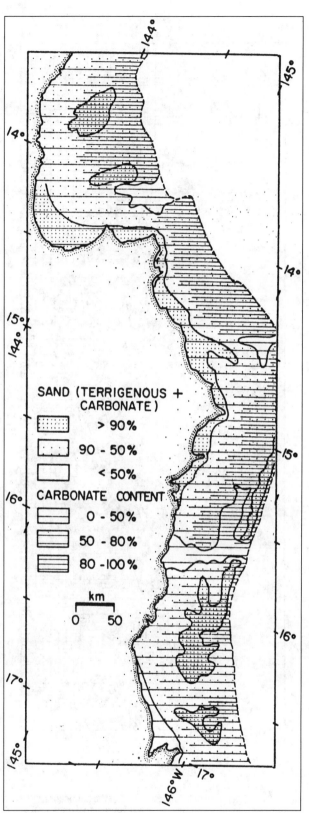

**Fig. 6–96.** Texture and carbonate content of the surface sediments on the relatively narrow northern part of the Queensland shelf, eastern Australia. From Ginsburg and James, 1974.

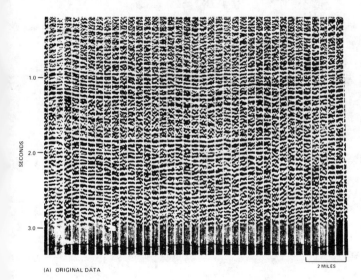

(A) ORIGINAL DATA

**Fig. 6–95.** Carbonate buildups both low platformlike banks and pinnacles are interpreted on basis of (1) Seismic facies change ham continuous parallel rejectors to mainly reflection-free to very discontinuous reflection zone, and (2) one to two cycles of onlap of overlying units onto buildups. From Bubb and Hatlelid, 1977, reprinted by permission.

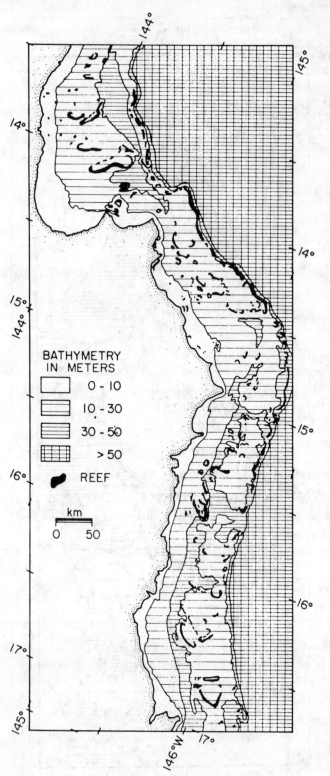

BATHYMETRY
IN METERS

| | |
|---|---|
| | 0 - 10 |
| | 10 - 30 |
| | 30 - 50 |
| | > 50 |
| REEF | |

km
0        50

**Fig. 6–97.** Bathymetry of the northern part of the continental shelf off Queensland, eastern Australia. From Ginsburg and James, 1974.

PREDOMINANT
CARBONATE GRAINS

MOLLUSC
FORAMINIFERA
CRYPTOCRY STALLINE GRAINS
H    HALIMEDA
P    PTEROPOD
CORAL, CORALLINE ALGAE, HALIMEDA
REEF

km
0        50

NOT KNOWN

200m

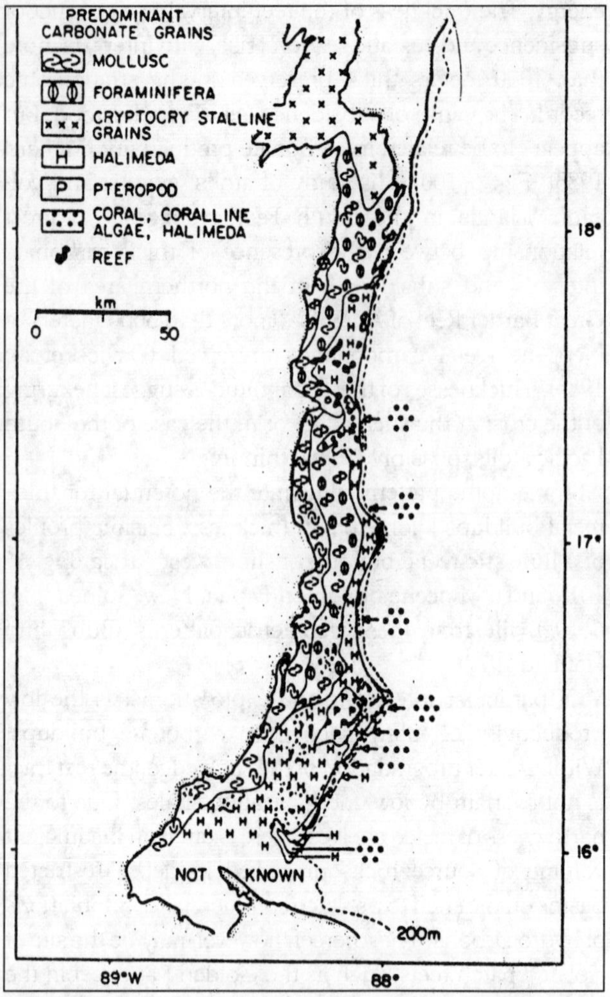

**Fig. 6–98.** Predominant carbonate grains in surface sediments on the continental shelf off Belize, British Honduras. From Ginsburg and James, 1974.

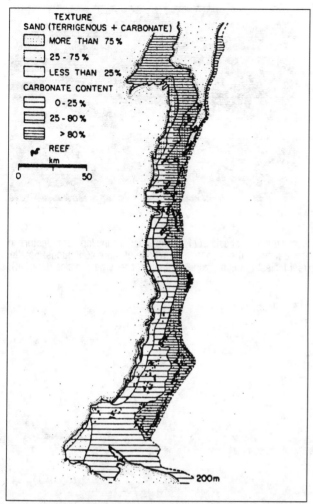

**Fig. 6–99.** Texture and carbonate content of surface sediments on the continental shelf off Belize, British Honduras. From Ginsburg and James, 1974.

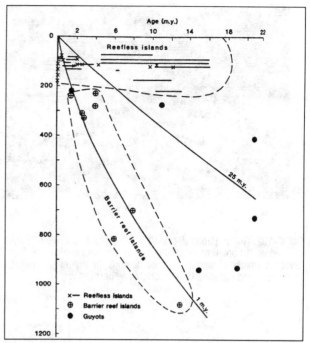

**Fig. 6–100.** From Menard, 1983.

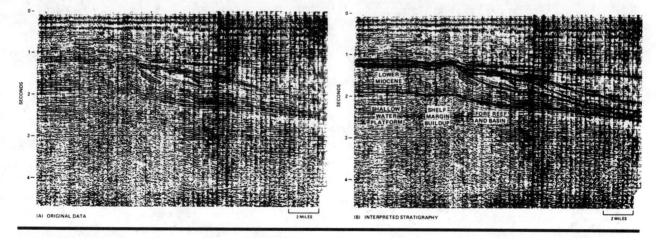

**Fig. 6–101.** Offshore West Africa (12-fold CDP Aquapulse® data). Shelf-margin carbonate buildup can be seen by (1) reflection from top and front of buildup, (2) onlap of cycles onto buildup, (3) change from continuous, parallel reflectors into discontinuous reflectors, (4) numerous diffractions, (5) drape over buildup, and (6) abrupt changes in dip of reflectors. Wells encountered series of Mesozoic shelf-margin buildups along eastern Atlantic continental margin off Africa. Buildup displayed on this line is interpreted as Late Jurassic. From Bubb and Hatlelid, 1977, reprinted by permission.

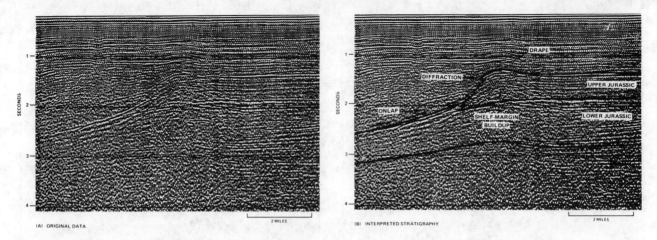

**Fig. 6-102.** Gulf of Papua (6-fold CDP dynamite data). Shelf-margin reef is interpreted on basis of (1) abrupt changes in slope of reflectors at shelf edge, (2) onlap of slope and basin units onto shelf edge, and (3) seismic facies pattern change from zone of more continuous, parallel reflectors (backreef-lagoon) to discontinuous, nearly reflection-free zone (reef facies) to thinner zone of dipping, convergent cycles (forereef). From Bubb and Hatlelid, 1977, reprinted by permission.

tract. These patch reef and mounded debris flows intervals often are onlapped and buried by finer-grained, deeper-water biogenic shales. In these cases shelf progradation and subsidence may result in sufficient burial depths for hydrocarbon generation, but preservation of organic rich shales is still a problem, unless the shelf margin is restricted due to mounding or to tectonic events.

## SUMMARY

Shelf models have been poorly defined. They represent a complex of processes that cannot be directly observed. The suggestion of seven differing siliciclastic and carbonate models is only a beginning of a synthesis of these response patterns. Certainly the range of processes observed requires unifying themes that can lead to the interpretation of the origin of diverse stratigraphic patterns. Little is known of response patterns from Holocene shelves; therefore, ancient stratigraphic patterns must be inferred from limited objective data. The seven models are sequences presently recognized. Petroleum exploration must reflect the interrelation of blanket, or shelf-edge, permeability patterns with the presence of intrashelf and intracratonic basins containing source rocks. These frameworks are common in the stratigraphic record. Exploration for siliciclastic or biochemical ramp sequences require an understanding of shelf geometry but may lead to large hydrocarbon accumulations, in the range of tens of billions of barrels of recoverable reserves.

## REFERENCES FOR FURTHER STUDY — CHAPTER 6

Allen, J.R.L., 1980, Sand waves a model of origin and internal structure: Sediment. Geol., v. 26, p. 281–328.

de Boer, P.L., A. van Gelder, and S.D. Nio, eds., 1988, Tide-influenced Sedimentary Environments and Facies: Dordrect, Holland, D. Reidel Publish. Co., 530 p.

Ginsburg, R.N., and N.P. James, 1974, Holocene carbonate sediments of Continental shelves, in Burk, C.A., and C.L. Drake, eds., The Geology of Continental Margins: New York, Springer-Verlag, p. 137–155.

Johnson, H.D., 1978, Shallow siliclastic seas, in Reading, H.G., ed., Sedimentary Environments and Facies: New York, Elsevier, p. 207–258.

Knight, R.J., and J.R. McLean, eds., 1986, Shelf Sands and Sandstones: Canadian Soc. of Petrol. Geol., Mem., no. 11, 347 p.

Murris, R.J., 1980, Middle East Stratigraphic evolution and oil habitat: Am. Assoc. Petrol. Geol. Bull., v. 64, p. 597–618.

Nummedal, D., O.H. Pilkey, and J.D. Howard, eds., 1987, Sea-level Fluctuations and Coastal Evolution: Soc. Econ. Paleont. and Mineral. Spec. Pub. no. 41, 267 p.

Sellwood, B.W., 1986, Shallow-marine carbonate environments in Reading H.G., ed., Sedimentary Environments and Facies: New York, Elsevier, p. 283–342.

Stride, A.H., ed., 1982, Offshore Tidal Sands—Processes and Deposits: New York, London, Chapman and Hall, 213 p.

Tillman, R.W., and C.T. Siemers, eds., 1984, Siliciclastic Shelf Sediments: Soc. Econ. Paleont. and Mineral., Spec. Pub. no. 34, 268 p.

Wilson, J.L., 1973, Carbonate Facies in Geologic History: New York, Springer-Verlag, 471 p.

# 7 EVAPORITE SEDIMENTOLOGY

Facts do not cease to exist. . . because they are ignored.   A. Huxley

## INTRODUCTION

The interpretation of the depositional history of evaporite units is difficult because few Holocene
examples are available to use as a basis for interpreting
either boundary conditions or sedimentary processes.
Response characteristics from many ancient sedimentary
units can be determined, but these do not provide sufficient information to interpret water depths, dynamic
balance between sea-water inflow and precipitation, rates
of accumulation, and post-depositional replacement and
alteration of deposited units. The balance between evaporation and inflow of waters of lesser concentration
appears to be critical, but the chemical or physical basis
for equilibrium has not been determined.

From geomorphic and paleogeographic patterns, it
appears that evaporite units occur in three differing
settings:

▶ Lagoonal or restricted seas and bays marginal to the
coastline,
▶ Partially restricted topographic basins filled or partially filled with brine, and
▶ Shoreline units associated with sabkhas.

Each of these associations reflects different boundary
conditions and depositional processes (Table 7–1). Inferences concerning these controls must be based upon
areal and vertical patterns of sedimentary units rather
than experimental or actualistic examples. Models based
upon inferences rarely can be used to predict sedimentary responses, but they are useful for comparisons.

## BOUNDARY CONDITIONS

The area of an evaporite sedimentary unit is an important indication of the ratio of evaporation to inflow. In
the progradational shoreline model of sedimentation, the
evaporite unit results in a continuous zone across a larger
portion of a shelf with limited topographic controls. In
lagoonal and restricted topographic basins, the geometry
of the topographic surface and the area of the deposited
sedimentary unit are interrelated. The boundary conditions of surface topography, tectonic history or sea-level
rise, and climate appear to be the major controls for these
types of evaporite units. In the shoreline or sabkha unit,
the control is primarily diagenetic and relates to climate
and concentration of capillary waters by evaporation.

**Table 7-1**
**Evaporite Models**

| | Restricted Topographic Basin | Lagoonal | Sabkha |
|---|---|---|---|
| Origin of brine | Inflow related to head created by evaporation | Flooding and complete evaporation | Capillary action related to proximity to the sea |
| Thickness | Thickness to 100's of m | Thin evaporite cycles maximum thickness is 10's of m | Thin to thick, depending on tidal flat cycles; typically 10–100 m |
| Mineralogy | Depends upon pattern of restriction: possibly monomineralic or proportions similar to sea water | Proportions the same as in original brine; halite to gypsum in sea water: 30/1 | Usually restricted to gypsum and anhydrite, possibly traces of halite and other mineral phases |
| Pattern of deposition | Minerals distributed in response to pattern of inflow | Related to topography of the lagoon | Related to supratidal units, possibly blanket |
| Bedding | Laminated to massive | Irregular, lensoid, disrupted | Nodular |
| Depositional history | Occurs during periods of onlap in marginal topographic basins | May occur in any stratigraphic unit without regard to depositional cycle | Occurs during offlap and results from exposure of tidal-flat sequences |

Most Holocene restricted basins and ancient evaporite units are ovate or elongate in form. Ovoid areas are strongly associated with marginal basins and topographic depressions; the entrance to the sea is limited by carbonate banks, bioherms, or marginal uplifts (Fig. 7–1). Elongate basins are commonly related to block-faulted plate margins either parallel or transverse to the continental margin. Restriction is associated with topographic uplifts at the terminus of faulted grabens. In both cases the geometry of the evaporite unit is controlled by the elevation of the basin floor and its relation to sea level. The geometry of the connection to the open sea appears to be critical to the development of an evaporite unit.

Thickness of evaporite sections reflects the topography of the basin or depression. Changes in topography or elevation during deposition cause changes in the geometry of the evaporite unit. In Holocene restricted basins or ancient sedimentary units, there is little suggestion of a dynamic equilibrium between depositional patterns and topographic responses. Sedimentation appears to be passive, and evaporite units fill the available space unless the developmental framework is modified by changes in entrance geometry or sea level.

## Tectonic and Eustatic Controls

The elevation and geometry of the inlet are the primary controls for restriction in evaporite basins. Consequently, systematic changes resulting from tectonism or eustatic controls profoundly affect evaporite sedimentology. Most evaporite units within topographic depressions or in marginal basins occur during periods of rising sea level. The last stage of offlap during the closing portion of a sequence exposes the continental surface. Tectonic and erosional forces modify this surface; grabens, basin subsidence, and erosional features provide the topographic framework that controls evaporite sedimentology. During rising sea level, these irregularities form restricted basins, troughs, and lagoons. The thickness of an evaporite sequence reflects the interrelation of topography and sea-level rise. Since rising sea level during the first stages of a sequence is systematic and progressive, a dynamic equilibrium is developed between evaporite deposition and sea level. In restricted basins with little sediment supply and the necessary climatic boundary conditions, thick evaporite sequences can be developed.

The tectonic control for evaporite deposition relates directly to patterns of plate tectonics and development of intracontinental and marginal grabens and basins. The tectonic history of these basins and rising sea level are the dominant controls for formation and preservation of evaporite units.

Evaporite units occur at the beginning of sequences in the marginal Williston basin at the close of the Mississippian, at the close of the Permian in the marginal Permian basin of Western Texas, in the middle Jurassic in the Gulf of Mexico, and in the Cretaceous Aptian Epoch of coastal graben basins in Brazil and Angola. Miocene evaporites of the Mediterranean also reflect a period of rising sea level. The associations between tectonic and eustatic changes are characteristic of many evaporite associations. These evaporite units represent a rapid fill of basins and are not formed in response to sediment accumulation.

## Climatic Controls

Without an equilibrium between influx of fresher waters and evaporation, a dynamic balance cannot exist. Cli-

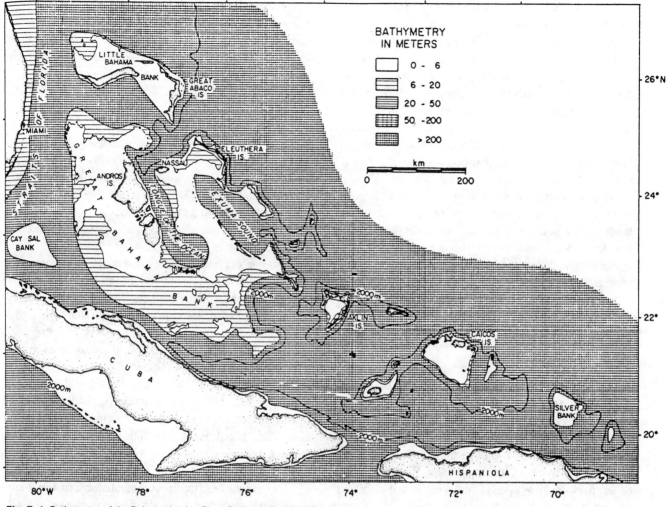

**Fig. 7–1.** Bathymetry of the Bahama banks. From Burk and Drake, 1974.

mate controls in conjunction with patterns of surface drainage and the influx of oceanic water provide the framework for evaporite sedimentation. Without excess evaporation, concentrations of three to ten times that of normal sea water could not develop. Evaporite units are positioned in relation to large continental areas, specific latitudes, and mountain systems. Development of thick evaporite sequences requires a combination of many of these boundary conditions. The difficulties in interpreting development of evaporite sedimentary sequences may rest with our limited understanding of these paleogeographic controls.

Not all evaporite sequences must be associated with deserts; simply a balance between evaporation and inflow is required. Rain shadows, continental margins with little fresh water runoff, subtropical latitudes, and continental interiors can be the sites of evaporite sedimentation. The common occurrence of evaporite sequences in ancient stratigraphic sections suggests multiple causes for their formation. Climate is an essential control, but optimum climatic conditions are common to many areas and times.

These are reflected throughout the stratigraphic record by evaporite deposits.

## Diagenetic Controls

Many evaporite units are related to diagenetic alteration of carbonate sediments. Connate waters saturated with evaporite mineral phases may be present in carbonate sediments, and progressive concentration may yield evaporite minerals as pore and fracture fillings, nodules, and replacement of other mineral phases. Close proximity of brines to subaerially exposed sediment surfaces can result in continuing brine concentration. In areas of low rainfall or periodic droughts, surface evaporation of vadose water may be reflected in evaporite deposition. The importance of this process is evident from study of Holocene coastal sedimentary units in many areas of the world. Boundary conditions responsible for these evaporite units include climate, proximity to brines, and characteristics of the deposited sediment.

The stratigraphic importance of these evaporite deposits has not been fully evaluated. Their thickness, association with carbonate sedimentary units, and the dynamic progradational history are interrelated. The separation of these responses and their relation to boundary conditions and to sedimentary processes can be inferred by comparing sedimentary characteristics from Holocene examples to similar characteristics preserved in ancient sedimentary units. The extent of diagenetic alteration is not simply related to a single process or boundary condition. The response may range from a few evaporite crystals or nodules to massive evaporite beds. Replacement may be selective; it may produce bedded or laminated evaporite units that cannot be simply related to the depositional framework. The determination of climatic, tectonic, and eustatic boundary conditions may provide a basis for interpreting these stratigraphic units.

## SEDIMENTARY PROCESSES AND RESPONSES

Evaporite sedimentology combines aspects of chemical and physical processes. The solubility of mineral phases in brines ranging from sea water to concentrations of more than 100 times sea water was thoroughly investigated (Braitsch 1971). The sequence of precipitation of evaporite minerals from progressively concentrated sea water is a much more difficult problem. Physical aspects of temperature, the rates of removing specific mineral phases from the brine, the density and tidal currents transporting brines of differing concentrations into contact with nonequilibrium mineral phases, and other such processes are reflected in the mineralogic and textural aspects of evaporite units. Also, at earlier stages of concentration, biochemical processes may strongly influence ion concentrations and mineral equilibria. These aspects cannot be simply related to develop depositional models to account for areal and vertical mineralogic patterns within an evaporite unit.

Depositional processes are related to patterns of density currents and to flow produced by differences in elevation of the water surface across restricted basins. These processes can be utilized to account for areal variations in mineralogy within evaporite units. Other processes related to convection and convective overturn, diffusion, and settling of crystals may also be important controls in evaporite sedimentology. The possible significance of each of these processes can be evaluated in relation to depositional patterns developed in ancient stratigraphic units.

Processes that affect the diagenetic formation of evaporite units include transport of water through the deposited sediment by capillary forces and ground water flow. Chemical equilibria among the brine and mineral phases within the sediment and evaporation are process controls for deposition. These processes are not easily interrelated but their products can be described in ancient stratigraphic units.

## Flow Patterns in Response to Evaporation

The single most important process is the flow produced by progressive evaporation. On a shallow shelf, lagoon, or restricted basin, an inflow of marine waters results from the higher head at the basin or shelf margin. The amount of inflow is directly proportional to the amount of evaporation that occurs over the surface of the restricted area. The slope of the water surface can be modified by onshore or offshore winds, tidal action, and surface drainage or groundwater flow from the land area, yet an equilibrium must be maintained. The greater the evaporation or distance from the inlet area, the higher the resultant brine concentration. This current or flow pattern requires that all of the dissolved constituents in the water be removed within the basin. Since the system is dynamic, even if the mean concentration of the brine in the basin does not reach saturation for individual evaporite mineral phases, the elemental components contained in the brine must still be removed.

Three flow responses are possible:

▸ Flow from the inlet across the basin with progressive concentration of the brine. Sedimentary patterns are areally distributed in response to evaporation. Distal portions of the basin are the site of deposition from the most concentrated brines. Thus, the depositional sequence and the areal patterns are the inverse of solubility.

▸ Evaporation and concentration of brines on the basin margin reflected in flow of denser brines toward the center or deeper portion of the basin. The depositional pattern is concentric with more soluble constituents deposited in the center of the basin.

▸ A dynamic system with reflux of denser brines flowing under the incoming marine waters. If the system is open or partially open, some of these denser brines may be lost from the basin system. This model could result in a simple monomineralic depositional pattern or simple or complex vertical cycles.

Progressive evaporation combined with flow away from the ocean can occur in many basin types. Data from the Red Sea graben during one portion of the year illustrates the pattern of surficial flow (Braitsch 1971, 252) (Fig. 7–2). With a net annual evaporation of 150 centimeters, a seasonal increase in salinity from 27.4 to 29.1% must result in the removal of 192 grams of halite per liter of brine (Briggs 1958). This type of calculation illustrates that within a dynamic system, salts must be removed from the basin. However, the mechanism responsible for this removal is not so easily determined. Also, the pattern of sedimentation, mineralogy of the evaporite deposits, and other sediment characteristics may be highly variable. Another example of this model is the Rhine graben (Richter-Bernburg 1972, 284) (Fig. 7–3).

## SEDIMENTOLOGIC THEMES

**B**oundary conditions and sedimentary processes are interrelated in a dynamic manner. The basin shape in

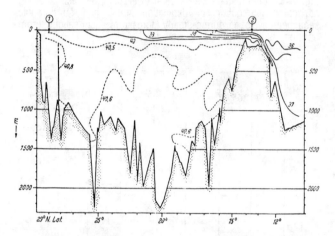

**Fig. 7-2.** The salt content of the Red Sea in %, June 1958. (1) Gulf of Aquaba and (2) Straits of Bab el Mandel in the Gulf of Aden. From Braitsch, 1971.

part controls the pattern of flow within the basin, which in part controls the pattern of evaporation. The pattern of evaporation is reflected in the areal distribution of evaporite sedimentology and the developmental history. These aspects of the sedimentology can be associated in a number of paleogeographic settings including the shallow marginal lagoon or bay, the partially restricted topographic sedimentary basin, and the coastal zone. In each setting, deposition is modified by changes in boundary conditions, which may be recognized by stratigraphic relationships or internal characteristics within the evaporite unit. Evaporite sedimentary models can be usefully related to these three patterns (Table 7–1).

### Restricted Basin Model

The developmental history of thick, laminated, and monomineralic evaporite sequences is a particularly difficult stratigraphic problem since, again, no Holocene examples are available to provide an actualistic basis for interpretation. Processes of evaporite sedimentation, however, can be inferred from evaporation and flow patterns from Holocene restricted basins. Oceanographic studies of restricted seas provide the basis for developing models for restricted basin evaporite deposition. Within the framework of rising sea level and an excess of evaporation over water inflow, topographic basins must be filled with evaporite sequences. These boundary conditions are

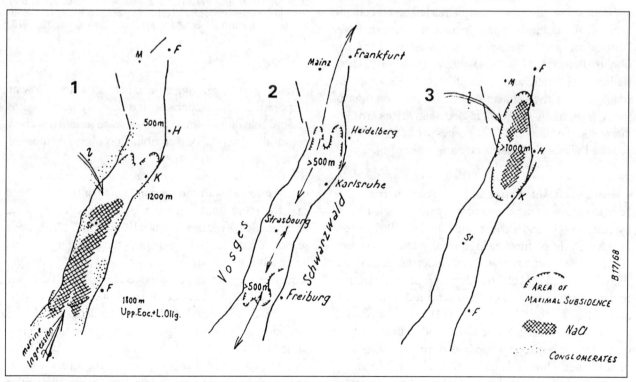

**Fig. 7-3.** Upper Rhine graben during (1) Upper Eocene-Lower Oligocene, (2) Middle-Upper Oligocene, and (3) Miocene. From Richter-Bernberg, 1972.

common throughout the stratigraphic record and result in basinal evaporite sequences.

A simplified illustration of a possible physical model was suggested by King (1947) and utilized by Scruton (1953) (Fig. 7–4). This model can be used to explain the complete removal of all dissolved ions within an evaporite basin, and it suggests that the total composition of deposited mineral phases must be proportional to those present in sea water. The complete removal of all phases in a salt basin is illustrated by the Permian Zechstein of western Germany (Schmalz 1979) (Fig. 7–5). Deposition includes, in the most distal extent of the basin complex, potassium and magnesium salts.

The development of a partial restriction allows reflux of the more concentrated brine back into the ocean. This reflects a dynamic flow system, and equilibrium results in the basin filled with a single evaporite minerals phase. The relative height of the sill, the pattern of wind or tidal action, and the geometry combine to produce differing models (Brongersma-Sanders and Groen 1970) (Fig. 7–6). These models illustrate the complexity of the system and the possible variations in sedimentary responses. Since the change in boundary conditions is slow with respect to the sedimentation rate, a coherent facies pattern may be developed.

The depositional response may be similar to that described by Hite for the Paradox basin (Hite 1970, 71) (Figs. 7–7 and 7–8). In these cases, the restriction may be reflected in part by the sedimentology with the biochemical deposition at the basin margin related to reflux of the brine. Other examples might include the Jurassic lower Portlandian and Triassic in northwestern Germany (Richter-Bernburg 1972, 282–3) (Fig. 7–9 and 7–10). Three well-studied stratigraphic examples of thick evaporite sequences are the Permian Castile-Salado formation of the Permian basin, the Permian Zechstein of western Germany, and the Silurian A-1 Evaporites of Michigan. These provide the model for restricted basin associations.

***Permian Basin Patterns.*** The Permian basin has been the focus of many years of study with classic reports by King (1947) and Adams and Rhodes (1960). More detailed study of the developmental history has been made on the Castile formation by Anderson et al. (1972) (Figs. 7–11a and 7–11b). The interpretation of the lamination as varves provides a detailed depositional time scale and history for the fill of the basin. The ratio of $NaCl$ to $CaSO_4$ is approximately one to four rather than thirty to one, as would be the response from complete evaporation of sea water. Thus, this basin must have had a strong reflux of saturated brine back into the ocean (Adams and Rhodes 1960). The other characteristic of these sediments

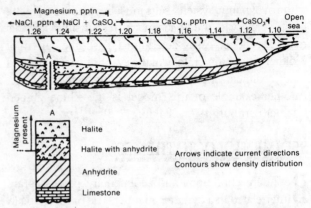

**Fig. 7–4.** Longitudinal cross section of evaporation basin showing horizontal segregation of deposition zones and stratigraphic sequence developed during steady increase of basin salinity. From Scruton, 1953, reprinted by permission.

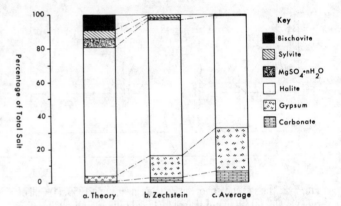

**Fig. 7–5.** Relative abundances of major evaporites (a) as they should be produced by total evaporation of normal sea water, (b) as they occur in Zechstein deposit, and (c) as they occur in average evaporite deposit. Adapted from Borchert and Muir, from Schmaltz, 1979, reprinted by permission.

is lamination, suggesting a periodicity of water inflow or evaporation. This response could possibly be related to seasonal change in temperature, surface runoff, direction of prevailing winds, or a combination of one or more of these aspects.

In the Salado formation, the proportion of evaporites is nearly reversed with the predominance of halite. In this case the most soluble mineral phase, sylvite, is found near the edge of the basin (Jones 1972) (Fig. 7–12). A cross section across the basin illustrates the changing depositional pattern for one of the sylvite zones (Fig. 7–13). This sequence appears to be developed in a restricted basin with little reflux. Thus, the change from a partially open to a closed basin developed during the depositional history and must reflect changing boundary conditions. The same sedimentary response could be developed by a number of different changes, including the fill of the topographic depression, changing sea level, or changes in the geometry of the inlet area.

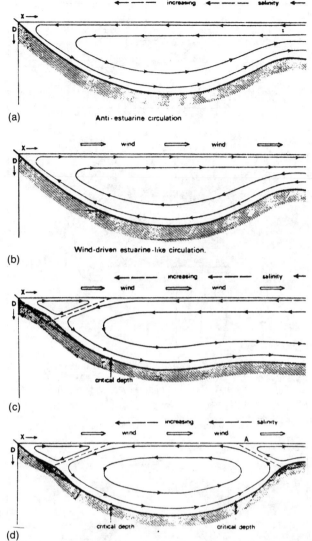

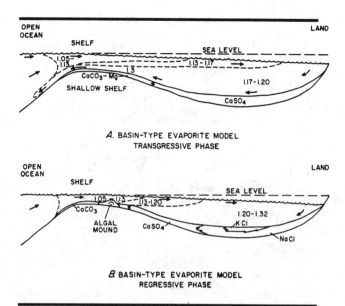

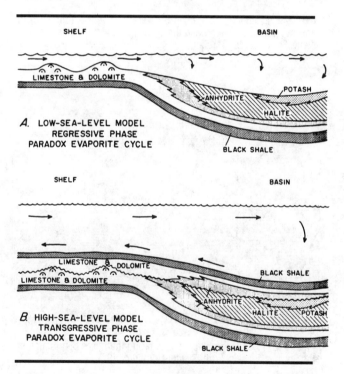

**Fig. 7-6.** *(a)* Circulation in basin with horizontal salinity gradient, wind effect negligible; *x* indicates axis and direction of opening of basin. *(b)* Circulation in basin without horizontal salinity gradient, wind blowing toward opening. *(c)* Circulation in basin with horizontal salinity gradient and counteracting wind; water depth above the sill larger than critical depth. *(d)* Circulation in basin with horizontal salinity gradient and counteracting wind; water depth above sill smaller than critical depth; upswelling at *A*. From Brongersma-Sanders and Groen, 1970.

**Fig. 7-7.** Models of a barred evaporite basin during (A) the transgressive phase (high sea level) and (B) the regressive phase (low sea level). Numbers representing water densities are approximate. From Hite, 1970.

**Fig. 7-8.** Depositional model of the Paradox basin during *A*, the regressive phase, and *B*, the transgressive phase showing relationships of various basin and shelf facies. Wavy line represents disconformity. Symbol in the shelf limestone and dolomite represents algal mound development. From Hite, 1970.

241

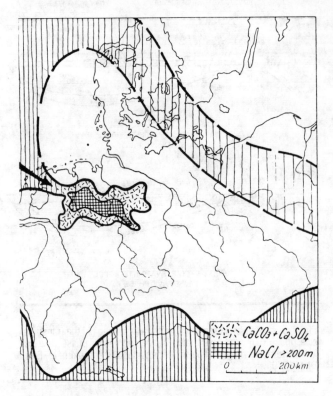

Fig. 7-9. Paleogeographic map of lower Portlandian (Jurassic) in northwestern Europe showing the saline basin in northwestern Germany. From Richter-Bernburg, 1972.

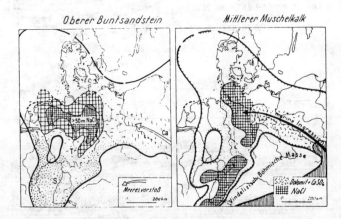

Fig. 7-10. Paleogeographic maps showing the facies distribution in the German Triassic. From Richter-Bernburg, 1972.

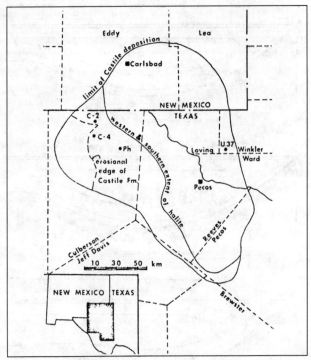

Fig. 7-11a. Index map showing the location of cores used in this investigation. *C2* = Cowden No. 2 well; *C4* = Cowden No. 4 well; *Ph* = Phillips No. 1 well; *U37* = Union University 37 No. 4 well. From Anderson et al., 1972.

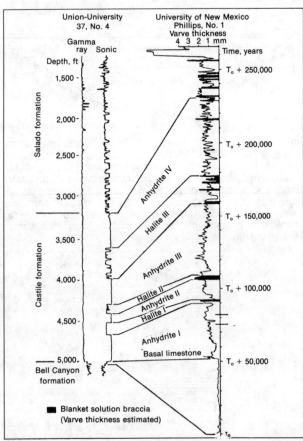

Fig. 7-11b. Correlation between gamma ray and sonic logs of the Union University well and smoothed calcite-anhydrite couplet thickness in the Phillips No. 1 core. Couplet thicknesses are estimated for halite units. From Anderson et al., 1972.

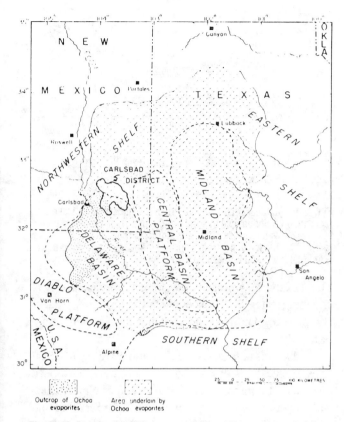

Fig. 7-12. Distribution of Ochoa evaporites with respect to subsidiary structures in the Permian basin, southwestern U.S. From Jones 1972.

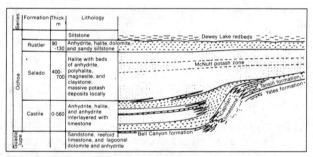

Fig. 7-13. Schematic of Ochoa evaporite formation in potassium-rich sections of the Permian basin, southwestern U.S. From Jones 1972.

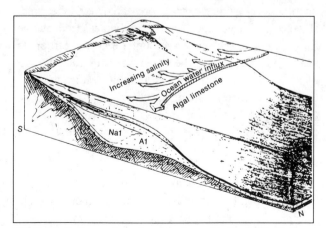

Fig. 7-14. Facies section of carbonate reef in Zechstein 2. From Richter-Bernburg 1972.

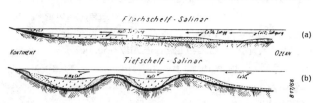

Fig. 7-15. (a) Shape of saline formations on shallow shelf areas (mega sabkha). (b) The same formations in deep shelf areas with the local and regional influence of shallow water ridges. Fractionated precipitation of different chemical phases. From Richter-Bernburg, 1972.

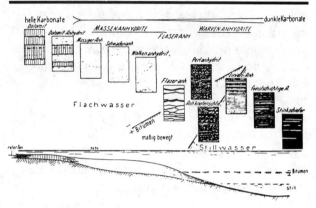

Fig. 7-16. Facies of the anhydrite near the margin and at the bottom of a basin. Bright and thick development in shallow water. Thin bedded and darkened by bituminous substances, also in low thickness in deep water. From Richter-Bernburg, 1972.

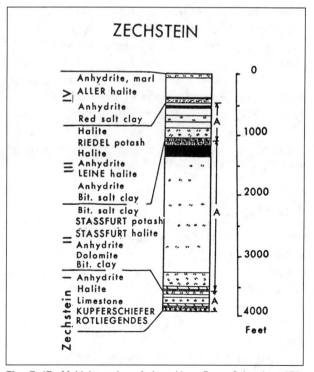

Fig. 7-17. Multiple cycles of deposition. From Schmalz, 1979, reprinted by permission.

**North German Zechstein Basins.** The Permian Zechstein is similar to the later stages of deposition in the Permian basin of New Mexico. The model suggested by Richter-Bernburg (1972) is a partially open shelf with progressive evaporation producing the sedimentary responses (Figs. 7–14, 7–15, and 7–16). Multiple cycles of deposition are indicated (Schmalz 1979) (Fig. 7–17). These cycles may be the product of changing deposition on the margin of the basin resulting from the deposition of algal mounds or from longer-term climatic changes. Progressive concentration across the basin produces the deposition of more soluble evaporite mineral phases. The removal of metal ions in response to progressive concentration and the absence of reflux of the concentrated brines is also associated with the basin. The depositional pattern reflects primarily the continentward flow of brines resulting from increased evaporation in the direction of flow. The most soluble phases were deposited nearest to the shore. These depositional units may have been restricted by anhydrite (or originally gypsum) that built up on the shelf (Richter-Bernburg 1972, 277). The model may be related to differential entrapment of brines in local basins formed by progressive evaporation across a partially restricted shelf or embayment.

**Michigan Basin.** Progressive evaporation on the basin margin with possible dilution from ground water or surface runoff may cause denser brines to flow toward the center of the basin (Figs. 7–18a and 7–18b). Circulation in a basin is restricted by shallow water margins having convergent influx and radial reflux. The most concentrated surface brines are near mid-basin (Figs. 7–19 and 7–20).

Basin-center evaporate sequences are not necessarily controlled by circulation patterns; they may represent shallow water accumulations related to a relative rise in sea level. Their stratigraphic importance cannot be understimated. The topographic geometry of the basin, restriction, and source-rock deposition are all closely related (Fig. 7–21).

## Marine Lagoon Model

Many examples of marginal lagoons adjacent to larger bodies of saline water occur throughout the world today. These areas are restricted by depth, occur in areas of high evaporation, and reflect the flow of marine waters into the marginal lagoon.

Published examples include the Ojo de Liebre lagoon of Baja California, Mexico, (Phleger 1969) (Figs. 7–22a and 7–22b) and the Kara-Bogaz-Gol adjacent to the Caspian Sea in the Soviet Union (Valyashko 1968) (Fig. 7–23).

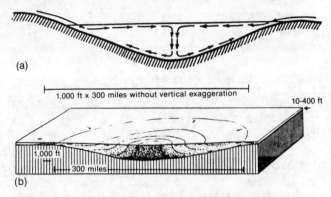

**Fig. 7–18.** (a) Ideal circulation pattern in a symmetrical salt lake with radial inflow uninfluenced by wind. (b) Circulation in a basin restricted by shallow water margins and having convergent influx and radial reflux. The most concentrated surface brines are near mid-basin. From Matthews and Egleson, 1974.

These examples are all shallow, have a complex history of changing water level, are periodically subject to partial or complete evaporation, and contain thin Holocene evaporite units.

Local playas or lagoons are common marginal to coastlines. Suggestions of leached evaporite units have been reported. The Jurassic Todilto formation of New Mexico was suggested to be a lagoonal deposit by Kirkland and Anderson (1960) (Figs. 7–24 and 7–25). The development of organic, carbonate, and evaporite laminae, which may reflect periodic influx of fresher water related to floods or periods of evaporation, is characteristic of many of these units.

## Sabkha Evaporite Model

Of the three evaporite models, the most controversial is the development of thick sequences of nodular and possibly bedded evaporites as a product of replacement in the coastal zone. This association of evaporites with coastal sabkhas or tidal flats is common to many Holocene depositional areas. The model is based principally on extensive work in the Arab-Persian Gulf, and carefully documented evidence was presented by Shearman (1966). The diagenetic process is characterized by the flow of brines from the sea through deposited carbonate sediments (Fig. 7–27). The common occurrence of nodular anhydrite within the stratigraphic record suggests that such a process may be important. Also, the deposition of evaporite mineral phases in restricted basins that do not contain bedded evaporites; for example, the Red Sea suggests the process-related balance between inflow and local evaporite precipitation.

Application of this model to ancient stratigraphic units was developed by Lucia (1972). Many sedimentary units throughout the stratigraphic record contain evidences of the presence of evaporites. For example, the Clearfork

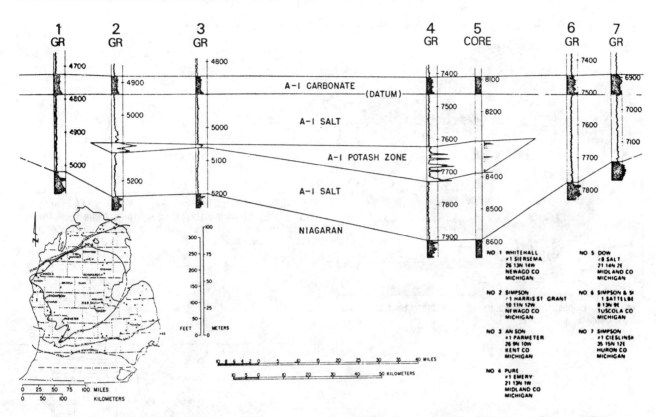

**Fig. 7–19.** West-east stratigraphic cross section of the *A-1 salt* and *A-1 basin* in the Michigan basin. From Matthews & Egleson, 1974.

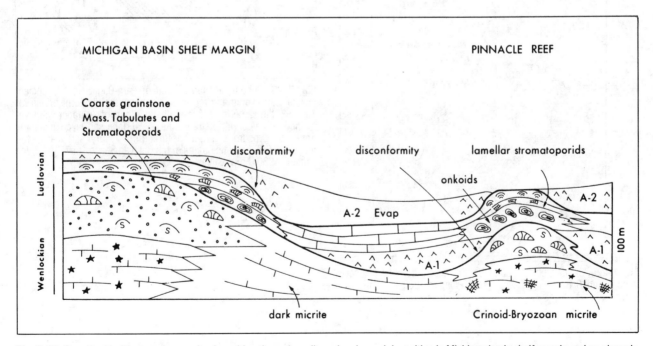

**Fig. 7–20.** Growth of buildups and evaporite deposition through cyclic and reciprocal deposition in Michigan basin shelf margin and on pinnacle reefs on the foreslope into the basin. From Wilson, 1975.

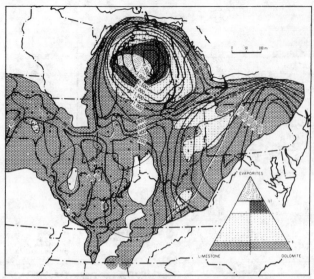

**Fig. 7–21.** Isopach and lithofacies of nonclastic, Silurian. From Sloss, Dapples, and Krumbein, 1960.

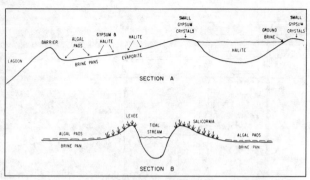

**Fig. 7–22b.** Diagrammatic cross section of brine and evaporite. Traverse A extends from main brine barrier to inner halite basin. Traverse B is section across tidal channel in brine basin. From Phleger, 1969, reprinted by permission.

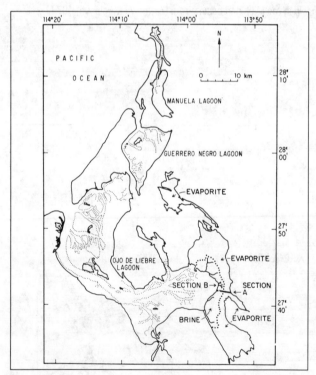

**Fig. 7–22a.** Ojo de Liebre lagoon area. Shows location of cross sections A and B from Fig. 7–22b. From Phleger, 1969, reprinted by permission.

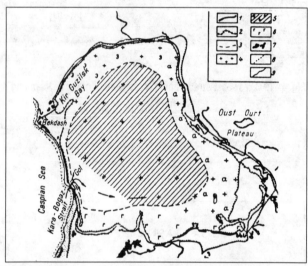

**Fig. 7–23.** Schematic map of Gulf of Kara-Bagaz-Gol. *1,* Ancient coastal earthworks and terraces; *2,* schematic divide of brines for 1930; *3,* the same for 1957; *4,* denuded part of an upper salt layer formed by halite with astrankanite; *5,* surface salt layer under the brine; *6,* gypsum salt marsh (drought); *7,* dry salt lakes, salt marshes; *8,* schematic outline of the Caspian Sea in 1957; and *9,* the same in 1930. From Valyasko, 1968.

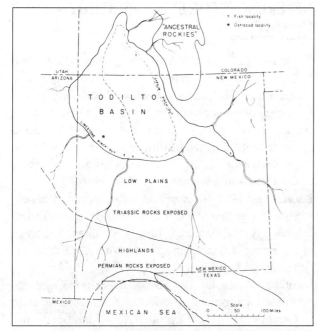

**Fig. 7–24.** Paleogeographic diagram of Todilto basin. Area of limestone deposition was 34,700 square miles. Gypsum deposition covered only about 12,000 square miles in center of basin. Weathering of Permian and Triassic strata in moist highlands on south may have provided source for salts deposited in basin. From Anderson and Kirkland, 1960, reprinted by permission.

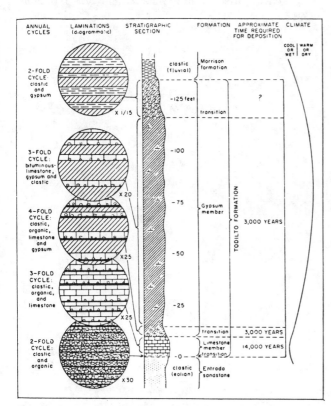

**Fig. 7–25.** Diagrammatic illustration of varved clastic-organic-evaporite (c-o-e) cycle. Regular procession from twofold cycle at base to fourfold cycle and back to twofold cycle at top. Clastic deposition persisted throughout. Gross lithologic changes were brought about by changes in individual laminae. From Anderson and Kirkland, 1960, reprinted by permission.

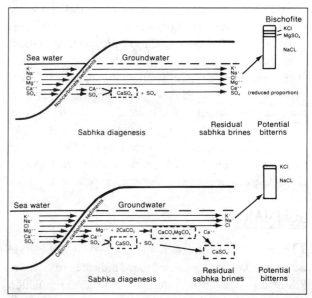

**Fig. 7–26.** (a) Schematic diagram showing trend of mineral and brine diagenesis in noncarbonate sabkhas. (b) Schematic diagram showing trend of mineral and brine diagenesis in carbonate sabkhas. From Shearman, 1966.

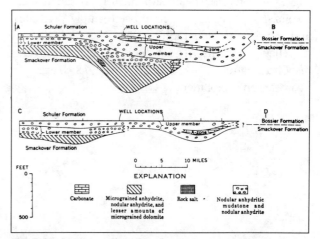

**Fig. 7–27.** Diagrammatic sections of the Buckner formation. Datum is top of Buckner formation. From Dickinson, 1974.

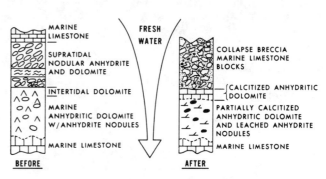

**Fig. 7–28.** Effects of fresh water on evaporite-carbonate shoreline deposits. From Lucia, 1972.

(Leonard) formation on the flank of the Midland basin and the Buchner formation (Jurassic) of the Gulf Coast basin (Fig. 7-27) illustrate relationships similar to those found on the Trucial coast of the Arab-Persian Gulf. Lucia (1972) suggested a typical stratigraphic pattern for these deposits (Fig. 7-28). This sequence is generalized from many ancient stratigraphic units and indicates a commonly preserved sequence suggesting the presence of diagenetic evaporites. Their presence prior to freshwater leaching must be inferred from mineralogic and textural characteristics preserved in the stratigraphic unit.

## SUMMARY

Evaporites are important indicators of hydrocarbon accumulations because they indicate restriction. Restricted, anoxic basins are the site of the accumulation of source rocks, and evaporate units may contain source rocks. Evaporite units are tectonically mobile, excellent hydrocarbon seals, and possibly reflect paleogeographic patterns reflecting unconformities and basinal history. For these reasons their presence in a stratigraphic section suggests appreciable hydrocarbon reserves exist in a reservoirs. Evaporite models have been simply presented and represent topographic, paleogeographic, and diagenetic patterns. These models reflect process-response patterns and can be used as bases for hydrocarbon exploration.

## REFERENCES FOR FURTHER STUDY – CHAPTER 7

Anderson, R.Y., Dean, W.E. Jr., Kirkland, D.W., and Snider, H.I., 1972, Permian Castile varved evaporite sequence, West Texas and New Mexico: Geological Society of American Bulletin, v. 83, p. 59–87.

Brongersma-Sanders, M., and Groen, P., 1970, Wind and water depth and their bearing on the circulation in evaporite basins: Third Symposium on Salt, Northern Ohio Geology Society, v. 1, p. 3–7.

Lucia, F.J., 1972, Recognition of evaporite-carbonate shoreline sedimentation: in recognition of ancient sedimentary environments: Tulsa, Oklahoma, Society of Economic Paleontologists and Mineralogists Spec. Pub. no. 17, p. 170–191.

Richter-Bernburg, G., 1972, Saline deposits in Germany: a review and general introduction to the excursions: *in* Geology of Saline Deposits: Paris, Hanover Symp. UNESCO Proc., p. 275–287.

Schmalz, R.F., 1969, Deep-sea evaporite deposition: A genetic model: Am. Assoc. of Petrol. Geol. Bulletin, v. 53, p. 798–823.

Shearman, D.J., 1966, Origin of marine evaporites by diagenesis: Inst. Mining Metal. Trans., v. 75, p. 208–215.

# 8 DELTAS

One main law which underlies modern progress is that, except for the rarest accidents of chance, thought precedes observation.

It may not decide the details, but it suggests the type. Nobody would count, whose mind was vacant of the idea of number.

Nobody directs attention when there is nothing he expects to see. The novel observation which comes by chance is a rare

accident and is usually wasted. For if there be no scheme to fit it into, its significance is lost. **A.N. Whitehead,** *The Function of Reason,* **1962**

## INTRODUCTION

Most land-derived sediment is deposited within the deltaic framework at the margin of cratonic plates. The problem of organizing observations about deltas has been the difficulty in determining three- and four-dimensional depositional responses on a sufficiently large scale to determine the actualistic controls. Study of surficial sediment patterns is not sufficient to understand the complex of physical, chemical, and biologic processes that operate within the delta framework. The effects of climate, topography, tectonic history, and alluvial, marine, and aeolian processes must be isolated before the history and pattern of deltaic sedimentology can be constructed.

Cored sections from modern deltas and the study of processes within Holocene deltas provide the observational basis for constructing stratigraphic models. Problems of scale, rates of change, climatic accidents, tectonic and eustatic variations, and the responses to specific processes have been interrelated (Wright and Coleman 1973; Coleman and Wright 1975). Physiography of deltas can be interpreted by integrating depositional

processes with boundary controls. Regression and transgression, tectonic controls of subsidence, growth faulting, and diapirism are the basis for constructing an interpretation of Holocene deltas and for interpreting and predicting stratigraphic patterns. This developmental approach to the determination of deltaic models has been possible only recently due to the accumulation of objective data from Holocene deltas.

Modern sedimentologic studies of deltas show the importance of both the boundary conditions and tectonic forces as the framework controlling deposition. Post-depositional responses and the effects of sedimentary processes must be understood as a basis for formulation of delta models.

Sedimentation generally is most rapid in the deltaic environment. This is conducive to the incorporation and preservation of large quantities of organic material. Rapid sedimentation under highly contrasting energy conditions produces abrupt lateral and vertical changes in sorting and lithology of both reservoir and non-reservoir facies.

All generally accepted conditions for hydrocarbon accumulation occur within the deltaic environment.

These include:

▶ Porous, reservoir-type strata,

▶ Abrupt horizontal and vertical lithologic changes,

▶ Abundance of rich, organic muds and silts, and

▶ Impermeable caprock shales and siltstones deposited under conditions of marine transgression; the latter is due to subsidence and compaction of the delta muds.

## DELTA SEDIMENTOLOGY

Study of modern deltas has provided insight into controls for depositional and stratigraphic responses. Simply examining the sedimentology of a portion of a distributary or the jet action at the mouth of the distributary is not sufficient to interpret either areal or vertical patterns of sedimentary units. The history of the delta is the important element. Processes must be understood if the developmental history is to be a basis for building delta models. The operation of these processes is partly controlled by boundary conditions. The interrelation of boundary conditions, tectonism, and sedimentary processes must precede analysis of specific geomorphic model elements of the delta. An understanding of these interrelationships forms the basis for stratigraphic delta models.

### Boundary Conditions

The physiography of the delta is functionally related to rates of sediment supply, shape of the depositional site, and relative subsidence. Each of these factors involves many possible variations. For example, differences in amount and proportion of suspension load and bedload and the frequency and magnitude of flood cycles are reflected in depositional patterns, sedimentary processes, and geometry of deposited units. Patterns of sediment transport are controlled by current, wave, wind, and tidal processes, which modify the delta physiography. The delta framework must include both the alluvial drainage basin and the marine platform. Questions concerning fetch, tidal period and amplitude, depth of the shelf platform, density of sediment-water mixtures, and longshore currents must be evaluated prior to the interpretation of the depositional patterns. Similarly, the alluvial valley responses must be related to latitude, climate, weathering, provenance, and history and pattern of sediment and fluid discharge before patterns of deposition within the delta can be interpreted and predicted.

The delta framework can be interpreted in relation to a physiographic framework exhibited in many Holocene deltas throughout the world. Six differing geomorphic components have been identified, and their relative importance related to dynamics of the physical processes

of the depositional site. Of the six components (1) alluvial valley, (2) upper deltaic plain, (3) lower deltaic plain, (4) marginal marine shelf, (5) marginal delta plain, and (6) marginal bay-basin—each one reflects delta processes (Fig. 8–1). For example:

▶ The width of the alluvial valley reflects subsidence or sea-level rise, flood periodicity and magnitude, and sediment transport.

▶ The distinction between upper and lower deltaic plains is related to distributary gradient, which reflects tidal influences, rates of sedimentation, nature of sediment load, and the rates of deltaic progradation.

▶ The marginal delta plain and the bay-basin components are related to longshore drift, sediment load, and marine transportation processes.

▶ The geometry of the subaqueous delta and its textural and depositional characteristics reflect the subsea topography, wave energy, tidal forces, and fluid densities.

The reconstruction of these process-related elements from depositional or stratigraphic characteristics cannot be based upon a single texture, structure, or biologic measurement; the broader implication of boundary conditions must be considered. The relation of offshore slope, wave power, and relative riverine influence may suggest the delta form and sand patterns (Wright and Coleman 1973, 397) (Table 8–1).

Description of these boundary conditions and the sedimentary process provides the basis for interpreting stratigraphic responses related to the geomorphic elements of deltas. The vertical and areal patterns of these elements form the basis for constructing stratigraphic models and the interpretation of ancient stratigraphic units (Table 8–2).

### Tectonic and Sea-Level Boundary Conditions

The geometry of the receiving basin, the areal and vertical extent of deltaic elements, and the pattern of shifting delta lobes are controlled by tectonism or sea-level changes. Each of these aspects is reflected in the delta morphology, the pattern of depositional units, and the internal sedimentary characteristics of stratigraphic units.

The geometry of the receiving basin can take many forms (Coleman and Wright 1975, 115) (Fig. 8–2) reflecting the tectonic framework of the deltaic area. Not only is the pattern of the delta controlled by this framework, but it also influences the physical processes. For example, Type I (Fig. 8–3) may be common to grabens related to separation of plates, and the topography may reflect tectonism rather than sedimentology. Tidal currents are a dominant control on the form and

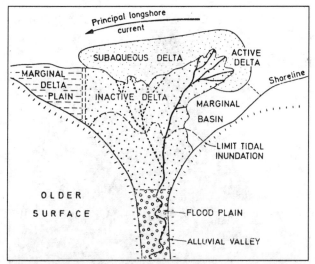

**Fig. 8-1.** Major physiographic components of a delta. From Gagliano and McIntire, 1968.

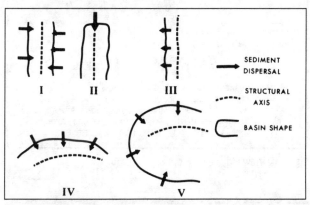

**Fig. 8-2.** Major configurations of receiving basins. From Coleman and Wright, 1975.

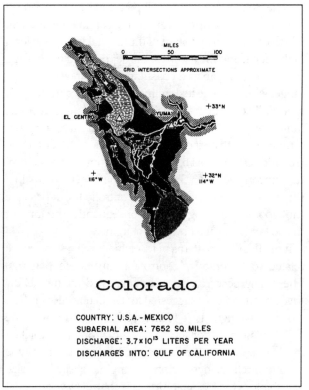

## Colorado

COUNTRY: U.S.A.- MEXICO
SUBAERIAL AREA: 7652 SQ. MILES
DISCHARGE: 3.7×10¹³ LITERS PER YEAR
DISCHARGES INTO: GULF OF CALIFORNIA

**Fig. 8-3.** Colorado River delta. From Smith, 1966.

**Table 8-1**
**Comparison of Wave and Riverine Delta**

| Offshore slope | Flat convex | | Steep concave |
|---|---|---|---|
| Relative nearshore wave power | Low | | High |
| Relative riverine influence | High | | Low |
| Configuration characteristics | Highly indented, extended distributaries } → | Smooth, arcuate, slightly protruding river mouths } → | Straight coastline, no protrusions |
| Delta plain landforms | Marsh, bays → | Subtle beach ridges and dunes alternating with marsh and mangrove } → | Barriers, beach ridge, and dunes dominate |
| Sand body | Shoestring sands, low lateral continuity } → | Moderate lateral continuity along localized bands } → | High lateral continuity; thick sand sheets |
| Sediment composition and sorting characteristics | Fine-grained, poorly sorted silts and sands; laterally, well-sorted channel sands } → | Alternating bands of moderately sorted dirty sands and poorly sorted silts and clays } → | Well-sorted clean sands |

From Wright and Coleman, 1973.

sedimentology of the delta. Type II reflects a graben-rift system that is transcurrent to the continental margin, which controls the position of the alluvial valley, the form of the delta, and the nature of the depositional processes. Estuarine and embayed coastlines can increase the tidal range, which is reflected in linear tidal ridges in the river mouth. The modern Irrawady River forms in a tectonic framework (Fig. 8–4) similar to the graben-rift system, but forming an embayed coastline. In addition, an aulacogen tectonic history possibly can have an important control (Dickinson 1974, 16). Specifically, the Niger delta is in a rift valley, and its distribution is controlled by a possible triple-point spreading center. Consequently, the form of the delta is controlled by plate margins.

Type III and IV patterns of receiving basin geometry are related to tectonically controlled subsidence patterns. These may occur on land, as suggested for the Yellow River, or offshore, as suggested for the Senegal River (Coleman and Wright 1975). Inland lakes or swamps of the deltaic plain or the offshore marine platform may thicken in response to subsidence in these depocenters. The Type V pattern reflects deposition in marginal basins; the size and shape of the topographic basin control wave and tidal processes and the history and pattern of subsidence are reflected in shelf and slope progradational elements (Fig. 8–5). These delta patterns are particularly well illustrated by the Mississippi delta in the Gulf of Mexico and the Danube and Dnieper in the Black Sea.

In combination, sea-level rise and subsidence control the thickness and patterns of environmental elements of the deltaic sequence. Tectonic and sea-level stability are reflected in thin depositional units in response to delta progradation across the marine platform producing a blanket stratigraphic unit. The rate of progradation is controlled both by the geometry of the receiving basin and rate of sediment supply. Progradation continues until an equilibrium is reached between sediment supply and either subsidence in the receiving basin or sediment dispersal. Under stable conditions, the margin of the shelf may be the only limit to progradation. In shallow interior basins, delta progradation may be followed by continued fill represented by alluvial sequences. These patterns, however, are strongly affected by subsidence or sea-level rise.

High rates of tectonic subsidence within the alluvial valley are reflected by shoreline migration toward the continent, thick alluvial sequences, and thin transgressive breaks. Similarly, subsidence within the delta produces laterally and vertically stacked delta lobes with little shoreline migration, a predominance of delta plain and marginal marine units, and local transgressive breaks between lobes. Subsidence on the continental shelf or marine platform is reflected by progradational lenses

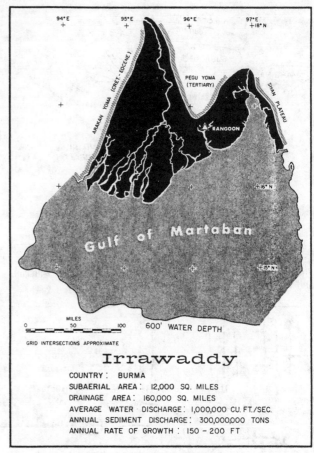

**Irrawaddy**

COUNTRY : BURMA
SUBAERIAL AREA : 12,000 SQ. MILES
DRAINAGE AREA : 160,000 SQ. MILES
AVERAGE WATER DISCHARGE : 1,000,000 CU. FT./SEC.
ANNUAL SEDIMENT DISCHARGE : 300,000,000 TONS
ANNUAL RATE OF GROWTH : 150 – 200 FT

**Fig. 8–4.** Irrawaddy delta. From Smith, 1966.

**Orinoco**

COUNTRY: VENEZUELA          DRAINAGE AREA: 340,000 SQ. MILES
SUBAERIAL AREA: 22,000 SQ. MILES          AVERAGE WATER DISCHARGE: 600,000 CU. FT./SEC.

**Fig. 8–5.** Orinoco delta. From Smith, 1966.

thickening to the margin of the shelf or basin, thick units of fine-grained clastics overlaid by blanket marginal marine sand units, and broad transgressive breaks in response to subsidence and to basin fill. Thus, both the vertical and areal patterns of delta units are related to tectonic boundary conditions.

The delta shape, the sedimentologic processes, and the vertical and areal pattern of sedimentologic units are related both to synsedimentary and post-sedimentary tectonism and to geomorphic boundary conditions. Three types of delta migration and progradation were identified by Coleman and Wright (1975, 113) (Fig. 8–6):

▶ Delta lobe switching, related to low slope, low wave, and tidal energies,

▶ Channel avulsion far up the delta plain with intermediate wave and tidal energies, and

▶ Alternate channel extension with high tidal or wave energies.

Types II and III result in a stable delta platform as a response to progradation into areas of low subsidence and steeper delta slopes and higher dispersion of clastic sediments. However, Type I reflects low energy and development of extensive units of prodelta clays. Thus, patterns of subsidence and topography of the depositional site control the geometry of deltaic stratigraphic units.

Examples of these patterns are illustrated by modern deltas. Type I is typical of the tectonic setting and history of the Mississippi delta (Fig. 8–7). Type II reflects the pattern of the Hwang Ho and the Ganges-Brahmaputra (Fig. 8–8). Type III is typical of the Danube and Nile deltas (Fig. 8–9). These modern examples are useful in determining the effects of tectonic and sedimentary boundary conditions on the vertical and areal patterns of depositional units.

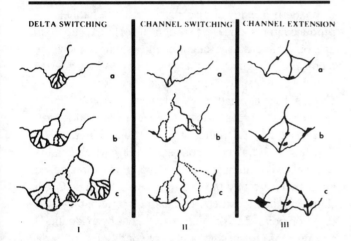

**Fig. 8–6.** Types of delta shifting patterns. From Coleman and Wright, 1975.

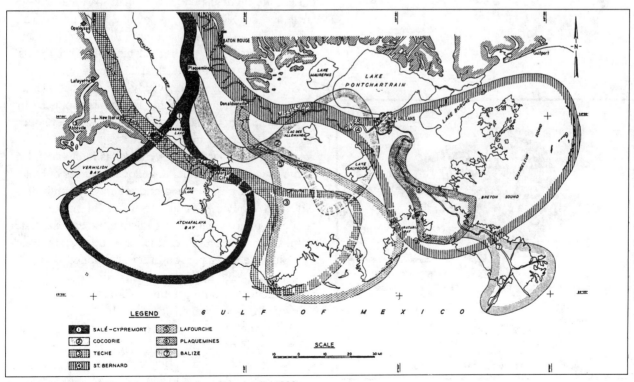

**Fig. 8–7.** Mississippi River deltas. From Kolb and Van Lopik, 1966.

**Fig. 8–8.** Ganges-Brahmaputra delta, From Smith, 1966.

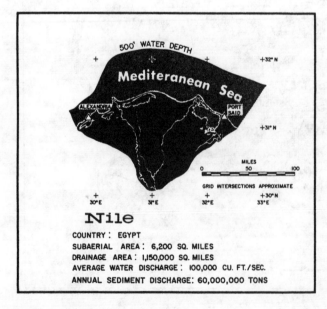

**Fig. 8–9.** Nile delta, From Smith, 1966.

## SEDIMENTARY PROCESSES

Within the framework of boundary conditions, riverine, tectonic, tidal, current, wind, chemical, and biologic processes, the areal and vertical pattern of depositional environments can be predicted.

### Riverine Processes

Sedimentary responses within the delta reflecting alluvial boundary conditions include the channel geometry and pattern of fill, texture of the deposited clastic detritus, and the areal patterns of geomorphic elements. Meandering, straight, and braided forms of channel geometry reflect the pattern of alluvial discharge. Braided channels are produced in response to a surplus of sediment supply in arid or arctic basins. Straight channels reflect a low gradient and a regular discharge pattern. A meandering channel demonstrates a higher gradient or a more discontinuous discharge pattern (Coleman and Wright 1975).

Three major types of lower deltaic-plain channel patterns have been identified (Colemen and Wright 1975, 109) (Fig. 8–10). Type A consists of seaward bifurcating channels with multiple mouths. Type B is characterized by many bifurcating but rejoining channels. Type C is a simple radiating pattern of channels originating at a common point. Each channel pattern has been identified as a response to specific boundary conditions and sedimentary processes. Conditions that produce the Type A pattern are high suspended sediment load and subsidence; and low slope, tidal range, and wave power. This pattern is typical of the Mississippi River (Coleman and Wright 1975). Type B pattern reflects a response to tidal action (greater than 2 meters), and steep offshore slopes as characterized by the Irrawaddy and Orinoco Deltas. The radiating channel pattern, Type C, is produced by high wave energy and steep offshore slopes, as developed in the São Francisco and Niger deltas.

Active channels from Holocene deltas rarely are filled with clastic detritus. Older abandoned courses of Holocene rivers have been cored and are mostly sand filled (Kolb and Van Lopik 1966). The process of sand fill may be similar to how channels are filled in the upper deltaic plain. Or marine processes, including tides and waves, may plug the lower portion of an abandoned channel, and the fill may contain fine-grained, poorly sorted detritus deposited during periods of flood.

Channel shapes are reflected in the pattern of fill. Braided channels form sheet-sand units and many depositional sequences. Straight channels form narrow shoestring sands. Meandering channels form a fining upward sequence in a broad alluvial valley.

Textural characteristics are dominantly controlled by

climatic aspects of the alluvial basin. The proportion of bedload to suspended load is controlled by weathering and the volume and pattern of water discharge. Erratic discharge is common to arid and polar climate regimes; this produces both a fan geometry and a predominance of bedload detritus. Alternately, stable channels reflect a more consistent discharge pattern and produce better sorting and a mixture of bedload and suspension loads. Very large rivers in regions of high temperature and rainfall transport more suspension load; the result is a meandering pattern in the alluvial valley. In the delta high suspended load is reflected by straight channels with little pattern of vertical size segregation and by broad distribution of fine-grained clastic detritus across the delta plain and the prodelta region (Coleman and Wright 1975).

The areal distribution of deltaic elements is controlled in part by the nature of the pattern of sediment-water discharge. Channel morphology and the pattern of channel fill are products of riverine processes. Also, the distribution of surficial elements of the delta is controlled in part by the flood history and development of natural levees.

The height and continuity of levees is a function of flood cycles. Extreme floods produce crevasse channels and deposition in swamps, lakes, and bays. The pattern of these features is in part controlled by the riverine process, but it is also affected by tectonic processes and other boundary conditions. An interplay exists between the volume of detritus transported by the river and the marine energy or power. An index of discharge effectiveness is reflected in the form of the delta. For example, the Mississippi River has a high index (sediment discharge/marine energy), and the response is development of long, linear sand bodies trending at high angles to the coast. The Senegal River has a low index, and the response is linear sand bodies parallel to the shoreline (Coleman and Wright 1975, 106).

## Marine Processes

These include waves, tides, and currents. These processes not only modify the form of the delta, but they are responsible for distributing sediments within the delta. Each major process is reflected in the sedimentary structures, bedding, textural characteristics, and their distribution.

Wave power plays a highly significant role in controlling the geometry and distribution of deltaic sands, their orientation, and their mineral composition (Coleman and Wright 1975, 110). High wave power redistributes the riverine sediment into strandplains, beach ridges, longshore bars, and offshore shoals. Sediments are winnowed, and metastable minerals are removed by abrasion and chemical action of the marine water. Low wave

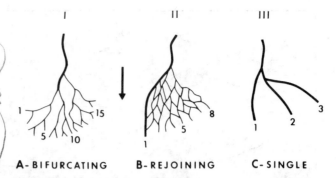

**Fig. 8-10.** Types of distributary channel patterns. From Coleman and Wright, 1975.

power allows the riverine process to dominate the form of the delta so that elongate fingers of coarse clastic sediment extend into the marine environment. These coarser lenses of sediment are often encased in fine-grained detritus. Sorting is good within channels, but the surrounding sheath of clastic detritus is poorly sorted and often disrupted by the indigenous biota.

Wave-power profiles were constructed for a number of deltas by Wright and Coleman (1973). There are three examples. One for low wave power is demonstrated by the Mississippi delta (Fig. 8-11). For high wave power, the Niger delta and the Sao Francisco are shown (Figs. 8-12a, 8-12b, and 8-13). The figures illustrate peak periods of discharge and wave power and the relative effectiveness of riverine discharge. Thus, the form of the delta, the relative significance of geomorphic elements, and the internal sediment characteristics are reflected by a balance between riverine discharge and wave forces. If the periods of high riverine discharge and high wave power coincide, the sediments are reworked as they are discharged from the delta. If they are out of phase, sediment may be widely distributed along the shoreline and into offshore areas. The size and internal characteristics of the delta reflect a response to both the timing and the relative strength of wave and riverine processes.

## Tidal Processes

Tidal processes are also effective in reworking riverine sediment, and they produce a markedly different pattern of deposition. Tidal currents are dominantly perpendicular to the shoreline, and the distribution of coarse sediment reflects this pattern. Linear offshore sand ridges, bell- or funnel-shaped distributary channel mouths dominated by sand shoals, channel meanders above the limit of tidal influence, and extensive tidal flats are all produced in response to medium to high tidal-range conditions. Internal characteristics of the deposited sedimentary units also reflect time-velocity asymmetry during the tidal cycle. Flood, ebb, or bidirectional cross-

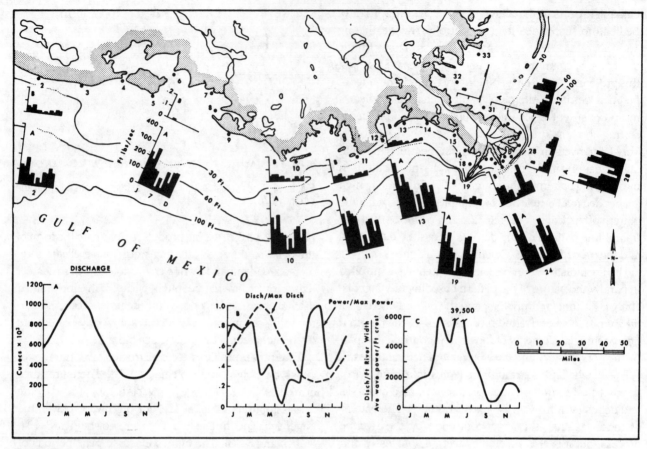

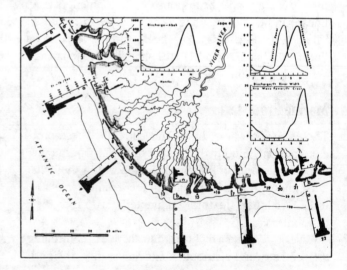

**Fig. 8–11.** Variations in morphology of major river deltas as functions of ocean waves and river discharge regimes. From Wright and Coleman, 1973, reprinted by permission.

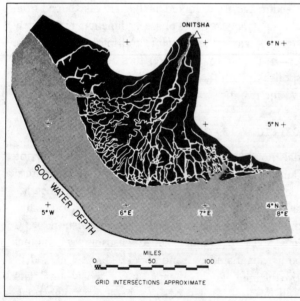

**Fig. 8–12a.** Niger delta. From Smith, 1966.

**Fig. 8–12b.** Energy patterns illustrate the dominant influence of wave processes, but the secondary tidal channel pattern can be seen. The distribution of energy throughout the year also influences the relative importance of individual depositional environments. From Wright and Coleman, 1973, reprinted by permission.

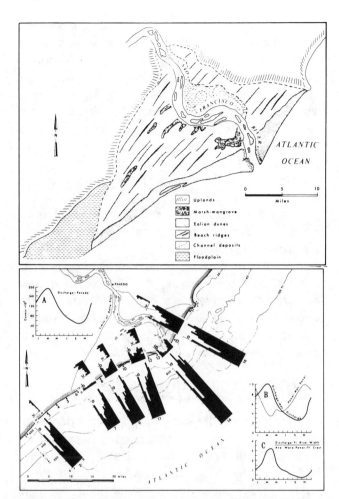

**Fig. 8-13.** Depositional environments in the São Francisco River delta. From Wright and Coleman, 1973 reprinted by permission.

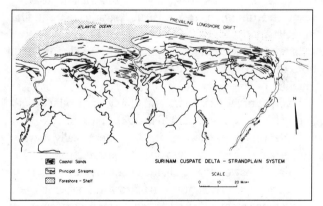

**Fig. 8-14.** Principal depositional environments, Surinam coast, northern South America. Inland coastal sands deposited as Pleistocene highly destructive deltas, fronted by modern accretionary strand plain sands. From Fisher, 1969.

stratification may result. In deltas with moderate to high tidal range, the pattern of sedimentation within the delta also is affected. During extreme tides—commonly developed during certain portions of the year or in response to storm conditions—the banks are crested and results in the formation of overbank sand sheets. Tidal processes modify the surface of the delta, channel morphology, offshore sand patterns and textures, sedimentary structures, and bedding patterns. Tidal forces can be recognized by these characteristics; they provide a means for evaluating the boundary conditions operating in a deltaic area.

## Marine Currents

In response to tidal, wind, and density forces operating in the offshore area of a delta, longshore drift systems develop. These currents redistribute sediments along the coast and produce strandplain depositional systems, barrier islands, cheniers, and marine shoal areas (Fig. 8-14). These are important in areas of high littoral drift formed in response to onshore winds, tides, and oceanic currents. The form of the delta, offshore textural patterns and bedforms, and the distribution of fine-grained, organically rich sediments is controlled by these processes. The biota also reflects the pattern of textures and organic detritus and can be used to interpret sedimentary processes and boundary conditions. The redistribution of the sediment by density currents is of major importance in many deltas. These currents can produce channels in the delta platform and transport sediment many hundreds of kilometers into the basin. Current also may flow back toward the mouth of the distributary and transport sediment into the distributary or concentrate sediment along the flanks of the distributary and help form subaqueous levees (Coleman and Wright 1975, 112). Density contrasts between fresh water, salt water, and sediment-laden fresh water are present at the delta mouth. The pattern of currents generated by riverine outflow, tidal inflow, and density stratification produces a complex of depositional responses.

## Wind Processes

Wind stress on the subaerial delta plain and on the deltaic platform produce waves and currents in the offshore portion of the delta. Thus, wind processes in the offshore area may not be easily separated from marine currents. Linear coastal dunes, sand sheets, fill of lakes and bays, and deposition on tidal flats are important in deltas in arid areas or where there are strong prevailing onshore winds. The combination of wind-driven waves, sediment reworking along the coast, and wind-transported detritus combines to produce characteristic delta patterns for example the São Francisco delta (Coleman and Wright 1975, 111) (Fig. 8–13).

## Chemical Processes

The removal of dissolved and colloidal load produced within the alluvial basin are important response patterns within the delta. Processes of adsorption, flocculation, biochemical reactions, and precipitation remove these phases from the alluvial system. Dissolved and complex ions of iron and silica are removed within the delta complex. Other ions of bicarbonate, sulfur, phosphate, and carbon also are removed by a combination of biochemical and inorganic processes within marine waters. After removal, these ions are reconstituted into stable mineral phases by chemical processes within the delta sediments. Thus, the delta contains characteristic mineral phases that can identify these stratigraphic units. Iron may occur as pyrite, limonite, siderite, hematite, or iron silicates or phosphates. All these phases reflect boundary conditions, chemical kinetics, and depositional history. From the pattern of these mineral phases, depositional aspects can be evaluated. Other reactions involving dissolved ions of calcium, magnesium, and sodium have longer residence times in ocean basins and are not concentrated in deltaic sediments unless there are unusual evaporation conditions.

Clay-mineral distribution within estuaries, bays, swamps, lakes, distributaries, and prodelta sedimentary units has been the focus of study. Principally, kaolinite occurs as a product of sedimentation or diagenesis in upper portions of the delta plain, distributaries, and restricted bays. Montmorillonite is dominant in the lower deltaic plain, open bays, and prodelta sediments, while chlorite (iron) and illite are more common in marginal marine sediments representing diagenetic processes in areas of lower rates of sedimentation. This pattern is highly variable, and local conditions within the delta (or post-depositional changes) may radically alter this distribution (Miller and Visher 1974).

## Biochemical and Biologic Processes

In areas of temperate or tropical climate regimes, the delta plain is the site of high organic productivity. Salt marshes, mangrove swamps, and delta plains produce thousands of metric tons of organic material per square kilometer per year. This material is transported and deposited as cuticle, humic acids, organic colloids, and many other organic compounds. It forms the base for a food chain that includes algae, bacteria, phytoplankton, protozoans, coelenterates, echinoderms, worms, mollusks, anthropods, and vertebrates. Biologic activity within the waters and sediments of the delta is pervasive and strongly modifies bedding patterns, preservation of organic materials, and the physical character of deposited fine-grained sediments. Biologic processes are partly responsible for the reduced density of prodelta clays and many post-depositional density-related tectonic processes. Sediment-biota interrelations have only been superficially studied, but patterns of trace fossils, preserved faunal remains, and chemical and biochemical reactions are attributable to the effects of the diverse biota within the delta.

The distribution of swamps and marshes and the subsequent formation of peats, lignites, and coals have been correlated to the delta regime (Cavaroc and Ferm 1968). The distribution of oyster banks, bay and lagoonal carbonates, and even reefs in the offshore have been correlated to the presence of deltas. Consequently, the biologic activity in the area of high nutrient supply, marine salinities, and equitable climates combines to produce a characteristic aspect of deltas.

## HOLOCENE DELTA MODELS

Boundary conditions control the pattern of delta sedimentology. The six elements shown in Figure 8–1 are useful as a basis for interpreting sedimentary responses. Each area of the delta reflects characteristic processes, and their relative importance determines the geomorphic framework for stratigraphic responses (Table 8–2). Delta history is reflected by the migration of these geomorphic model elements. Their recognition in ancient stratigraphic units provides the basis for interpreting the developmental history. Each of the model elements supplies the framework for analyzing the sedimentologic responses required to construct stratigraphic delta models.

Each of the four major types of deltas develop in relation to their principal depositional processes. These include:

▸ Riverine processes dominant, characteristically developed in interior seas and in marginal basins;
▸ Wave processes dominant, characteristically developed on marginal shelves open to oceanic basins;
▸ Tidal forces dominant, characteristically developed on tectonic coastal margins with narrow shelves, embayed coastlines marginal to restricted ocean basins; and
▸ Fan and high latitude deltas characterized by coarse fan-glomerates, braided stream deposits, often associated with rift valleys and foredeeps.

**Table 8-2**
**Delta Response Characteristics**

| | Riverine | Wave | Tidal | Fan-Arctic |
|---|---|---|---|---|
| Delta form | Elongate | Cuspate-lobate | Bell-shaped | Lobate |
| Dominant elements | Marginal basin | Marginal plain | Estuary and tidal channels | Marginal shelf |
| Channel geometry | Straight, low width/depth | Meandering, high width/depth | Braided moderate width/depth | Braided extreme width/depth |
| Depocenter | River channels; lower delta plain | Strandplain and offshore | Marginal shelf river mouth | Offshore |
| Migration history | Lobe switching and lateral stacking | Sheet progradation and vertical stacking | Channel switching; sediment bypassing into deep water | Sheet progradation plus extension of marginal plain |
| Vertical sequences Channel | Little change | Fining upward | Interbedded coarse and fine | Gravel bars fining upward |
| Interchannel | Bay-crevasse-levee-swamp | Multiple coarsening upward cycles | Fining upward tidal flat cycles | Little pattern |
| Marginal shelf | Coarsening upward; single cycle | Multiple coarsening upward cycles | Multiple coarsening upward cycles | Gravel bars |
| Marginal plain | Coarsening upward; single cycle | Multiple coarsening upward cycles | Multiple fining upward cycles | Coarsening upward cycles |
| Marginal basin | Shelf-bay-marsh-algal limestone | Thin coarsening upward and interbedded shell banks | Interbedded clay and shell banks | Interbedded peat and clay |

# RIVERINE DELTA MODEL

The delta pattern is particularly well illustrated by the Mississippi delta in the Gulf of Mexico. Under stable conditions the shelf margin is the only limit to progradation. Progradation is by lobe switching.

The shifting of delta lobes is a response to channel extension, and the development of lobe stacking is in response to the history of subsidence. The array of depositional units within the delta plain and the pattern of sedimentation in the prodelta area are interrelated to the outlined boundary conditions. The evolution of this type of delta can therefore be outlined.

## Upper Deltaic Plain

The upper deltaic plain is transitional from the alluvial valley into the area where marine effects are predominant. It represents in most modern deltas areas that were originally a part of the prodelta and lower delta plain. In response to progradation, processes of sedimentation reflect an increasing subaerial control and decreasing submarine processes that were common during the early progradational history. Surficial sedimentary processes and modification of the distributary channel pattern, both in active and abandoned channels, provide a basis for studying developmental processes and interpreting the resulting areal and vertical sedimentary patterns.

The limits between the alluvial valley and the upper deltaic plain are process controlled. The alluvial valley represents a unit developed in response to riverine processes of channel migration, overbank flooding, and valley fill by a single channel. At the point where the land surface approaches sea level, flood stage no longer produces a significant increase in gradient. Increases in discharge must be accommodated by multiple channels rather than a single floodplain channel. A bifurcating distributary pattern is the geomorphic response to this change (Fig. 8-15).

Distributary patterns in the upper deltaic plain reflect continuing subsidence, which results in the growth of natural levees and an interconnected pattern of smaller channels and crevasse splay deposits (Figs. 8-16, 8-17, and 8-18). The result is an overthickened channel sequence capped by a fining upward unit or a massive sand with little textural variation dominated by parallel laminae, rippled cross beds, and planar to trough-shaped sedimentary structures (Figs. 8-19 and 8-20).

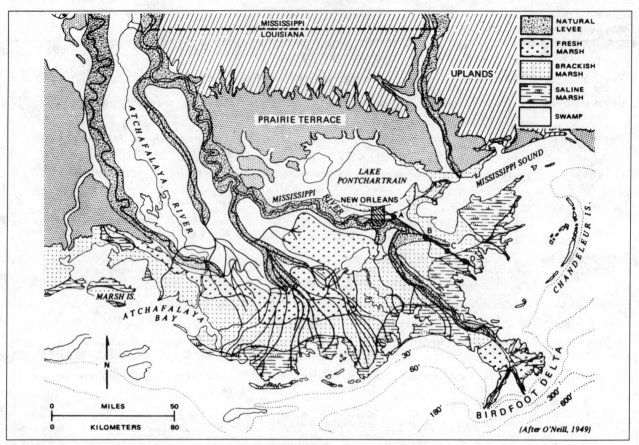

**Fig. 8-15.** Distribution of swamp and marsh environments on Mississippi deltaic plain. From Gould, 1970, reprinted by permission.

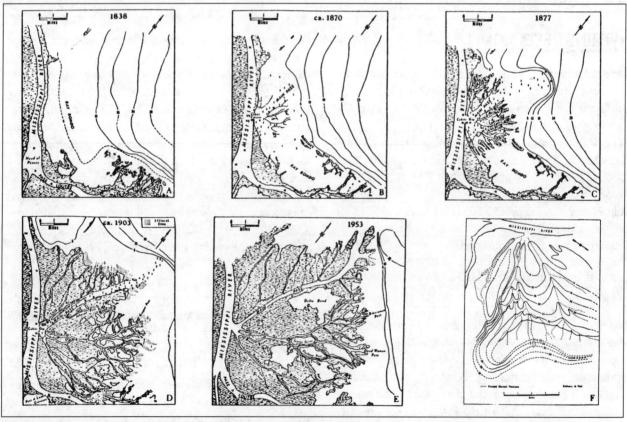

**Fig. 8-16.** Delta processes on the delta plain are controlled by processes of overbank flooding, and a history of crevasse splay, sedimentation, levee development, and bay fill results. From Wilder, 1959.

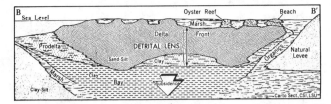

**Fig. 8-17.** Cross sections of crevasse model. From Coleman and Gagliano, 1964.

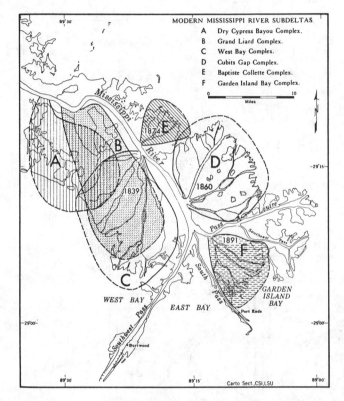

**Fig. 8-18.** Subdeltas of the Modern Birdfoot or Belize delta. Dates indicate year of crevasse breakthrough. From Coleman and Gagliano, 1964.

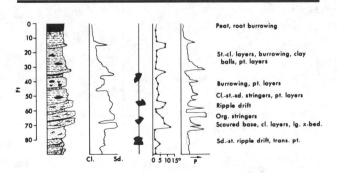

**Fig. 8-19.** Main channel sedimentary pattern. From Coleman and Prior, 1980, reprinted by permission.

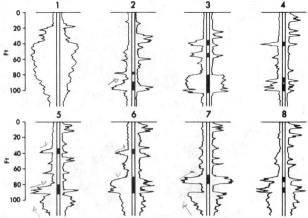

**Fig. 8-20.** Diagram illustrating characteristics of distributary channel-fill deposits. From Coleman and Prior, 1980, reprinted by permission.

## Lower Deltaic Plain

The transition from the upper deltaic plain has been attributed to hydraulic parameters; the transition into the marginal marine shelf also is gradational and reflects marine processes.

The lower deltaic plain is formed by a complex of processes related to crevasse splay deposition, interchannel bay-fill deposition, and subaqueous and subaerial natural levees (Figs. 8-21, 8-22a, 8-22b, and 8-23).

## Distributary Mouth Bar and Strandplain

Distributary mouth bar sedimentation is an important aspect of all types of deltas. Buoyant, frictional, and inertial forces are all developed in this complex environmental setting (Fig. 8-24). These patterns are reflected in a simple distributary mouth bar from the Mississippi delta (Fig. 8-25). Developmental history of such a bar was suggested by Coleman and Prior (Figs. 8-26, 8-27, 8-28, and 8-29).

Synsedimentary tectonism also is an important control for the sand distribution pattern. Subsidence of channel sands, growth faulting with sand deposition on the downthrown side, and channel patterns controlled by shale diapirism are important processes in the lower delta plain (Fig. 8-30). These processes are most important in deltas with strong riverine influence and high suspended sediment load. The Mississippi River is a well-documented example, and extensive studies have been published (Woodbury 1973) (Fig. 8-31) (Fisk 1961) (Fig. 8-32).

The final depositional environment that is often developed in front of riverine deltas is the area of gravity slumping (Fig. 8-33). In this setting the seaward face of the distributary mouth bar slides into deeper water. These

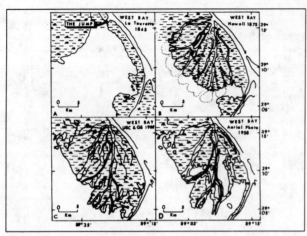

**Fig. 8-21.** History of bay-fill sedimentation, Mississippi delta. From Coleman and Gagliano, 1969.

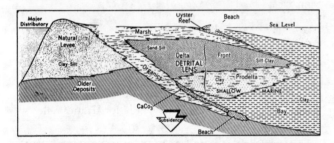

**Fig. 8-22a.** Cross section bay-fill sequence. From Coleman and Gagliano, 1964.

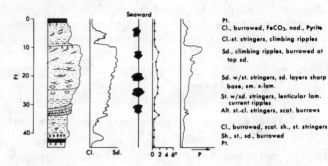

**Fig. 8-22b.** Vertical section. From Coleman and Prior, 1980, reprinted by permission.

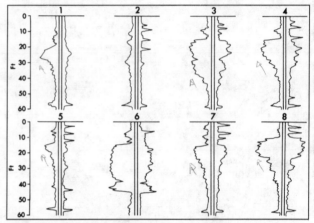

**Fig. 8-23.** Log shapes of lower delta plain, bay-fill sequences. From Coleman and Prior, 1980, reprinted by permission.

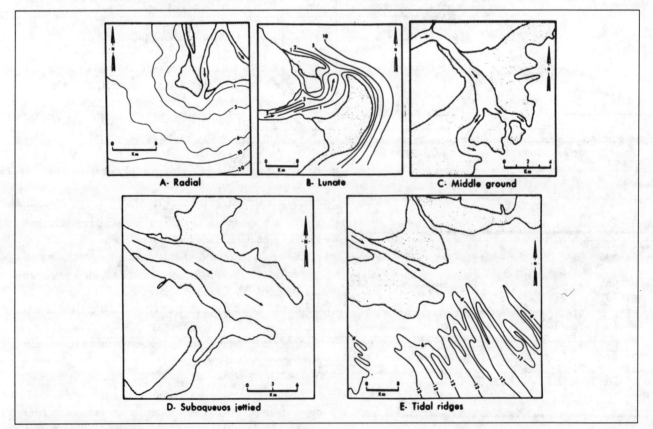

**Fig. 8-24.** Type of river mouth bars. From Coleman and Prior, 1980, reprinted by permission.

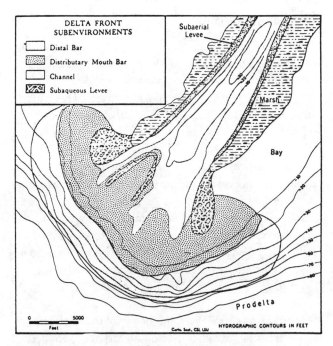

**Fig. 8-25.** Delta front environments, mouth of southwest pass, Mississippi River delta. From Coleman and Gagliano, 1965.

## DISTRIBUTARY MOUTH BAR

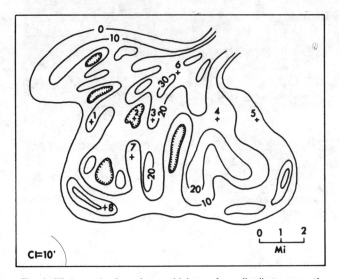

**Fig. 8-26.** Historic development of Southwest Pass bar finger. From Coleman and Prior, 1980, reprinted by permission.

**Fig. 8-27.** Isopach of sandstone thickness for a distributary mouth bar. From Coleman and Prior, 1980, reprinted by permission.

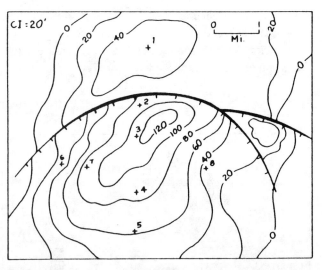

**Fig. 8-28.** Diagram illustrating the sedimentary characteristics of the distributary mouth-bar deposits. From Coleman and Prior, 1980, reprinted by permission.

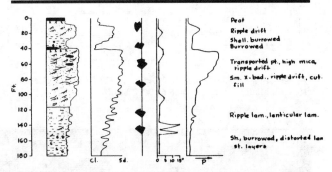

**Fig. 8-29.** Diagram illustrating the sedimentary characteristics of the distributary mouth-bar deposits. From Coleman and Prior, 1980, reprinted by permission.

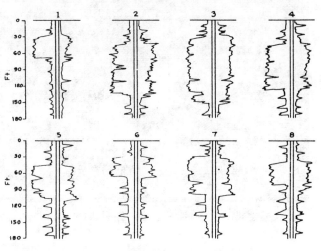

**Fig. 8-30.** Diagram illustrating the sedimentary characteristics of the distributary mouth-bar deposits. From Coleman and Prior, 1980, reprinted by permission.

slumps can produce local subaqueous scours, but they also may produce a complex hummocky topographic expression, or even coherent slide blocks that are many tens of meters in thickness and hundreds of meters wide.

## Depositional History

The progradational history is followed by a transgression resulting from compaction and subsidence (Fig. 8–34). The amount of subsidence possible is indicated by the warping of the Pleistocene subaerial surface. This pattern suggests that the causality of the subsidence is directly related to the presence of the Holocene depocenter, which is reflected through the older stratigraphic units. The relative subsidence of the depositional interface produced a shoreward movement of the strandline. This history is complex, with the first event the formation of a brackish marsh, then followed by a lagoonal environment, a transgressive blanket sandstone, and overlain by shelf sedimentary environments. This paleogeographic pattern is illustrated by the depositional history of the St. Bernard subdelta of the Mississippi River (Fig. 8–35). A cored section is located on the marginal levee of the Modern Mississippi River and does not reflect the transgressive history of either the St. Bernard or the Holocene birdfoot delta (Fig. 8–36). This section, however, can be used to develop a depositional model for Riverine Deltas. The geometry of the geomorphic elements is useful for identifying ancient stratigraphic patterns (Table 8–3).

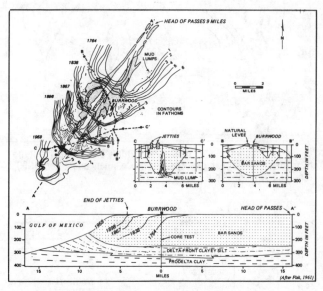

**Fig. 8–31.** Plio-Pleistocene delta system, Northern Gulf Coast Basin. From Woodbury, 1973, reprinted by permission.

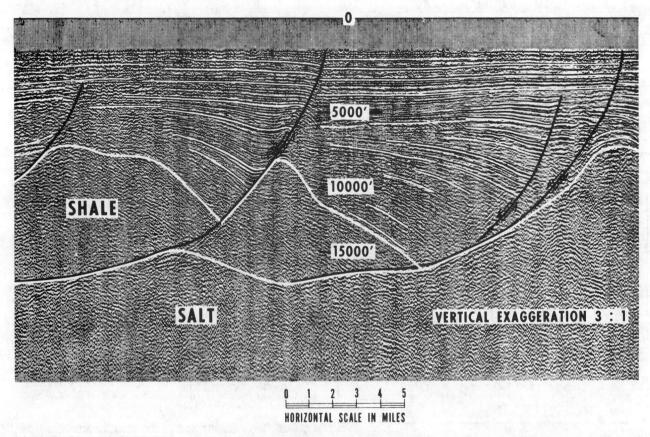

**Fig. 8–32.** Diagram illustrating sedimentary characteristics of slump deposits offshore of the deltaic sequence. From Fisk, 1961, reprinted by permission.

## Ancient Examples of Riverine Deltas

The Mississippi River delta has been used as the type model for sedimentation in epeiric seas and in restricted ocean basins. In these cases neither wave or tidal forces are strongly developed. Work by Wright and Coleman suggests a probable vertical sequence based upon their studies of deltas throughout the world (Fig. 8–37).

*Bartlesville Sandstone.* Examples of vertical sequences from the Bartlesville Sandstone are directly comparable to those of the Mississippi River. The interdistributary pattern is similar to bay-basin sedimentary patterns: the distributary section represents lower deltaic plain channel fill during avulsion, and the prodelta pattern is similar to that illustrated in Figure 8–35 (Fig. 8–38).

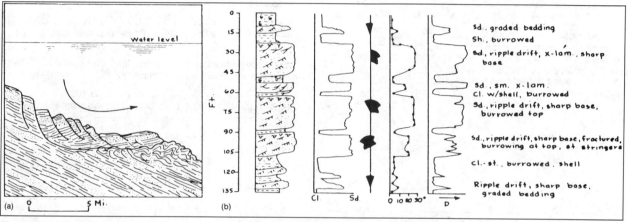

**Fig. 8–33.** (a) Diagram illustrating sedimentary characteristics of slump deposits offshore of the deltaic sequence. (b) Idealized core section of resulting sedimentary units. From Coleman and Prior, 1980, reprinted by permission.

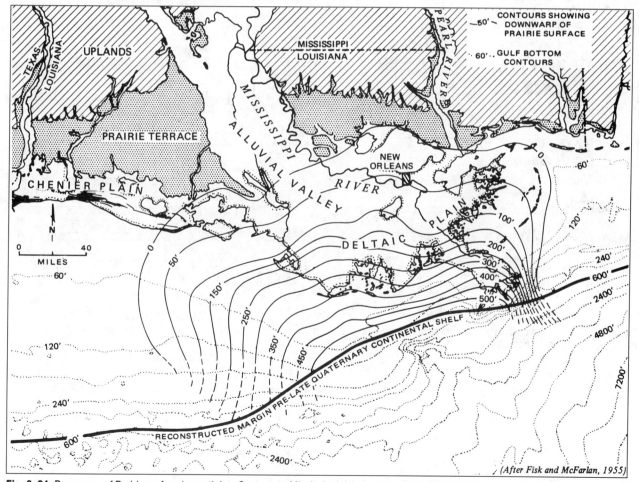

**Fig. 8–34.** Downwarp of Prairie surface beneath late Quaternary Mississippi deltaic mass. From Gould, 1970, reprinted by permission.

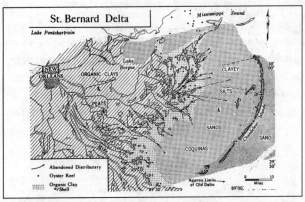

**Fig. 8-35.** The history of regression is followed by a transgressive period. This produces a vertical sequence that can be recognized in both Holocene and ancient stratigraphic sections. From Coleman and Gagliano, 1973.

**Table 8-3**
**Riverine-Dominated Delta Reservoirs**

Straight Distributaries, 10–40 m thick, width 1–5 kms, elongation greater than 10:1, flow units up to 5 m thick, 200 m wide, and 300 m long

Crevasse Splay interval, from 10–40 m thick, over areas up to 1000 kms², with 3–10 m thick reservoirs, width to 100 m and length to 3 kms, less than 100 m flow unit continuity

Bay fill sandstone units, from 3–10 m thick, ovoid in shape, with width to length ratio 1:2, flow units in top 1–3 m, continuous over areas of up to 5 kms²

Strandplain systems 10–25 m thick, 5–10 kms wide, and 20–60 kms long, reservoir flow units continuous up to 20 kms² in top 5 m

Top of sandstone interval shows less than 25% change in thickness relative to overlying event marker

Base of sandstone irregular, with as much as 100% change in interval thickness relative to an underlying event marker

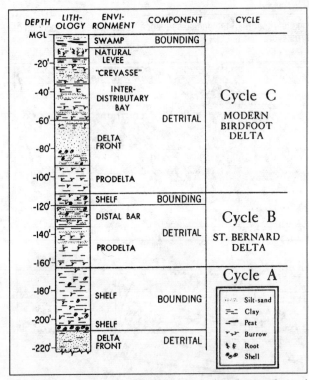

**Fig. 8-36.** A cored section illustrating the cycles of regression and transgression during the Holocene. From Coleman and Wright, 1975.

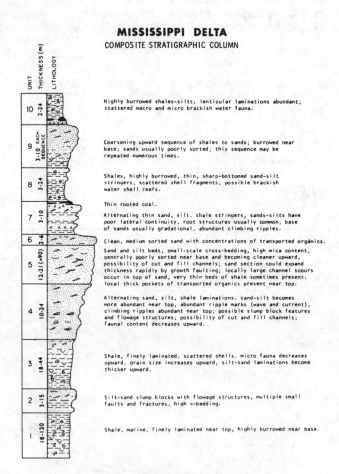

**Fig. 8-37.** Composite stratigraphic column of Mississippi River delta. From Coleman and Wright, 1975.

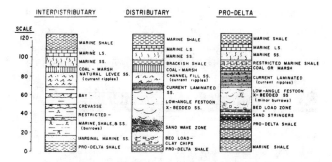

**Fig. 8-38.** Genetic sequences of lithologic units. These patterns characterize process-response sequences within individual areas of deltas. From Visher, Saitta, and Phares, 1971, reprinted by permission.

Paleogeography and channel patterns of the Bartlesville Sandstone are similar to the Mississippi delta. Note the presence of narrow channels, thicker sandstone development at points of bifurcation, thicker sandstone development in distributary mouth bars at the termination of channels, and bay fill. Some tidal influence is indicated in the outcrop at the very southern end of the map. This may also be reflected in the rejoining of some of the distributaries (Fig. 8-39).

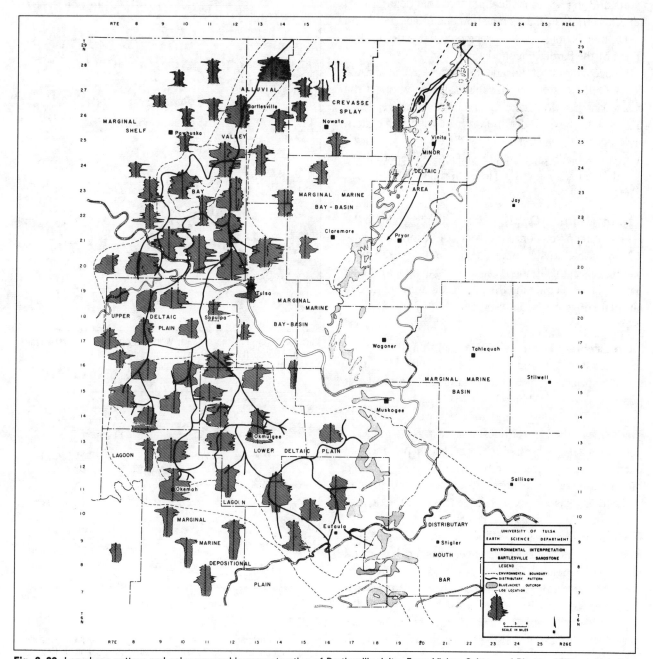

**Fig. 8-39.** Log-shape pattern and paleogeographic reconstruction of Bartlesville delta. From Visher, Saitta, and Phares, 1971, reprinted by permission.

The Bartlesville Sandstone contains the geomorphic elements described for the Mississippi delta. The alluvial valley, upper delta plain, and lower delta plain channel patterns are similar, and the bay-basin area, the marginal shelf, and the marginal depositional plain are recognizable (Fig. 8–39).

***Wilcox Sandstone.*** The Eocene lower Wilcox of southern Texas has been studied extensively by Fisher and McGowen and by Galloway. They show that a portion of the unit is directly comparable to the Mississippi delta system (Figs. 8–40 and 8–41). A pattern of growth faults and log sections illustrate that this delta system reflects vertical stacking on a short shelf (Figs. 8–42 and 8–43). It does not show the progradational patterns illustrated by the Bartlesville sandstone.

Paleogeography of Eocene delta systems along the Texas Gulf Coast show that the eastern portion of the coast contains elongate and highly constructive deltas (Fig. 8–44). The delta distributory pattern for the eastern style delta was characterized by Fisher and McGowen (Fig. 8–45). Galloway illustrates the log-shape pattern for the distributaries for the Holly Springs delta system (Fig. 8–46).

***Pennsylvanian Deltas.*** Other examples include the Pennsylvanian Red Fork Sandstone of Oklahoma (Fig. 8–47) and the Pennsylvanina Degonia and Hardinsburg Sandstones of the Illinois basin. This again illustrates the pattern of log shapes and the paleogeography of the deltaic plain (Figs. 8–48a and 8–48b).

***Cretaceous Examples.*** Production from these types of delta sandstone bodies may be very high. Compactional drape across the top of the sandstone unit makes it possible to identify channel patterns by detailed seismic analysis. The Muddy Sandstone equivalent to the D sandstone of the Denver-Julesburg basin is a significant petroleum reservoir. Structure, wavelet shape, and isopach patterns all are identifiable in wavelet-processed broad band data. The cross section across the channel indicates log shapes, vertical sequences, and the depositional controls. (Figs. 8–49 and 8–50).

A similar pattern is developed for the Dakota Sandstone, the J sandstone of the Denver-Julesburg basin, and illustrates the relative importance of textural patterns to the water saturation. Note the sequence of textures and the pattern of porosity and permeability (Fig. 8–51). The large pore-size distribution of the channel center provides for higher permeability. E-log patterns and production

indicate that the finer-grained distributary mouth bar and strandplain are nonproductive. The coarser-grained channel fill has higher permeability and larger pore-size distribution, and it is a commercial reservoir (Fig. 8–52).

# WAVE-DOMINATED DELTA MODEL

**W**ind-driven waves and oceanic circulation patterns result in development of longshore drift systems. Both suspended and bedload sediments are distributed

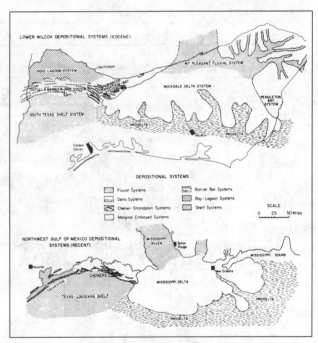

**Fig. 8–40.** Comparison of lower Wilcox (Eocene) and northwestern Gulf of Mexico (Holocene) depositional systems. From Fisher and McGowen, 1969, reprinted by permission.

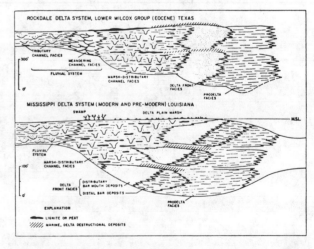

**Fig. 8–41.** Comparison of component facies of Mississippi delta systems (modified from partly hypothetical cross section of Coleman and Gagliano) and Rockdale delta system, Wilcox Group, Texas. From Fisher and McGowen, 1969, reprinted by permission.

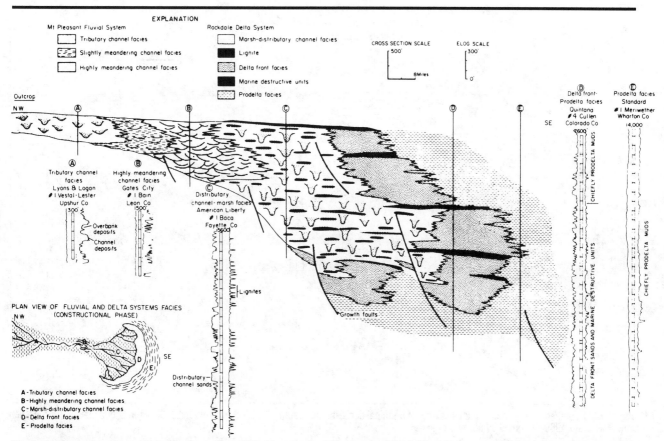

**Fig. 8–42.** Diagrammatic stratigraphic dip section of Mt. Pleasant fluvial system and Rockdale delta system shows relation and character of principal component facies. From Fisher and McGowen, 1969, reprinted by permission.

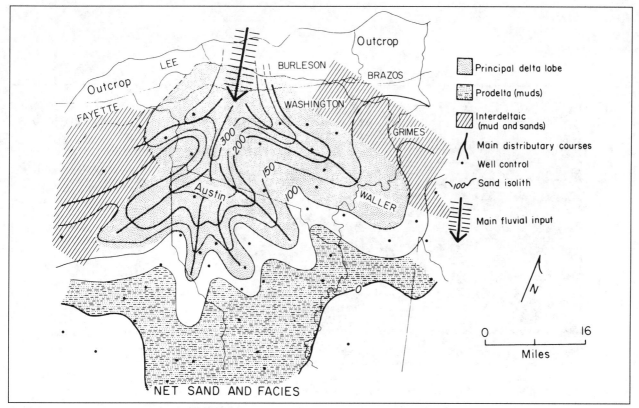

**Fig. 8–43.** Tertiary delta system from the Texas Gulf Coast. From Bruce, 1973, reprinted by permission.

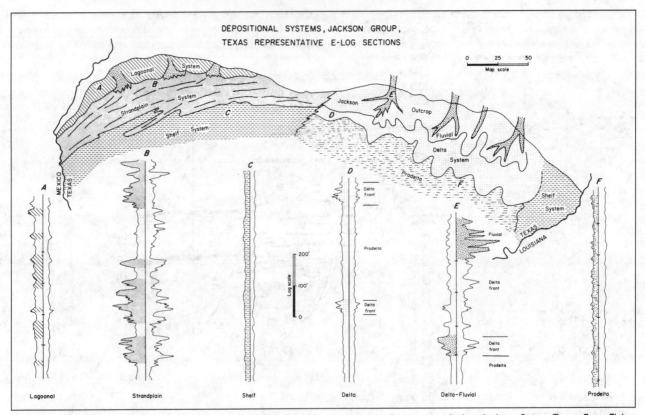

**Fig. 8-44.** Representative E-logs and log patterns of principal depositional systems and component facies, Jackson Group, Texas. From Fisher and McGowen, 1969, reprinted by permission.

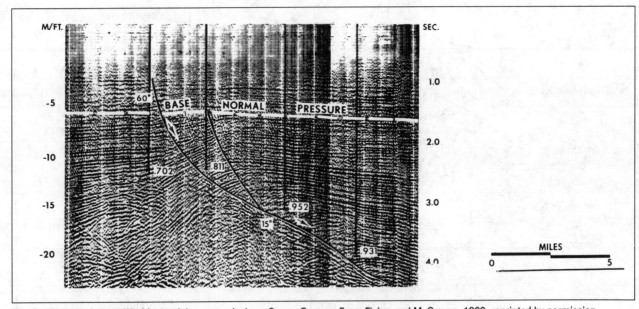

**Fig. 8-45.** Isopach map, Washington lobe, upper Jackson Group, Eocene. From Fisher and McGowen, 1969, reprinted by permission.

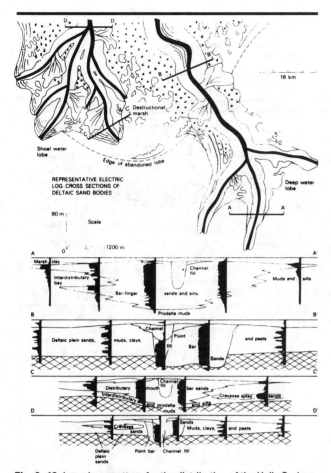

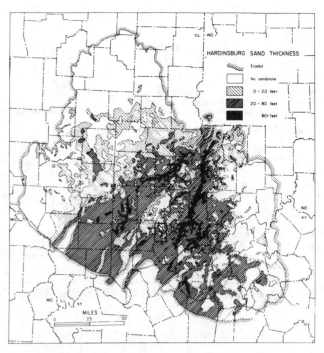

**Fig. 8–48a.** Total thickness of sandstone in Hardingsburg formation as determined from electric logs. From Swann, 1964, reprinted by permission.

**Fig. 8–46.** Log-shape pattern for the distribution of the Holly Springs delta system. From Galloway, 1968.

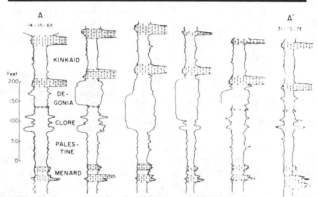

**Fig. 8–48b.** Electric log cross section of Degonia and adjacent formations. From Swann, 1964, reprinted by permission.

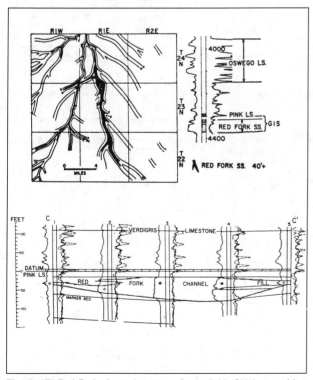

**Fig. 8–47.** Red Fork channel pattern, Ceres field, Oklahoma. After Scott, from Busch, 1974, reprinted by permission.

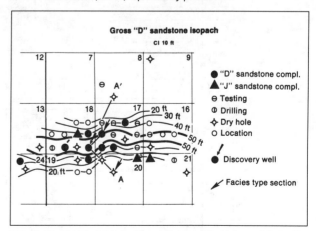

**Fig. 8–49.** Zenith field significant Dakota (Muddy) sandstone discovery. From Root, 1980.

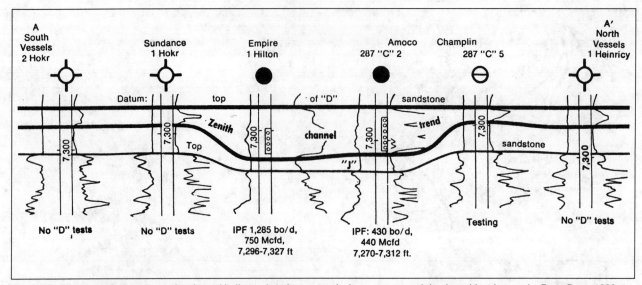

**Fig. 8-50.** The cross section across the channel indicates log shapes, vertical sequences, and the depositional controls. From Root, 1980.

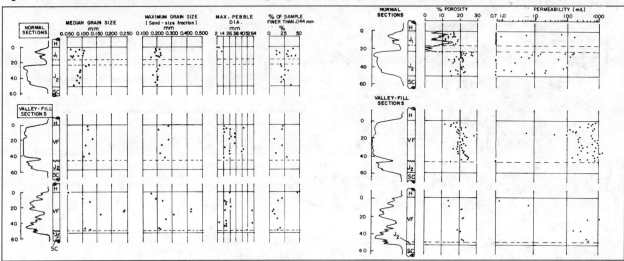

**Fig. 8-51.** The Dakota sandstone, or J sandstone, of the Denver Julesburg basin illustrates the relative importance of textural patterns to the water saturation. From Harms and Exum, 1966, reprinted by permission.

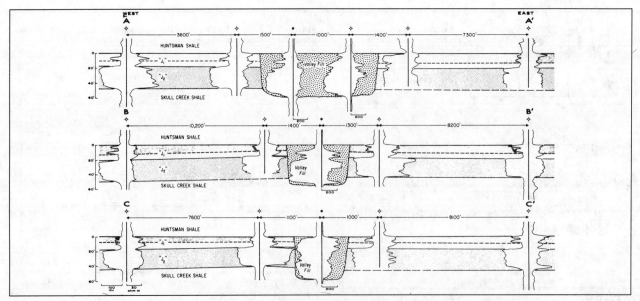

**Fig. 8-52.** Electric-log cross sections showing valley fill. From Harms and Exum, 1966, reprinted by permission.

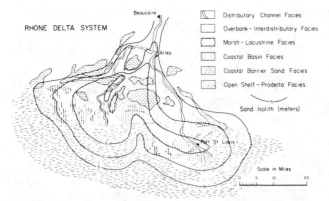

**Fig. 8–53.** Principal depositional environments, Rhone delta system (Holocene, southern coast of France). A wave-dominated high-destructive delta system. From Fisher, 1969.

by this drift. The bedload requires transport in the zone of shoaling waves to produce sediment accumulation in the marginal delta plain, but suspended load may reflect offshore current patterns. The dominant forces responsible for accumulating strandplain systems are those associated with waves. Wave height and power reflect water depth, fetch, and wind velocity. The effectiveness of alongshore transport is correlated with these aspects. The direction of the wind and its relation to times of maximum riverine sediment discharge also must be considered if a physical model is to be developed.

Marginal delta plains formed in areas of high wave and/or current power produce a strandplain dominated by beach ridges, reworked sheet sands, and windblown detritus. The Rhone delta is a modern example reflecting these processes (Fig. 8–53). Extensive interdeltaic strandplain systems can develop under conditions of strong coastal winds, high sand riverine discharge, and steep offshore gradients. Such a system has been suggested for the extensive longshore and barrier island depositional system from the Trinity River to the Rio Grande along the Gulf Coast. Sand is contributed by a number of rivers, including the Colorado, Brazos, Trinity, and Rio Grande. The primary deposition is in the extensive beach-bar system flanking the coast. Textural patterns are similar to the marginal marine shelf with a coarsening upward pattern, poorly sorted bioturbated marine platform and lower shoreface sands, and progressively better sorting upward into the upper shoreface and lower foreshore zones. Thickness is usually less than 10 meters in areas of high wave power. Areas close to the delta, where continued subsidence results from rapid sedimentation in the marginal marine delta plain, poorer sorting, and less bioturbation, are developed. Sedimentary structures are similar to those described for the marginal marine shelf but with small-scale current and wave ripples, planar truncating surfaces, and planar cross-bedding.

Wave-dominated deltas are characterized by restricted delta plain areas, meandering distributaries with broader fining upward channel trends, a dominance of strandplain environments, and minor bay-basin, levee, and marsh depositional patterns. The prodelta environment is characterized by sand waves or a single coarsening upward cycle of deposition. For example, the Nile and Rhone delta systems illustrate these patterns (Figs. 8–9 and 8–53).

This pattern is even more strongly illustrated by the Surinam delta. Note the dominance of meandering channels, strandplain development, and the low-energy marsh deposits separating individual beach ridges (Fig. 8–14).

The Niger delta is not as strongly wave-dominated and has a considerable tidal influence. The presence of barrier-bar patterns, prodelta coarsening upward cycles, and sand waves suggests this as a good example for many ancient deltaic reservoirs (Fig. 8–54). Energy patterns illustrate the dominant influence of wave processes, but the secondary tidal channel pattern can be seen. The distribution of energy throughout the year also influences the relative importance of individual depositional environments (Fig. 8–12a and 8–12b).

Seismic sections across the delta platform illustrate the Tertiary history of regression and transgression. Transgressive breaks are shown as reflectors due to the high impedance contrast between marsh deposits and regressive delta cycles reflecting sand deposition (Fig. 8–55). Depositional history is reflected by a seismic section typical of this type of delta; note the growth fault pattern in Figure 8–56.

The depositional processes of the Niger delta have been well described and interpreted. The surficial patterns can be related to the formation of stratigaphic sequences detailed by shallow core holes through the Holocene section (Fig. 8–57).

There are few examples of log-shape patterns through wave-dominated Holocene sedimentary sequences, but the wave-dominated bay fill pattern from the Mississippi is similar to those developed in the ancient examples. Similarly, wave-dominated delta front sequences from the Mississippi are similar to those indicated from the Niger delta (Fig. 8–58).

## Ancient Examples of Wave-Dominated Deltas

Coleman and Wright suggest that the São Francisco delta illustrates the vertical sequence that can develop in response to wave processes (Fig. 8–59). This entire sequence is not often preserved in ancient delta deposits. The rooted zone above the beach-distributary mouth bar is often the topmost unit preserved. The geometry and sequence of geomorphic elements can be summarized for comparison to ancient stratigraphic units (Table 8–4).

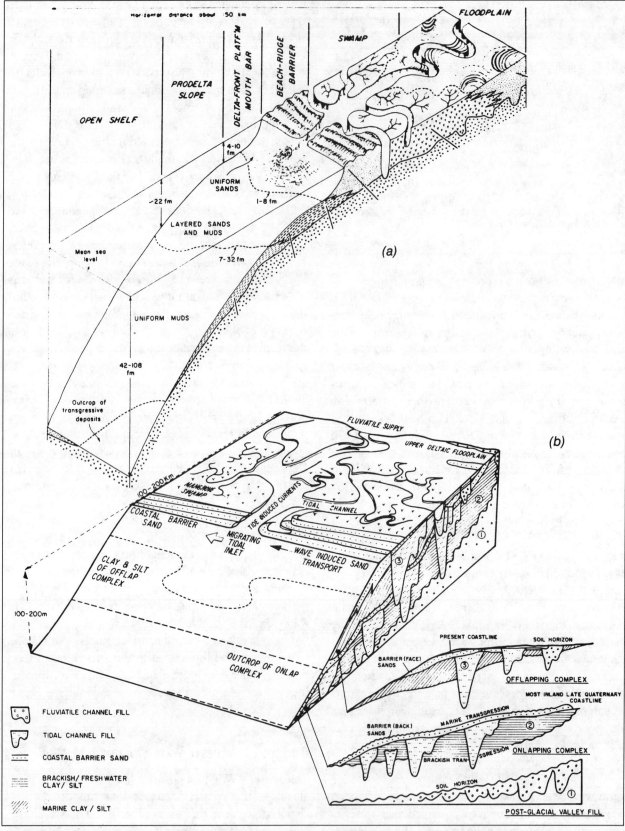

**Fig. 8-54.** A, Schematic illustration of the properties and relationships of principal sedimentary facies of the modern Niger delta. B, Lithofacies relations in the Late Quarternary Niger delta. From Allen, 1970 (A), and Oomkens, 1974 (B).

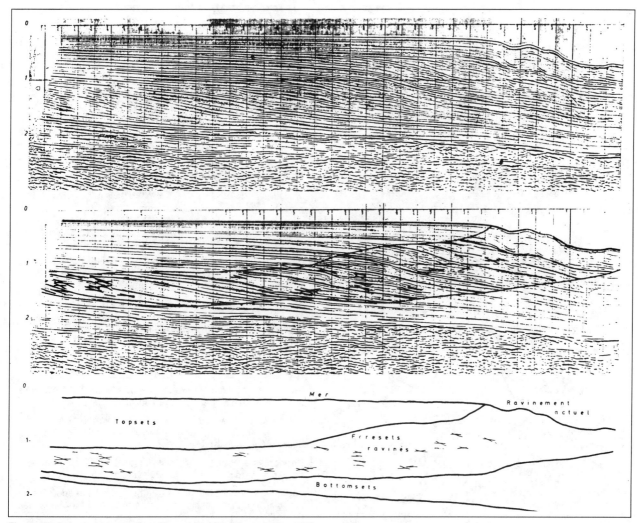

**Fig. 8-55.** Seismic cross section, Niger delta plain. From Dailly, 1975.

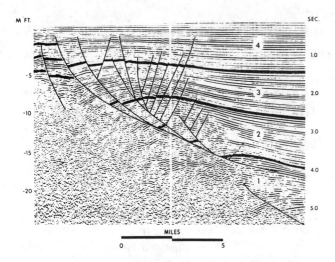

**Fig. 8-56.** Growth fault pattern of a wave-dominated delta system from the Gulf Coast. A similar pattern is developed in the Niger delta. From Woodbury et al., 1973, reprinted by permission.

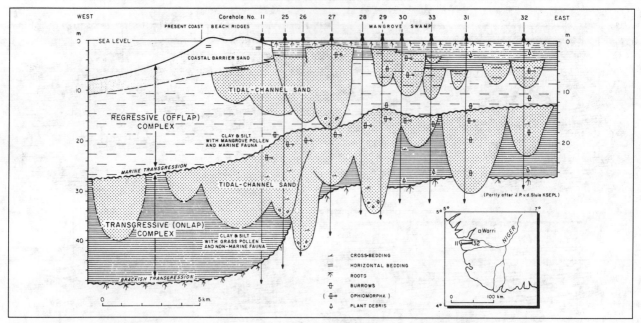

**Fig. 8–57.** Transgressive and regressive sequences in the Late Quarternary Niger delta. From Weber and Daukoru, 1975.

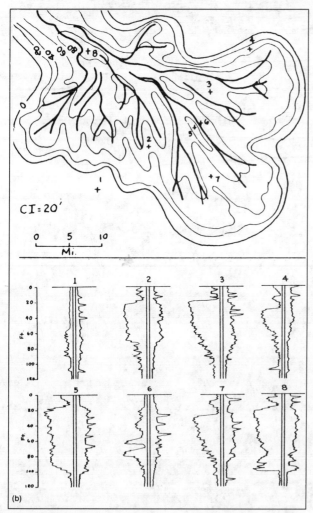

**Fig. 8–58.** *(a)* Geometry of a wave-dominated distributary mouth bar pattern. Similar in scale and pattern to those developed in the Niger delta. *(b)* Characteristic log shapes. From Coleman and Prior, 1980, reprinted by permission.

**Table 8-4**
**Wave-Domianated Deltaic Reservoirs**

> Interval is lensoid, 10–20 m thick, lobate in area, 40–160 kms wide and 100–400 kms long (parallel to shoreline)
>
> Distributaries, irregular in area, 10–20 m thick, 5–15 kms wide, and more than 40 kms long, reservoir flow units less than 5 m thick, 100 m wide, and 500 m long
>
> Strandplain system 5–20 m thick, length up 10 times width, with flow units up to 5 m thick, and 500 m long
>
> Lagoonal intervals, less than 10 kms² parallel to shoreline or irregular in shape, reservoir 1–10 m thick and flow units less than 100 m long

*Eocene Jackson Group.* As mentioned previously, the Eocene Jackson Group contained both a riverine depositional process pattern on the eastern portion of the Texas Gulf Coast and a wave-dominated pattern on the western portion of the Coast. The contrast in depositional pattern is readily apparent.

*Vicksburg Delta.* Few examples have been thoroughly documented in the literature, but a geomorphic pattern developed by Gregory (1966) and presented by Fisher et al. illustrates the general features for the Middle Vicksburg delta system (Fig. 8–60).

*Niger Delta.* Few deltas are exclusively the product of a single depositional process. The Niger delta, though

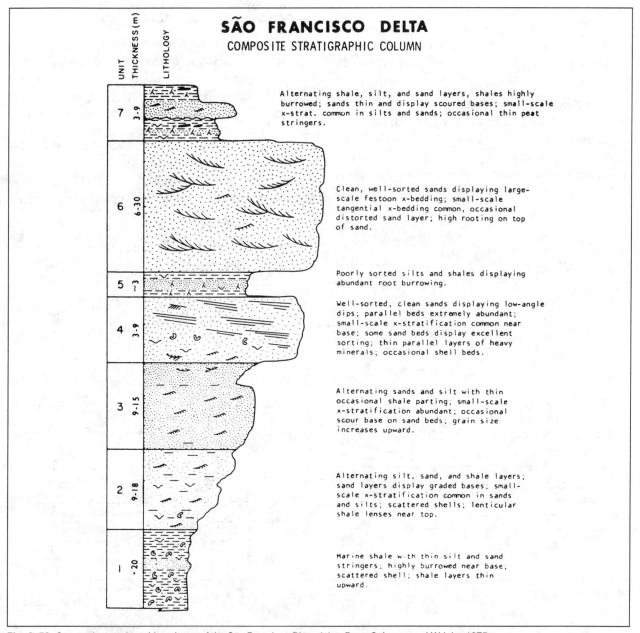

## SÃO FRANCISCO DELTA
### COMPOSITE STRATIGRAPHIC COLUMN

**7** (3-9) — Alternating shale, silt, and sand layers, shales highly burrowed; sands thin and display scoured bases; small-scale x-strat. common in silts and sands; occasional thin peat stringers.

**6** (6-30) — Clean, well-sorted sands displaying large-scale festoon x-bedding; small-scale tangential x-bedding common, occasional distorted sand layer; high rooting on top of sand.

**5** (3) — Poorly sorted silts and shales displaying abundant root burrowing.

**4** (3-9) — Well-sorted, clean sands displaying low-angle dips; parallel beds extremely abundant; small-scale x-stratification common near base; some sand beds display excellent sorting; thin parallel layers of heavy minerals; occasional shell beds.

**3** (9-15) — Alternating sands and silt with thin occasional shale parting; small-scale x-stratification abundant; occasional scour base on sand beds; grain size increases upward.

**2** (9-18) — Alternating silt, sand, and shale layers; sand layers display graded bases; small-scale x-stratification common in sands and silts; scattered shells; lenticular shale lenses near top.

**1** (20) — Marine shale with thin silt and sand stringers; highly burrowed near base; scattered shell; shale layers thin upward.

**Fig. 8–59.** Composite stratigraphic column of the São Francisco River delta. From Coleman and Wright, 1975.

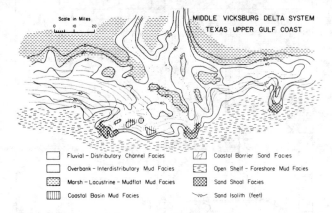

**Fig. 8–60.** Middle Vicksburg delta system, Upper Texas Gulf Coast. Ancient example of a highly destructive, wave-dominated delta system. From Fisher et al., 1969.

dominated by high-energy wave processes, also illustrates tidal influences. For example, a log cross section through a part of the Egwa field illustrates channel-fill sedimentary units (Fig. 8–61). A log section through a distributary channel fill shows the importance of the strandplain deposition (Fig. 8–62) (barrier bar and barrier foot units).

***Pennsylvania Foster Sandstone Delta.*** The Morrowan Spiro sandstone from the Arkoma basin shelf margin is a well-documented example (Figs. 8–63 and 8–64). The log cross section illustrates the relation of the meandering distributaries to strandplain coarsening upward intervals and the presence of carbonates in high-energy, near shore environments along depositional strike (Fig. 8–65). The paleogeographic pattern illustrates the close association both laterally and vertically between contrasting depositional association of subgraywacke channel sandstones and quartzose shelf sandstones.

***Other Examples.*** The development of a strike parallel quartzose sandstone body at the edge of the Morrowan continental shelf, the Cretaceous of Utah, Colorado, and Wyoming, the Woodbine of northern Louisiana, and Desmoinsian units on the northwestern shelf of the Anadarko basin are examples of these types of deltas.

# TIDAL PROCESS DELTA MODEL

The dominance of tidal processes in the delta framework changes in fundamental ways the relative importance and pattern of the geomorphic elements typical for all deltas (Table 8-2). Process patterns of alternate channel fill, interchannel tidal estuary, and tidal flat channels, all lead to a complex of delta plain channel patterns of varying geometries (Smith 1966) (Fig. 8–68). The changing stream gradients due to the tidal cycle produces meandering and/or rejoining channel systems across most of delta plain. Also the flood-ebb tidal cycle concen-

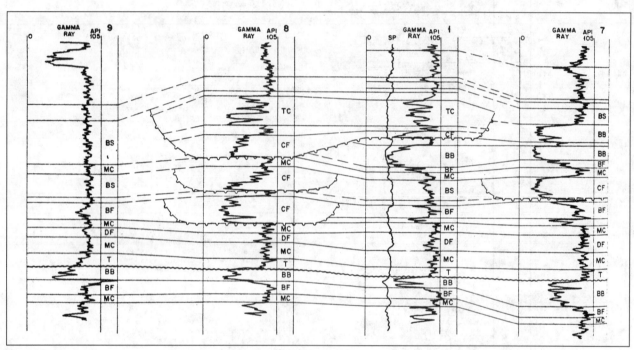

**Fig. 8–61.** Log section of Egwa field. From Weber and Daukuro, 1975.

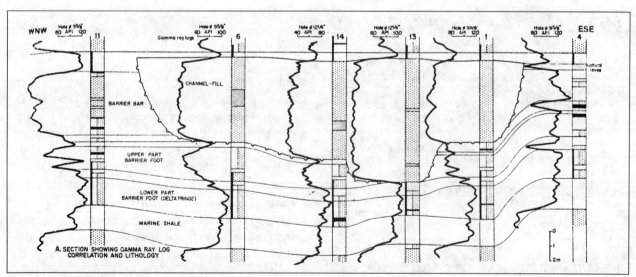

**Fig. 8–62.** Log section distributary channel fill. From Weber and Daukuro, 1975.

trates coarser detritus in major channels, leading to bell shaped channel mouths, and thick channel lag deposits. In addition, ebb tidal flow produces an extensive marine blanket of tidal sand ridges oriented perpendicular to the shoreline as illustrated by the Yellow River Delta (Fig. 8-68). As indicated for the Ganges-Brahmaputra (Fig. 8-8), the shelf edge shows the development of a submarine canyon. The tidal process distributes sediment across the lower delta plain, the continental shelf, and down the shelf slope. Deposition of sediment is not localized into a single depocenter; consequently, the geometry of sedimentary sequences is controlled by depositional processes and not by the history of subsidence.

A number of Modern examples of depositional response patterns have been reported. The Altamaha River estuary of south Georgia illustrates the aforementioned patterns of meandering and rejoining tidal and riverine channel patterns, the bell shape of the channel mouth, and the offshore linear sand ridge geometry (Visher and Howard 1974) (Fig. 8-69). It also illustrates the coarse sand lag deposition in deeper channels both on and offshore. A similar study of the dynamics of tidal delta sedimentation was conducted on the Ord River delta by Coleman and Wright (1975) (Fig. 8-70). Their study suggested a typical vertical sequence of depositional units, reflecting seaward progradation of the delta

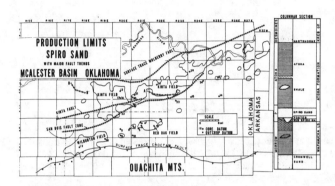

**Fig. 8-63.** Spiro sand production and main fault trends in McAlester basin. Kinta and San Bois faults are Atokan-age growth faults. Choctaw and Mulberry faults are post-Atokan. Spiro production shows include submarginal as well as produced and abandoned wells. From Lumsden et al., 1971, reprinted by permission.

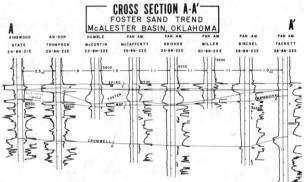

**Fig. 8-64.** Cross-section A-A', Foster sand trend number II. Line of section is shown. Note lens shape of Foster sand channeled in sub-Spiro shale. In other areas, Foster cuts deeply into Wapanucka limestone. From Lumsden, Pittman and Buchanan, 1971 reprinted by permission.

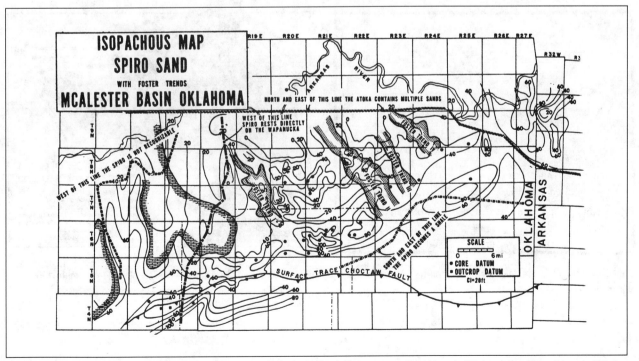

**Fig. 8-65.** Isopach map of Spiro sand with four Foster sand trends outlined. Note east-northeast trend of Spiro and northwest trend of Foster. From Lumsden, Pittman, and Buchanan, 1971, reprinted by permission.

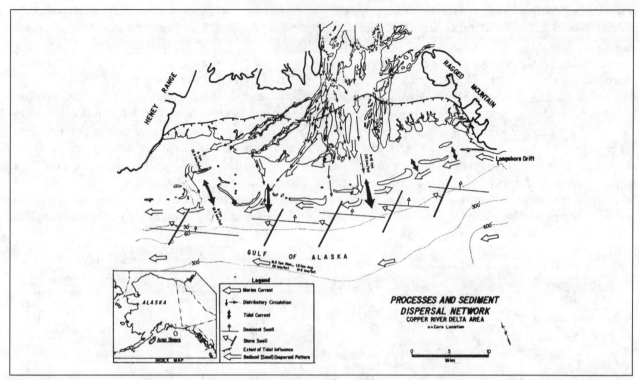

**Fig. 8–66.** Summary diagram showing processes active in the Holocene delta of the Cooper River, Alaska, and the resultant longshore transport paths of bed-load sediment. Suspended load sediment (very fine sand to clay) spreads farther out onto the shelf and is transported westward by the intruding marine current. The delta plain includes lagoonal sand flats and marsh, which are influenced by tides, and prominent marginal coastal barrier bars. From Galloway, 1976.

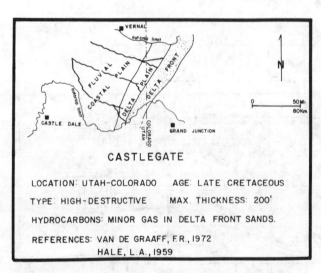

**Fig. 8–67.** Pattern of fan delta. Reprinted courtesy Houston Geological Society.

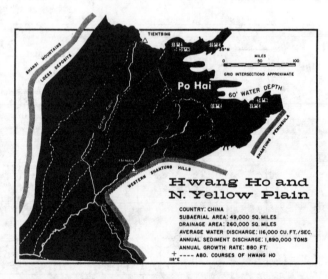

**Fig. 8–68.** Hwang Ho and North Yellow plain. From Smith, 1966.

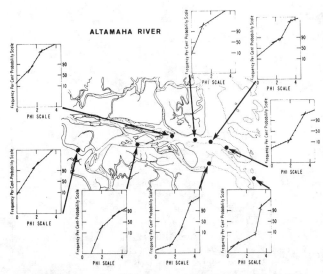

**Fig. 8–69.** Pattern of texture changes within the estuary. From Visher and Howard, 1974.

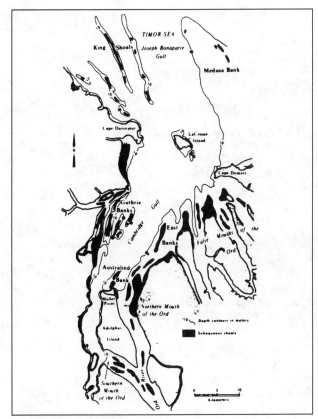

**Fig. 8–70.** Linear tidal ridges and marginal shoals in the mouth of the river. From Coleman and Wright, 1975.

platform (Fig. 8–71). This sequence of depositional units may be usefully applied for the recognition of ancient tidal delta intervals.

Another more extensive study was conducted on the Mahakam Delta of east Kalimantan, Indonesia, by Allen and others (1974). They present a general geomorphic pattern of the delta (Fig. 8–72); detailed core studies through the lower delta plain (Fig. 8–73); a reconstructed vertical profile of sedimentary units (Fig. 8–74); a strike cross section showing the history of alternate channel deposition producing an interweaving of delta plain and marine depositional units (Fig. 8–75); and also a dip section suggesting a history of stacking of delta intervals in response to eustatic changes in sea level (Fig. 8–76). These illustrations provide a basis for comparison to reservoir sequences of giant fields from the Mahakam Delta and other tidally deposited deltaic stratigraphic sequences (Table 8–5).

Indications from many Holocene tidally dominated shelves is that both submarine canyons and tidal sand ridges are common to areas offshore of tidal deltas. A review of the distribution of tidal sand ridges by Belderson (1986) (Table 8–6) indicates such an association, with sand waves and ridges offshore many rivers from around the world, including the Ord River, northwest shelf of Australia; the East China Sea, offshore the Yangtzee and Yellow Rivers; and Malacca Delta off west Malaysia. In the East China Sea occurrence these tidal ridge system are thought to be relict (Chang-shu and Jia-song 1988) (Figs. 8–77 and 8–78), but the pattern may be stratigraphically important. Also associated with tidal dominated shelves are submarine canyon systems. These systems transport sediment from the shelf down the slope and are proximal to submarine fans. Examples of canyons offshore deltas include the Ganges-Brahmaputra, Irrawady, Mahakam, and Amazon deltas. The stratigrahic importance of this association has been indicated by the occurrence of submarine fans, and by drilling on the shelf offshore each of these delta systems. Many more examples can be cited, including the Orinoco, Mackenzie, Parana, and Indus deltas.

## Ancient Examples of Tidal-Dominated Deltas

The paleogeographic and depositional complexity cited for sedimentary intervals associated with tidal delta systems may indicate the reason for the difficulty experienced by stratigraphers in the recognition of these delta systems in the stratigraphic record. Exploration success for hydrocarbon accumulations also has been slow to develop, possibly due to the lack of localized depocenters, and the complex array of depositional processes and patterns producing petroleum reservoirs (Table 8–5).

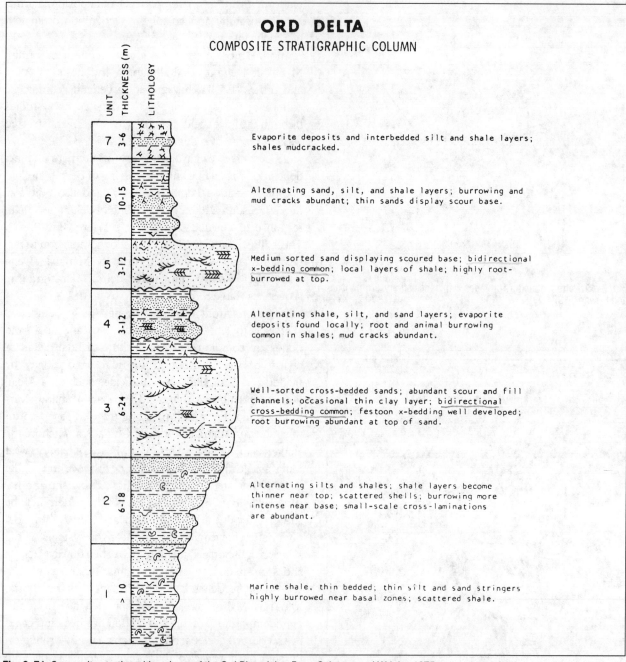

## ORD DELTA
### COMPOSITE STRATIGRAPHIC COLUMN

UNIT | THICKNESS (m) | LITHOLOGY

**7** — 3-6
Evaporite deposits and interbedded silt and shale layers; shales mudcracked.

**6** — 10-15
Alternating sand, silt, and shale layers; burrowing and mud cracks abundant; thin sands display scour base.

**5** — 3-12
Medium sorted sand displaying scoured base; bidirectional x-bedding common; local layers of shale; highly root-burrowed at top.

**4** — 3-12
Alternating shale, silt, and sand layers; evaporite deposits found locally; root and animal burrowing common in shales; mud cracks abundant.

**3** — 6-24
Well-sorted cross-bedded sands; abundant scour and fill channels; occasional thin clay layer; bidirectional cross-bedding common; festoon x-bedding well developed; root burrowing abundant at top of sand.

**2** — 6-18
Alternating silts and shales; shale layers become thinner near top; scattered shells; burrowing more intense near base; small-scale cross-laminations are abundant.

**I** — > 10
Marine shale, thin bedded; thin silt and sand stringers highly burrowed near basal zones; scattered shale.

**Fig. 8-71.** Composite stratigraphic column of the Ord River delta. From Coleman and Wright, 1975.

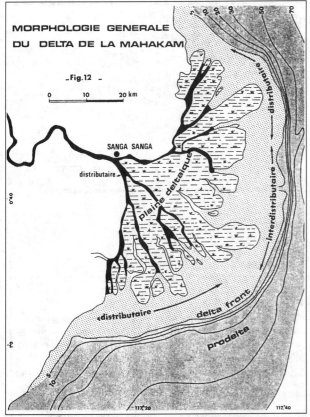

**Fig. 8-72.** General morphology of Mahakam delta. From Allen, Laurier, and Thouvenin, reprinted courtesy Total Compagnie Française de Pétroles, 1974.

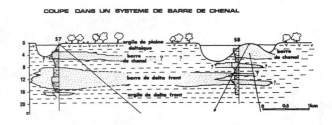

**Fig. 8-73a.** Geometry based upon boreholes from lower delta plain. From Allen, Laurier, and Thouvenin, reprinted courtesy Total Compagnie Française des Pétroles, 1974.

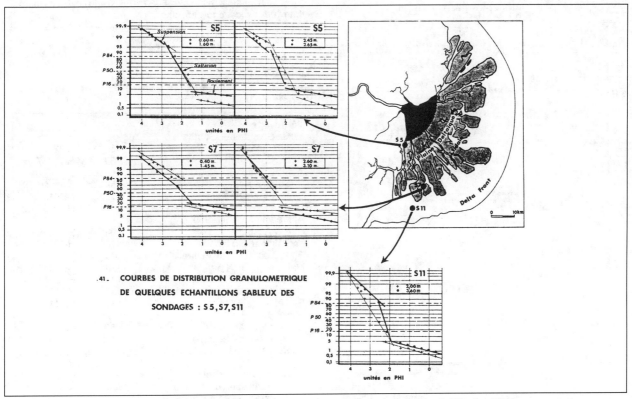

**Fig. 8-73b.** Grain size curves from Holocene cores, reprinted courtesy Total Compagnie Française de Pétroles, 1974.

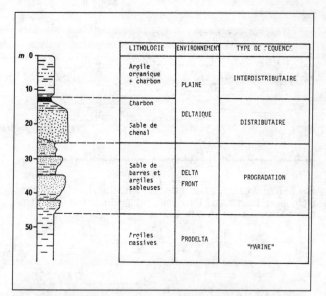

**Fig. 8-74.** Vertical sequence of sedimentary units. From Allen, Laurier, and Thouvenin, reprinted courtesy Total Compagnie Française des Pétroles, 1974.

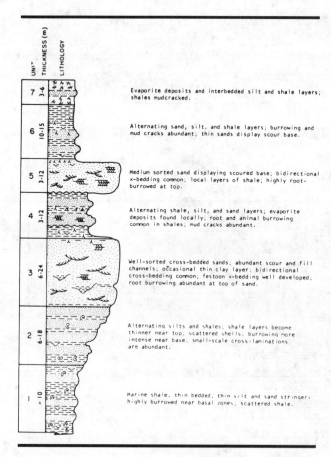

**Fig. 8-75.** Pattern of interdeltaic depocenters. From Allen, Laurier, and Thouvenin, reprinted courtesy Total Compagnie Française des Pétroles, 1974.

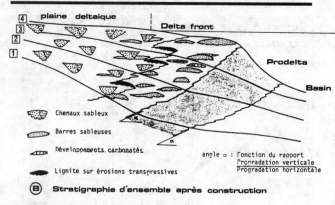

**Fig. 8-76.** Stratigraphic cross section. From Allen, Laurier, and Thouvenin, 1974.

### Table 8-5
### Tide-Dominated Delta Reservoirs

Interval more than 100 m thick, lensoid in shape, over an area of greater than 2000 kms²

Distributary reservoirs, 10–40 m thick, 2–10 kms wide, 10–50 kms long, with gradational base over 1–3 m, lateral pichouts greater than 1 km wide, flow units up to 5 m thick, greater than 100 m wide, and 500 m long

Tidal flat reservoirs, 1–8 m thick, up 10 kms² in area, stacked in an interval up to 100 m thick, flow units 1–5 m thick, highly irregular in shape, and less than 100 m long

Marine platform reservoirs, 2–10 m thick, up to 100 m wide, .1–2 kms long, and stacked in interval 10–50 m thick, over an area greater than 1000 kms², flow units 1–5 m thick, up to 100 m wide and 1 km long

Shelf edge channeled, 1–10 kms wide and greater than 20 m deep, with multiple reservoirs, 5–30 m thick, and flow units more than 200 m wide, and 1 km long

### Table 8-6
### Mean Spring Near-surface Peak Tidal Current Speeds Worldwide*

| Worldwide | | |
|---|---|---|
| White Sea Funnel (Southern Barents Sea) | 160 | HD2270 |
| Korea Bay (North Korea) | 130 | HD1257 |
| Approaches to Seoul (South Korea) | 150 | Kang (1969) |
| SW coast South Korea | 130 | HD913 |
| Cambridge Gulf (NW Australia) | 150 | HD/AUS 32 |
| Approaches to King Sound (NW Australia) | 90 | HD1206 |
| Van Diemen Gulf (NW Australia) | >70 | HD/AUS 308 |
| Endeavour Strait (N Australia) | 100 | HD2321 |
| Malacca Strait (N Sands) | 130 | HD3945 |
| Malacca Strait (S Sands) | 130 | HD3946 |
| Gulf of Cambay (W coast of India) | 130 | HD1486 |
| Gulf of Kutch (W coast of India) | 130 | HD43 |
| Bay of Bengal | 130-160 | HD814 |
| Head of Persian Gulf (muddy) | ≈50 | HD2884 |
| Georges Bank (NE USA) | 80-140 | NOAA 13009 |
| Nantucket Shoals (NE USA) | 60 (near-bed) | Swift et al. (1981) |
| Entrance to Para River (Brazil) | 160 | HD2186 |
| Valdes Peninsula (Argentina) | 200 | HD3067 |
| Approaches to Bahia Blanca (Argentina) | 90 | HD1331 |
| Cabo Blanco (S side Gulf of San Jorge, Argentina) | 100 | HD3106 |

*Information mainly from British Admiralty charts (HD).

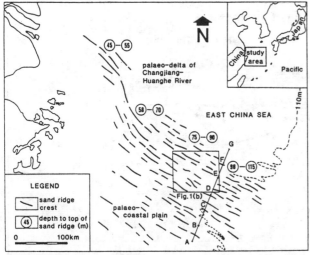

**Fig. 8-77.** Tidal sand ridges on the shelf of the East China Sea. From Chang-Shi and Jia-song, 1988.

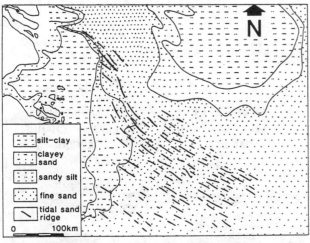

**Fig. 8-78.** The types of surficial sediments on the east China shelf (data from the Bureau of Marine Geol. Survey and the Second Inst. of Oceanogr.) From Chang-Shi and Jia-song, 1988.

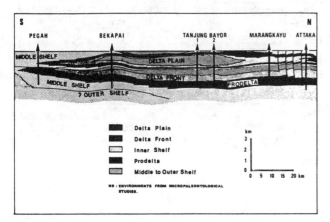

**Fig. 8-79.** Geological section along the Bekapai-Attaka trend, Mahakam Contract Area. Note the two main paleodeltas separated by a marine transgression. DeMatharel et al., 1980.

The Mahakam Delta is an obvious example of deposition controlled by tidal processes. Subsurface information presented by Verdier and others (1980) indicates that the depositional history is similar to that illustrated by Holocene depositional processes. Regressive-transgressive cycles are stacked and reflect patterns of intervals associated with eustatic cycles (Fig. 8-79). Within a single sequence associated with alternate channel depositional sequences, correlation of individual sands between wells is very difficult (Fig. 8-80). Typical log sequences of the delta plain and offshore intervals are helpful in placing a well within a single depositional setting (Fig. 8-81). Log sections of typical sedimentary sequences illustrate the stacking of individual reservoir units (Fig. 8-82). A north-south cross section through the giant Handil field illustrates the complexity of the horizontal and vertical stacking of reservoir sandstone units. Hydrocarbon distribution within the field is a complex interrelation between basement faulting, occurrence of bounding sealing stratigraphic units, and the geometry of individual reservoir sandstone units (Fig. 8-83). The structural pattern has little relation to depositional history of the delta, and contrary to most other delta systems, exploration is principally based upon the structural history related to basement faulting (Fig. 8-84). Field development, however, must be based upon a detailed analysis of the characteristics and the geometry of reservoir and sealing depositional units.

***Booch Delta.*** The Pennsylvanian Booch sandstone interval from southeastern Oklahoma illustrates the pattern to be expected from a tidal delta (Busch 1953, 1973) (Fig. 8-85). Field work by the author indicates that stratigraphic units, both below and above this interval, show evidence of response patterns indicative of tidal influence, with a probable tidal range from 2 to 3 meters (Visher 1968). The depositional history indicates an unconformity surface underlying the Booch interval, and the thicker sandstone sequence reflects valley fill deposition in an estuarine setting.

***Permian Basin.*** Outcrop and subsurface analysis of the Upper Permian from Sydney basin also illustrates marine shelf to channel vertical sequences, abundant coal units bounding sedimentary sequences, and coarse channel fill sandstone units with a rejoining channel pattern (Connolly and Ferm 1971) (Fig. 8-86).

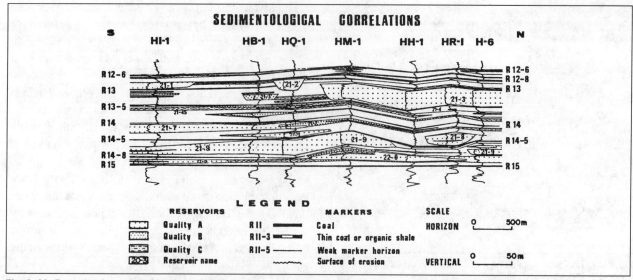

Fig. 8–80. Example of correlation within the Handil field based on the continuity of coal beds. From Verdin, 1980.

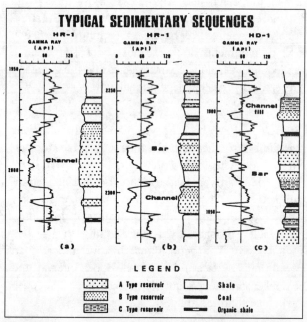

Fig. 8–81. Typical sedimentary sequences: *(a)*, thick channel unit in a complete sequence, *(b)*, bar sequence overlying a thinner channel unit, and *(c)*, series of uncomplete sequences, no thick reservoir development. Handil Field AAPG 1980, Verdier et al., 1980.

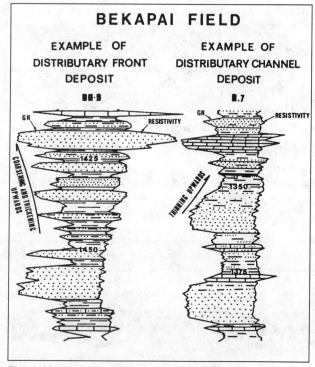

Fig. 8–82. Diagram showing examples of distributary front and distributary channel deposits, Bekapai field, Mahakam delta. From DeMatharel et. al, 1980, reprinted by permission.

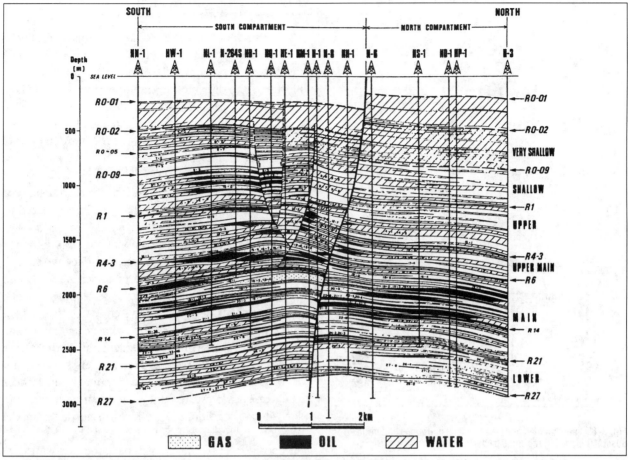

**Fig. 8-83.** Schematic north to south cross section of Handil field indicating the multiple reservoirs and the distribution of the hydrocarbons. From Verdin, 1980.

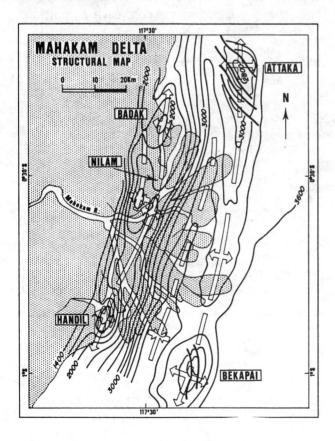

**Fig. 8-84.** Structural contours showing north-to-south structural trend of the Mahakam Delta area. Note Bakapai field at the southern end of the trend and Attaka field at the northern end. From De Matharel, 1980, reprinted by permission.

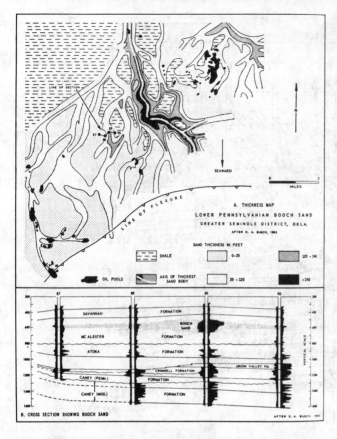

**Fig. 8–85.** Isopach map and cross section of Booch sandstone. From Busch 1953; 1973, reprinted by permission.

# ARCTIC-FAN DELTA MODELS

Analysis of deltaic processes and boundary conditions allows the subdivision of deltas into natural groupings. Study of modern deltas (Coleman and Wright 1975) illustrates the difficulty in forming a rigorous classification of delta types. Definitions and classifications need to change as new information is acquired, and as usefullness of a classification changes. The history of the definition of arctic, fan, and braid deltas was recently reviewed by McPherson and others (1987). This discussion focuses on the need to define the process basis for each delta type before developing a generalized model. Principal to this, for the arctic-fan delta depositional system, is the tectonics of the source and depositional sites. Discussion of coastal plain depositional systems illustrated the usefullness of interrelating grain size, channel gradient, and shoreline depositional processes for developing a genetic classification of depositional systems. Consistent with this classification, the arctic-fan delta system represents deposition of coarse-grained sediment transported across the coastal and delta plain by high gradient streams, and deposited in tectonically mobile marginal basins. These characteristics provide a unique array of boundary and depositional conditions to produce a useful deltaic model (Wescott and Ethridge 1980).

Deposition of fan deltas may be associated with foreland, lacustrine, extensional, and wrench-faulted basins. The geometry of the delta system reflects these depositional settings. Provenance reflects both tectonic and

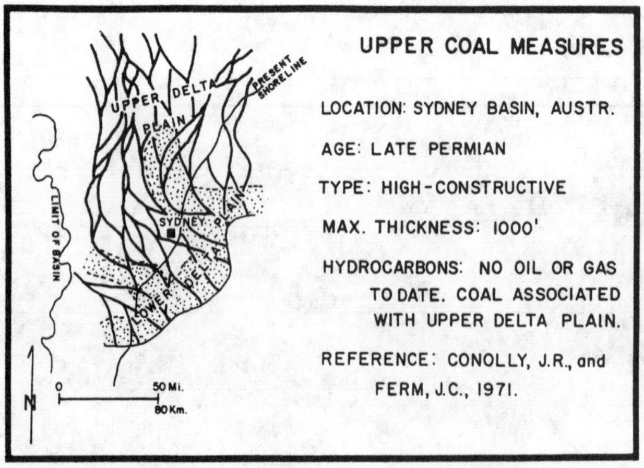

**Fig. 8–86.** Upper coal measures. From Houston Geological Society, 1975, after Connolly and Ferm, 1971.

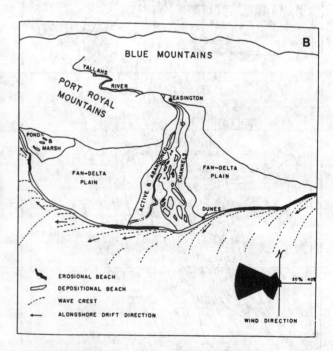

**Fig. 8–87.** Interpretive sketch map of Yallahs fan delta. From Wescott and Ethridge, 1980.

climatic controls, but more importantly the gradient between source and depositional setting provides a unifying theme. Braided rivers with stream gradients of meters/kilometer (McPherson et al. 1987) dominate transportational and depositional processes, producing extensive sheet sandstone and conglomeratic intervals. Deltas may range in size from a few hundred to several thousand square kilometers. Channel sequences may range from a few meters to tens of meters in thickness. Interchannel areas may be dominated by marsh, splay, or sheetflood deposits, producing an array of differing depositional sequences.

Few examples of Holocene fan deltas have been reported upon in detail, but two examples may be useful: 1) the Yallahs delta of Jamaica, a small coarse-grained delta (Wescott and Ethridge 1980) (Fig. 8–87); and 2) a more extensive study of the Copper River delta of southwest Alaska (Galloway 1975; 1976) (Fig. 8–88). The latter study is supported by cores, and textural data, and provides a useful model for comparison to ancient, humid, fan delta sequences (Fig. 8–89). This delta contains an area of more than 150 square kilometers, and the delta plain is dominated by marsh deposits. The lower delta

plain and the marine platform are tidally influenced, typically ranging from 2 to 3 meters, but offshore sand bars are parallel to the shoreline, reflecting wave and current processes.

Few other examples of Holocene fan deltas have been described to aid in the development of a process/response fan delta model (Nemec and Steel 1988). The suggestion of geometric patterns for fan delta reservoirs is based both upon understanding of Holocene examples, and many ancient stratigraphic examples (Table 8–7). Inference concerning a fan delta origin for many of these examples is based upon texture of the detritus, the blanket nature of the stratigraphic interval, and the paleogeographic and the tectonic depositional framework.

## Ancient Fan Deltas

Many summary articles and books are available describing ancient fan delta systems. These include recent books by Ethridge and Wescott (1984) and Nemec and Steel (1988). The research emphasis on fan deltas is apparently a result of their identification as producing reservoirs in strata ranging in age from Paleozoic, to Ter-

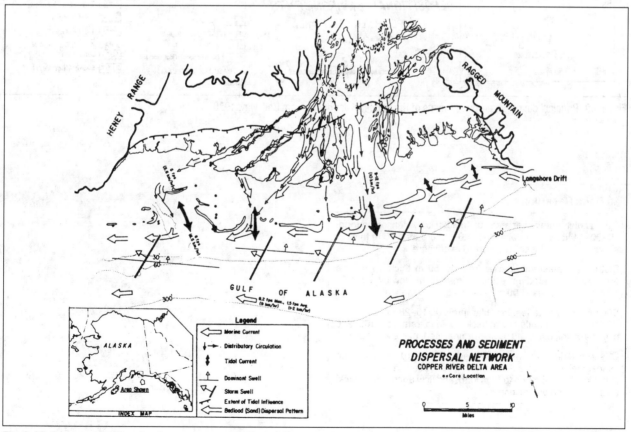

**Fig. 8–88.** Summary diagram showing processes active in the Holocene delta of the Copper River, Alaska, and the resultant longshore paths of bed load sediment. Suspended load sediment (very fine sand to clay) spreads farther out onto the shelf and is transported westward by the intruding marine current. The delta plain includes lagoonal sand flats and marsh, which are influenced by tides, and prominent marginal coastal barrier bars. From Galloway, 1975.

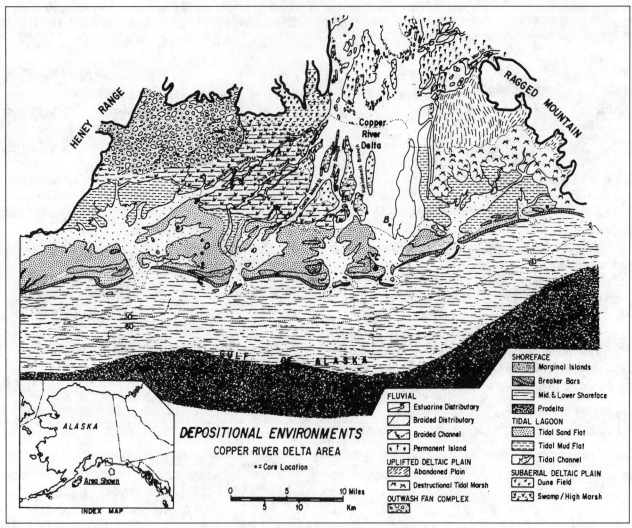

**Fig. 8–89.** Principal depositional environments of the Copper River delta. From Galloway, 1976.

**Table 8-7**
**Fan Delta Reservoirs**

Interval distributed over areas up to 3000 kms², lensoid in form, 10–30 m thick upper delta plain, 20 to more than 50 m thick at shoreline, and thins to less than 10 m offshore

Distributary reservoir interval from 10–50 m thick, 2–20 kms wide, and elongated more than 5 times width, flow units are 5–10 m thick, less than 300 m wide and 500 m long

Between distributaries, irregular intervals 10–20 m thick, containing reservoir units 1–5 m thick, ovoid in form from 100–200 m in diameter, enclosed by fine-grained sediments and coals

Offshore delta platform, over areas of 500–1000 kms², with sandstone sequences 5–20 m thick, and reservoirs up to 5–10 m thick, 100 m wide, up to 1 km long, and flow units 1–5 m thick, 50 m wide, and 500 m long

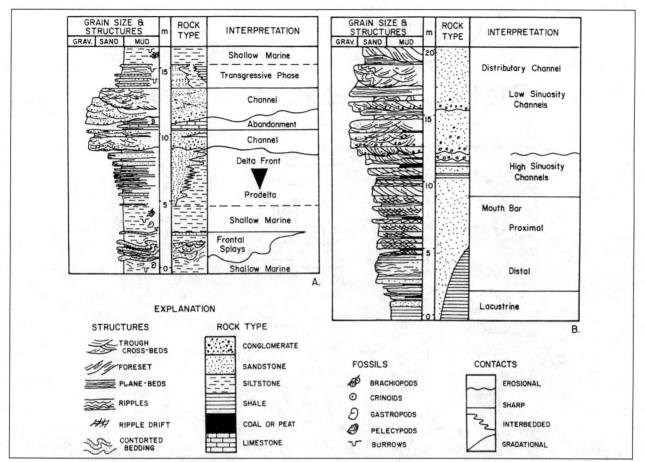

**Fig. 8–90.** Detailed vertical sequences through shelf-type fan-delta deposits. *A,* Pennsylvanian deposits of the Taos trough, New Mexico. *B,* Devonian, lacustrine fan-delta deposits, Hornelen basin, western Norway. From Ethridge and Wescott, 1984.

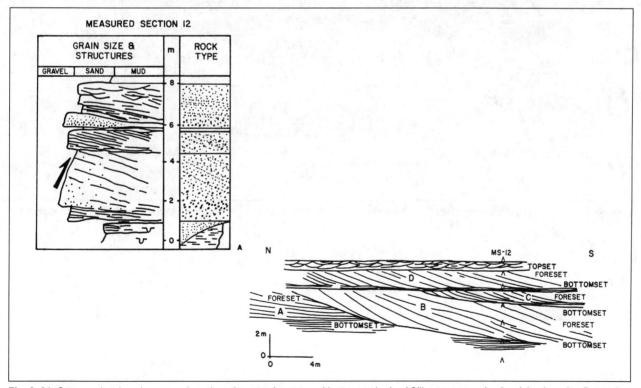

**Fig. 8–91.** Outcrop sketch and measured section of topset, foreset, and bottomset beds of Gilbert-type, marine fan-delta deposits, Pennsylvanian, Taos Trough, New Mexico. From Ethridge & Wescott, 1984.

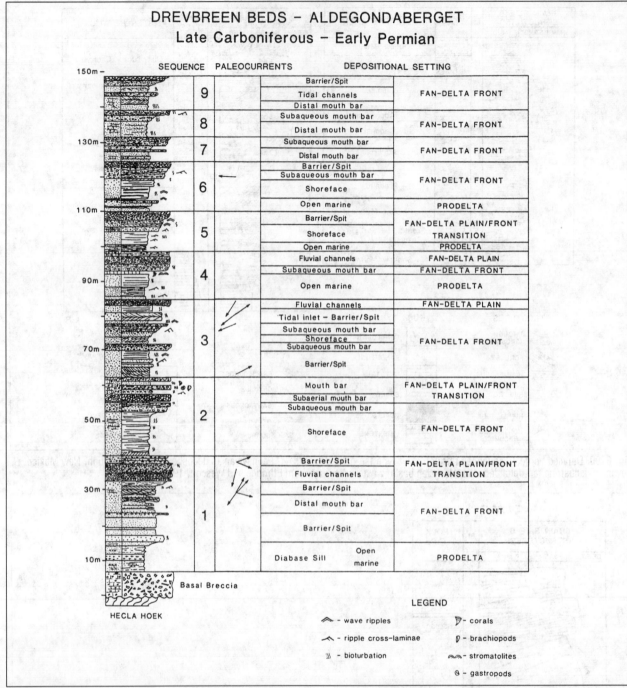

**Fig. 8-92.** A generalized stratigraphic profile of the Reinodden formation at Aldegondaberget showing the organization of the succession into nine sequences. From Kleinspehn et al., 1984.

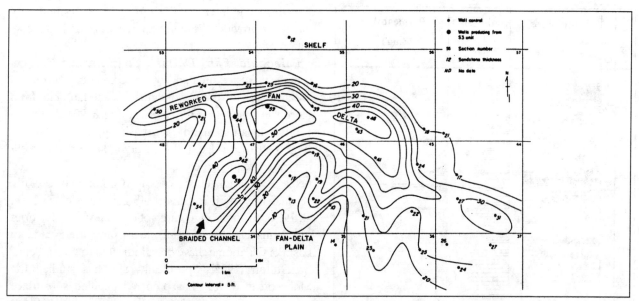

**Fig. 8-93.** Isopach of S3 sandstone showing extensive reworking of distal fan along strike. From Dutton, 1982, reprinted by permission.

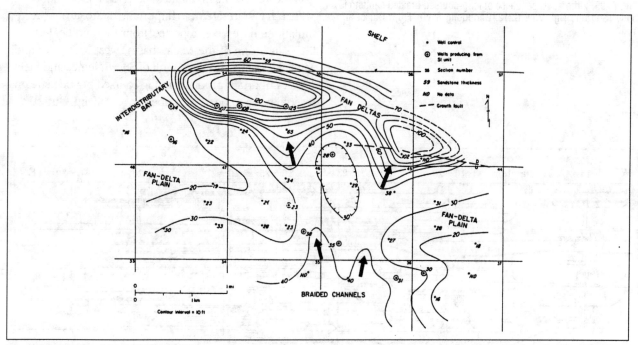

**Fig. 8-94.** Isopach of S1 fan-delta sandstone, showing braided channels which fed fan lobes basinward of underlying algal mounds. From Dutton, 1982, reprinted by permission.

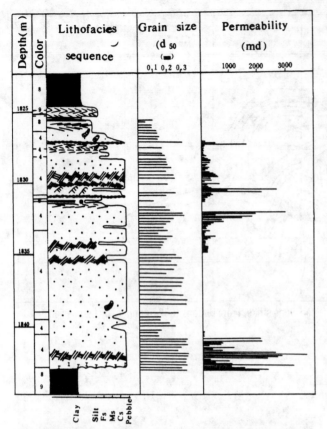

**Fig. 8-95.** Lithofacies sequence of long-coursed braided sandstone body (Subzone NgI, Well G212, Gongdong field). From Qiu et al., 1987.

tiary Neogene intervals. A short review of a number of examples provides the observational basis for their recognition.

***Paleozoic Fan Deltas.*** Characteristic vertical sequences of textures and other sedimentary attributes are available for the Devonian lacustrine fan delta from Hornelan basin of western Norway (Fig. 8-90) and for the Pennsylvanian foreland Taos basin from New Mexico (Fig. 8-91) (Ethridge and Wescott 1984). Another well-described sequence of fan deltas is from the Drvbreen beds: Aldegondagebert, of Late Carboniferous—Early Permian of western Spitsbergen (Fig. 8-92) (Kleinspehn et al. 1984). In addition research of the many oil fields from the Pennsylvanian of north Texas, and the Anadarko basin of Oklahoma, have been indicated to be fan deltas by Dutton (1982). Specifically, she has studied the Mobeetie field, and has prepared detailed subsurface maps of the reservoir geometry (Figs. 8-93 and 8-94).

***Mesozoic Fan Deltas.*** Resrvoir intervals from the giant oil fields of Jurassic age from the North Sea have been suggested to be of a fan delta origin (Harms and McMichael 1983). These fields reflect a high continuity of reservoir intervals, a high percentage of sandstone in the intervals, and the high permeability of fan delta systems as indicated by the high rate of petroleum production.

***Tertiary Fan Deltas.*** Important hydrocarbon accumulations have been reported from the middle Tertiary "K" sandstone of the Malay basin, west Malaysia, and Miocene clastics from the Gulf of Suez (Nemec and Steel 1988). Important producing long coursed braided stream reservoirs have been described in detail by Qui et al.

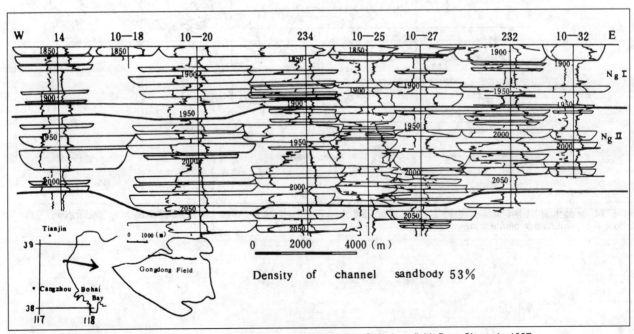

**Fig. 8-96.** Correlation section of Ng, the long-coursed braided sandstone bodies, Gongdong field. From Qiu et al., 1987.

(1987). These reservoir intervals overlie the lower Ecocene unconformity surface in the Shengli area adjacent to the Gulf of Bohai, China (Fig. 8–95). Stratigraphic sections illustrate patterns of bounding fine grained units, and percentages of sandstone within depositional intervals, which are characteristic of fan delta systems (Fig. 8–96). The high rates of hydrocarbon productivity also supports the fan delta depositional origin of these intervals. Yet another example has been described from the Pliocene of the foreland Apennine basin, from Spain (Ricci Lucchi et al. (1981).

These oil field studies illustrate the recent emphasis on recognition of fan deltas in the stratigraphic record. This emphasis has led to new research studies and publication, which will be useful for refining and expanding current knowledge on the tectonic, climatic, and paleogeographic framework for the development of fan delta systems.

## SUMMARY

Deltaic sedimentary patterns are most important in planning and developing an exploration program for hydrocarbon distributions. Each of the four models suggested show differing depositional patterns, and hydrocarbon accumulations are controlled by differing structural and stratigraphic associations. Stratigraphic traps can be associated with strand plain depositional systems. Shelf riverine deltas are associated with regional structural and topographic patterns, while hydrocarbon accumulations associated with fan deltas, or tidal deltas, suggest a structural or erosional channel framework for traps. Deltas are probably one of the more important depositional frameworks resulting in commercial hydrocarbon accumulations, but their genetic interpretation is of paramount importance for structuring an exploration program.

## REFERENCES FOR FURTHER STUDY — CHAPTER 8

Coleman, J.M., and D.B. Prior, 1980, Deltaic sand bodies: Am. Assoc. of Petrol. Geol., Continuing Education Course Note Series, no. 15, 171 p.

Coleman, J.M., and L.D. Wright, 1975, Modern river detlas variability of processes and sand bodies, *in* Broussard, M.L., ed., Deltas, Models for Exploration: Houston, Texas, Houston Geol. Soc., p. 99–149.

Koster, E.H., R.J. Steel, eds., 1984, Sedimentology of gravels and conglomerates: Canadian Soc. of Petrol. Geol. Mem., no. 10, Calgary, Canada, 441 p.

Nemec, W., and R.J. Steel, eds., 1988, Fan Deltas, Sedimentology and Tectonic Settings: Bishopbriggs, Scotland, Blackie and Son Ltd., 464 p.

# 9 SLOPE AND BASIN SEDIMENTOLOGY

*. . . we cannot pretend to explain everything which appears: and that our theories, which necessarily are imperfect are*

*not to be considered as erroneous when not explaining everything which is in nature, but only when they are found*

*contrary to or inconsistent with the laws of nature, which are known, and with which the case in question may be*

*properly compared.* **J. Hutton,** *Theory of the Earth,* **1795**

## INTRODUCTION

The geomorphology and depositional responses on the continental shelves and in marginal basins has been the focus of oceanographic research. Data collected in these areas have been useful in developing and understanding sedimentologic processes and provides objective information for developing stratigraphic models. The physiography, current and density flow processes, and the geometry of the deposited units provide the framework for interpreting sedimentary units. Direct observation by submersibles and scuba divers and indirect measurement and sampling combine to provide the objective information on processes and sedimentologic responses. Information on bedforms, textures, distribution of faunas, and rates and patterns of deposition is available for comparing Holocene areas and interpreting ancient stratigraphic units.

## BOUNDARY CONDITIONS

Geometry of the shelf, the shelf slope, and the configuration and topography of the basin floor reflect and produce sedimentary responses. Sediment provenance, whether related to delta progradation, volcanic activity, or tectonic forces, is of major importance in interpreting sedimentary responses. Tectonic and sea-level changes, rates and patterns of uplift, and earthquake frequency and intensity are related to plate tectonic forces and boundary conditions. Sedimentary responses are an interrelation of these aspects. Their importance must be evaluated as a part of the model approach to interpreting continental slope and marginal basin sedimentary sequences.

### Shelf-Slope-Basin Geometry

The development of continental shelves is a complex process that demonstrates a balance between depositional and erosional forces. Deposition on the continental slope, the associated rise, and the basin floor reflect shelf processes. Progradational shelves are dominated by a surplus of sediment supply. The related slope and basin reflect this supply, but slope processes produce differing topographic expressions. The presence of slumping, canyons, and growth faults can change the topography of the slope and basin margin. The continental rise may be modified by diapiric structures, coherent slumps, and depositional fans. These processes indicate that the continental slope, rise, and basin floor are not passive features, but reflect many diverse processes.

Topographic boundary conditions must be understood in light of the many possible processes that operate at plate margins. Canyons reflect the complex relationship among sedimentation, density currents, provenance, and ocean basin processes. The topography of submarine canyons has been studied in detail, but much remains to be learned of their origin, maintenance, and fill. Canyons are a primary control for physical processes operating at continental margins, yet few studies have interrelated boundary conditions, physical processes, and sedimentologic responses. These aspects are an essential part of models relating sedimentology to development of stratigraphic units.

A common topographic pattern at the base of continental slopes is a hummocky and seismically chaotic area. The origin of the topography may reflect depositional processes, but it also may be in response to post-depositional diapiric or tectonic forces.

Other topographic aspects, including the gradient of the continental slope, the area and form of the continental rise, and the geometry of the basin floor, reflect a complex of processes. These features and the geometry of associated sedimentary units can be used as the basis for interpreting eustatic, sedimentary, and tectonic boundary conditions.

## Clastic Provenance

The sedimentation rate is a principal control for shelf, slope, and basinal depositional processes. In areas of high rates of clastic supply, the result of deltaic sedimentology, volcanic activity, or vertical uplift may be debris flows, small or large landslides, or active canyons. To know which of these alternatives is important depends upon other boundary conditions.

The increase in the slope angle by progradation of the shelf may result in landslides or growth faults with diapiric responses in the area of continental rise. Patterns of regression and transgression across the shelf are reflected in the topography of the slope, rise, and basin floor. One common response of coastal progradation to the continental shelf margin is the development of submarine canyons and fans.

Tectonism, reflected by either uplift and formation of scarps or by volcanic activity, strongly affects the rate of clastic supply and the mineralogy and texture of the detritus. These processes may produce an unstable shelf slope, and the sedimentology may be dominated by

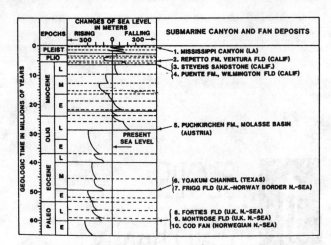

**Fig. 9–1.** Hydrocarbon-bearing submarine canyon and fan deposits that closely correspond to low sea level. From Shanmugan and Moiola, 1982.

either debris flows or landslides. The scale of the downslope gravity movement may range from debris flows containing a few hundred cubic meters to slumps involving tens of cubic kilometers. Rates of clastic supply may be high, but the depositional site often may be only at the base of the slope or on the basin floor. Sedimentologic responses reflect both the provenance and the mechanics of transportation.

Provenance patterns may be similar, but the form of the slope, continental rise, and basin floor may result from, or be a result of, many diverse processes.

## Tectonic and Eustatic Controls

The shelf width, the presence of a marginal trench, or patterns of block faulting and basinal highs are primary sedimentary controls. The geometry of deposited units, the directions and patterns of paleocurrents, and sedimentary characteristics that reflect provenance are measurable aspects useful for interpreting tectonic boundary conditions. Facies and isopach patterns provide information on the history of uplift, eustatic sea-level changes, or subsidence, but these interpretations require detailed time-rock correlations.

Systematic changes in relative sea level are important to sedimentology. A narrow or emergent shelf reflects a low sea level, and slope and basin sedimentology is more closely related to riverine and subaerial processes. At the beginning of a sequence, prior to significant onlap, rivers discharge at or below the edge of the continental shelf. Subaerial erosion to a depth of 200 to more than 500 meters initiates the formation of canyons in the shelf margin. Once formed, these canyons act as a funnel to channel sediments across the shelf and down the slope and produce a submarine fan. This pattern is common to basal portions of sequences (Fig. 9–1).

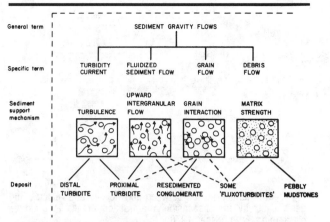

**Fig. 9-2.** Classification of subaqueous sediment gravity flows. From Middleton and Hampton, 1973.

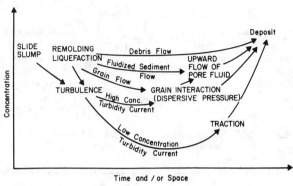

**Fig. 9-3.** Hypothetical evolution of a single flow, either in time or in space. From Middleton and Hampton, 1973.

The sedimentology of slopes and basins are similarly affected by periods of transgression that reflect rising sea level. The primary depositional site for most Holocene clastic debris is at or near the strandline. As this depocenter moves towards the continent, less sediment is available for accumulation on the shelf, slope, and the marginal continental rise.

Developmental history is the primary observational basis for interpreting the results of eustatic changes. Geophysical profiles may be used to provide the depositional framework and to suggest the time-stratigraphic patterns.

## SEDIMENTARY PROCESSES AND RESPONSES

The primary process in the transportation and deposition of clastic detritus on the shelf, rise, and basin floor is gravity flow. The nature of the sedimentary response depends upon the properties of the flow and the texture of the sediment. These aspects can be determined from properties of the deposited sediment.

### Gravity Flows

A gravity flow simply is the flow of a denser fluid through a less dense fluid. In sedimentology, the difference in density is in response to either a dissolved or an entrained sediment load. Movement of the fluid is entrained by currents generated by wind, wave, or gravitational forces. Most fluids of higher density reflect gravity, and usually this force produces a descending density current. The properties of the current can be stated in terms of density, velocity, acceleration, and viscosity adhesive strength, and flow thickness. These aspects correspond to laminar flow, turbulence shear, and dispersive stress, which can

be related to aspects of sediment transport and deposition. The mechanics of sediment transport by a gravity flow were categorized into four types by Middleton and Hampton (1973) (Fig. 9-2). This useful classification interrelates the nature of the flow, the mechanism of sediment transport, and aspects of the deposited sediment.

The analysis of the mechanics of transport and deposition from sedimentary characteristics has been the focus of much research (Hampton 1972, Middleton and Hampton 1973, Lowe 1976a and 1976b). Theoretical and laboratory studies provide the framework for interpretation, but few studies of Holocene depositional units are available. The physical characteristics of sediment flows have not been measured. Much of the information on gravity flows was derived from analysis of deposited units. Extensive laboratory study of gravity flows was conducted by Middleton (1967), Komar (1972), and Hampton (1972). Four processes of sediment transport were described: (1) debris flow, (2) grain flow, (3) fluidized flow, and (4) turbidity currents. They are physically related and may also be temporally related during the initiation, transport, and deposition of a sedimentary unit (Fig. 9-3). The admixture of clay was suggested to be the critical aspect in the depositional response (Lowe 1976a). An admixture of as little as 10% clay significantly modifies the physical properties of the gravity flow and the depositional response.

Critical to the interpretation of sediment transport and depositional patterns is the understanding of the physical nature, and sedimentary processes associated with gravity flows. Lowe (1979, 1982) has continued his research on gravity flows and their mechanics. Additional work by Fisher (1983) relates to the historical changes developed during the period of a single flow event. A summary article by Potsma (1986) attempts to define

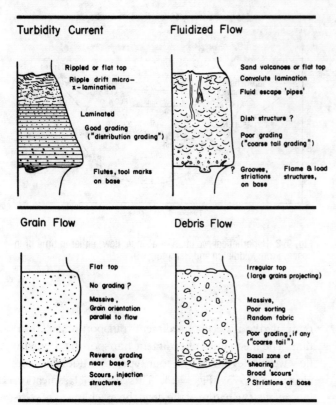

**Fig. 9–4.** Sequence of structures in hypothetical single-mechanism deposits. From Middleton and Hampton, 1973.

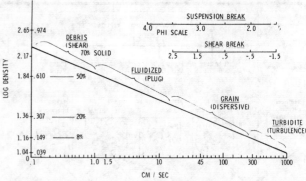

**Fig. 9–5.** Potential relation between flow density and velocity. Suggests causality for suspension and traction breaks in grain-size distribution. From Visher, 1978b.

properties of gravity flows and derive a possible genetic interpretation of their state at the time of sedimentary accretion. He postulates that three aspects of flows may be important: (1) laminar or turbulent, (2) cohesive or cohesionless, and (3) high and low density. The interrelation of these aspects defines eight theoretical flows, but of this only four are commonly developed. He utilizes internal sedimentary structures and bed thickness as criteria for interpreting the physical nature of the flow. This classification, however, does not recognize the useful subdivision of liquified, fluidized, and grain flows. (Fig. 9–4)

If textural aspects are utilized, the truncation and joining of separate textural populations or modes allows the recognition of both the transportational and depositional processes. These aspects have been shown to vary for differing types of deposited units (Visher 1969; Sagoe and Visher 1977). Speculation on the density and velocity of flows and their relation to internal breaks in the grain size distribution was suggested by Visher (1978b) (Fig. 9–5). In this postulated relation, the shear break may reflect coarse tail truncation associated with changes in velocity, shear stress, and flow density. The suspension break could also relate to flow density, as well as to turbulence, viscosity (static and eddy), grain density, and grain shape (Sagoe and Visher 1977) (Fig. 2–124).

In addition, the vertical sequence of textures, and the density, position, orientation and the shape of grains and

clasts in the deposited unit may reflect the nature of sediment transport and process of sediment accretion. Also internal sedimentary structures reflect turbulent, grain, fluidized, and debris flows as suggested by Middleton and Hampton (1973) (Fig. 9–4).

The admixture of clay to a gravity flow is significant in terms of boundary conditions, provenance, initiation of the density current, and tectonics of the basin. Sedimentary fill and transportation in a canyon and fan system reflect high energy and winnowing of the sediment. Deposition at the edge of the continental shelf by volcanic, tectonic, or progradational processes produce an admixture of clay. The transport of sand in a high-energy canyon setting and the transport of mixed sand and clay in an unstable shelf-margin setting constitute a dynamic difference in the nature of the clastic detritus and the mechanics of sediment transport. Turbidity currents are clay-rich turbulent suspensions, and transport is by slumping, landslides, and turbidity currents. Fluidized flow and grain flow are grain-supported traction flows, and transport is by avalanching, gravity flow, sediment creep, and bed shear. The difference between turbulent-supported, clay-rich suspensions and flows supported by dispersive stress resulting from grain-to-grain contacts is critical.

Current patterns in submarine canyons has been the focus of study for many years. Work at the Scripps Institution by Shepard and coworkers has produced information on the tidal relation of both up-canyon and down-canyon flow patterns (Shepard et al. 1979) (Fig. 9–6). Flow velocities have been measured in tens of centimeters/second, with velocities near the bed nearly the same as those many meters above the sediment water interface (Fig. 9–7). Current velocities sufficient to transport medium to fine sand have been measured. Sediment transport by diurnal tides produces suspension and dispersion of the fine-grained matrix, resulting in a sand supported fabric.

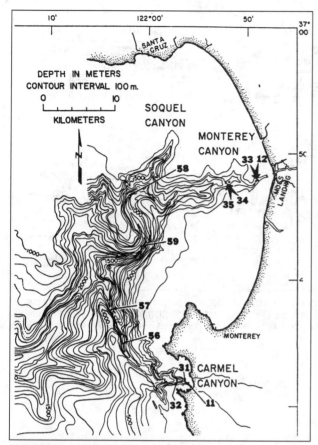

**Fig. 9-6.** Contours of Monterey and Carmel Canyons; note the meandering course of Monterey Canyon. The current-meter stations of both canyons are indicated. From Shepard et al., 1979, reprinted by permission.

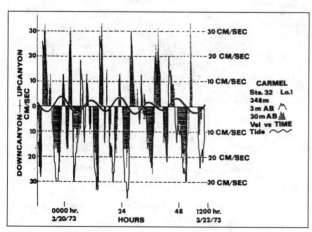

**Fig. 9-7.** Examples of remarkably similar time-velocity curves at 3 and 30 m above bottom at the 348-m station in Carmel Canyon. Agreement is good both in times of current direction change and in speed of currents. This canyon is unique in having such good agreement of records from overlying current meters. Shepard, 1979, reprinted by permission.

Transport of sand by gravity flows would thus be primarily by grain and fluidized flows. These current patterns were not developed in shallower portions of the canyon, less that 200 meters deep. In these areas the canyon heads fill with non-winnowed clastic detritus, and mass flow may be the transport mechanism down-canyon.

## SEDIMENTOLOGIC THEMES

Boundary conditions and sedimentary processes appear to reflect two fundamentally differing depositional settings and developmental histories. Passive or block-faulted plate margins reflect eustatic changes, progradational events on the shelf, and formation of equilibrium shelves, submarine canyons, and fans. Tectonically active plate margins are reflected by volcanic activity, little evidence of eustatic changes, and abrupt changes in sedimentology both in space and time. Sedimentation in the first instance is the dominant control for the pattern of depositional units. The topographic patterns, sedimentary processes, and geometry of depositional units are interrelated. In the latter setting the sedimentology is controlled by topographic features produced by tectonic forces. Sedimentary events are initiated by earthquakes, and the pattern of depositional units is tectonically controlled. The two sedimentologic models produce very distinctive depositional characteristics. Differences occur in internal characteristics, the sedimentary sequence, and geometry of the deposited units. These distinctions are not universal to all slope, rise, and basin floor sedimentary units. Obviously, canyons and fans are present at some time or place along active plate margins, and faulting, earthquakes, and volcanic activity are present at some place or time along passive plate margins. However, model characteristics are useful in distinguishing these responses when and where they occur.

## CANYON AND FAN MODEL

Both Holocene and ancient canyons and fans have been the focus for extensive research. The development of detailed studies of Holocene depositional fan systems has provided the basis for understanding of depositional patterns and history. This knowledge base reflects the diversity of tectonic, topographic, and sedimentologic boundary conditions. The development of a single model from such a diverse data base presents a fundamental problem. The principal control, however, is focused on supply and a point source for the canyon system. Recognition of shelf edge canyons, with proximal to distal fan deposition across the fan, provides the basis for simpli-

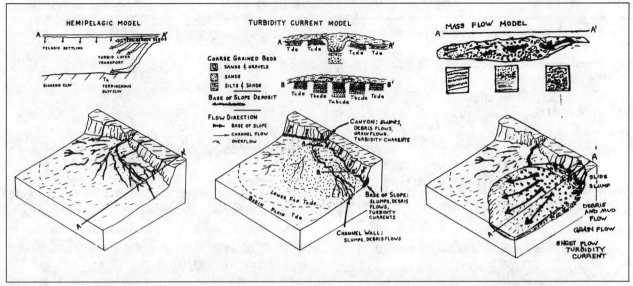

**Fig. 9–8.** The major deep-sea fan sedimentary processes plan view show expected sediment dispersal patterns of turbid layer turbidity current and mass transport flows. Turbidity current cross sections A-A¹ and B-B¹ indicate (1) hypothetical distribution of sediment in the flow and (2) pattern of sedimentary structures (T$_{CDE}$, etc.) in deposits laid down by a single flow. Mass transport cross-section A-A¹ shows hypothetical proximal (A¹) to distal (A) processes and deposits. From Nelson and Kulm, 1973.

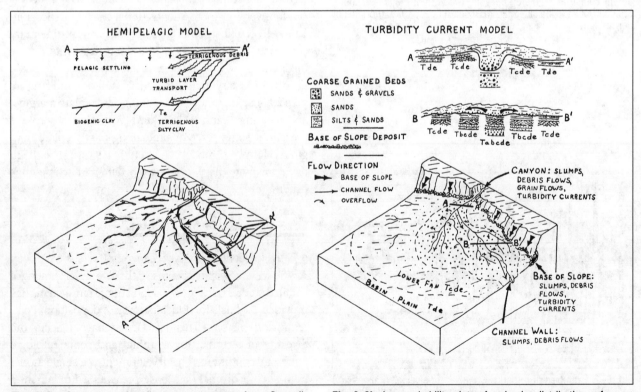

**Fig. 9–9a.** Sequences of sedimentary structures from Cascadia Channel compared with complete turbidite sequence (Bouma 1962). Percentage occurrence of Cascadia Channel sequences are indicated. Ninety percent of the sequences were classified. From Griggs and Kulm, 1973.

**Fig. 9–9b.** Log-probability plots of grain-size distributions of representative types of Astoria Fan. Massive clean sands and gravels and massive muddy sands and gravels are from upper fan valleys. Bouma AB and CD sands are from the middle and lower fan, and hemipelagic muds are from throughout the fan. From Nelson and Kulm, 1973.

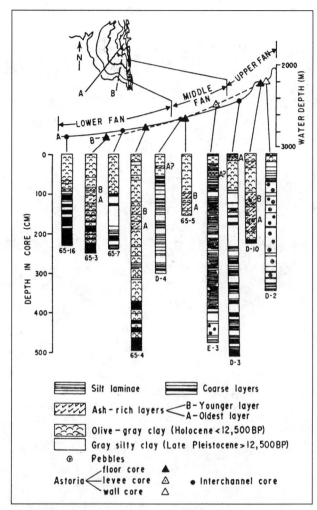

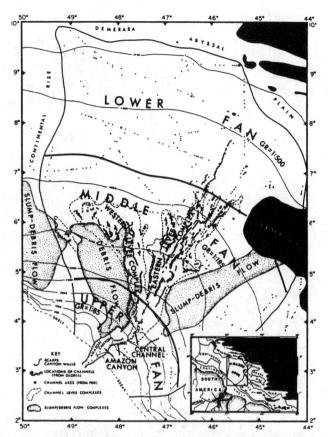

**Fig. 9-10.** Lithology and stratigraphy at representative locations on Astoria Fan. From Nelson, 1976.

**Fig. 9-11.** Distributary channels and other morphologic features of Amazon deep-sea fan revealed by GLORIA survey. Bathymetry in corrected metres. Channel trends mapped from GLORIA sonographs and conventional PDR ecograms (*black dots* = channel axes, *GR* = gradient). From Damuth et al., 1983.

fying the complex of depositional processes and response patterns.

## Holocene Depositional Pattern

A particularly well studied fan system is from the Oregon coast, the Astoria fan (Nelson and Kulm 1973; Nelson 1976) (Fig. 9-8). Cores suggest that the depositional system does not compare to a simple Bouma sequence (Fig. 9-9a). Of particular importance, the textural patterns of depositional facies from channel to distal fan suggest that transportation reflects a history of winnowing, with grain flow deposition dominant across the fan (Fig. 9-9b). These papers indicate the importance of the geometry of the depositional site (Fig. 9-10), and depositional controls on sedimentary patterns. Coriolis forces cause the development of higher levees on the right side of downslope channels in the northern hemisphere (Fig. 9-8). Fan morphology has been extensively studied by deep tow mapping systems, with extensive GLORIA mapping on the Amazon (Damuth et al. 1983) (Fig. 9-11)

and Rhone fan systems (Droz and Bellarche 1985) (Fig. 9-12). High frequency seismic data has outlined some of the depositional patterns across modern fans, but interpretation lacking cores is always equivocal (Droz and Bellarche 1985) (Fig. 9-13). Other processes related to debris flows, contour currents winnowing of fan intervals after deposition, and depositional events related to storms and tectonic events present a complex of depositional facies that rarely can be associated to a specific depositional event. The model must represent equilibrium conditions, not unique events; consequently, a few areal patterns, based upon core data, provide some of the information required to develop a depositional model.

Bourcart, utilizing both topography and cores, provides an actualistic pattern of sedimentation off the south coast of France (Bourcart 1964) (Fig. 9-14). Also Gorsline and Emery (1959) show a pattern of textural and bedding geometries in the San Pedro, and Santa Monica basins, off the California coast (Fig. 9-15). These older studies have not been paralleled by newer actualistic studies of sediment dispersion across Holocene depositional fans.

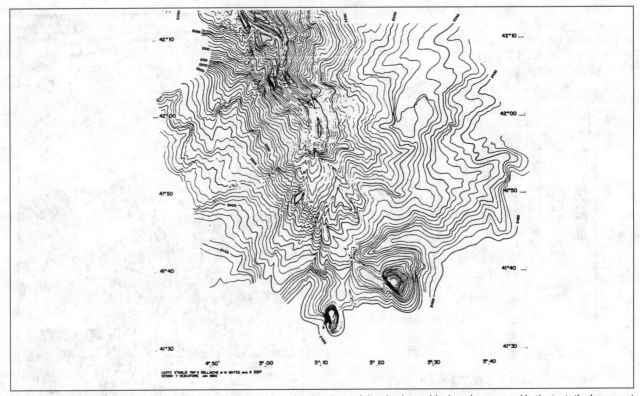

**Fig. 9–12.** Bathymetric chart of part of the Rhone deep-sea fan, showing three of the physiographic domains exposed in the text: the lower part of the canyon system, marked by very steep slopes; the upper fan characterized by a sinuous leveed main channel; and the mid-fan, where the main channel is subdivided into numerous smaller distributary channels. Note the presence of piercing salt domes, disturbing the fan structure. From Droz and Bellarche, 1985, reprinted by permission.

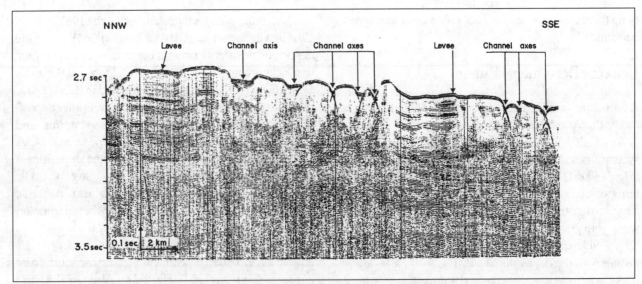

**Fig. 9–13.** Longitudinal sparker profile 85-85′ (3,000 joules) along main channel axis. Profile traverses several channel meanders. Well-stratified, regularly bedded reflectors on left and right (with smooth sea-floor relief) represent levee deposits, whereas central diffracting facies overlain by rugged sea floor represent channel axis filled with coarse sediments. From Droz and Bellarche, 1985, reprinted by permission.

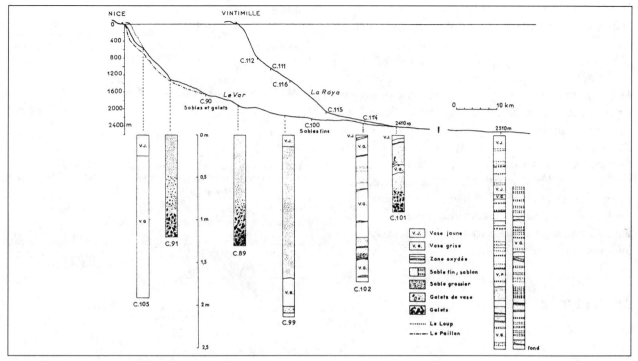

**Fig. 9–14.** Longitudinal profile of the submarine valley of Nice and its effluents. From Bourcart, 1964.

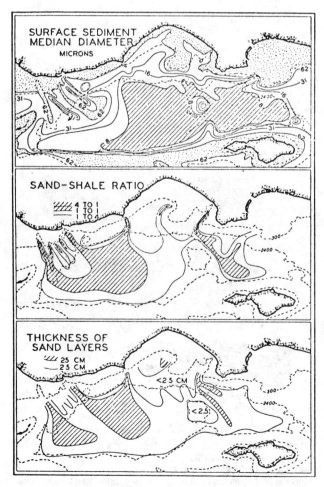

**Fig. 9–15.** Modern example of California basin. From Gorsline and Emery, 1959.

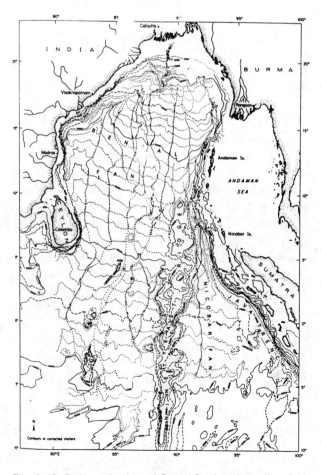

**Fig. 9–16.** Bathymetric chart of Bengal Fan based on all available soundings. Fan valleys are indicated by dotted lines. Contour interval is variable. From Curray and Moore, 1971.

The importance of these depositional systems to stratigraphy is illustrated by the scale of the Bengal Fan from the Bay of Bengal (Curray and Moore 1971) (Fig. 9–16). Leveed channels, kilometers in width, and up to 100 meters deep cross, a fan more than 2500 kilometers long. This depositional system forms a subaqueous fan with topographic relief of more than 5000 meters. Seismic and gravity data suggest that an additional 5000 meters may be preserved beneath the fans' topographic expression. Stratigraphic intervals of this magnitude must represent significant preserved intervals in the subsurface. Identifying these intervals, utilizing the limited core information presently available, has presented substantial difficulties in the application of the fan model to the interpretation of depositional systems.

## Ancient Fan Patterns

Of principal importance to the identification of submarine fans in the stratigraphic record is the determination of the topographic, tectonic, and historic framework of the submarine fan setting. Comparison may be made using Holocene fan settings and geometries, but for stratigraphic intervals it requires either detailed log and core sections or mapping of stratigraphic intervals utilizing outcrop or seismic information. Based upon both Holocene and ancient fan systems, Table 9–1 suggests some of the geometric characteristics useful for identifying submarine fan systems.

The concept of deeper water fluidized flow producing thick stratigraphic intervals was developed in the 1950's. This information was applied to the interpretation of stratigraphic intervals throughout the world. Work on basins on the California borderland (Gorsline and Emery 1959) (Fig. 9–15) were related to depositional processes and patterns in Tertiary intervals from the Los Angeles and other California Basins (Conrey 1959). On the Atlantic continental margin, earlier work by Erickson, Ewing, and Heezen (1952) related submarine canyon history to transport and deposition of sand in deeper water. The turbidity current hypothesis was extensively applied to the interpretation of the origin of flysch deposits in orogenic areas by Bouma (1962).

Application of the turbidity current concept to the interpretation of extensive outcropping depositional systems provided the detailed geometric, paleocurrent, and sedimentologic data needed for comparison to other stratigraphic intervals. Work by Kuenen and others (1957), Stanley (1961), Bouma (1962), and Stanley and Bouma (1964) on the Annot sandstone of uppermost Eocene and lower Oligocene age of the Maritime Alps of southern France has been a useful case study. Geometric

**Table 9-1**
**Submarine Fan Reservoirs**

Near shelf edge sandstone intervals 20–100 m thick, and 1–10 kms wide

In proximal-medial slope areas the interval thickens to more than 100 m, and ranges from 20–100 kms wide

Fining and thinning upward sandstone sequences, 5–50 m thick, irregularly distributed both areally and vertically, interbedded with horizontally bedded fine-grained siltstone and shales

Sandstone sequences, 5–50 m thick, are channelized and are elongated with width to length greater than 1:10, and may be radial downslope or parallel to shelf edge

In distal rise area the interval is less than 200 m thick, and is distributed over an area greater than 2000 kms²

Distal areas dominated by fining and thickening upward sandstone bedding sequences .5–5 m thick, interbedded with shales and silstone units.

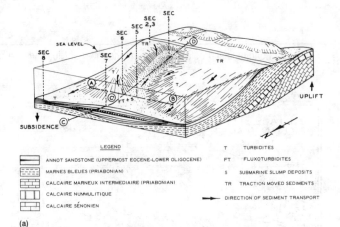

(a)

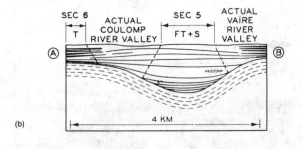

(b)

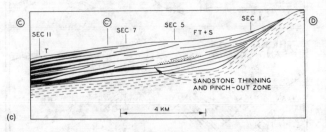

(c)

**Fig. 9-17.** Interpretations of the depositional origin of the Annot sandstone formation in the vicinity of Annot (Basses-Alpes). The block diagram (vertical scale exaggerated) attempts to show the interrelation between lithologic variability, depositional environment, mechanisms of sediment transport, and structural mobility. From Stanley and Bouma, 1964.

aspects as indicated in Figure 9–17 are consistent with geometric studies of Modern submarine fan intervals (Stanley and Bouma 1964). The geometrical patterns were based upon a synthesis of paleocurrent, isopachous, and outcropping sandstone geometries. Detailed descriptions of measured sections (Lanteaume and others 1967) illustrate the internal sedimentary patterns and depositional sequences for an upper fan-valley channel fill (Fig. 9–18). A number of facies of this fan can be identified and the detailed paleogeography reconstructed. The geometry of sedimentary units is illustrated (Fig. 9–17). This geometry, particularly the onlap of deposited units up the continental slope as the fan progrades, is typical of other ancient stratigraphic examples. The Annot sandstone reflects deposition related to the upper Eocene sea-level lowstand (Fig. 9–1), and the submarine canyon is related to downcutting of the shelf edge by subaerial processes, and the bypassing of clastic sediments across the narrow continental shelf (Fig. 9–17a).

Few studies integrating stratigraphic patterns with internal characteristics of sedimentary units are available. Patterns of bedding units and their facies distribution

were suggested by Mutti (1974) for a number of ancient deep-sea fan intervals. He suggests a model interrelating areas of deposition and a possible onlapping depositional sequence (Fig. 9–19). However this sequence pattern may be inverted or consist of a complex interbedding of fan facies that reflect changing depocenters and patterns on the fan surface (Galloway and Brown 1973) (Fig. 9–20).

A similar study of fan deposition was conducted by Conrey on the Los Angeles basin (1959). Continued work on the basin produced a textural and isopachous pattern for the Pliocene Repetto formation in the Los Angles

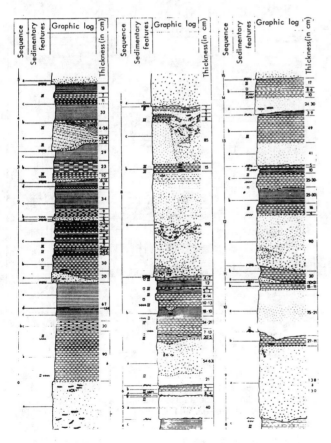

Fig. 9–18. Graphic log showing lower part of the section observed between Col de l'Orme and Baisse de la Cabanette in the Peira-Cava region. Note large proportion of poorly graded, or nongraded, massive pebbly sandstone and chaotic units between shales and thinner units. The thicker units are interpreted as upper fan-valley channels; the finer and overbank as interchannel deposits. From Lanteaume, Beaudoin, and Campredon, 1967.

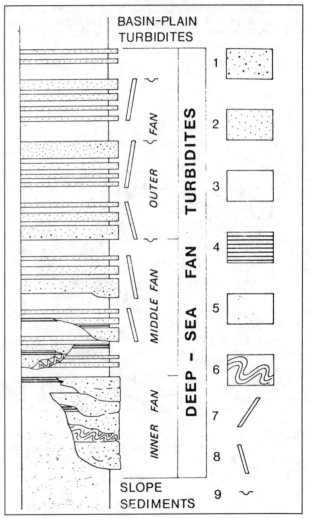

Fig. 9–19. Depositional model for ancient deep-sea fan deposits. (1) Thick-bedded, graded and/or crudely laminated, coarse-grained sandstone and conglomerate; (2) thick-bedded, graded sandstone and varying amounts of finer-grained, current-laminated deposits, including mudstone; (3) mudstone and parallel bedded, thoroughly current-laminated sandstone and siltstone; (4) thin, lenticular, and discontinuous beds of fine to coarse-grained sandstone exhibiting grading and poorly developed current laminae; (5) so-called hemipelagic mudstone; (6) chaotic sediments; (7) thickening upward cycle; (8) thinning upward cycle; and (9) shallow channels (less than 2 m). From Mutti, 1977.

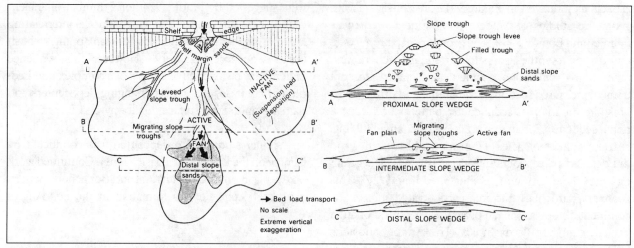

**Fig. 9-20.** Slope submarine fan model illustrating general processes and resulting composition that typify these deepwater depositional systems. Successive fans may offlap or may onlap, depending on a sustained or diminishing sediment supply, respectively. Submarine fan deposits may stack in vertical or superposed manner if subsidence rates exceed sediment supply, thus producing an uplap system. From Galloway and Brown, 1973, reprinted by permission.

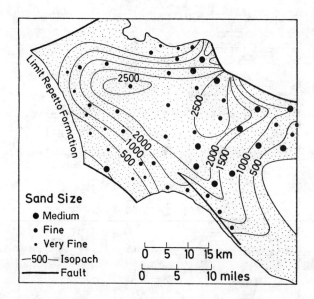

**Fig. 9-21.** Thickness of sandstone and grain size in Lower Pliocene Rapetta formation, Los Angeles basin, California. From Pettijohn, Potter and Siever, 1972.

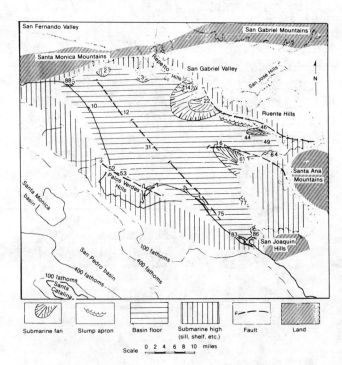

**Fig. 9-22.** Paleogeographic map of Los Angeles basin during early Pliocene. The paleophysiographic data are superimposed on a map showing modern subaerial and submarine features. From Conrey, 1967.

Basin (Conrey 1967) (Fig. 9–21). This fan interval is also related to a lowstand in sea level (Fig. 9–1). A paleogeographic map of the basin margin based upon both subsurface and outcrop patterns suggests the development of submarine fans adjacent to a narrow shelf marginal to the basin (Fig. 9–22).

The identification of canyon-fan depositional systems in the stratigraphic record often must be based on limited subsurface information. Seismic data often can provide the framework for identifying these depositional systems. Brown and Fisher (1977) illustrate patterns developed by Holocene and ancient slope systems (Fig. 9–23a), and compare these to seismic reflection patterns (Fig. 9–23b). These geometries can be related to structural and paleogeographic settings typically developed on continental margins. Diapiric and block faulted margins similar to that illustrated by the Sardinian coastal area may preserve fan systems (Wezel and Savelli 1981) (Figs. 9–24 and 9–25). Also shelf paleogeography and depositional processes may contribute to the history and pattern of submarine fan development (Brown and Fisher 1980)

(Figs. 9–26 and 9–27). Sea level and sedimentologic controls also may be important in the transport and erosion of deltaic and shelf sedimentary intervals (Brown and Fisher 1980 after Moore and Asquith 1971) (Fig. 9–28). Both offlap and onlap patterns are common in seismic record sections, and their interpretation requires careful data processing (Brown and Fisher 1980) (Figs. 9–29 and 9–30). The submarine canyon is only preserved in the uppermost portion of the shelf margin, and is characterized by chaotic, or discontinuous reflection (bedding) patterns (Figs. 9–31 and 9–32).

## Hydrocarbon Accumulations

Submarine fans are important reservoirs for hydrocarbons. Depositional processes produce sandstone intervals with good porosity, lateral changes in pore size distributions, and potentially closely associated reservoirs and source rocks. Prediction of canyon and fan systems is facilitated by understanding the boundary conditions required for their development. Continental

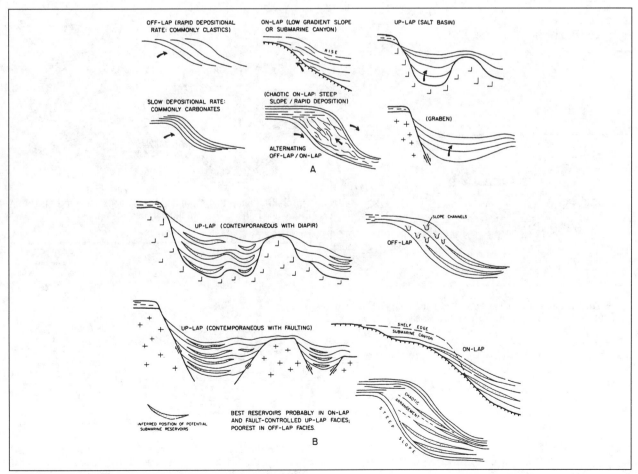

**Fig. 9–23.** Seismic reflection configurations and inferred distribution and geometry of submarine-fan reservoirs. *A,* Typical reflection patterns displayed by Holocene and ancient slope systems. *B,* Inferred sand distribution and stratigraphic trap potential based on seismic-facies interpretation. From Brown and Fisher, 1977, reprinted by permission.

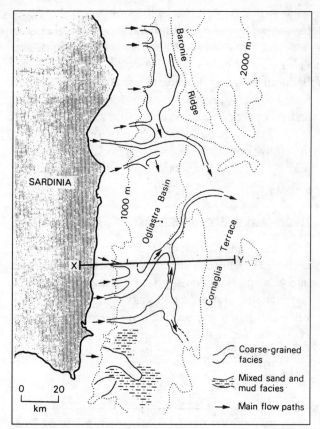

**Fig. 9–24.** From Wezel, Savelli et al., 1981.

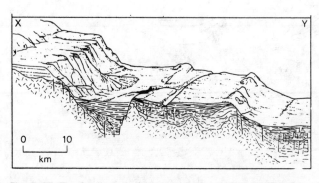

**Fig. 9–25.** The Sardinnia faulted slope-apron and marginal basin in the Tyrrhenian Sea west of Italy. From Wezel, Savelli, et al., 1981.

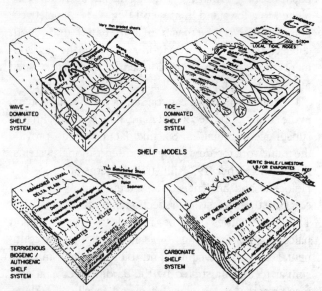

**Fig. 9–26.** Conceptual shelf models illustrating biogenic/authigenic shelf, tide-dominated shelf, and carbonate shelf systems. From Brown and Fisher, 1980, reprinted by permission.

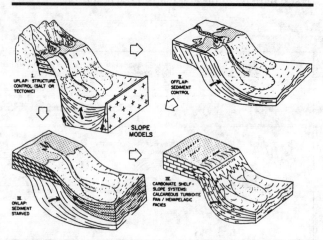

**Fig. 9–27.** Conceptual slope and basin depositional models. From Brown and Fisher, 1980, reprinted by permission.

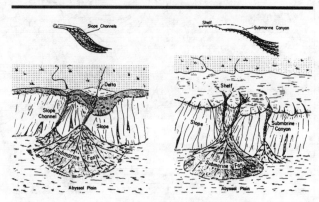

**Fig. 9–28.** Schematic comparison of constructional (active) offlap slopes and destructional (passive) onlap slopes. Model indicates that these types of slopes infer different sediment-supply mechanisms, rates/volumes of sediment input, and quality of organic matter. From Brown and Fisher, 1980, after Moore and Asquith, 1971, reprinted by permission.

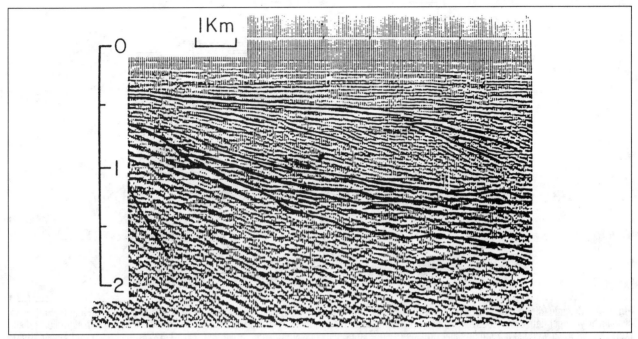

**Fig. 9–29.** Seismic (dip) profile of a passive-margin basin, offshore Brazil. Well-defined upper Cretaceous continental-rise sequence onlaps (to left) erosional unconformity. Younger prograded slope and associated mixed carbonate and clastic platform sequences represent late Cretaceous and Tertiary deposition. High-amplitude clinoforin reflections are generated by hemipelagic drapes within slope that prograded to the right. Local onlap wedges are longitudinal sections through submarine canyons. Intersecting strike section, see Fig. 9–30. From Brown and Fisher, 1980, reprinted by permission.

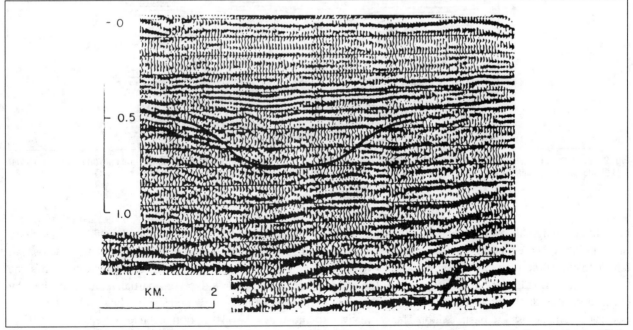

**Fig. 9–30.** Seismic (strike) section of a passive-margin basin, offshore Brazil. Line intersects a submarine canyon eroded into clastic upper-slope facies. Canyon is not outlined by a basal reflection since there is little velocity contrast between the fill and slope deposits. Reflection terminations, diffractions, and a reflector-free seismic facies help confirm the canyon. Special processing results in better definition. Intersecting dip section, see Fig. 9–29. From Brown and Fisher, 1980, reprinted by permission.

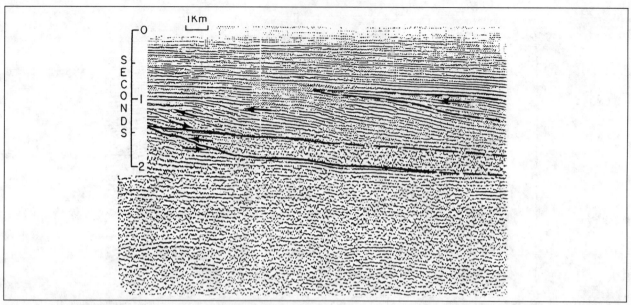

**Fig. 9-31.** Seismic (dip) profile of a passive-margin Brazilian basin. Note eroded older (mid-Cretaceous) carbonate platform and marine onlap overlain by younger (late Cretaceous and Teritiary) platform that exhibits oblique reflections (left) grading basinward (right) into sigmoidal reflections. Small onlap wedges are longitudinal profiles of submarine canyons. Intersects strike line, Fig. 9-32. From Brown and Fisher, 1980, reprinted by permission.

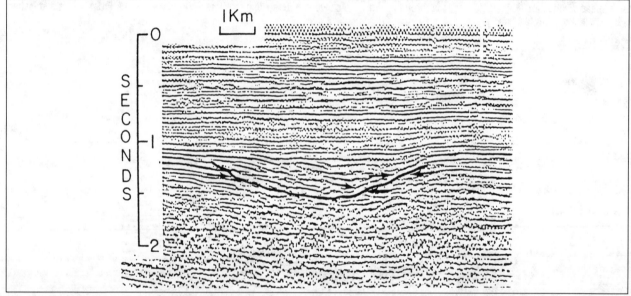

**Fig. 9-32.** Seismic (strike) profile of Teritiary carbonate platform in passive-margin Brazilian basin. Note submarine canyon eroded into lower part of shelf and underlain by slope. Intersects dip line, Fig 9-31. Reprinted by permission from Brown & Fisher.

margins and marginal basins associated with deltaic sedimentary intervals are the sites of submarine canyon and fan depositional systems during times of sea-level fall. Canyons are formed by subaerial erosion into the shelf edge, and the depocenter for terrestrial detritus is downslope and across the basinal floor. The Pennsylvanian Red Oak sandstone reflects this depositional history (Vedros and Visher 1978) (Fig. 9-33). The stratigraphic pattern is illustrated by a cross section across the fan (Fig. 9-34). Both sandstone isolith and interval isopachous maps indicate a channeled fan depositional geometry (Figs. 9-35 and 9-36). Detailed cross sections across fan channels indicate the pattern of deposition, with some channels filled and others partially partially filled by sandstone (Figs. 9-37 and 9-38). The well log motif map illustrates the distribution of the reservoir sandstone (Fig. 9-39). Continued fan deposition throughout the middle Atoka reflects tidally influenced deltaic sedimentation on a narrow marginal shelf. Development and preservation of multiple canyon heads cutting into the shelf edge

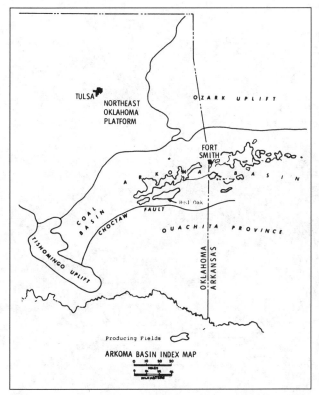

**Fig. 9-33.** Arkoma basin index map. From Vedros and Visher, 1978.

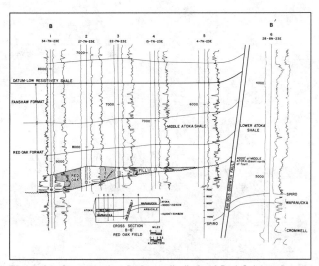

**Fig. 9-34.** Cross-section $B^N$-$B^1$, updip limit of Red Oak sandstone. From Vedros and Visher, 1978.

reflects the history of a sea-level lowstand at the base of the Atoka, high rates of clastic deposition derived by major river systems, and tidal processes during the deposition of the Atokan Series. This sandtone unit has produced in excess of one trillion cubic feet of gas, and with additional development this reservoir should produce an additional one trillion cubic feet.

Another giant gas accumulation in a submarine fan interval is the Frigg field from the Eocene of the North Sea (Heritier and others 1980) (Fig. 9-40). A structure map

from seismic reflections from the top of the fan illustrate the mounded buildup of the fan (Fig. 9-41). Patterns of internal seisimc reflections, together with core and sample data from wells suggest the depositional history of the fan (Fig. 9-42). Detailed log sections illustrate the geometry of sandstone units from proximal to distal across the fan (Fig. 9-43). In excess of nine trillion cubic feet of gas may ultimately be produced from this fan system.

Two examples of fan deposition from a forearc basin are developed in the Sacramento and San Joaquin Valleys. Upper Miocene Stevens sandstone fan intervals are developed in response to the Upper Miocene sea-level fall (Fig. 9-1). A cross section across the area shows multiple stacked sandstone units (MacPhearson 1978) (Fig. 9-44). Other figures illustrate the vertical stacking of channel sandstone sequences, and the areal distribution of channel intervals across the submarine fan (Figs. 9-45 and 9-46). Oil fields are developed on a basement arch, with a combination of stratigraphic traps and seals associated with growth faults. A series of submarine canyons have been mapped in the Sacramento Valley associated with sea level lowstands (Almgren 1978) (Fig. 9-47). A well-log cross-section illustrates the presence of canyons downcutting through a series of interbedded shelf sandstone and shale sedimentary intervals (Fig. 9-48). Mapping of the canyon system shows a canyon up to 20 kilometers wide and up to 600 meters deep across the marginal shelf (Fig. 9-49a and 9-49b). Vail et al. (1977a) have suggested that the Lower Eocene sea-level fall amounted to at least more than 300 meters. The amount of downcutting illustrated by the Meganos canyon could be considered as supporting evidence for more than this amount of sea-level fall. Outcrop examination of the basal channel fill by the author revealed imbricated and rounded cobbles and quartz pebbles, suggesting greater than 760 meters of relative fall in sea level. A schematic paleogeographic framework for this depositional setting was suggested by Moore (1966) (Fig. 9-50). Gas fields are associated with permeability traps developed on the flanks of the dominantly shale-filled submarine canyon systems.

Similar canyon systems have been identified from the United States Gulf Coast associated with major sea-level falls. The Oligocene Hackberry canyon-fill system illustrates the presence of canyon neary 500 meters deep, with an associated shelf edge growth fault (Halbouty and Barber 1961) (Fig. 9-51). The seismic section across this canyon system illustrates the downcutting and onlap fill of the canyon system by interbedded sandstones and shales (Brown 1984) (Fig. 9-52). The lower Eocene sea-level fall also is reflected on the Gulf Coast by the development of the Yoakum and Hardin shale-filled can-

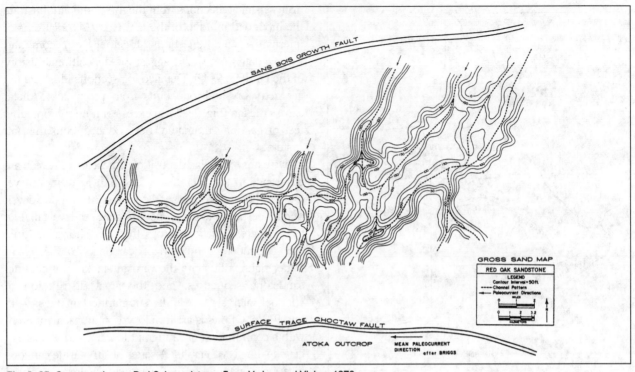

**Fig. 9-35.** Gross sand map, Red Oak sandstone. From Vedros and Visher, 1978.

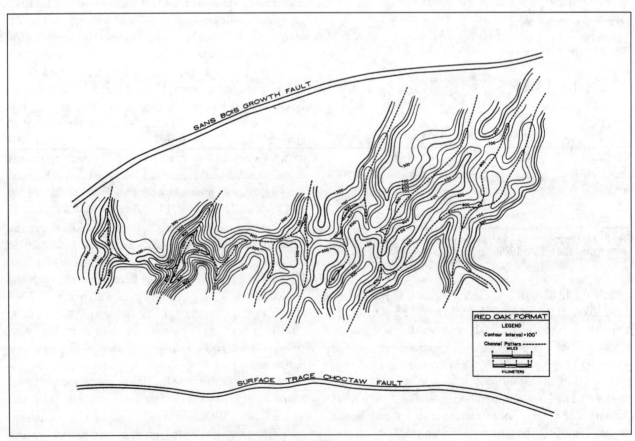

**Fig. 9-36.** Red Oak sandstone format map. From Vedros and Visher, 1978.

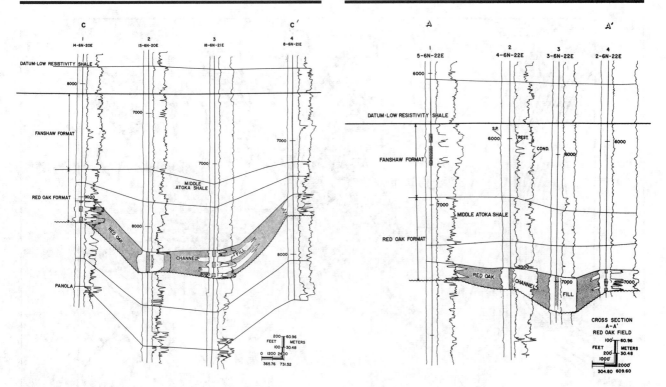

**Fig. 9–37.** Cross-section $C^N$-$C^1$, profile across mid-fan channel complex. From Vedros and Visher, 1978.

**Fig. 9–38.** Cross section $A^N$-$A^1$, profile across lower end of upper fan valley. From Vedros and Visher, 1978.

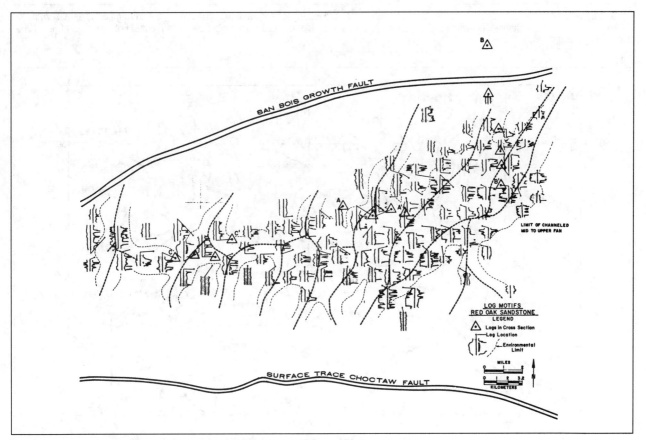

**Fig. 9–39.** Log motif map, Red Oak sandstone. From Vedros and Visher, 1978.

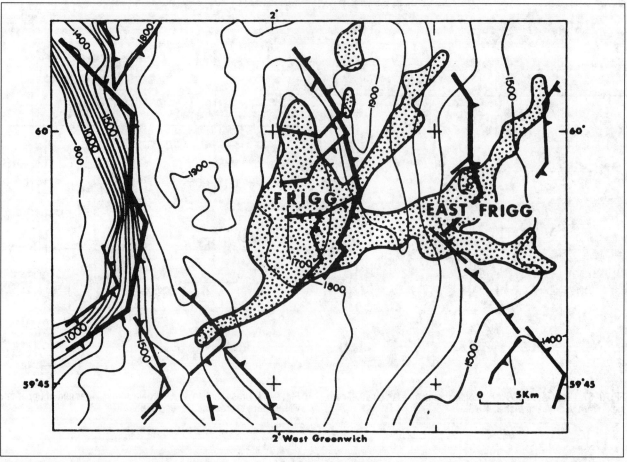

**Fig. 9–40.** Pre–Cimmerian structure (C.I. = 100 msec one-way time) and extent of Frigg sand (*dotted*). Axis of main Frigg field is parallel with axis of deep structure. Feeder channel follows deep fault zone which parallels old northeast-southwest Caledonian faults. From Heritier et. al., 1980, reprinted by permission.

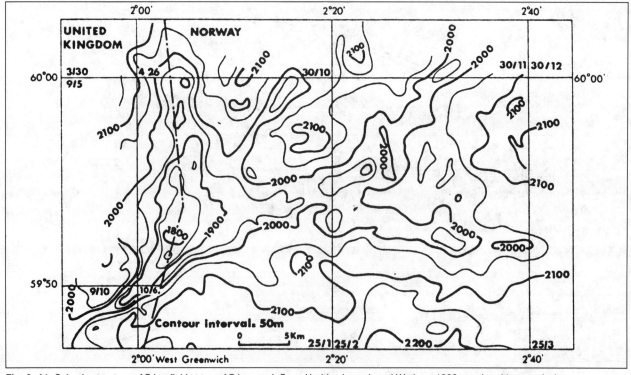

**Fig. 9–41.** Seismic structure of Frigg field at top of Frigg sand. From Heritier, Lossel, and Wathne, 1980, reprinted by permission.

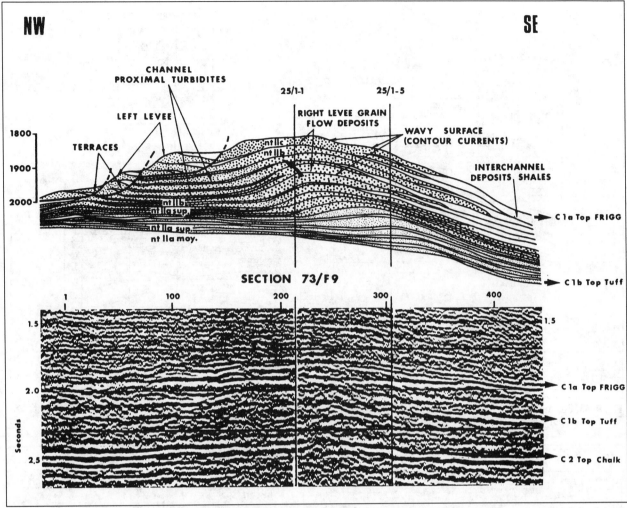

**Fig. 9–42.** Sedimentologic interpretation of typical Frigg seismic section crossing wells 25/1-1 and 25/1-5. Vertical exaggerations is 2x. Seismic "flat spot" is well defined. From Heritier et al., 1980, reprinted by permission.

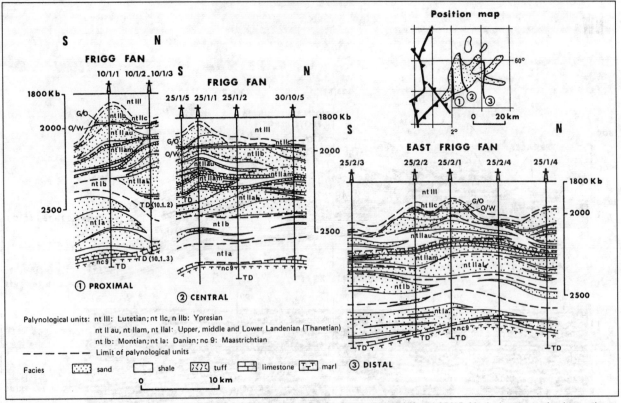

Palynological units: nt III: Lutetian; nt IIc, n IIb: Ypresian
nt II au, nt IIam, nt IIal: Upper, middle and Lower Landenian (Thanetian)
nt Ib: Montian; nt Ia: Danian; nc 9: Maastrichtian
----- Limit of palynological units

Facies [sand] [shale] [tuff] [limestone] [marl]

0         10 km

**Fig. 9–43.** Sedimentologic interpretation of typical Frigg seismic section crossing wells 25/11 and 25/15. Vertical exaggeration is two times. Seismic flat spot is well defined. From Heritier, Lossel, and Wathne, 1980, reprinted by permission.

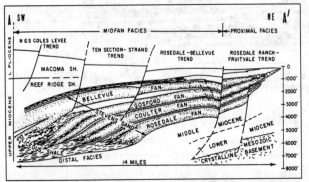

**Fig. 9–44.** Cross section A-A¹ showing upper Miocene turbidite fans and related facies, Bakersfield arch, southern San Joaquin Valley, Kern County, California. Section parallels direction of transport. From MacPhearson, 1978, reprinted by permission.

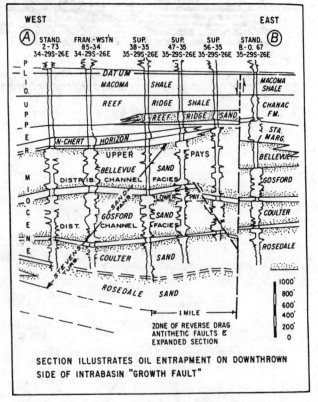

SECTION ILLUSTRATES OIL ENTRAPMENT ON DOWNTHROWN SIDE OF INTRABASIN "GROWTH FAULT"

**Fig. 9–45.** Stratigraphic section across Bellevue field. From Mac-Phearson, 1978, reprinted by permission.

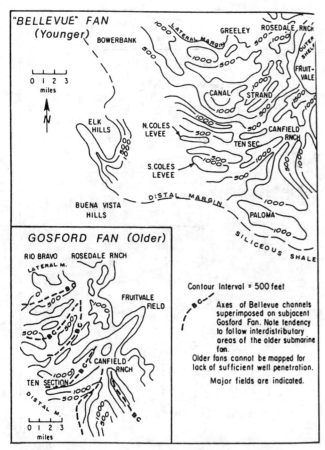

**Fig. 9-46.** Isopachs of Bellevue and Gosford turbidite fans on Bakersfield arch. From MacPhearson, 1978, reprinted by permission.

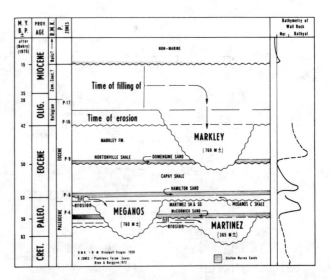

**Fig. 9-47.** Schematic chart showing relationship of Tertiary caryons to periods of erosion and cycles sedimentation. Nore the thickness of the fill in each canyon is something something something and positions of zone are tentative. From Almgren, 1987.

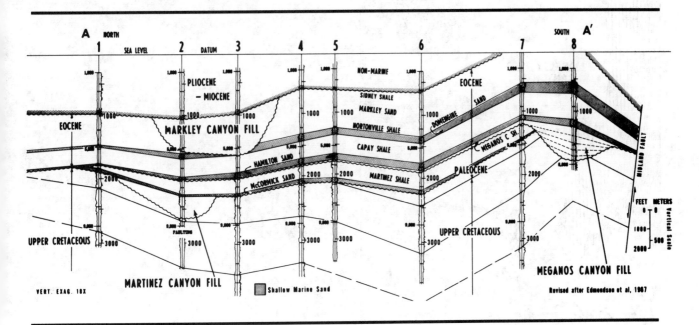

**Fig. 9-48.** Stratigraphic cross-section A-A', showing relationship of Tertiary canyon fills to associated cycles of sedimentation. From Almgren, 1978.

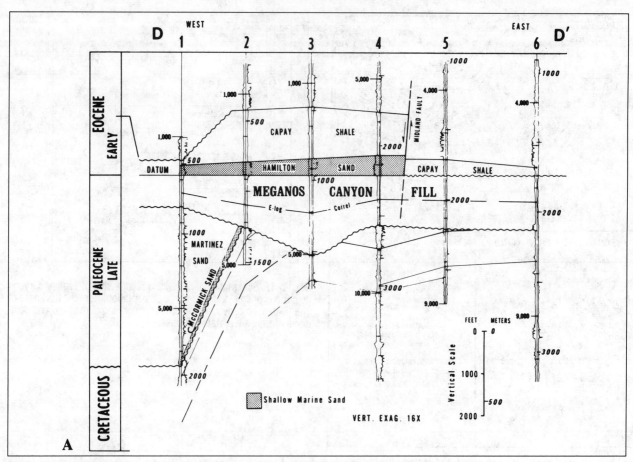

**Fig. 9–49a.** Meganos canyon stratigraphic cross section. Fig 9–49b shows line *D-D'* perpendicular to canyon axis.

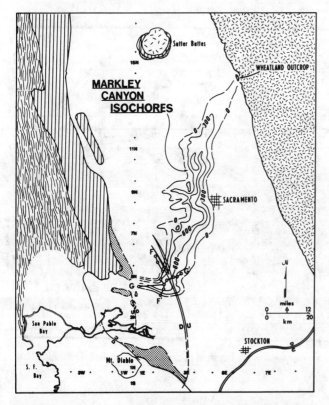

**Fig. 9–49b.** Map showing location of Meganas Canyon with approximate isochores of the shale fill in meters. From Almgren, 1978.

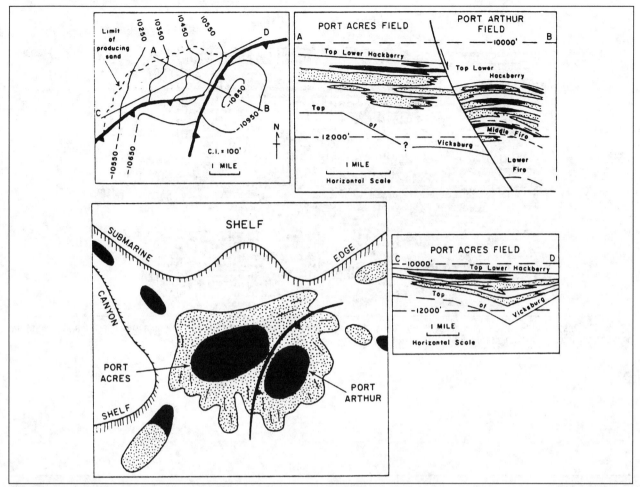

**Fig. 9-51.** Stratigraphic and structural traps in Oligocene Hackberry canyon-fill sands, Port Acres–Port Arthur field, Gulf Coast Basin, Texas. Reactivated faults and rollover provide combined structural and stratigraphic traps in deep-water sands. From Brown, 1984, modified from Halbouty and Barber (1961).

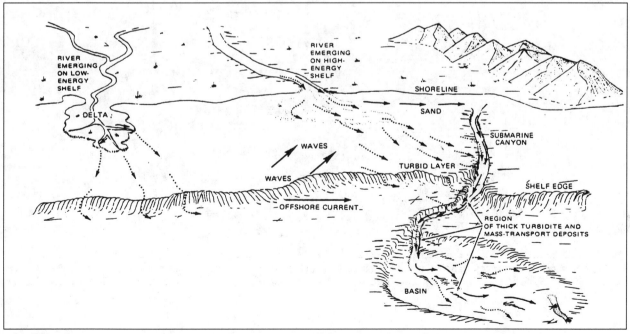

**Fig. 9-50.** California coastal paleogeographic maxxx. After Moore 1966, from Sangree et al., 1978.

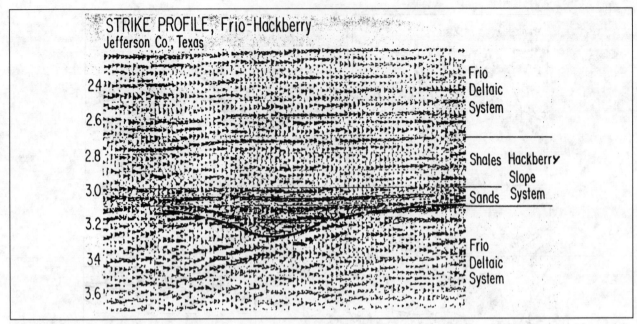

**Fig. 9–52.** Seismic profile across an Oligocene Hackberry submarine canyon, Gulf Coast of Texas. Note onlap-fill reflections at base of the canyon and almost transparent shale. From Brown, 1984.

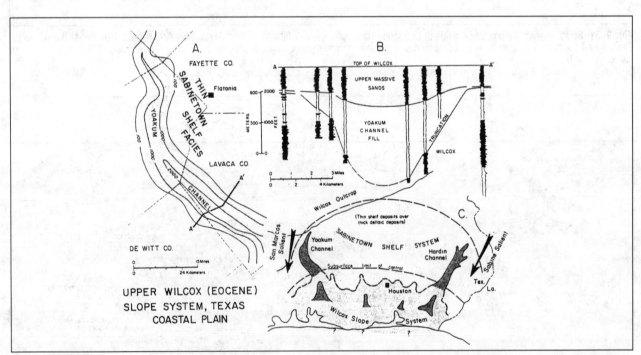

**Fig. 9–53.** Stratigraphic setting of Eocene Yoakum and Hardin canyons, Gulf Coast Basin of Texas. Canyon eroded lower Wilcox deltaic deposits during a period of regional transgression. Canyon was filled with prodelta sediments during upper Wilcox delta progradation. From Hoyt, 1959.

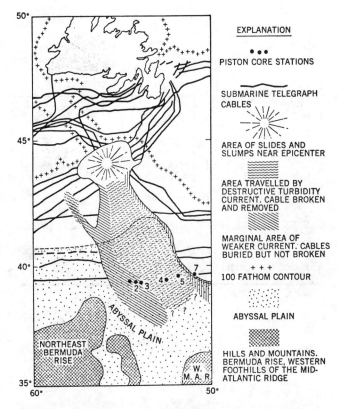

**Fig. 9-54.** The 1929 Grand Banks turbidity current. Sediment cores obtained downslope of the earthquake epicenter contain graded, muddy sand that was deposited by the turbidity current. This catastrophic flow broke many submarine telegraph cables and deposited sand, silt, and silty mud over an area of about 200,000 sq mi. From Heezen and Hollister, 1971.

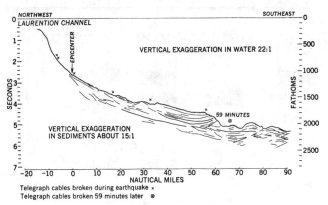

**Fig. 9-55.** Grand Banks slump. This enormous upturned block of sediment slid downslope during the 1929 Grand Banks earthquake, severing all adjacent submarine cables. This profile is a tracing of the seismic reflection records made along a profile from the Laurentian Channel across the continental slope and continental rise south of the Grand Banks. All cables crossing the area of the slump were broken at the instant of the earthquake. From Heezen and Hollister, 1971.

yons (Hoyt 1959) (Fig. 9–53). The depth of the canyon is indicated to be greater than 600 meters, which is greater than that suggested by Vail et al. (1977a) for the post-Ypresian sea-level fall, and may indicate a combination of downcutting and shelf subsidence. Again oil is trapped against the flank of the canyon fill. These relationships suggest that many marginal shelf systems, dominated by deltas reflecting tidal processes, may have similar paleogeographic patterns, with shelf sandstone intervals truncated, producing traps adjacent to shale-filled shelf and slope canyons.

## TURBIDITY CURRENT MODEL

Separation of gravity flows into the two differing types ( (1) those associated with grain flows and fluidized flows and (2) those associated with turbulent flow with varying fluid densities) produces fundamental differences in flow geometry and texture of the deposited sediments. The concept of turbidity currents producing graded beds was suggested in 1950 by Kuenen and Migliorini.

### Holocene Depositional Patterns

In 1967 Heezen and Hollister reported on an unusual depositional event associated with the 1929 Grand Banks earthquake (Heezen & Hollister 1971). A nearly actualistic experiment was possible by the timing of the earthquake and the successive rupture of a series of submarine cables (Figs. 9–54). A tracing from a seismic line suggests the initiation of mass movement along a slip plane, and the simultaneous rupture of a series of cables, was followed by a break of a cable 59 minutes after the earthquake shock (Fig. 9–55). Subsequent cable breaks occured over a period of more than 13 hours after the earthquake. They postulated that a density current was initiated by the earthquake, and the current traveled at an initial velocity of 55 knots, subsequently decreasing to 12 knots, when the last cable was ruptured (Fig. 9–56). Cores taken from deposits generated by the flow are indicated on Fig. 9–54, and record the generation of clay-rich graded beds.

The turbidity current concept was extensively applied to the interpretation of ancient deep water deposits (Ten Haaf 1959) (Fig. 9–57). However, the distinction between muddy-graded flysch sequences and winnowed grain supported bedding units was not emphasized until work by Middleton and Hampton (1973) (Figs. 9–58 and 9–59). The importance of this distinction may be fundamental to the interpretation of both Holocene and ancient deeper water depositional systems. Klein (1975), reporting on cores of Holocene trench sediments, recognized graded bedding sequences (Figs. 9–60 and 9–61), similar to those

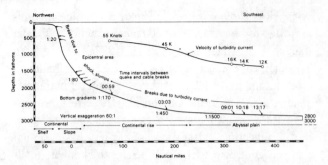

**Fig. 9-56.** Record of submarine cable breaks. From Heezen and Hollister, 1971.

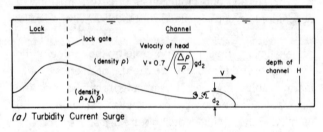

(AFTER TEN HAAF, 1959)

**Fig. 9-57.** Inferred process-response pattern for a turbidity current sequence. Reprinted courtesy E. Ten Haaf, 1959.

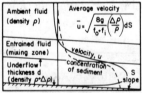

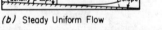

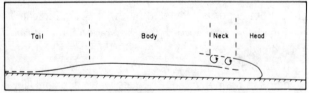

*(d)* Schematic Subdivision of a Turbidity Current

**Fig. 9-58.** Hydraulics of turbidity currents. *(a)* Turbidity current surge, as observed in a horizontal channel after releasing suspension from a lock at one end. The velocity of the head, $v$, is related to the thickness of the head, $d_2$, the density difference between the turbidity current and the water above, $\Delta p$, the density of the water, $p$, and the acceleration due to gravity, $g$. *(b)* Steady, uniform flow of a turbidity current down a slope, $s$. The average velocity of flow, $u$, is related to the thickness of the flow, $d$, the density difference, and the frictional resistance at the bottom, $f_0$, and the upper interface, $f_i$ *(c)* Flow pattern within and around the head of a turbidity current. *(d)* Schematic division of a turbidity current into head, neck, body, and tail. From Middleton and Hampton, 1973.

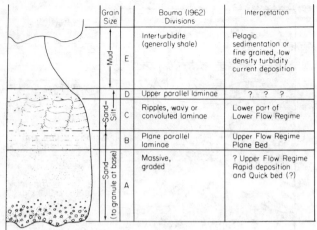

**Fig. 9-59.** Ideal sequence of structures in a turbidite bed. From Middleton and Hampton, 1973.

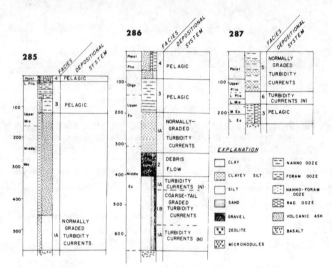

**Fig. 9-60.** Comparison of vertical distribution of sedimentary facies and associated depositional systems for sites *285*, *286*, and *287*; N = Normally graded. From Klein, 1975.

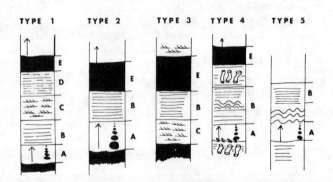

**Fig. 9-61.** Sedimentary sequences encountered in facies 1 cores at sites 285 and 286. *Type 1* is typical of Bouma sequence. *Type 2* consists of intervals *A*, *B*, and *E* of Bouma sequence. Graded interval is coarse-tail graded. *Type 3* differs from Bouma $T_{c-e}$ sequence in that the overlying parallel-laminated interval is coarser or of same grain size as interval *C*. *Type 4* contains convoluted interval within *B* interval, which is overlain by an interval of sandy silt containing dewatering pipes. *Type 5* consists of interval of antidune wave forms between *A* and *B* interval of Bouma. From Klein, 1975.

described by Ten Haaf and Bouma (1962). Patterns of sedimentary structures and textures are indicated similar to those for mud-supported flysch deposits described from many ancient trench systems. Also, work by Pilkey and others (1980) illustrated extensive depositional units similar to those described by Heezen and Hollister across abyssal plains marginal to the North American continent (Fig. 9–62). Volumes of sediment for these graded units has been interpreted to be from 1010 to 1012 cubic meters for individual flows (Pilkey et al. 1980).

## Ancient Examples

Stratigraphic patterns of turbidity current and flysch deposits have been mapped in detail throughout the world. Their association with geosynclinal basins and tectonically active plate margins has been conclusively demonstrated. These mapped patterns demonstrate that sedimentology can be used as the basis for a stratigraphic model. A useful pattern of internal change in a sedimentary unit was documented by Enos (1969) for the Gaspé Peninsula in Quebec, Canada (Fig. 9–63). The volume of a single turbidity current unit, the areal pattern of change in the sequence of structures and textures, and the relation of the depositional sequence to basin geometry are indicated in Table 9–2. This pattern is useful for comparison to other turbidity current sequences and to infer boundary conditions and sedimentologic patterns. The stratigraphic interpretation of ancient turbidity current and flysch units requires comparison to other ancient depositional sequences. Facies frameworks can be suggested by examining paleocurrent patterns, changes in texture and sequence of sedimentary structures, and determining the paleogeographic setting for the stratigraphic section. These internal and mapped relationships are sufficient to identify the depositional process, but they are not useful in predicting the developmental history or boundary conditions of flysch-filled basins.

The close association of flysch and turbidite sedimentary sequences to tectonic processes requires that a developmental model include aspects of plate margin tectonic history. The association of melange, wildflysch, and slump blocks with folding and faulting suggests that an interrelation of these relationships is required before a developmental model can be suggested.

Flysch or turbidite sequences are not commonly associated with petroleum accumulations. Porosity is usually filled by the clay, which provided the increased density and transport velocity of the flow requisite to produce turbulence and the development of a graded suspension of detrital clasts. The nature of a single flow, however, may change in time and/or space, and the resulting depositional unit may show both vertical and areal changes in

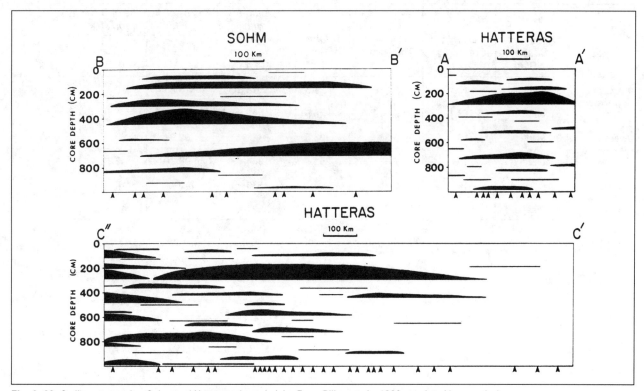

**Fig. 9-62.** Sedimentary units, Sohm and Hatteras abyssal plain. From Pilkey et al., 1980, reprinted by permission.

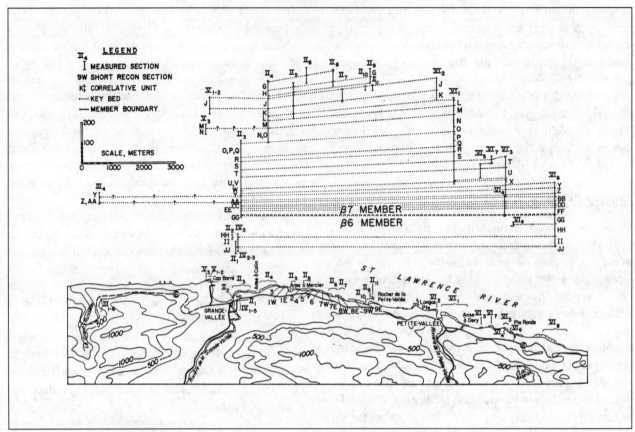

**Fig. 9–63.** Correlation of turbidity current units, Gaspé Peninsula, Canada. From Enos, 1969.

**Table 9-2**
**Estimated Volumes of Graywacke Beds and Turbidity Currents**

| Type of Bed | Assumed Shape in Plan | Length Corrected, $10^3$ m | Thickness, m | Volume of Bed, $10^6$ m³ | Volume with Argillite Correlation,[1] $10^6$ m³ | Volume of Turbidity Current,[2] $10^9$ m³ | | |
|---|---|---|---|---|---|---|---|---|
| | | | | | | $\Delta\rho = 0.01$ | $\Delta\rho = 0.10$ | $\Delta\rho = 0.60$ |
| Average graywacke β7 | 3:1 elliptical | 13.1 | 0.218 | 9.8 | 14.3 | 3.85 | 0.39 | 0.064 |
| | circular | 13.1 | 0.218 | 29.5 | 42.3 | 11.5 | 1.15 | 0.19 |
| Average graywacke β1 | 3:1 elliptical | 19.1 | 0.88 | 83.6 | 121 | 32.9 | 3.27 | 0.55 |
| | circular | 19.1 | 0.88 | 251 | 364 | 98.2 | 9.82 | 1.6 |
| Thickest graywacke β1 | 3:1 elliptical | 19.1 | 3.85 | 366 | 530 | 142 | 14 | 2.4 |
| | circular | 19.1 | 3.85 | 1,097 | 1,591 | 430 | 43 | 7.2 |

From Enos, 1969.
[1] An estimated 31 percent of each current deposit is argillite in addition to the graywacke volume.
[2] The density contrast, $\Delta\rho$, = current density – water density. A density of 1 g/cm³ is used for water.

texture and sedimentary features (Middleton and Hampton 1973) (Fig. 9–3). Consequently, the turbidity current model of sedimentation may not always be uniquely associated with catastrophic flow events in trenches and across abyssal plains.

## REFERENCES FOR FURTHER STUDY — CHAPTER 9

Bouma. A. H., G.T. Moore, and J.M. Coleman, eds., 1978, Framework, Facies, and Oil-Trapping Characteristics of the Upper Continental Margin: Am. Assoc. Petrol. Geol., Studies in Geology, no. 7, 326 p.

Doyle, L.J., and O.H. Pilkey, eds., 1979, Geology of Continental Slopes: Soc. Econ. Paleont. and Mineral., Spec. Pub. no. 27, 374 p.

Lowe, D.R., 1979, Sediment gravity flows: their classification and some problems of application to natural flow and deposits, *in* Doyle, L.J., and O.H. Pilkey, eds., Geology of Continental Slopes: Soc. Econ. Paleont. and Mineral., Spec. Pub., no. 27, p. 75–82.

Lowe, D.R., 1982, Sediment gravity flows II, depositional models with special reference to the deposits of high density turbidity currents: Jour. Sed. Petrol., v. 52, p. 279–298.

Potsma, G., 1986, Classification for sediment gravity-flow deposits based on flow conditions during sedimentation: Geology, v. 14, p. 291–294.

Stanley, D.J., and G. Kelling, eds., 1978, Sedimentation in Submarine Canyons, Fans, and Trenches: Stroudsburg, Pennsylvania, Dowden, Hutchinson, & Ross, 395 p.

Stanley, D.J., and D.J.P. Swift, eds., 1976, Marine Sediment Transport and Environmental Management: New York, John Wiley & Sons, 602 p.

# CHAPTER

# 10 BASIN ANALYSIS

Science is a self-correcting and self-enlarging system. It aims to unify experience. It creates patches of organized knowledge in the vast

expanse of human ignorance. The patches of knowledge grow, and may fuse to form more comprehensive patterns. The trend is clearly

towards an eventual single organization of conceptual thought, holding all aspects of experience in its web of relations, uniting all the

separate patches of knowledge into one living and growing body of organized understanding. **J. Huxley,** *The Humanist Frame,* **1961**

## INTRODUCTION

The most significant level of integration of stratigraphic information is at the basinal level. Depositional history can be determined at the chronozone level, and the interrelation of structural patterns with depositional and synsedimentary tectonic patterns is possible. The stratigraphic record of basinal fill is the basis for interpreting the causality of petroleum generation and accumulation.

The history of basin analysis has included stages from description to development of geometrical relationships, to classifications related to cratonic positions, and to the history of basin fill. This progression from Dana's work to Aubouin to Marshall Kay's treatise on geosynclines (1951) has been usefully applied to interpretation of stratigraphic units. With the availability of a three-dimensional pattern of wells and measured sections developed in the subsurface exploration for hydrocarbon accumulations, a basis was developed for regional facies studies of time-stratigraphic intervals. Sloss, Dapples, and Krumbein published an atlas of facies maps that illustrated the patterns of stratigraphic variation (1960). A tectonic control for interpretation of these maps was developed and was usefully applied to the genetic interpretation of stratigraphic intervals.

During the 1960s, observations were made on the depth, first motion and frequency of earthquakes, the magnitude and polarity of magnetic anomalies, heat flow patterns, and the topography of ocean basins. In 1968 Isacks et al. suggested a model of sea-floor spreading. In 1968 the concept of transform faulting was suggested. In 1970 an important synthesis of these data by Dewey and Bird suggested the form of a plate tectonic model. This model has provided the missing link for the interpretation and understanding of the genesis of basins. Many types of data from stratigraphy, paleontology, volcanology, seismology, and igneous petrology could be interrelated and used to understand the causality of both basin origin and developmental history. New syntheses, including geodynamics, paleobiogeography, and paleomagnetics are currently the focus of research symposia and books. It is now possible to determine from direct stratigraphic observations the nature of a basin, to suggest additional observations for interpreting structural and stratigraphic history, and with limited data to predict the types of patterns that developed.

# BASIN ANALYSIS

Basins are relatively permanent depocenters that persist for periods of tens to hundreds of millions of years. Patterns of change are predictable if an understanding of aspects of plate tectonics is part of the analysis (Fig. 10–15).

Basins reflect a developmental history related to plate tectonics, changes in heat flow, eustatic changes, and sedimentation. Plate separation, uplift, and convergence are in response to changes in crustal thickness and crustal and mantle densities. Crustal-mantle relationships from high heat flow, passive oceanic crustal areas, and areas of rapid sea-floor spreading illustrate these relationships.

One of the principal needs for stratigraphers is to determine the origin and history of sedimentary basins. Basins are the site for accumulation of the bulk of the sedimentary deposits preserved near the surface of the earth. Recognition of the importance of such depositional sites has long been a focus of study. Dana (1873) formulated the geosynclinal concept in an attempt to explain the thick stratigraphic sequences associated with the Appalachian Mountain system. Continued research was carried out on the interrelation of unconformities, tectonic events, and stratigraphic response patterns (Termier 1902) (Shuchert 1923) (Stille 1913, 1936, 1940). Wise (1974) (Fig. 1–40) suggested these authors' influence in developing the worldwide occurrence of depositional sequences. Others instrumental in the elaboration of the geosynclinal concept were Kay (1947, 1951) and Aubouin (1959, 1965). In addition, Krumbein and Sloss (1951, 1963) developed the concept of the tectonic control for sedimentation, emphasizing paleogeography, basins, and tectonic history. Coupled with this work directed towards stratigraphy was parallel work on tectonics, with the interpretation of the structure of the Alps, the Indonesian Archepelego, the Appalachian and Ouachita orogens, and the complex terranes of Trinidad. Much of this work was objective, based upon detailed field mapping, but also unifying concepts relating to causality of subsidence, uplift, thrusting, and uplift of deeply buried metamorphic facies from granulites and granites to blueschist and eclogites required an understanding of tectonic processes lacking to earlier observers.

The accumulation of magnetic, topographic, heat flow, seismic, and earthquake data from ocean basins during the 1950s and 1960s led to the unifying model concept of plate tectonics. An outline of the history of the development of the plate tectonic model was described by Frankel (337–353) in Schneer (1979). A summary of this development is presented in Chapter 3, p. 138. The importance of this concept to the orgin of sedimentary basins cannot be overstated. As described in the chapter

on models, models are not scientific "truth," but are only useful tools or devices for synthesis of data. New observations in the last decade have changed our perspective on causality of basin development and history. Deep crustal drilling, COCORP and BIDEP seismic profiling, topographic and gravity data derived from a satellite observation platform, thermal models of the mantle derived from a worldwide seismological network, new heatflow data from ocean basins, and integrated seismic, magnetic, gravity, and field mapping from orogenic terrances have provided new information requiring synthesis. Other information on matching of exotic terranes using new age dates, paleontologic, and facies data have led to the development of new histories and architectures for the interpretation of mobile belts. Much of the newer literature is highly subjective and simply suggests only possible mechanisms to explain many of the observed relationships.

At the 1987 meeting of the International Union of Geodesy and Geophysics meeting in Vancouver, Canada, many papers were presented on current research relative to earth history and structure. Of particular note was a

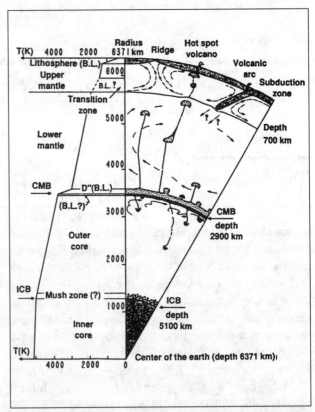

**Fig. 10-1.** Schematic diagram of core-mantle coupling. Hot instabilities emitted from D″ in the lowermost mantle rise through the mantle, leading to TPW and lithospheric activity (hot-spot volcanism, faster root-mean-square plate velocities, lithospheric polar wander). Cold instabilities emitted from the top of the core may eventually destabilize the main geomagnetic field and cause reversals. Temperature distribution (in kelvins) is outlined on the left side (74). Abbreviations: *CMB*, core-mantle boundary; *ICB*, inner-outer core boundary; *B.L.*, boundary layer. From Courtillot and Basse, 1987.

summary keynote lecture presented by Peter J. Wylie of the Californian Institute of Techonology on Magma Genesis and Plate Tectonics. Published symposia volumes from this conference will provide an updating of our expanding data base on the origin of sedimentary basins. A newer synthesis of earth structure, similar to that illustrated by Wylie, was published by Courtillot and Besse (1987) (Fig. 10-1). The development of geodynamic models of core, mantle, and crustal interactions is a highly subjective research area. Each new observational data base requires a new schema to interpret the data (after Humphreys, Clayton, and Hager from Kerr 1984) (Fig. 10-2). Fundamental concepts such as subduction, obduction, thin-skinned tectonics, and three dimensional geometry of triple point junctions are presently under serious objective review. Recognition from many types of observation that the upper mantle at a depth of approximately 650 kilometers is probably separated from the lower mantle. This has required a redirection of research on patterns of heat flow in the mantle. Tomography of seismic velocities within the mantle has indicated a complex pattern of thermal anomalies (Dziewonski et al. 1984, 1987). Only in the upper mantle near the asthenosphere boundary is there a strong correlation with plate boundaries. The concept of heat flow from the core-mantle boundary, providing unstable high heat areas at the base of the upper mantle may lead to rising hot spots, thinning of the lithosphere, and the formation of spreading centers. Supporting the dynamic nature of upper mantle processes is the presence of major differences in sea level, or geoidal elevation, across ocean basins (Canby 1983). These changes in the geoid may lead to gravity spreading, and vertical instability in crustal elevations. The presence of spreading centers, clearly indicated by topographic highs and sea level or geoidal highs, possibly produce gravity influenced sinking of lithospheric slabs into the asthenosphere. Earthquake data (Pennington 1983) illustrate the geometry of the dipping slab into the upper mantle (Fig. 10-3). The section illustrated at Tohoku, Japan, shows a double band of earthquake foci, and is thought to represent the upper and lower lithospheric boundary, with the lithosphere approximating 30 kilometers in thickness.

Postulated lithosphere-asthenosphere patterns at oceanic plate spreading boundaries suggest the effects of higher heat flow, patterns of intrusive and extrusive ophiolitic igneous rocks (Lewis 1983) (Figs. 10-4 and 10-5). Similar lithospheric thinning may also develop crustal blisters under cratonic plates, or at cratonic-oceanic plate boundaries. This complex of possibilities may lead to development of three types of plate boundaries, divergent, convergent, and transform. Matching of magnetic strips on either side of a spreading center has long been recognized, and these magnetic reversal events have been dated, and provide a history of the opening of ocean basins, and even the history of microplate movement (Rabinowitz et al. 1983) (Figs. 10-6 and 10-7).

Observation of earthquake foci at convergent oceanic-cratonic margins suggests the formation of a number of models possibly related to the subduction angle of the lithospheric plate, the heat flow, or the sedimentation history (Seeley and Dickinson 1977) (Fig. 10-8). This model has been the focus for the interpretation of continental margin seismic transects, and the depositional history of

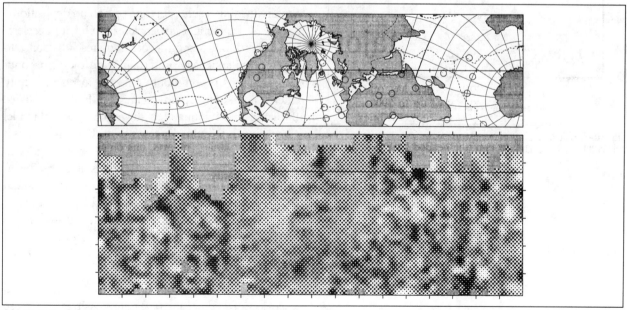

**Fig. 10-2.** Looking down to the earth's core. From Kerr, 1984.

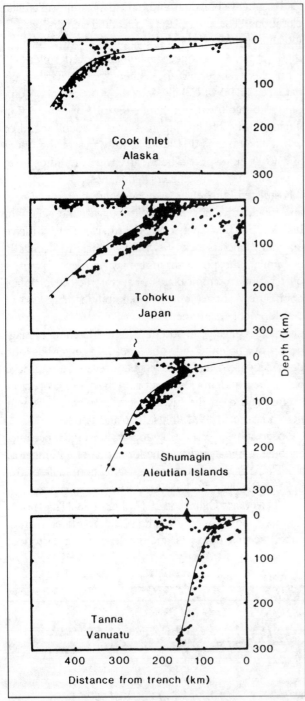

**Fig. 10–3.** Role of shallow phase changes in the subdictions of oceanic crust. From Pennington, 1983.

forearc basins (Scholl and Marlow 1974) (Fig. 10–9) (Biddle and Seeley 1983) (Fig. 10–10).

## Origin and History of Basins

A relative simple basin classification relative to plate tectonics was suggested by Halbouty et al. (1970) and Klemme (1975) (Fig. 10–11). This classification suggests a relation to cratonic plates, with simple, extensional and foreland basins on the craton; and, on cratonic margins, marginal, extensional, mobile belts, and deltaic depositional centers. One of the more important contributions to the origin and history of basinal development has been the suggestion that higher heatflow produces an uplift, due to lower density of the aesthenosphere (McKenzie 1969; Sleep 1971). Bott, utilizing this concept, developed two possible response patterns that may be responsible for basinal development (Bott 1979). Simple uplift and erosional truncation of lighter crustal material may result in erosion of lighter crustal material and gravity sliding of the lighter lithospheric crust away from the thermal anomalay (Fig. 10–12). A second possibly is the thermal effects of metamorphism resulting in increased density of the lower crust, or underplating of the crust by the addition of intrusives. Spreading related to magma plumes may also affect temperatures in the lithosphere, with the response reflecting changed viscosities between the shallow, more ductile region of the lithosphere, and increased temperature allowing flow and spreading in the underlying more ductile lithosphere. Graben response patterns possibly would be related to ductile flow in response to the uplift (Bott 1979) (Fig. 10–13). Often continental margins response patterns cannot be simply related to thermal heat flow anomalies. Eustatic and sedimentary loading result in marine onlap and depositional wedges of sediment of lower density. Gravity acceleration, to produce a smooth geoid, requires subsidence to produce similar density profiles. Flexure folding of continental margins is commonly reflected in unconformity patterns and stratigraphic thicknesses (Bott 1979) (Figs. 10–14 and 10–15) Pitman has suggested a thermal model to explain coastal subsidence patterns. Utilizing a causal mechanism related to the growth of mid-ocean ridges due to the development of spreading centers, it is possible to model the response on continental margins (Pitman 1979) (Figs. 10–16 and 10–17). The displacement of sea water by changes in subsea elevation of the subsea crustal surface should produce changes in water depths on continental margins. Sea-floor elevations, as indicated by depth measurements across sreading centers and ridges, reflect the processes described above. Both shrinking and expansion of mid-ocean ridges are possible, and these can be related to spreading rates. The response in terms

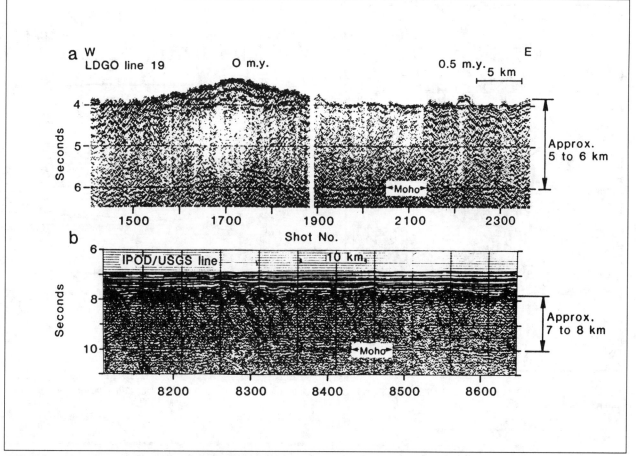

**Fig. 10–4.** Multichannel seismic reflection profiles from *a,* the East Pacific Rise at 9°30′,(8) and *b,* Atlantic crust 135 million years (*m.y.*) old. The seismic Moho occurs at a two-way travel time of about 2.1 seconds in both profiles. There is no evidence of an increase in travel time under the East Pacific Rise axis due to low-velocity magma. From Lewis, 1983.

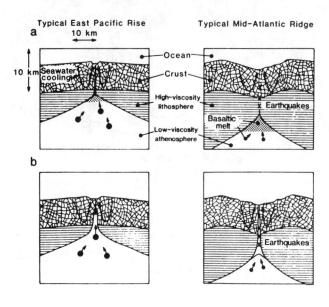

**Fig. 10–5.** *a:* Diagram of the model of ocean crust formation in which the crust is formed by episodic injection of basaltic dikes from a mantle reservoir and the Moho is a petrologic boundary. This model requires that the amount of basaltic melt is independent of spreading rate and all of the melt is available to form crust. *b:* Diagram of the model of ocean crust formation that invokes diapiric intrusion of mantle containing basaltic partial melt. The depth to the Moho in this model is controlled by the depth to which hydrothermal circulation and metamorphism extend—if the thickness of the basaltic section is less than about 5 km. This model accounts for the fairly uniform depth to the Moho—independent of spreading rate and mantle temperature—the absence of steady-state crustal magma chambers, the gravity data on the East Pacific Rise, and the seismic data. From Lewis, 1983.

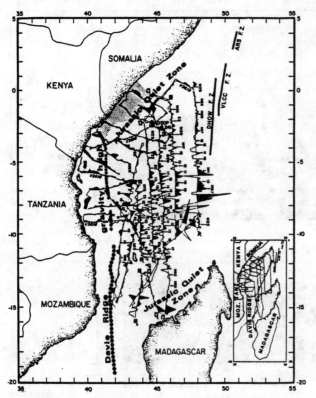

**Fig. 10-6.** Magnetic anomaly profiles plotted normal to ships' tracks. From Rabinowitz et al., 1983.

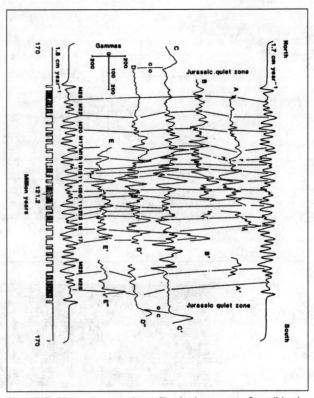

**Fig. 10-7.** Magnetic anomaly profiles in the western Somali basin. From Rabinowitz, Coffin, Falvey, 1983.

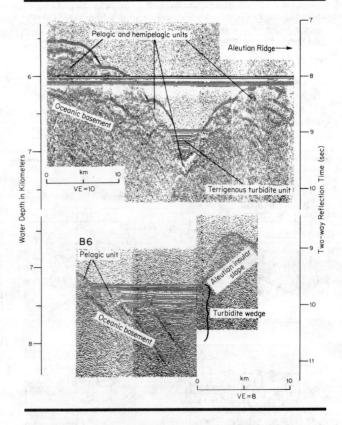

**Fig. 10-9.** Acoustic reflection profiles of central (B6), or underthrust; and western, or nonunderthrust, segments of Aleutian trench. From Scholl and Marlow, 1974.

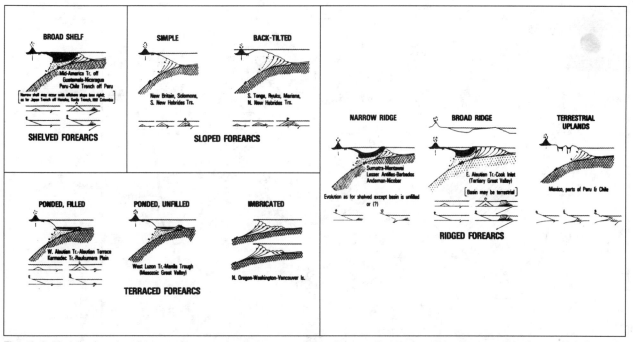

**Fig. 10-8.** Models of modern forearcs. Conceptual evolutionary cartoons shown at the bottoms of the diagrams. Question marks indicate the unknown nature and position of the massif-subduction complex boundary, which is a current subject of JOIDES research. From Seeley and Dickinson, 1977, reprinted by permission.

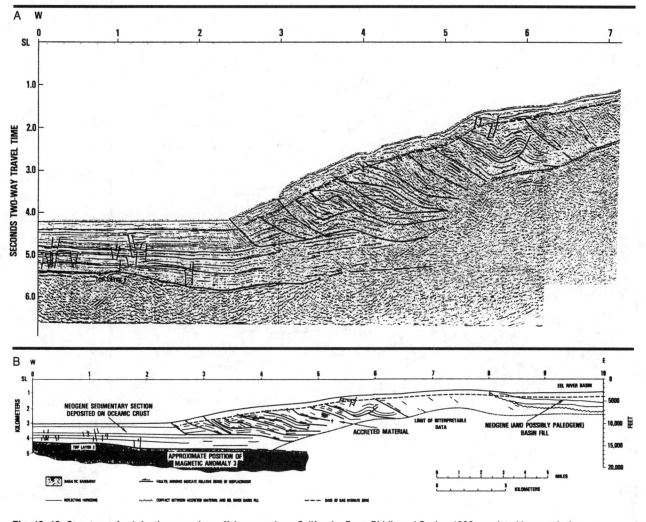

**Fig. 10-10.** Structure of subduction complex, offshore northern California. From Biddle and Seeley, 1983, reprinted by permission.

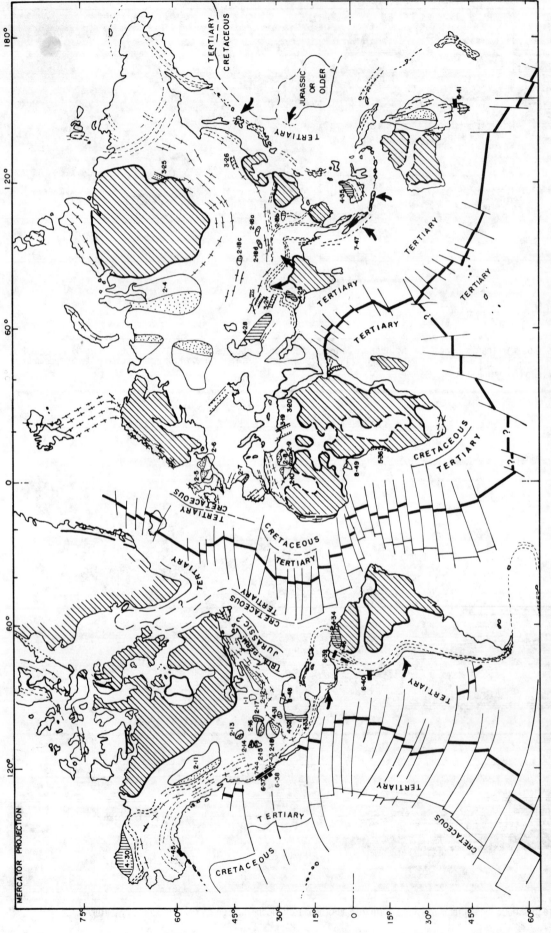

**Fig. 10–11.** Basins reflect a developmental history related to plate tectonics charges heat flow, eustatic changes, and sedimentation. From Halbourty et al., 1970.

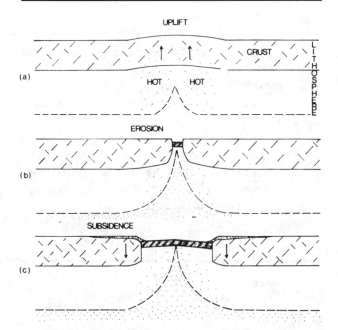

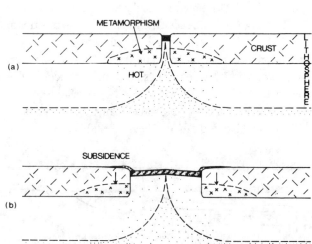

**Fig. 10-12.** *A,* The basic thermal hypothesis. *(a)* Uplift following heating of the lithosphere; *(b)* initiation of new ocean accompanied by erosion of continental uplifted region, causing crustal thinning; and *(c)* subsidence of continental margins as underlying lithosphere cools. Note the nearly vertical edges of continental crust predicted by this model.

*B,* The thermal hypothesis omitting the rift stage. *(a)* Heating the continental lithosphere prior to and during split causes metamorphic transition of lower crust from greenschist to amphibolite facies, raising the mean crustal density, *(b)* Subsidence of continental margins as underlying lithosphere cools. From Bott, 1979, reprinted by permission.

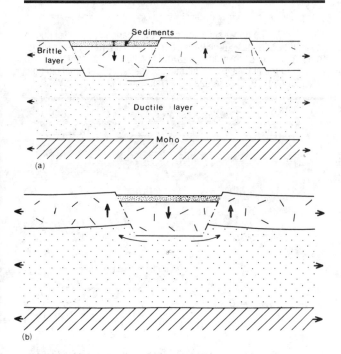

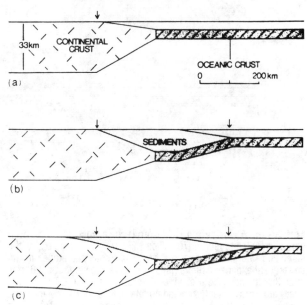

**Fig. 10-13.** The mechanism of graben formation by wedge subsidence affecting the upper continental crust with outflow in the lower crust. *(a)* Subsidence compensated by horst uplift and *(b)* subsidence compensated by elastic upbending. From Bott, 1979, reprinted by permission.

**Fig. 10-14.** The gravity loading hypothesis: *(a)* the initial situation prior to loading, assuming a pre-existing 200-km-wide transition between oceanic and continental crust; *(b)* the result of local Airy sediment loading, assuming density of the sediments of 2450 kg/m³; *(c)* the result of flexural loading, assuming that the lithosphere has a flexural rigidity of 2x1022 Newton meters and that densities are as in *(b).* From Bott, 1979, reprinted by permission.

of strandline movement on the shelf can be modeled for the variations in changes of ridge volume (Figs. 10–18 and 10–19). This can then be shown diagrammatically for shoreline position at specific time periods, and these can be associated with time-rock bedding plane terminations derived from seismic sections. In addition, depositional models can be constructed, with paleogeographic interpretations, in relation to strandlines, rates of sea-level rise, progradational sequences, and the displacement of magnetic strips. These events can be correlated to sea-level falls indicated by biozone events, and have been utilized by Haq and others (1987) to establish one of the response patterns for a time framework useful for dating seismic sequences.

A model adapted from Falvey (1974) suggests the following history in response to these changes in measured spreading rates.

▸ Driving subsidence is a continuous function related to a density increase in the crust due to cooling. This is a linear function.
▸ Sedimentation is controlled by receptor capacity, which is related to sea-level rise. This is a discontinuous function. There is a positive feedback mechanism—more sea-level rise, more receptor capacity, and more deposition.

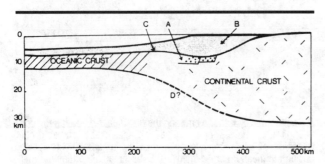

**Fig. 10–15.** Section across a passive margin showing some of the characteristic features of sediment and crustal structure: *A,* pre-split graben sediments; *B,* post split sediments associated with flexural subsidence; *C,* the proglematic position of the continent-ocean crustal contact beneath the sediments; *D,* the apparent gradational contact between deep continental and oceanic crust. From Bott, 1979, reprinted by permission.

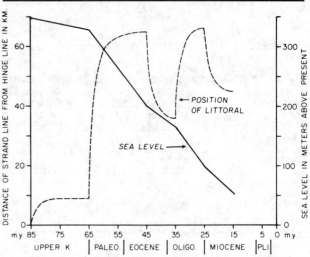

**Fig. 10–17.** The upper solid line gives the change in sea level due to the change in ridge volume for the period from 85 million years (*m.y.*) to 15 m.y. The dashed line gives the position of the shoreline with respect to the hingeline as a function of the rate of sea-level change. From Pitman, 1979, reprinted by permission.

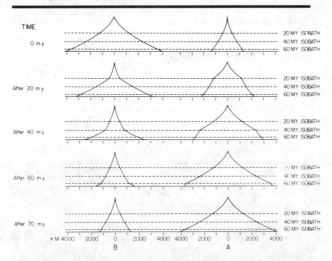

**Fig. 10–16.** *A,* At the top is the profile of a ridge that has been spreading at 2 cm/year for 70 million years (*m.y.*). At time 0 m.y. the spreading rate is increased to 6 cm/year. Sequential stages in the consequent expansion of the ridge profile are shown: first at 20 m.y. after the spreading rate change; then at 40 m.y., at 60 m.y., and finally at 70 m.y., after the spreading rate change. At 70 m.y. the ridge will be at a new steady-state profile; the cross-sectional area at this time will be three times what it was at 0 m.y. *B,* At the top of the profile of a ridge that has spread at 6 cm/year for 70 m.y. At 0 m.y. the spreading rate is reduced to 2 cm/year. The sequential stages in the subsequent contractions of the ridge are shown. At 70 m.y. after the change in spreading rate, the ridge will be at a new steady-state profile; the cross-sectional area will be one-third of what it was at 0 m.y. From Pitman, 1979, reprinted by permission.

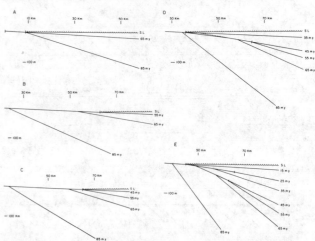

**Fig. 10–18.** Hypothetical stratigraphic sections are shown for a sequence of stages from 85 million years (*m.y.*) to 15 m.y. The distance from the hingeline is given in kilometers. The only variable is the rate of sea-level fall given by the slope of the sea-level curve in Figure 10–17. From Pitman, 1979, reprinted by permission.

▸ If the crust is 34 kilometers thick, with a density of 2.91, and the lithosphere is 10 kilometers thick, with a density of 3.35, a change of 1.0 to 1.2 HFU (Heat Flow Units) (1 μcalory/square centimeters/second) will cause a rise of the crustal surface of 2 kilometers. With erosion of 2 kilometers the surface elevation would be 210 meters (200 x 2.91 = 179 x 3.25.)

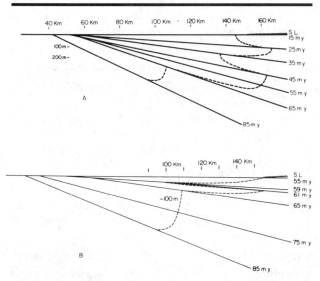

**Fig. 10-19.** *A,* Same as in Figure 10–18, but in this case *s* = 1 cm/1,000 years. Note that the pattern of pinchout and truncation of the various sedimentary wedges is the same as Figure 10–18, but the position of the shoreline marked by the heavy-dash bed line looping up through the section is pushed seaward. *B,* As in Figure 10–19A s = 1 cm/1,000 years, but in this case *s* is reduced to zero for the interval from 61 million years (*m.y.*) to 59 m.y. and then increased back to 1 cm/1,000 years. From Pitman, 1979, reprinted by permission.

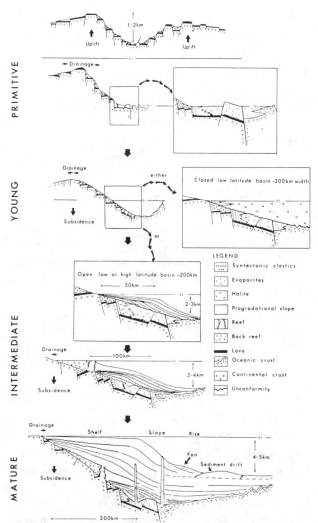

**Fig. 10-21.** Basin evolution in response to depositional history and subsidence. From Roberts and Caston, 1975.

**Table 10-1**
**Basin Classification**

| | Relation to Plates | | |
|---|---|---|---|
| Tectonic Style | Cratonic Plate | Marginal to Cratonic Plate | Oceanic Crust |
| Epeirogenic downwarp | Cratonic basin (simple) Type 1 | Marginal basin (open) Type C | Abyssal plain (passive) |
| Extension | Graben basin (rifted) Type 3 | Marginal basin (block faulted) Type 5 | Midocean ridge (volcanic pile) |
| Vertical—Type 1 (drape) | Intermontane basin (high-angle reverse) Type 2A | Progradational wedge (deltas) Type 8 | Fan and continental rise (accretion) |
| Vertical—Type 2 (gravity) | Intermontane basin (low-angle thrust) Type 2 | Marginal basin (closed by marginal uplift) Type 4A | Fore-arc trench (subduction) |
| Horizontal and vertical movement | Fault block basin (conjugate shear) Type 6 | Transform fault basin (mobile belt) Type 7 | Fracture zone (transform) |

▸ Similarly a change in mantle density to 3.35 will cause subsidence of 380 meters (380 x 1.03 + 1620 x 3.35 = 2000 x 2.91.

▸ Metamorphism can add to subsidence—a change in density, an increase from 2.01 to 3.09 will increase receptor capacity and sediment cover will increase temperature. Thus, thermal metamorphism will produce an increased density.

▸ This system has a lag proportional to heat conductivity, which results in a lag of 15 x 10[6] years between expansion and subsidence [roughly equivalent to a second order sequence of Vail (1987)].

The lag indicated is a maximum for changes in heat flow. If the asthenosphere boundary is quasi-liquid, the viscosity of the mantle at this level could be low enough to allow flow rates of centimeters per year. This would result in subsidence to maintain the constancy of gravitational acceleration for the interval from the surface of the geoid down to the asthenosphere.

Basin history may also be related to depositional rates (Schwab 1976) (Fig. 10–20). A history for basins at extensional plate boundaries was suggested by Roberts and Caston (1975) (Fig. 10–21). Initial extension, related to spreading, may produce a basin with basaltic flows and intrusions, or may produce a topographic basin that is a receptor for lacustrine sequences, fan-deltas, or low sinuosity stream deposition. Continued subsidence related to sequence-related sea-level rises, may produce at low latitude an evaporitic sequence of sedimentation. Continued subsidence with additional sequence-related sea-level rises maintains a basin with narrow shelf margins and a topographic basin available to receive deeper

water sedimentary sequences. Continued expansion of the shelf platforms, subsidence, and opening to ocean basins may result in either deeper water or shelf carbonate depositional sequences. These sequences may show strong reefal development near the shelf margin. Additional subsidence with continued clastic detrital provenance results in a history of marginal basinal depositional sequences, or coastal prograding clastic sequences. This tectonic-depositional history is of major importance on passive continental margins, and is the control for marginal basinal sedimentary sequence development throughout the stratigraphic record. Examples are easily recognized in the Mesozoic of the North Sea, the Atlantic margin of North America, the late Mesozoic and Tertiary margins of the Gulf Coast of North America, and similar passive margins from South America, Africa, India, and China (Roberts and Caston 1975).

## BASIN CLASSIFICATION

Many attempts have been made to classify basins simply in relation to their tectonic setting and history. These attempts led to classifications that were virtually unusable, since every basin had a somewhat different tectonic history. The first step was to suggest one of five differing plate tectonic settings: (1) divergent margin basins, including intracratonic rifts, divergent margin wedges, aulacogens, and failed rifts; (2) convergent margin basins related to volcanic arcs, including trenches, forearc, backarc, and retroarc basins; (3) basins associated with transform margins, and megashears; (4) basins generated during the process of continental colli-

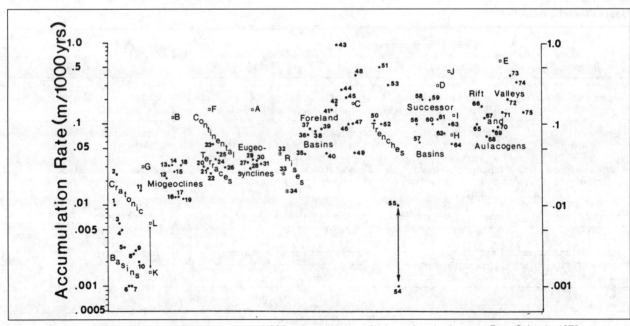

**Fig. 10–20.** Rate of sediment accumulation in metres per 1,000 yr (logarithmic scale) for various basin types. From Schwab, 1976.

sion and suturing, including foreland, foredeep, or peripheral basins, and some graben and wrench-faulted basins; and (5) cratonic basins whose development is not obviously related to plate margin tectonics (Dickinson 1974). Such a process framework for basin classification presupposes that all factors relevant to basin origin and history are well known, but it neglects the unknown factors related to sedimentation, heat flow, and changing tectonic history.

In an attempt to develop a usable classification, Bally (1980) suggested a hierarchy of controls important to basin formation and history. In his classification scheme, some 27 different basin types and sub-types were thought necessary to explain worldwide observed basinal patterns. An even more complicated pattern for basin classification was published by Kingston and coworkers at Exxon (1983), suggesting at least six levels of possible basinal variation, which could provide sufficient detail so that every basin in the world could be uniquely classified.

The goal of classification is to make complex patterns of variation simple, and to generate natural groupings that can be usefully used for comparison. If the classification has an objective genetic basis, rather than a descriptive subjective framework, it does lead to a predictive framework for interpretation. Such a classification was suggested by Halbouty et al. (1970) and by Klemme (1975). They suggested that only 8 general basinal frameworks could provide the framework for a basinal classification scheme:

- Type 1 — Simple cratonic basins;
- Type 2 — Foreland cratonic basins with or without horizontal thrusting;
- Type 3 — Successor or rifted extensional cratonic basin with or without wrench faulting;
- Type 4 — Cratonic margin depocenters on thinned cratonic crust, with or without horizontal thrusting;
- Type 5 — Extensional basins on cratonic passive margins;
- Type 6 — Basins at active plate boundaries, including forearc, intra-arc, and retro-arc basins, with wrench faulting cutting perpendicular to the cratonic margin;

- Type 7 — Basins at active plate boundaries with wrench faulting parallel to the cratonic margin;
- Type 8 — Progradational deltaic depocenters.

Klemme related these basin types to the worldwide distribution of giant hydrocarbon accumulations (Fig. 10–11) (Halbouty et al. 1970).

The expansion of the Klemme classification to include tectonic style, and the relation of basins to crustal type provides the basis for a genetic intepretation for depositional and internal structural history of basinal types (Table 10–1). The advantages of such a classification is that similarities, and not differences between basins, are emphasized; themes of sedimentation can be suggested that are present in natural groupings of basins; and a tectonic history, such as that proposed by Roberts and Caston (1975) (Fig. 10–21), has predictive value. In addition, unconformity, stratigraphic, and structural patterns may be related to the occurrence of giant hydrocarbon accumulations for each basinal type.

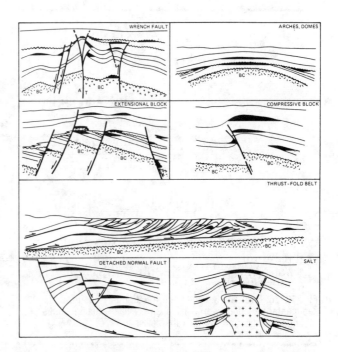

**Fig. 10-22.** Structural styles related to basin history. From Harding and Lowell, 1979, reprinted by permission.

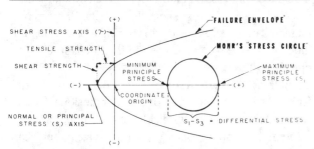

**Fig. 10-23.** Graphical representation of a stress field by a stress circle and a failure envelope. From Meissner, 1978.

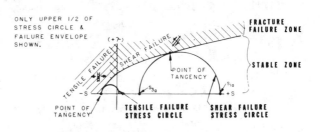

**Fig. 10-24.** Graphical representation of stress fields producing shear and tension fracture failure. From Meissner, 1978.

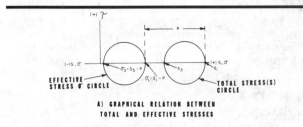

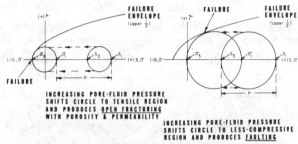

**Fig. 10-25.** (a) Graphical relation between total and effective stresses and (b) graphical representations of fracture failure produced by increasing pore pressure. From Meissner, 1978.

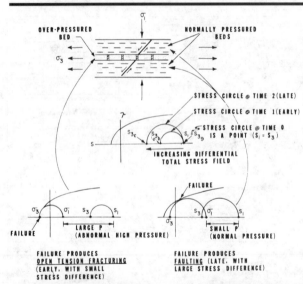

**Fig. 10-26.** Both faulting and open fracturing can be produced in a sedimentary section with a rising stress field affecting beds with differing pore-fluid pressure. From Meissner, 1978.

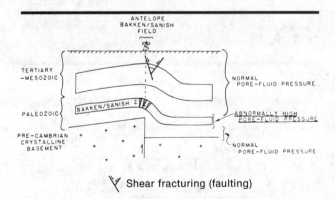

**Fig. 10-27.** Changes in failure response as a function of pore pressure. From Meissner, 1978.

## Structural Styles

Examination of seismic sections has revealed that structural styles are developed in a few relatively simple patterns (Fig. 10-22). These patterns also can be shown to reflect responses characteristically developed in each of the eight basin models presented (Table 10-1). Forces involved are extension with the development of arches and domes, extensional block, and diapiric salt (or low-density shale); compression with the development of compressive block and wrench faulting; or gravity sliding associated with thrust-fold belts or detached normal fault patterns (Harding and Lowell 1979) (Fig. 10-23). Some of these may be combinations with wrench faulting resulting in both extensional and compressional responses, the extensional block reflecting extension along strike-slip faults, and mounded uplifts associated with diapirism. The wrench-fault pattern is developed in convergent margin and successor basins (Type 6 and 7 basins). Arches and domes are present in simple basins (Type 1) and in foreland basins (Types 2 and 4). Extensional blocks are present in Type 3 and 5 basins. Gravity thrusting is present in both foreland and marginal basins (Types 2 and 4). Detached normal faulting and diapiric structures are present in marginal basins and in a few pull-apart basins (Types 4, 5, and 8). This observational synthesis is useful for interpreting both the origin and pattern of structures in each type of basin.

## Fracture-Failure Controls

Some information is available on the local generation of stress fields and on the mechanisms of failure. The Mohr's stress ellipsoid suggests that forces can be modeled using the three mutually perpendicular vectors of the stress field. Sigma 1, 2, and 3 represent the principal, intermediate, and minimum stress vectors. Failure can be related to the absolute difference in magnitude of the stress between Sigma 1 and 3. Failure is a property of the rock system and the magnitude of the stress differential (Meissner 1978) (Fig. 10-24). The pore fluid pressure is one of the more important controls for the stress differential between Sigma 1 and 3. The magnitude of this difference controls the nature of the rock system failure whether tensional or by shear (Fig. 10-25). Incremental increases in the stress field will terminate only when the differential stress increases to the fracture-failure envelope (Fig. 10-26).

Failure must occur in the quadrant defined by the differential stress and be either by shear or tensional failure. Generation of the stress field may be due to forces external to the pore geometry and fluid system, but it can be due to local changes in permeability with fluid expan-

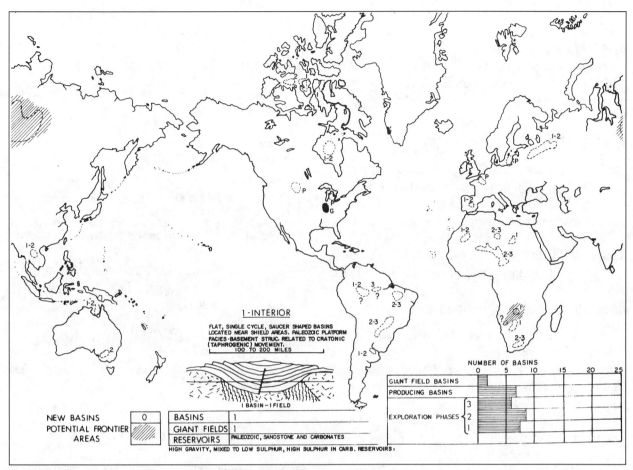

**Fig. 10-28.** Cratonic basins and their exploration stages. Reprinted courtesy H.D. Klemme.

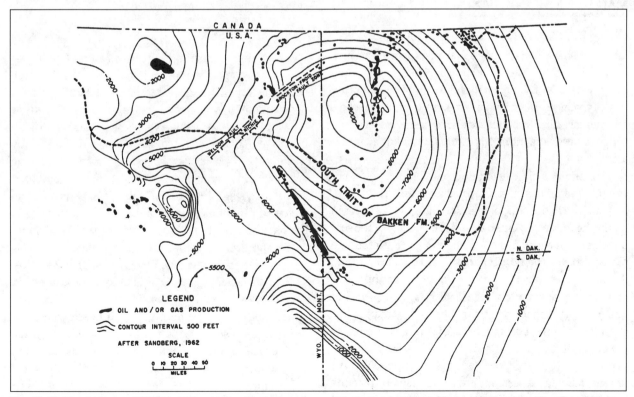

**Fig. 10-29.** Structure map of Williston basin; structure contours on base of Mississippian strata and limits of Bakken formation. From Meissner, 1978.

sion and compaction leading to an increasing stress field (Fig. 10–27). Depending upon permeability and the rigidity of the pore geometry, failure can be the result of increasing pore fluid pressure.

Much still must be learned concerning the causality of the origin of stress fields in the earth and the responses that lead to basin formation and synsedimentary and post-depositional structural history. Measurement of pressure, strain accumulation and release, and other geodynamic measurements are a focus of continuing research. If the results of these processes can be observed in the stratigraphic record and appropriate models developed to interrelate observations, it may be possible to utilize stratigraphic patterns to interpret the origin of responses in sedimentary basins. This base then can be used to predict developmental patterns with limited data.

## Basin Models and Exploration Strategies

The systematic analysis of the eight basin types described by Klemme can lead to understanding the origin of the basin, the distribution of hydrocarbons, and the development of exploration strategies. Each of the basin types are described, related to the appropriate model, and related to the pattern and history of hydrocarbon accumulations. The stratigraphic response patterns are the bases for this analysis.

## SIMPLE CRATONIC BASINS

Simple cratonic basins, Klemme's Type 1, are flat, saucer-shaped basins up to 200 miles across and generally located on Precambrian shields. They contain multiple cycles of sedimentation broken by unconformities. Typically, they contain Paleozoic and Mesozoic sedimentary fill. Structures are due to basement tectonics with the development of drape folds, broad domes, and arches.

Basin origin is difficult to understand at this point, but several possibilities exist. Basins may be related to inhomogeneities in the crust or mantle; changes may occur in mantle heat-flow patterns causing subsidence; or the sedimentation may cause the initiation of the basin and be the control for the developmental history. Determining the causality of basin formation simply from the stratigraphic response pattern is difficult, but a sequence of time-rock facies maps suggest that the sedimentation history is an important control, if not the primary control for the origin of the basins. Basin centers appear to migrate in relation to deltaic depocenters, to subside in relation to density differences resulting from sedimentary patterns (producing changing receptor areas), and subsequently, to control the patterns of basin fill.

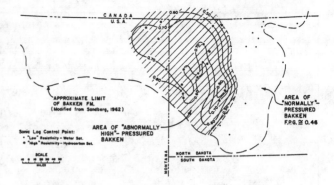

**Fig. 10–30.** Aerial distribution of the pore-fluid gradients in the Bakken formation and areas of normal and abnormally high pressure. Note coincidence of low electrical resistivity with normal pressure and high resistivity with abnormally high pressure. From Meissner, 1978.

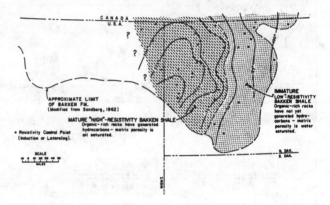

**Fig. 10–31.** Areas of "high" and "low" electrical resistivity in Bakken shales, with subsurface isotherm contours and interpreted area of source-rock "maturity". From Meissner, 1978.

Recognition of basin types requires an understanding of the depositional controls and the tectonic and depositional history. These aspects can be determined from detailed stratigraphic studies, but information of this type is not universally available. A suggested worldwide distribution of simple cratonic basins was developed by Klemme (1983). He also suggested basins containing giant oil fields (more than 500 million barrels) and their exploration stage (Fig. 10–28). This synthesis is not complete and cannot include all of the possible variations in basins throughout the world. However, it is a useful model approach for the focus of further studies.

From Klemme's experience and other sources, it is possible to characterize the distribution of hydrocarbons in simple cratonic basins. These basins contain both carbonate and sandstone reservoirs with possibly an even division in reserves. Source rocks include biogenic marine shales, argillaceous carbonate muds, and carbonaceous shales and coals. The thickness of sedimentary fill in these basins rarely exceeds 4,000 meters. With a normal temperature gradient of 0.7°– 1.0°C/30 meters, only a small portion of the basin has

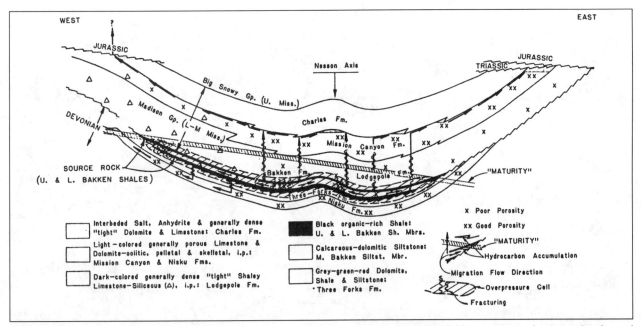

**Fig. 10-32.** Schematic east-west section across the Williston basin showing source-rock maturity, fluid over-pressure, fracture, migration and hydrocarbon accumulation patterns in the Bakken formation and adjacent units. From Meissner, 1978.

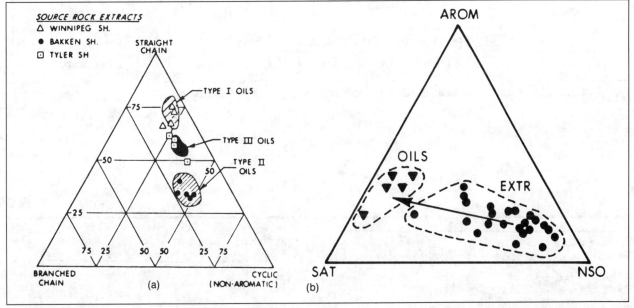

**Fig. 10-33.** (a) Distribution of straight chain, branched, and cyclic paraffins in the $C_4$-$C_7$ fraction of Williston basin crude oils and source rock extracts. From Barker, reprinted by permission. (b) Distribution of saturate, aromatic, and NSO compounds in Jurassic crude oils and source rock extracts from the Parentis basin, France. From Barker, 1978, reprinted by permission.

attained sufficient temperatures for hydrocarbon generation. Consequently, oil reserves are low, approximating 18,000 barrels/cubic miles.

Basin sequences reflect changes in relative sea level with the same unconformities recognized on continental margins. The placement of these basins on the craton removes them from the initial effects of onlap during each sequence. Sedimentation may commence later, and subaerial exposure may commence earlier. Consequently, unconformities are more persistent, are more frequently developed, and more strongly control the depositional history.

The largest trap is structural and contains 50% of the basin's giant field reserves. Gas appears to generate late, often occurs in the basin center, and is associated with carbonaceous shales and coals. Reservoir seals are shales and evaporites with stratigraphic traps on the basin margin.

## Williston Basin Example

The Williston basin is a useful example of stratigraphic and structural patterns. A structure map on the top of a

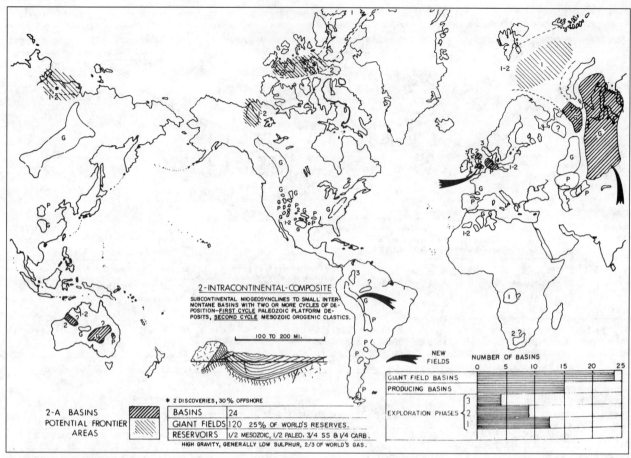

**Fig. 10–34.** Type 2 basins and their hydrocarbon potential. Reprinted courtesy H.D. Klemme, 1979.

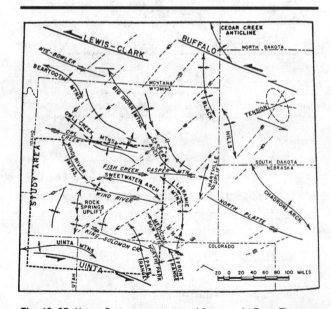

**Fig. 10–35.** Upper Cretaceous structural framework. From Thomas, 1971.

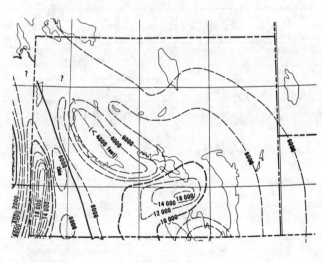

**Fig. 10–36.** Pre-Tertiary-Tectonic framework. From Weimer and Haun, 1960.

major source rock interval provides a time-rock datum that can be used to interpret paleogeography and the pattern of post-depositional basinal subsidence (Fig. 10–29). Both the preserved distribution of the source rock and the geometry of structural features in and around the basin are indicated. Temperature and pressure gradient maps define areas of abnormally high resistivity (Figs. 10–30 and 10–31). These areas can be interpreted to reflect the presence of hydrocarbons in the pores of a source rock. A cross section through the basin indicates the patterns of temperature, overpressured shale, unconformities, and potential seals (Fig. 10–32). By using a simple model reflecting the presence of a basement fault, it can be suggested that strain induced by overpressure and by synsedimentary movement along the fault trace indicates that tensile rupture should occur in the overpressure zone (Fig. 10–27). This model causes hydrocarbons to move from the overpressured source rock up along vertical fractures into porous and permeable strata. Hydrocarbons should be trapped by the impermeable evaporitic caprock in areas of low pressure at the crest of structures and in updip stratigraphic pinchouts. Confirmation of this model can be seen by the comparison of molecular composition of hydrocarbon fractions from the source rock and the oil reservoirs (Fig. 10–33). Some compositional changes do occur, but these can be attributed to differences in molecular properties between saturated normal paraffins and the nitrogen, oxygen, and sulfur compounds.

## Summary of Simple Cratonic Basins

The outlined example can be developed and applied in a similar fashion to the Illinois, Michigan, Paris, Baltic, Chad, and Congo basins. Other basins in Africa and South America are currently being explored, and hydrocarbons have been found. The basin model outlined here can be usefully applied.

Exploration for the single largest structure in the center of the basin can provide a high probability for exploration success. Source-rock patterns can be evaluated and related to depositional history and permeability patterns within the basin. The history of petroleum generation, migration, and entrapment can be the basis for structuring an exploration program for hydrocarbon reserves.

## CRATONIC FORELAND BASINS

This type of cratonic basin reflects the presence of a faulted or uplifted flank. The couple developed interrelates subsidence with a complementary synsedimentary uplift. This change causes a strong increase in clastic provenance, which results in the devel-

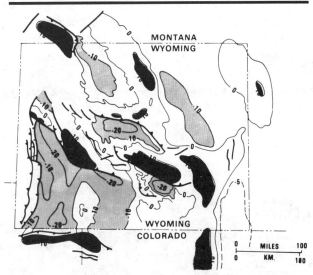

**Fig. 10–37.** Differential structural relief as a product of wrench-fault shearing. From Sales, 1968, reprinted by permission.

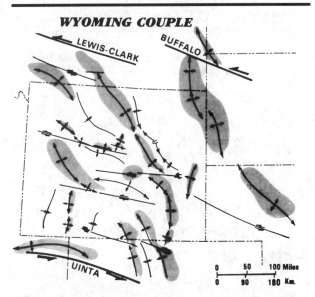

**Fig. 10–38.** Uplift and basinal axes produced in response to shear and basinal sedimentation. From Thomas, 1971.

opment of an alluvial, deltaic, and prodeltaic depocenter and a foredeep adjacent to the marginal uplift.

This type of basin persists for tens or hundreds of millions of years. It continues to subside, and may accumulate sedimentary thicknesses in excess of 10,000 meters. It is the habitat of 25% of the world's giant field reserves (Klemme 1983) (Fig. 10–34). Sedimentary fill is dominated by clastics, but reefs and evaporites may occur prior to closure by faulting. Two types of basins may be developed: one with a single-faulted uplift which closes the basin, and the other with continued uplift resulting in the development of gravity sliding with compressional deformation of the sedimentary fill.

The mechanism of foreland deformation is important for interpreting the origin and pattern of structures containing hydrocarbons and the distribution and patterns of unconformities that control depositional patterns and sequences. Crustal studies by COCORP and mapping of surface structural patterns suggest that the cratonic basement is involved in the development of high-angle reverse faulting in the sedimentary fill. The geometry of these faults in relation to deep crustal patterns and the stress distribution within the crust is presently unknown. However, the response pattern that is developed can be used to predict patterns of faulting and the development and timing of formation of structural traps. Similarly, the pattern of gravity sliding and the formation of compressional features is related in the depositional history.

Source rock potential is increased due to the increased depth of burial, higher heat flow associated with the marginal uplifts, and the presence of fluvial-deltaic depocenters. Basin closure provides the possibility of restriction with an increased preservation of the organic fraction in shales. In a personal communication to the author in 1979, Klemme suggested that source rock potential is approximately 120,000 barrels/mile. Possibly due to the presence of a deltaic type of sedimentary regime, the oil is primarily high gravity and dominated by paraffinic and aromatic fractions. Giant and supergiant fields are developed in large anticlinal structures and in unconformity traps. Seals are mostly shale. Locally reservoirs formed during early stages of development may be sealed by evaporites.

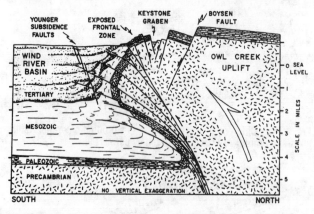

**Fig. 10–39.** Diagrammatic view of Owl Creek uplift with keystone faults and gravity slumps. (After Wise, 1963) From Lowell, 1985.

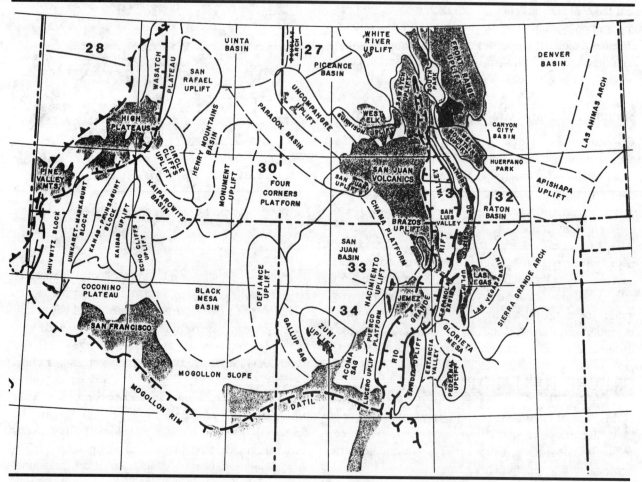

**Fig. 10–40.** San Juan basin Volcanic field. From Gries & Dyer, 1987.

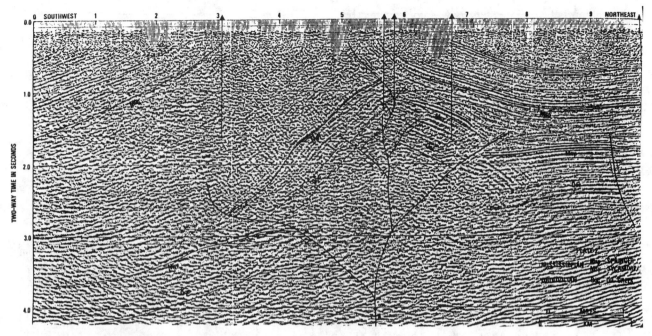

**Fig. 10–41.** Foreland basin. From Harding and Lowell, 1979, reprinted by permission.

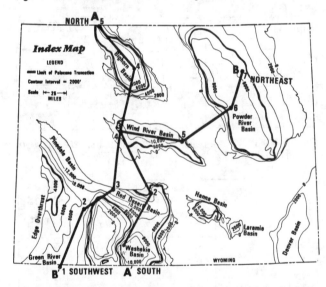

**Fig. 10–42.** Post-depositional structural history produced truncation of Paleocene-Eocene depositional cycles. *A-A'*, and *B-B'*, are lines of correlation.

## Simple Foreland Basin Example

The structural framework of the foreland area needs to be established as a boundary condition for interpreting the later structural developments. The tectonic framework must be related to preexisting structural trends and the stress distribution within the crustal plate. Often this is not possible, but Thomas (1971) suggested a wrench-fault pattern that could explain stratigraphic and structural features within the Wyoming foreland (Fig. 10–35). The structural-stratigraphic platform, upon which the system was activated, is reflected in the isopach map of the underlying Upper Cretaceous (Fig. 10–36). This illus-

trates that the basinal elements were present in southcentral Wyoming, the Hanna basin and in the prograding fan deltas of the rising Rocky Mountain system in southwestern Wyoming. The only uplift within the foreland area reflected in the isopach map is in west-central Wyoming in the Wind River uplift. This framework represented the base for later tectonic and depositional patterns. These patterns are reflected in the structural relief presently defined in the area (Sales 1968) (Fig. 10–37). The mechanics of the formation of this pattern can be suggested by a strike-slip, wrench-fault tectonic system (Fig. 10–38). High-angle reverse-fault patterns were confirmed in the sedimentary fill by field mapping, seismic cross sections, and drilling. The high angle pattern of faults marginal to uplifts has long been confirmed by both seismic and drilling data (Wise 1963) (Fig. 10–39). The overthrust of Precambrian basement ramps over marginal basinal stratigraphic intervals. The history of these structures, and the pattern of basins and foreland uplifts across the Colorado and Wyoming have been the subject of study (after Wise 1963, from Lowell 1985). The origin of vertical and/or compressive uplift is still a matter for debate. Newer information (Chapin 1985) suggests that the foreland was subject to more than 100 kilometers of strike-slip movement associated with the uplift of the Colorado Plateau. Volcanic terranes marginal to the uplift suggest a hot spot, or rising mantle plume (Gries et al. 1987) (Fig. 10–40), which could have driven the wrenching. This pattern is suppportive of the suggestions of Stone (1969) as illustrated by Thomas (1971) (Figs. 10–35 and 10–38). The wrenching style in foreland basins is illustrated by the Cement Field of the Anadarko basin (Harding and Lowell 1979) (Fig. 10–41).

Correlation between depocenters was complicated by later uplift, but sedimentary sequences are preserved in the centers of foreland basins (Fig. 10–42). Log correlation suggests two cycles of deposition commencing with fluvial-deltaic environments and terminating with deeper-water black shale units (Fig. 10–43). This depositional pattern is confirmed by the paleogeographic pattern (Fig. 10–44) and a regional cross section of sedimentary facies (Fig. 10–45). Sedimentary sequences developed within this time period have been suggested (Vail 1987) (Fig. 10–46). This pattern is confirmed by examining seismic record sections and mapping uncon-

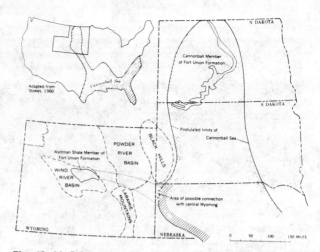

**Fig. 10-44.** Paleogeographic pattern of the Cannonball Sea. From Kiefer, 1965.

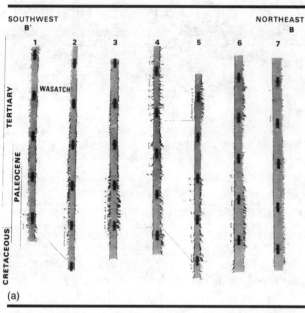

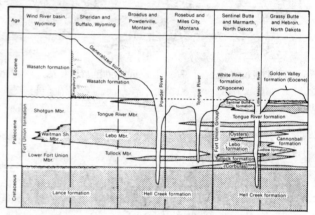

**Fig. 10-45.** Black shale pattern, North Dakota, Montana and Wyoming. From Brown, 1962.

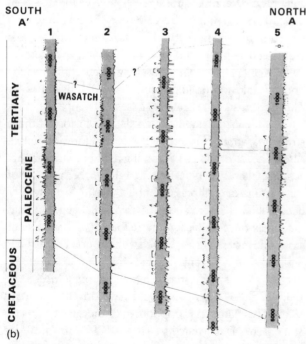

**Fig. 10–43.** *(a) Stratigraphic cross section A-A', (b)* Stratigraphic cross section *B-B'.*

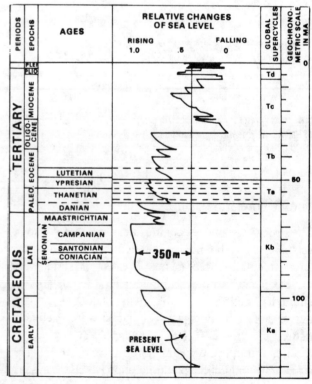

**Fig. 10–46.** Eustatic changes. From Vail, 1987.

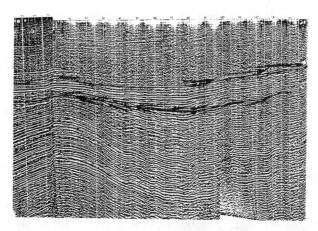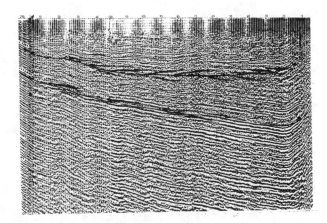

**Fig. 10–47.** *(a)* Green River basin. *(b)* Red Desert basin. Pattern of truncation and onlap are shown for Cretaceous and post-Danian unconformity surfaces. From Stearns et al., 1975.

formity surfaces by terminations of time-rock reflections (Stearns et al. 1975) (Fig. 10–47). Age relationships are based upon palynological zonation within the fluvial-deltaic depositional framework.

Paleogeographic patterns can be interpreted from a combination of well log and outcrop data. Well log and outcrop sections can be correlated and stratigraphic information synthesized (Figs. 10–48 and 10–49). All such sections can be similarly treated, and the tectonic controls for each stratigraphic sequence reconstructed from isopach maps and the paleogeographic patterns from faunal, facies, and paleocurrent studies. The basal sequence (Upper Cretaceous to mid-Paleocene) indicates the presence of an inland sea with local deltaic depocenters (Fig. 10–50). Petroleum occurrences are related to several of these foreland depocenters, but others remain to be explored. The upper sequence reflects a much stronger control in the development of foreland structural features. It reflects a paleogeographic fill of the foreland area by continental sedimentary units (Fig. 10–51). This pattern illustrates coastal areas where continuing sea-level rise led to the stacking of coastal marsh environments—coals more than 40 meters thick are developed—and the development of permeable riverine channels that drain uplifted crustal blocks. These channels are the site for the precipitation of uranium and thorium minerals of commercial significance.

Stratigraphic analysis of foreland basins can be constructed in other areas and basins using this approach. The stratigraphic record shows response patterns that are controlled by the tectonic history and stratigraphic boundary conditions.

## Deformed Foreland Basins

Complete foreland depositional systems may undergo more complex tectonic histories. Depocenters may reflect shelf progradation and development followed by uplift of the outer shelf area and formation of a foreland depocenter on the cratonic interior. This sequence pattern may involve sedimentary thickness of more than 10,000 meters. Restricted basins may be developed during shelf progradation and during the uplift of the outer shelf.

The intrashelf and intracratonic basins are topographic basins. Restriction due to driving subsidence can lead to the accumulation of thick biogenic and prodeltaic shales with source rock potential. The Alberta basin suggests such a framework (Fig. 10–52). Uplift that closed the basin produced topographic relief that caused pressure gradients (Fig. 10–53). The presence of unconformities that extend to the bottom of the foredeep can be documented by seismic and stratigraphic data (Fig. 10–53). The basin becomes a dynamic system of water influx, hydrocarbon expulsion from overpressured source rocks, migration of fluids through permeable unconformity sandstones, which lead to hydrocarbon accumulations reflecting changes in the relative permeability of oil and water in structures and stratigraphic traps. This mechanism can account for more than one trillion barrels of oil accumulation in low-pressure areas marginal to the foredeep depocenter. A second tectonic stage may result in increased burial depths, increased pressure gradients, and formation of large structures near the center of the depocenter. This stage is related to continued uplift of the basin margin and leads to the development of gravity sliding on surfaces of lowered shear strength. Such surfaces are developed in response to increased heat flow, reduction of permeability in massive shale intervals, continuing subsidence due to sedimentary and tectonic loading, and increased pore pressures. Increased pore pressures cause the development of zones of low shear strength and produce gravity slide patterns (Mudge 1970) (Figs. 10–54 and 10–55).

Improved seismic processing and drilling to confirm a seismic interpretation has established the overthrust

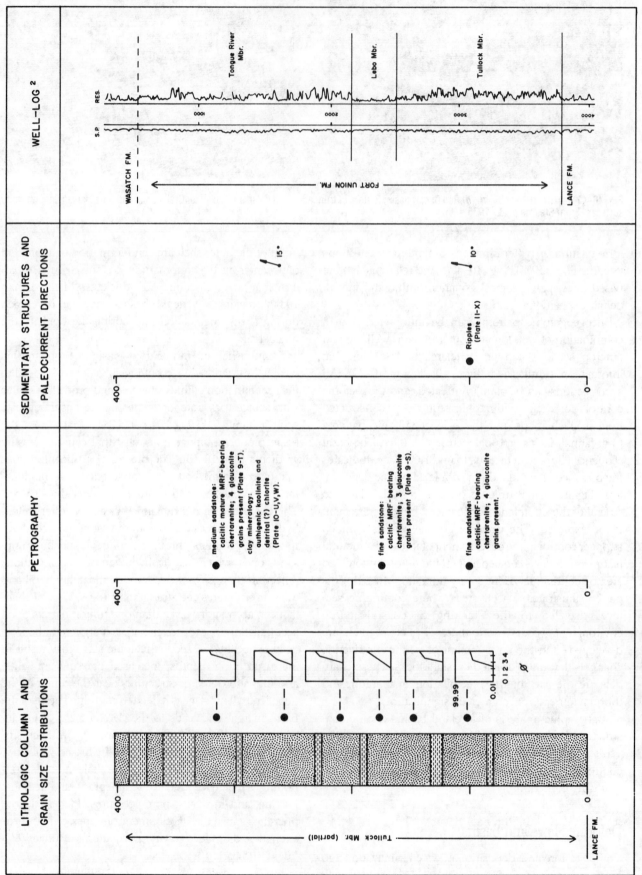

**Fig. 10–48.** Moorcroft section, Powder River basin. Reprinted courtesy R. Lyday.

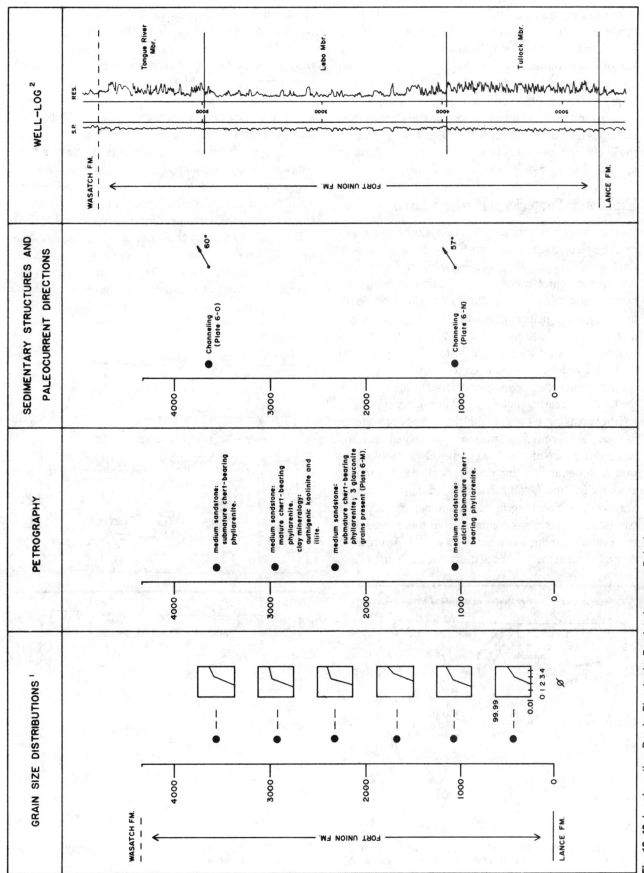

**Fig. 10-49.** Lynch section, Powder River basin. Reprinted courtesy R. Lyday.

structural pattern. Many possible structural solutions are possible, but with drilling, an interpretation to the depth of the drilling may be confirmed (Williams and Dixon 1985) (Figs. 10–56 and 10–57). The preparation of an interpreted crustal section, however, is still subjective without well control as a basis for processing and interpreting the seismic information (Fig. 10–58). This section shows the thrust plane occurs at a higher stratigraphic interval, and there is little evidence that the Darby thrust underlies the Absaroka thrust (Fig. 10–57) in the area of the Ryckman Creek and Painter Reservoir fields (Frank et al. 1982) (Fig. 10–59).

## Exploration Strategies in Foreland Basins

Patterns of unconformities and depositional sequences must be recognized. These provide the basis for correlation, interpretation of paleogeographic histories, and patterns of source rocks and permeable pathways for hydrocarbon migration. Structural features can be identified by surface mapping, gravity and magnetic maps, migration patterns from salinity, and pressure patterns within the basin. Temperature history is important to determine which units have been source rocks. Studies of the maturation stage possibly could lead to an estimate of the total hydrocarbon potential in a frontier basin.

Other areas in the world with these types of patterns include the western Siberian basin, the Volga basin, the Anadarko basin of Oklahoma, the Permian basin of western Texas, and the Erg basins of North Africa. Frontier areas include the Arctic Islands of Canada, the foreland area south of Spitsburgen, and a number of Australian basins. Exploration has commenced in these areas with hydrocarbon production established in the Devonian of Australia and in the Jurassic of the Canadian Arctic.

## CRATONIC EXTENSIONAL BASINS

The causality of cratonic extension is still under debate. Extensional basins occur as successor basins overlying orogenic zones; they occur in areas of rifted triple-point spreading centers; and they are associated with wrench-fault tectonic patterns. In each of these instances, the initial stages of basin development reflect higher heat flows and typically the presence of basaltic intrusions and flows. It suggests that the initiation of spreading is due to vertical uplift in response to a high temperature mantle plume. Rifting would be in response to gravity sliding away from the central uplift. Movement could be radial, producing a triple-point spreading center or linear, producing intracratonic graben patterns.

The basinal geometry must reflect the geometry of the uplift, the history of wrench-fault tectonics, and the direction and duration of spreading. The aulocogen is the primary structural style preserved on the cratonic plate.

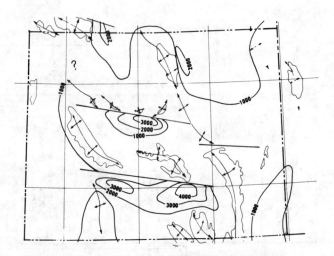

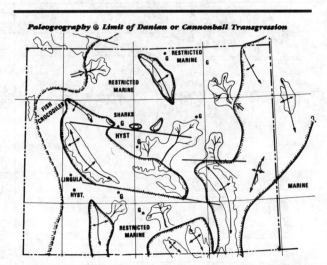

**Fig. 10–50.** (a) Isopach at limit of Danian or Cannonball transgression. (b) Paleogeography at limit of Danian or Cannonball transgression.

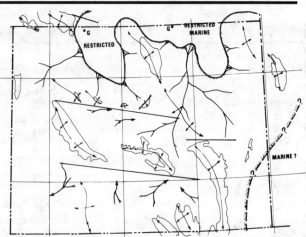

**Fig. 10–51.** (a) Isopach for upper Tongue River transgression. (b) Paleogeography for upper Tongue River transgression.

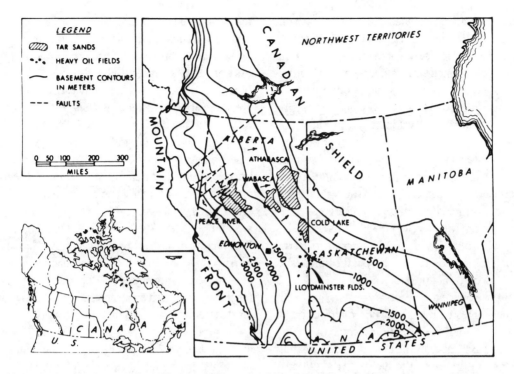

**Fig. 10-52.** Alberta basin gathering area. From Momper, 1978, reprinted by permission.

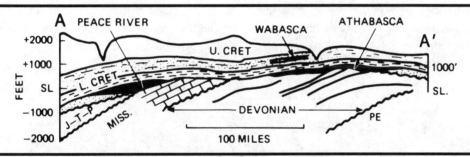

**Fig. 10-53.** Diagrammatic east-west cross section showing geological setting of Lower Cretaceous oil sands, Athabasca-Peace River. From Hitchon, 1969.

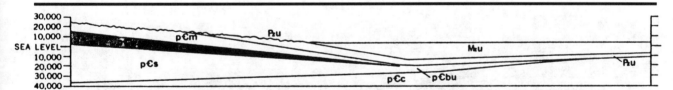

EVENT B.—End of Mesozoic sedimentation. Uplift to the west began during Late Jurassic, and continued periodically with some igneous activity through the Cretaceous. The thickest Mesozoic sediments accumulated in a narrow elongate basin along the east side of the highland

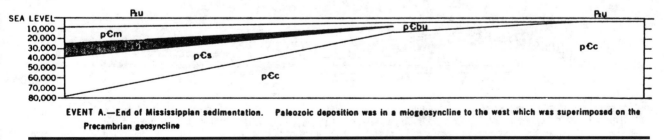

EVENT A.—End of Mississippian sedimentation. Paleozoic deposition was in a miogeosyncline to the west which was superimposed on the Precambrian geosyncline

**Fig. 10-54.** Diagrammatic reconstruction of some events before, during, and after the early Tertiary orogeny. Horizontal scale is the same as the vertical scale. From Mudge, 1970.

These basins contain mostly Mesozoic and Tertiary fill, but there are Paleozoic basins of this type. The Anadarko basin of Oklahoma reflects some of these aspects during early stages of development and fill. The internal structure reflects differentiation of horst blocks and grabens with driving subsidence. They are topographic basins through the early stages of development and fill. These basins are restricted and have an early history of salt deposition and later periods of biogenic shale deposition. Some may be open to the ocean, but depocenters occur on cratonic crust. Both carbonate and clastic sedimentary fill are developed, and both are excellent reservoirs.

Unconformities are well developed within these basins in response to sea-level falls. Depositional patterns produce stratigraphic traps in the center of the basin, and patterns of onlap across both depositional mounds and structural uplifts produce the necessary seals.

Klemme has suggested the worldwide distribution of basins of this type (Fig. 10–60). He indicates there is a wide distribution of these cratonic basins, and more than a dozen basins are presently producing hydrocarbons. Some overlap in classification is indicated due to lack of understanding of stratigraphic and structural patterns and causality of basin formation. Hydrocarbon productivity is 140,000 barrels/mile, and there are basins with

reserves in excess of 20 billion barrels, e.g., North Sea and Sirte basins. Oil generation is in deep areas between horst blocks, and migration occurs along fractures or unconformity surfaces. Oil can be waxy and low gravity (Reconcavo basin) or high gravity paraffinic (Sirte and North Sea basins) (Klemme 1983).

Basin history has been characterized by Harding (1984) by an initial extension followed by an early depositional history, tilting of stratigraphic sequences followed by unconformities, and finally a broad sag develops which extends beyond the initial extensional limits of the basin (Fig. 10–61). History may change in response to the amount of extension, which has been characterized by Beta factors of spreading, indicated by [1], or 100%, to as much as [4] (McKenzie 1978). If spreading exceeds [4], the cratonic crust ruptures, and an ocean basin develops, resulting in a passive cratonic marginal basin, with a differing structural and stratigraphic history.

A simple example of a spreading center with limited amount extension perpendicular to the craton margin is illustrated by the Reconcavo basin of the Brazilian continental margin (Ghignone and de Andrade 1970) (Fig. 10–62). A typical pattern of blanket fluvial sedimentation on the floor of the extensional basin, with marginal fan deltas on the faulted margin, with a central uplift

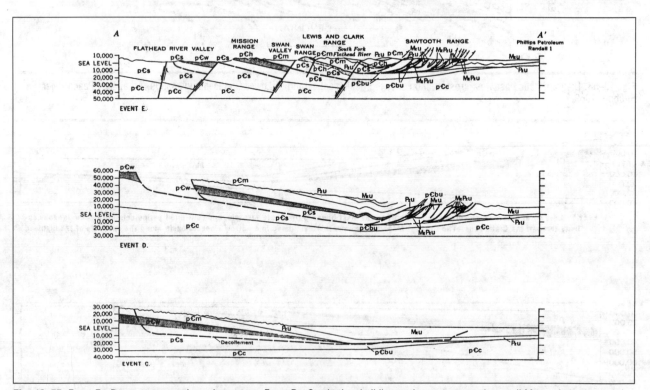

**Fig. 10–55.** Event E—Present topography and structure. Event D—Gravitational gliding to the east across the small Mesozoic basin, creating folds at the toe and piling one thrust plate upon another. The gap at the pull away zone is strictly diagrammatic. The amount of uplift shown is the maximum that may have occurred during this period. Event C—Renewed uplift of the western area during early Paleocene with continued erosion. Abnormal fluid pressures very likely at depths as much as 25,000 feet where the decollement formed. From Mudge, 1970.

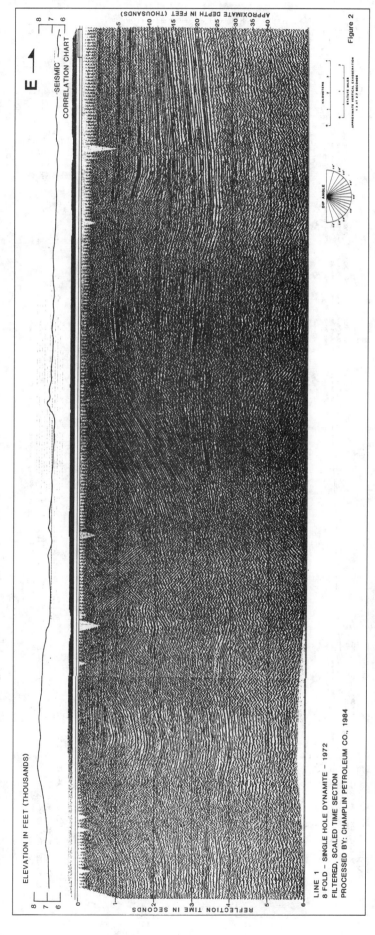

**Fig. 10-56.** Seismic section, western Wyoming overthrust. From William and Dixon, 1985.

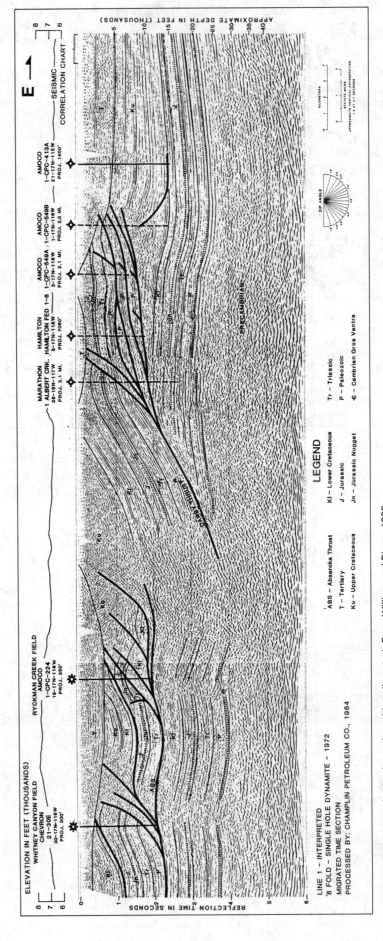

**Fig. 10–57.** Interpreted seismic section with well control. From William and Dixon, 1985.

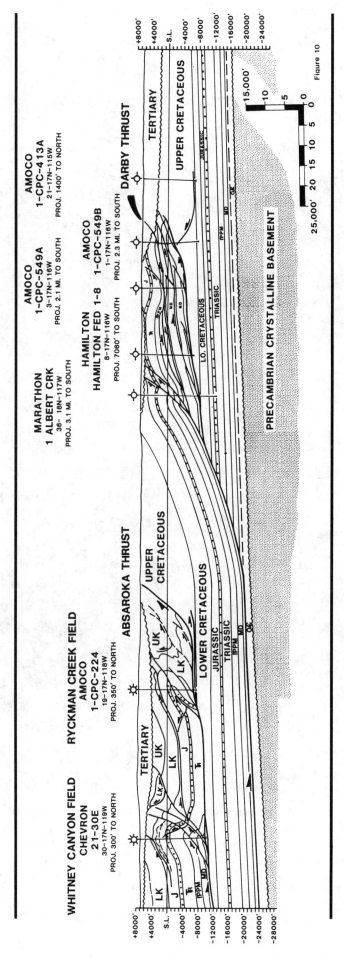

**Fig. 10-58.** Overthrust model. Williams and Dixon, 1985.

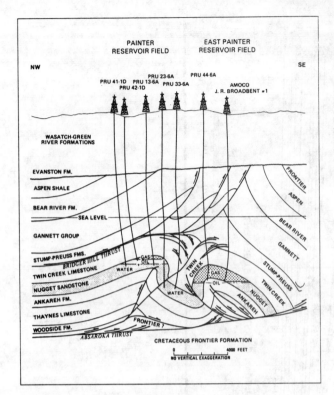

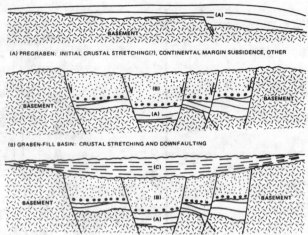

**Fig. 10–61.** From Harding, 1984, reprinted by permission.

**Fig. 10–59.** Northwest-southeast structure section through the Painter Reservoir and East Painter Reservoir fields. In these and other Wyoming-Utah thrust-belt fields it is critical that reservoir rocks in the upper plate of the Absaroka thrust be in contact with subthrust Cretaceous rocks that are the source of oil and gas. After Frank et al., 1982 From Lowell, 1979.

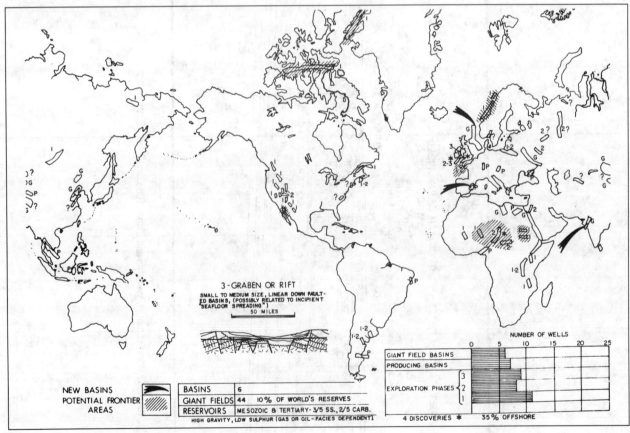

**Fig. 10–60.** Cratonic basins. Reprinted courtesy H.D. Klemme.

remaining after the principal rifting period, is developed. This basin reflects minimum extension. No basaltic flows were initiated by crustal thinning, and the depositional history was limited to progradational fill of the rift valley (Fig. 10–63). Driving subsidence was reflected in deeper water fan and fluidized sediment deformation in the Ilhas member near the spreading center. Production is from the basal fluvial Sergi formation and from lower Cretaceous deltaic and prodeltaic clastic intervals. The depositional history is illustrated by the dip section down the basin axis (Fig. 10–63). The southern portion of the basin is filled with gravity-transported fine-grained sandstone deposited in a mounded fan on the floor of the topographic basin. Growth faulting is indicated by micropaleontologic correlations. Production in the basin is from all sandstone intervals, with more than one billion barrels of recoverable hydrocarbon reserves. Exploration is for both structural and stratigraphic traps. Seals are from prograding prodelta shales, and source rocks are from onlapping deep basin shales present in the center of

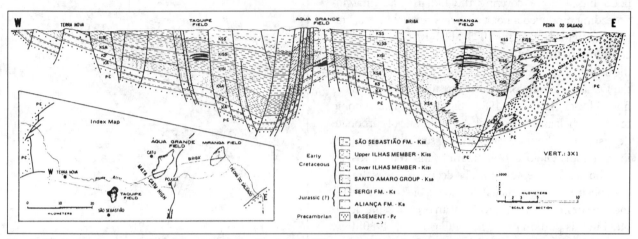

**Fig. 10–62.** Regional cosss section of Reconcãvo basin, Brazil. After Ghignone and dé Andrade, 1970, fig. 4). From Lowell, 1985.

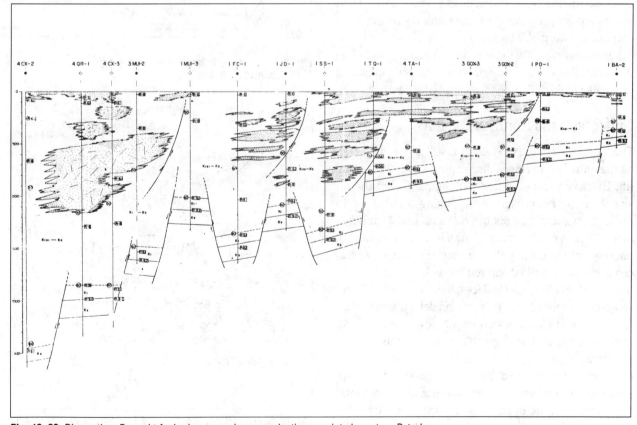

**Fig. 10–63.** Dip section. From dé Andrade personal communication, reprinted courtesy Petrobras.

the restricted basin. The source rock reflects a continental provenance and a restricted marine basin, producing a Type 1 source rock with the kerogen dominantly of a terrestrial and algal origin, resulting in a very waxy crude.

## Cambay Trough and Basin

In a larger basin initiated at a triple point spreading center, the Cambay rift valley is an aulocogen ocurring perpendicular to the Indian plate. The basin history started with latest Cretaceous rifting and spreading, with flood basalt flows covering the Indian plate margin. Spreading produced a horst-graben structure on the cratonic margin (Chowdary 1975) (Fig. 10–64). Spreading continued to the southwest, and resulted in formation of the Indian ocean. An intrashelf basin formed, resulting in the Dahanu depression, flanked by uplifted and rotated horst blocks (Basu et al. 1980) (Fig. 10–65). The cratonic marginal shelf and horst blocks were the site of carbonate deposition, and during low stands of sea level, especially in the middle Oligocene, submarine fans were deposited into the Dahanu depression. The deep basin was the site of deposition of more than 2500 meters of sediment, with the development of source rocks, with overpressure developed in this portion of the basin (Basu et al. 1980) (Fig. 10–66). Overpressure caused fluid expulsion and migration of hydrocarbons up into Eocene carbonate reservoirs developed on the crest of horst blocks. The Bombay High is a supergiant hydrocarbon accumulation with potential ultimate recovery of 2 billion barrels of oil. Other major fields are associated with carbonate reservoirs, including the South Bassein and Bassein structural accumulations on smaller faulted highs to the south of the Bombay High.

## Gulf of Suez and North African Rifts

The Red Sea and Gulf of Suez rift is still active, with high temperature saline brines discharging along the central rift. Extension has produced faulting parallel to the rift axis (Harding 1984) (Fig. 10–67). Rifting commenced in the Upper Cretaceous, but the rift valley has been punctuated by synsedimentary faulting and a series of unconformities during the Tertiary. Pre-Upper Cretaceous clastics are tilted and remain as high horst blocks (Harding 1984) (Fig. 10–68). Horst blocks provided Cretaceous clastics for alluvial fans and fan deltas, and Tertiary carbonate and clastic reservoirs are developed as fault, unconformity, and stratigraphic traps on the crest of structural highs (Fig. 10–69). Seals are associated with onlapping fine-grained sediments associated with sequence unconformities, and the sea level lowstand associated with the upper Miocene has produced a salt sequence within the rift basin (Fig. 10–70). Source rocks

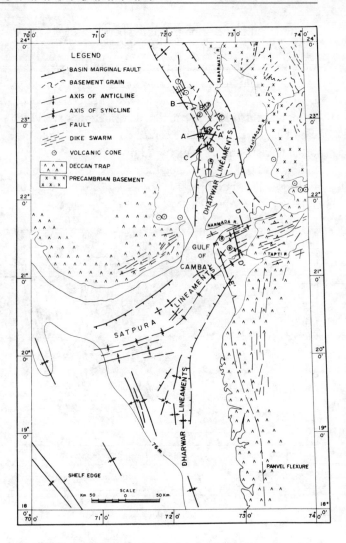

**Fig. 10-64.** Tectonic lineaments in Cambay basin. From Chowdary, 1975, reprinted by permission.

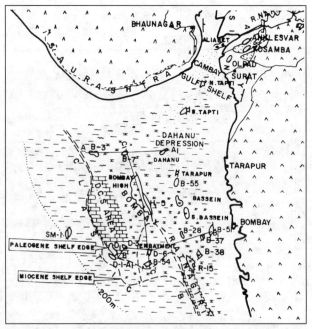

**Fig. 10-65.** From Basu, Banerjee, and Tamhane, 1980, reprinted by permission.

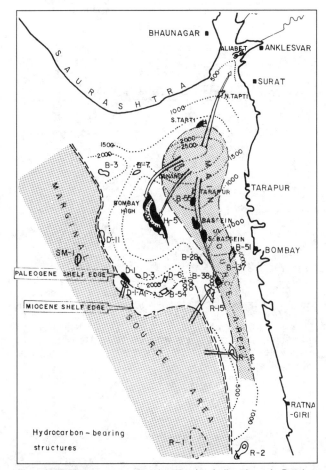

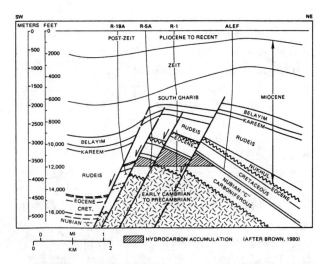

Fig. 10-66. Source area migration trends of oil and gas in Bombay offshore basin India. From Basu, Banerjee, and Tamhane, 1980, reprinted by permission.

Fig. 10-68. Cross section across Ramadan oil field, Gulf of Suez, showing hydrocarbon trap in pre-graben-fill reservoirs (Nubian "C" sands) that is controlled by graben-age normal faults. Associated flexing forms narrow rollover at high side of block boundary and broader syncline in low side. From Harding, 1984, reprinted by permission.

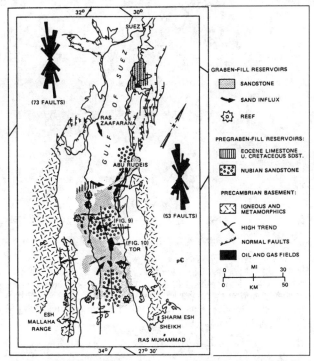

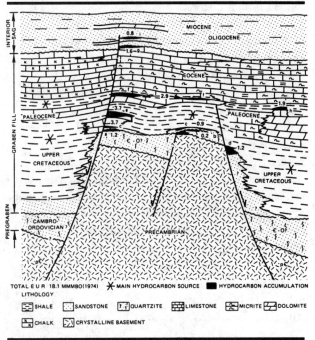

Fig. 10-67. Oil fields and distribution of productive zones in Gulf of Suez. Numbers of surface faults with orientations within each 5° geographic quadrant are plotted in rose diagrams for northeast and southwest margins of graben. From Harding, 1984, reprinted by permission.

Fig. 10-69. Schematic diagram of cross-sectional character of traps and stratigraphic level and lithology of hydrocarbon-producing reservoirs in northwest portion of Sirte basin, Libya. Succession of basin types is indicated at left margin. Estimated ultimate recoverable reserves (EUR) are for total liquids and are stated in billions of barrels of oil (MMMBO). From Harding, 1984, reprinted by permission.

occur in Upper Cretaceous to Miocene sedimentary sequences flanking uplifts, and migration is principally associated with faults and unconformities (Fig. 10–70). Similar depositional histories have been developed for fields in the Western Desert of Egypt and in Libya. Pre-Upper Cretaceous clastics are productive from faulted structures in both basins. The Libyan extensional basin was the site of continued sedimentary fill which resulted in Paleocene and Eocene carbonate reservoir sequences with hydrocarbon accumulations on the updip flank of tilted horst blocks, and Oligocene clastics in the supergiant Gialo field developing on the margin of the basin (Harding 1984) (Fig. 10–71). Source rocks were deposited in troughs adjacent to uplifts and provided the source for more than 3 billion barrels of recoverable oil in the basin.

## North Sea Grabens

The post-Hercynian structure provided the framework for extensional centers and rifts. Major wrench faults separated contrasting terranes accreted during the pre-Permian tectonic history (Ziegler 1978) (Fig. 10–72). A map of Permian and Mesozoic tectonic elements reflects this Hercynian structural framework (Fig. 10–73). Basins are indicated by an isopachous map of the interval from the Hercynian unconformity to the Holocene (Fig. 10–74). North Sea structure also reflects the complex history of basinal development and fill, with extension producing tilted horst and graben structures, deep topographic basins, and relatively high platform areas (Harding 1984) (Fig. 10–75). A similar pattern of synsedimentary faulting, regional unconformities, and the formation of a broad sag during the Tertiary is developed. The tectonic history of northwestern Europe illustrates the pattern of unconformities, tectonic events, and stratigraphic sequences (Ziegler 1978) (Fig. 10–76). Vail's sea-level curve is shown for comparison, with unconformities developed at times of sea-level fall. Correlation of tectonic events and depositional history is of prime importance to the interpretation of the structural history of North Sea basins. The origin of specific structures is still open to discussion. An example from the north German salt basin illustrates the development of stable platform structures, salt cored anticlinal structures, and wrench-faulting (Ziegler 1983) (Fig. 10–77). Due to the presence of salt deposits throughout the central basinal area of the North Sea graben, it often is not possible to obtain seismic information at the level of the pre-Permian basement to determine its involvement in the development of anticlines and horst blocks. Wrenching along strike-slip faults separating differing terranes illustrated on the post-Hercynian basement map (Fig. 10–72) has been observed in outcrop and is probable within basinal areas.

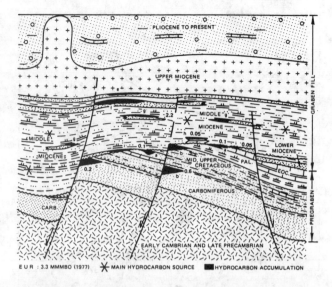

EUR : 3.3 MMMBO (1977)   ✳ MAIN HYDROCARBON SOURCE   ■ HYDROCARBON ACCUMULATION

**Fig. 10–70.** Schematic diagram of major hydrocarbon accumulations and their estimated ultimate recoverable reserves (*EUR*) in Gulf of Suez, Egypt. Trap types and reservoir lithologic characteristics have been synthesized from published data (see text). Hydrocarbon reserves are for liquids only. Trap types discussed in text indicated by *a-c*. Salt indicated by + , marl by-. From Harding, 1984, reprinted by permission.

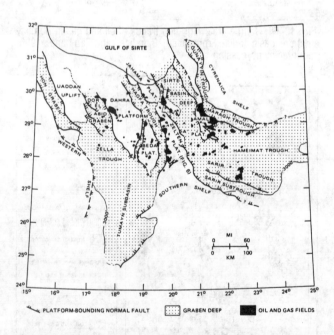

⌐ PLATFORM-BOUNDING NORMAL FAULT   ▣ GRABEN DEEP   ■ OIL AND GAS FIELDS

**Fig. 10–71.** Major tectonic elements Sirte basin, Libya. Five major northwest-trending grabens have formed an unusually wide basin complex in northwest portion of basin and provide several sites for maturation of hydrocarbons. From Harding, 1984, reprinted by permission.

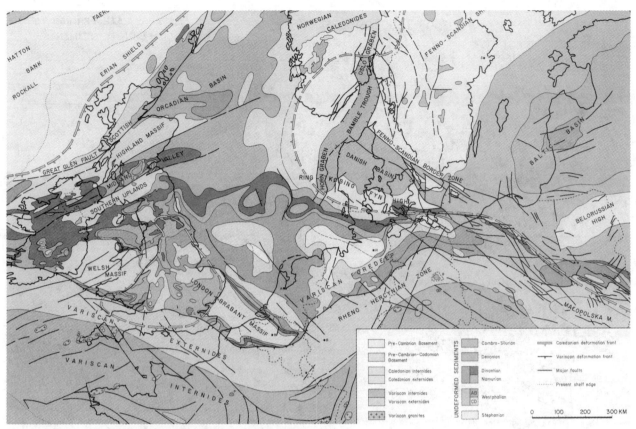

**Fig. 10-72.** Post Hercynian tectonic framework. From Ziegler, 1978.

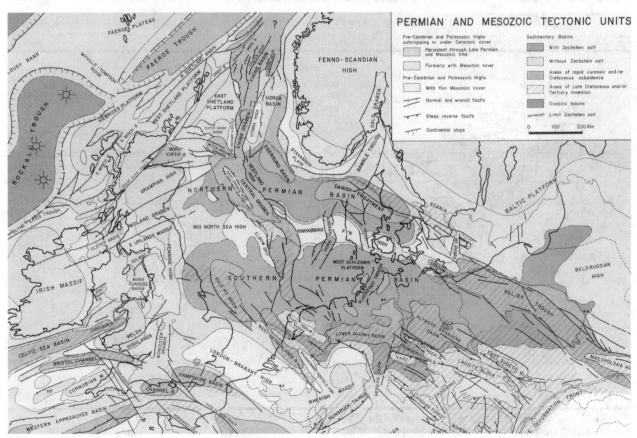

**Fig. 10-73.** Axes of inverted basins of northwest Europe with Upper Cretaceous paleogeography; Upper Cretaceous has been eroded from inversion structures. Inversion is also known west of U.K. and north of S. German high and Bohemian massif. From Ziegler, 1978.

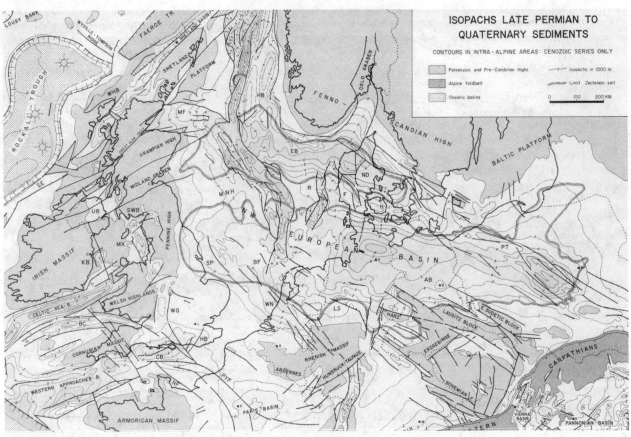

**Fig. 10–74.** Basinal areas derived from a post-depositional isopachous map. From Ziegler, 1978.

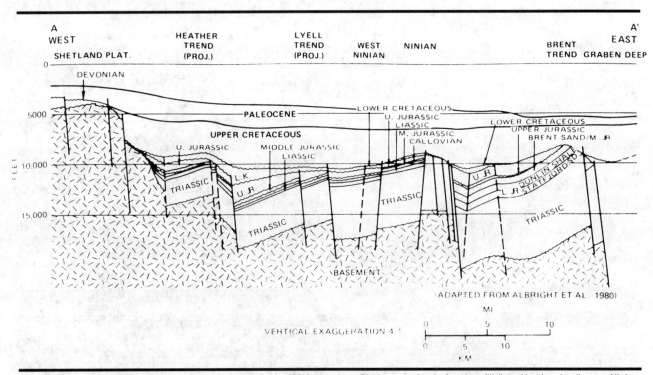

**Fig. 10–75.** East-West cross section across northwest flank of Viking graben. Platforms at level of graben fill (i.e., Heather, Lyell, west Ninian, Ninian, and Brent) are rotated westward into large normal faults located at each platform's western boundary. Structural relief within lower sag fill has flexure stype expressed by Upper-Lower Cretaceous contact. After Harding, 1984, reprinted by permission.

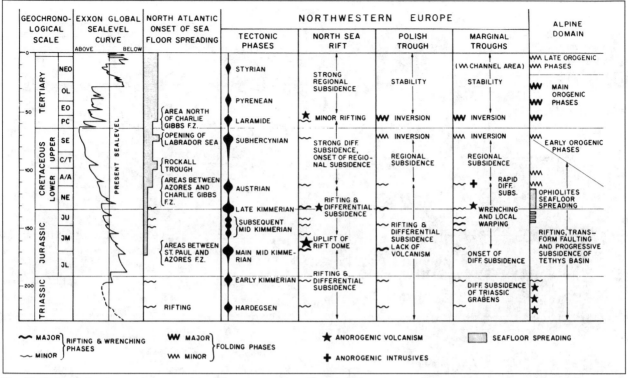

**Fig. 10-76.** Unconformity patterns, tectonics, and sequences. From Ziegler, 1978.

A series of isopachous and paleogeographic maps have been published by Ziegler (1978). These maps illustrate the depositional and structural history of the North Sea basinal area. The earliest map shows the paleogeographic and isopachous framework of the basal Permian Rotliegend formation (Fig. 10–78). At this stratigraphic level, basinal subsidence is indicated, but little evidence is developed for rifting. Coaser clastics equivalent to the Salt interval are reservoir rocks, with a complex depositional history ranging from continental to marine. Subsequent stratigraphic intervals in the Permian and Triassic indicate the continued development of salt deposition, but again with little evidence for rifting. By later Jurassic, isopachous and facies patterns suggest the beginning of active rifting (Fig. 10–79). Marginal clastic shelves flank the beginning development of the Viking Graben and the Danish-Polish furrow. Lower Cretaceous deposition reflects continued subsidence and rifting (Fig. 10–80). By Upper Cretaceous time a broad deep basin is developed, dominated by deposition of pelagic carbonates (Fig. 10–81). The Tertiary was a period of fill by progradational deltaic and submarine fan units (Fig. 10–82). These formed supergiant hydrocarbon accumulations, with the Paleocene Forties field having recoverable reserves of 2 billion barrels, and the Eocene Frigg field expected to produce in excess of 7 trillion cubic feet of gas (Berg 1982) (Fig. 10–83).

Unconformity patterns, rifting and subsidence, rotation of fault blocks, and the patterns of depositional onlap all led to the development of patterns of petroleum migration (Harding 1984) (Fig. 10–84). The cause of high-angle faulting is related to extension, but as in other aulocogens it appears to be basement controlled with some strike-slip faulting (Williams et al. 1975) (Fig. 10–85). Permeable pathways associated with faults and unconfonformities interrelate areas of overpressured source rocks to structurally high lower-pressured reservoirs. The principal source rock is the uppermost Jurassic Kimmeridgian biogenic shale, which onlapped upper Jurassic reservoirs on faulted anticlinal structures. Continued rifting, synsedimentary basement faulting, and unconformities all led to the accumulations of hydrocarbons in Cretaceous and Tertiary reservoirs.

## Summary

As can be seen in these examples, tectonic-depositional systems are interrelated. Causality for subsidence can be attributed to low-density depositional units and lithosphere thinning. Horst blocks are tilted in response to contined extension, and basinal areas undergo driving subsidence in response to changes in heatflow, lower rates of sedimentary fill, and to extension. Wrenching may produce local compressive uplifts or inversions

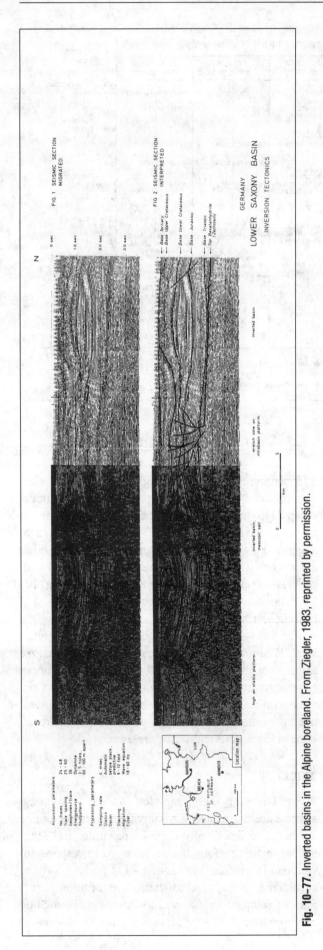

**Fig. 10–77.** Inverted basins in the Alpine boreland. From Ziegler, 1983, reprinted by permission.

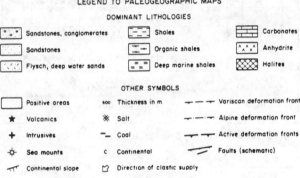

**Fig. 10-78.** Basal fill of post-Hercynian surface—clastic and evaporite basin. From Ziegler, 1978.

### LEGEND TO PALEOGEOGRAPHIC MAPS

#### DOMINANT LITHOLOGIES

| | | |
|---|---|---|
| Sandstones, conglomerates | Shales | Carbonates |
| Sandstones | Organic shales | Anhydrite |
| Flysch, deep water sands | Deep marine shales | Halites |

#### OTHER SYMBOLS

| | | | |
|---|---|---|---|
| Positive areas | 500 Thickness in m. | — — Variscan deformation front |
| ★ Volcanics | ✳ Salt | Alpine deformation front |
| + Intrusives | — Coal | Active deformation fronts |
| ✳ Sea mounts | c Continental | Faults (schematic) |
| Continental slope | Direction of clastic supply | |

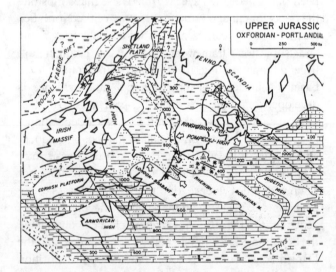

**Fig. 10-79.** Paleogeography of northwestern Europe during the Upper Jurassic. From Ziegler, 1978.

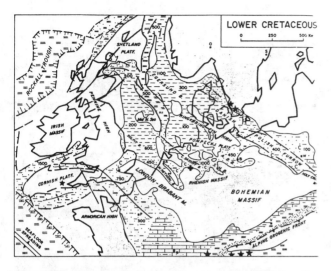

**Fig. 10-80.** Paleogeography of northwestern Europe during the Lower Cretaceous. From Ziegler, 1978.

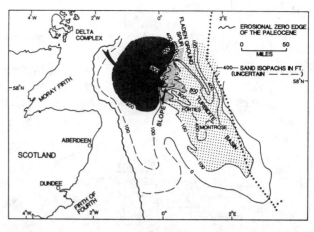

**Fig. 10-83.** Isopach of Paleocene sands and depositional environments, central North Sea. From Berg, 1982, reprinted by permission.

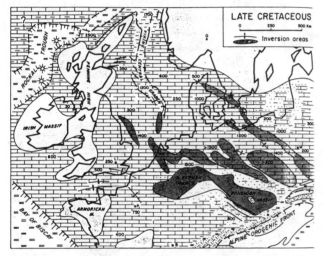

**Fig. 10-81.** Late Cretaceous carbonate patterns. From Ziegler, 1978.

**Fig. 10-82.** Tertiary patterns of subsidence and fill. From Ziegler, 1978.

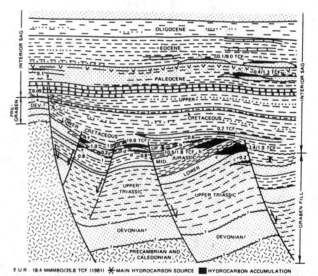

**Fig. 10-84.** Schematic diagram of major hydrocarbon accumulations and their estimated ultimate recoverable reserves (*EUR*) in Viking graben, North Sea. Reserves include gas stated in trillions of cubic feet (*TCF*). Trap types discused in text indicated by *a-h*. Several traps related to salt flowage at far south end of graben are excluded. Volcanic tuff indicated by *v*. From Harding, 1984, reprinted by permission.

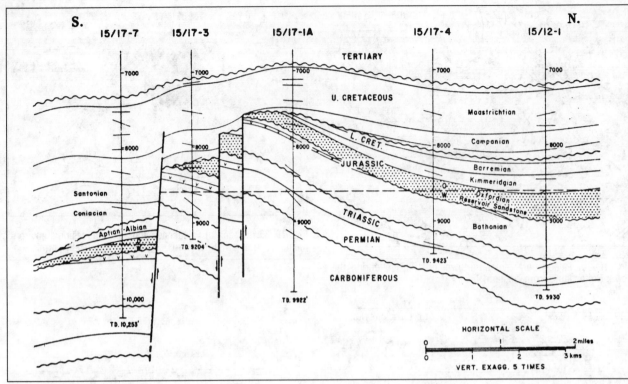

**Fig. 10-85.** Piper field illustrating stratigraphic and structural history. From Williams, Conner, and Peterson, 1975, reprinted by permission.

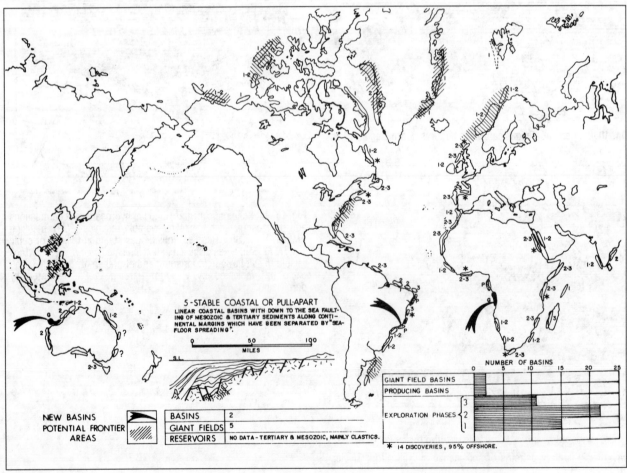

**Fig. 10-86.** Continental margin extensional basin. Personal communication. Reprinted courtesy H.D. Klemme, 1979.

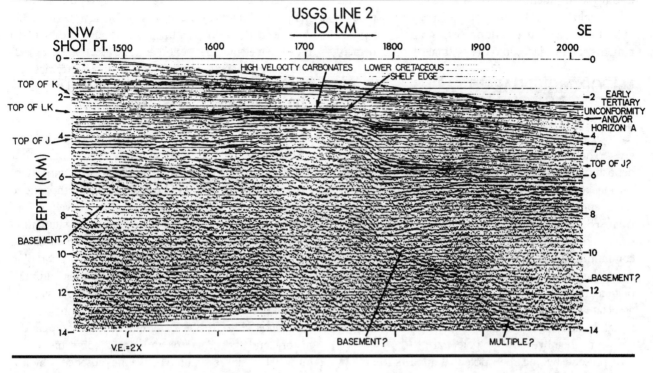

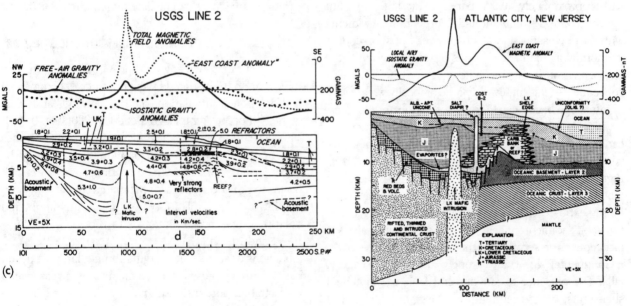

**Fig. 10-87.** (a) Depth section for line 2 used velocity analyses every 1 km. The tops of the Upper Cretaceous and Lower Cretaceous at SPN 1700 are at 1.8 and 2.6 km, respectively. (b) Velocity sections summarize interval velocity data sampled every 3 km along profiles and then averaged over 2050-km wide bands. (c)Schematic interpreted cross section along CDP line 2 through wide ocean-continent transition zone associated with the Baltimore Canyon trough off New Jersey. From Grow, Mattlock, and Schlee, 1979, reprinted by permission.

during the subsidence history. Most basins, therefore, are self-actualizing depositional receptor areas with a predictable history of deposition. Stratigraphic response patterns can be placed into this basinal tectonic theme.

## MARGINAL PULL-APART BASINS

Extensional tectonic patterns may extend across the shelf and involve the continental rise and oceanic crust. The change in crustal thickness and responses to depositional load are important differences for hydrocarbon accumulations. These basins may evolve from rifted cratonic basins and result in cratonic marginal basins. Separation of this type of basin from the cratonic extensional basin type is often difficult. In a personal communication in 1979, Klemme suggested that these basins are common along passive or pull-apart cratonic plate margins (Fig. 10–86). Only a few giant oil fields are associated with this basin type, but more than 48 additional basins have not been sufficiently explored to determine their hydrocarbon potential.

At present, hydrocarbon potential appears to be low, with approximately 40,000 barrels/cubic miles in giant hydrocarbon accumulations. Temperature gradients within these basins start at a high level, but soon cool to subnormal as sedimentary units prograde onto the abyssal plain. Source organic material is from the continent; consequently, it is highly gas prone or locally waxy. Only in intraslope basins can there be restriction and the preservation of marine biogenic organic detritus. Reservoirs can be either carbonate or clastic depending upon provenance patterns. Typically, these basins do not have high depositional rates, and the sedimentary sequence tends to be thin. In addition, continental slopes are commonly exposed to oceanic currents, and long residence times for sedimentary deposits at the depositional interface reduces preservation of the organic fraction.

Continental slopes are strongly affected by sea-level falls. The presence of canyons and fans, resedimented shelf units, contour currents resulting in truncation of deposited sequences, and gravity slumping, all lead to complex unconformity patterns in preserved stratigraphic sequences.

Hydrocarbon generation and migration must be from continental-rise sedimentary sequences exposed to these processes or must occur in intraslope basins.

### North Atlantic Coastal Margin

Continental margin exploration requires synthesis of stratigraphic and geophysical information. The history of basinal extension and subsidence is reflected in stratigraphic response patterns. The interpretation of seismic profiles is aided by velocity analysis, and correlation to magnetic and gravity surveys (Grow et al. 1979) (Fig. 10–87). As can be seen from this figure it often is difficult to pick the basement reflection and the location of the cratonic margin, and to recognize older unconformities and the geometry of depositional sequences. Drilling of exploratory stratigraphic test wells aids in the interpretation of shallower reflection patterns. Much of the sea-level history that has been documented by Vail and co-workers has been established by study of cratonic margins, but variations in heat flow, history of cratonic margin extension, rates and history of deposition, and lithospheric changes all are important controls for basin development. Recent wells in the area of the Baltimore submarine canyon have established the presence of the Mesozoic shelf edge reef development indicated by the stratigraphic interpretation (Fig. 10–87c). Often many wells and multiple seismic profiles are required before the history of basinal development and its petroleum potential can be ascertained (Fig. 10–88). Determining basin temperature history and the geometry and stratigraphic history of intrashelf basins is required before the presence of hydrocarbon source rocks can be determined (Fig. 10–89).

### Other Examples of Extensional Cratonic Margins

An example of a shelf edge carbonate buildup at the cratonic margin is illustrated by the presence of commercial hydrocarbon accumulations off the Palawan Islands west of the Philippine archipelago (Beddoes 1980) (Fig. 10–90). This Tertiary example is significant in that it illustrates that if intrashelf basins are developed, there may be unconformity surfaces extending down into intrashelf basins to provide avenues for petroleum migration, stratigraphic traps available for petroleum accumulation, and onlapping fine grained intervals to provide seals (Figs. 10–91 and 10–92).

If extensional cratonic margins contain large intrashelf basins with the presence of sediments that are associated with stratigraphic intervals that are dominated by biogenic sediments, these settings can be the site for giant hydrocarbon accumulations. Such is the case for Jeanne D'Arc basin off eastern Canada. This basin is more than 10 kilometers long, and 6 kilometers wide, with up to 14 kilometers of sedimentary fill (Grant et al. 1986) (Figs. 10–93 and 10–94). An interpreted seismic profile and a well section illustrates the extensional nature of the structure with both listric and antithetic faulting (Arthur et al. 1982) (Figs. 10–95a and 10–95b). Possibly of even greater importance is the the stratigraphic history, with the development of biogenic upper Jurassic sediments within the basin, which are important source rocks throughout the world (Fig. 10–96).

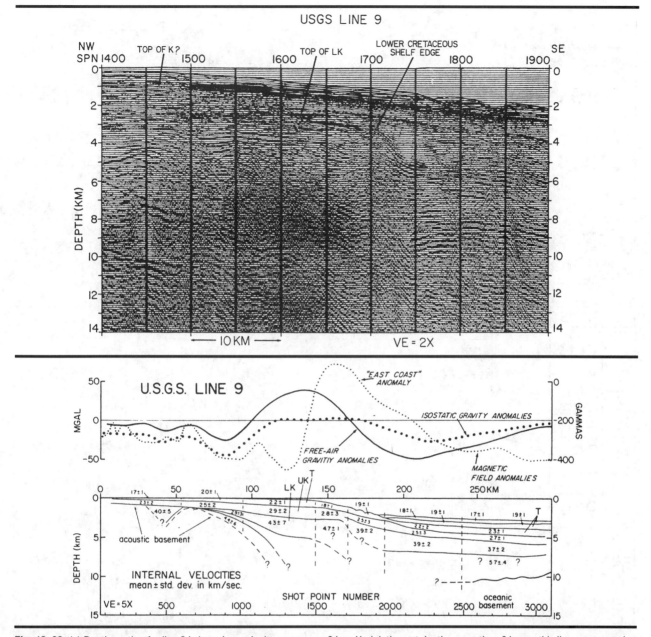

**Fig. 10–88.** (a) Depth section for line 9 is based on velocity scan every 3 km. Undulations at depths more than 6 km on this line are somewhat influenced by noisy data and scatter in the velocity valves. The Lower Cretaceous shelf edge is probably about 3 km deep at SPN 1700. (b) Geophysical interpretation, USGS Line 9. From Grow, Mattlock, and Schlee, 1979 reprinted by permission.

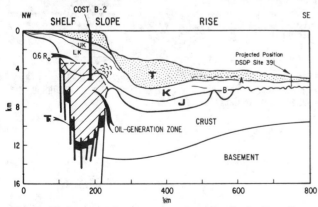

**Fig. 10–89.** Source-rock patterns in an intraslope basin. From Dow, 1979, reprinted by permission.

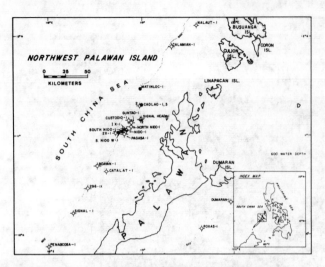

**Fig. 10-90.** Well location map for northwest Palawan Island. From Beddoes, 1980.

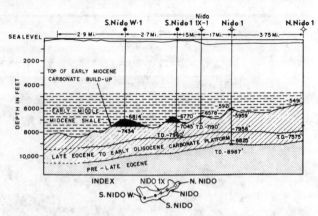

**Fig. 10-92.** Diagrammatic cross section of Nido reef trend. From Beddoes, 1980.

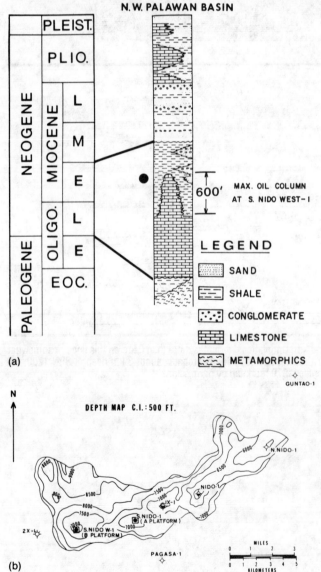

**Fig. 10-91.** (a) Generalized stratigraphic column for northwest Palawan basin. (b) Top of early Miocene carbonate contour map, Nido Reef trend. From Beddoes, 1980.

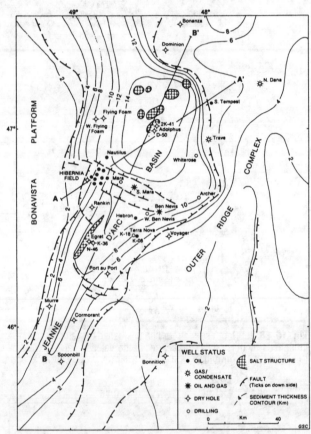

**Fig. 10-93.** Diagrammatic structure and sediment thickness map of the Jeanne d'Arc basin. Contours in km. From Grant, MacAlpine, and Wade, 1986, reprinted by permission.

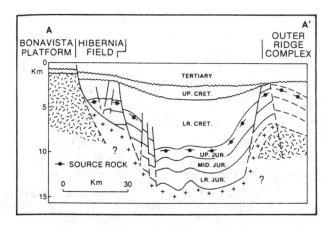

**Fig. 10-94.** Schematic transverse section of the Jeanne d'Arc basin. From Grant et al., 1986, reprinted by permission.

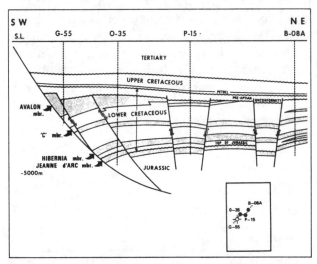

**Fig. 10-95a.** Southwest-northeast structural geologic cross section across the Hibernia structure. From Arthur, Cole, Henderson, and Kushnir, 1922.

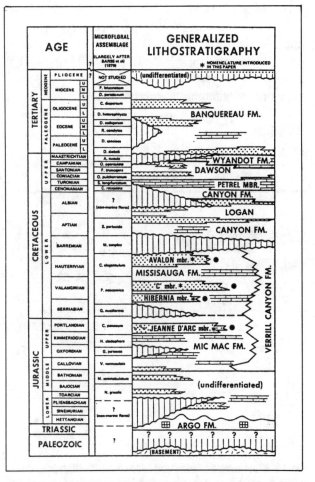

**Fig. 10-96.** Generalized lithostratigraphic chart for the East Newfoundland basin. From Arthur, Cole, and Henderson, and Kushnir, 1982, reprinted by permission.

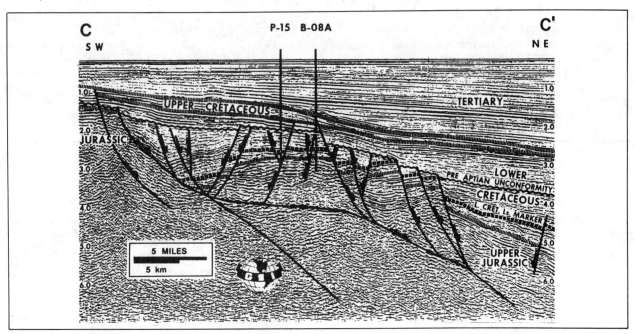

**Fig. 10-95b.** Migrated time seismic section which passes near the Chevron et al. Hibernia P-15 well and the Mobil et al Hibernia B-08A well. From Arthur, Cole, Henderson, Kushnir, 1982, reprinted by permission.

The opening of the south Atlantic also produced restricted basins with giant hydrocarbon accumulations. Initial separation commenced with deposition of basal Cretaceous stratigraphic intervals, and continued separation salt deposits were associated with the lower Aptian sequence unconformity (Campos et al. 1975) (Fig. 10–97). Giant fields were developed on the African margin (Brink 1974) (Fig. 10–98) and the South American margin (Dailly 1982) (Fig. 10–99). Seismic sections illustrate both listric

and antithetic fault patterns from the Campos basin (Fig. 10–100) and the Sergipe-Allagoas basin (Dauzacker 1985) (Fig. 10–101). On both sides of the South Atlantic the principle source rocks underlie the Aptian salt horizon, with hydrocarbon accumulations both below and above the salt horizon. Migration along faults and unconformity surfaces are important to the accumulation of giant fields.

A final example illustrates the importance of interpreting the depositional sequence history. Offshore Guyana, the spreading history was slightly later and was not coupled with the formation of a large intrashelf basin associated with salt deposition (Lawrence and Coster 1985) (Fig. 10–102). The sequence history is similar to that indicated by Vail, with lower Aptian, mid-Cenomanian, Santonian, post Danian, post-Ypresian, early Miocene, and late Miocene unconformities (Fig. 10–103). The history reflected by these unconformities provides a basis for evaluating the maturation and migration histories required for hydrocarbons accumulations (Fig. 10–104). A map of the Time-Temperature Index index illustrates the reason for the absence of major hydrocarbon occurrences in the basin (Fig. 10–105).

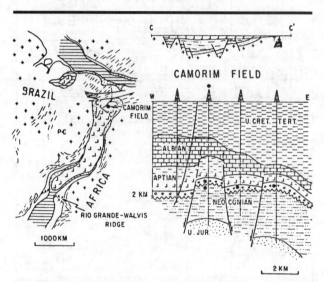

**Fig. 10–97.** External basin with salt deposition. After Campos, Muira, and Reis, 1975.

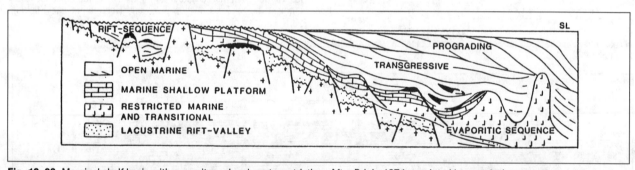

**Fig. 10–98.** Marginal shelf basin with evaporite and carbonate restriction. After Brink, 1974, reprinted by permission.

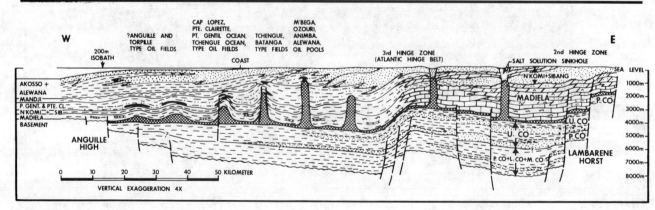

**Fig. 10–99.** From Dailly, 1982, reprinted by permission.

## Exploration Strategies

Seismic exploration for drape folds and shelf-edge carbonate banks and reefs is required. Topography of the depositional area at the time of reservoir development is critical. As has been shown for all basins, unconformity analysis is required to interpret migration pathways, geometry of depocenters, and direction and magnitude of onlapping source rocks and seals. Geochemical analysis is required to determine the depth for hydrocarbon generation, the volume of potentially hydrocarbon-producing marine source rocks, and the distribution of overpressured biogenic shale within intrashelf and intraslope basins.

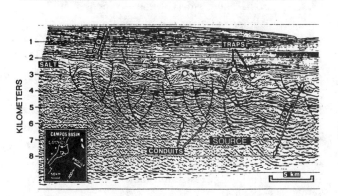

**Fig. 10–100.** Seismic profile, Campos basin. From Dauzacker.

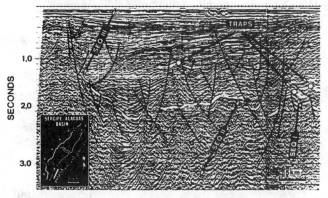

**Fig. 10–101.** Seismic profile, Sergipe-Alagoas basin. From Dauzacker, 1985.

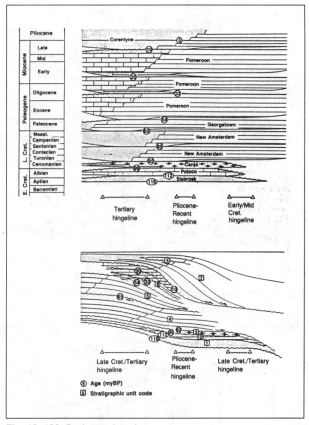

**Fig. 10–103.** Basin stratigraphy

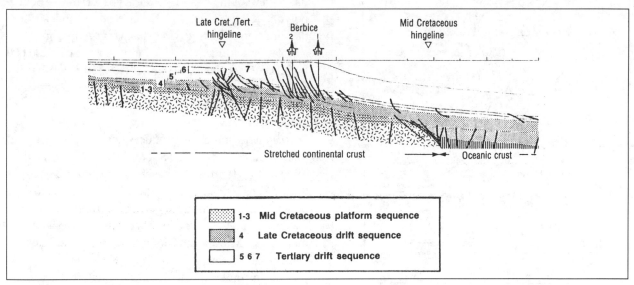

**Fig. 10–102.** Basin profile From Lawrence and Coster, 1985.

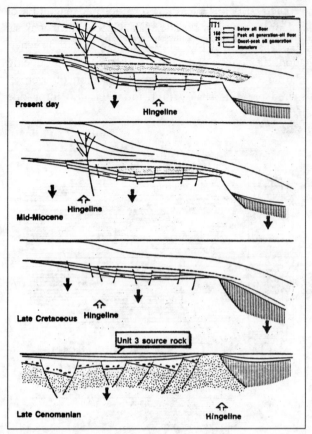

**Fig. 10-104.** Maturation/migration history. From Lawrence and Coster, 1985.

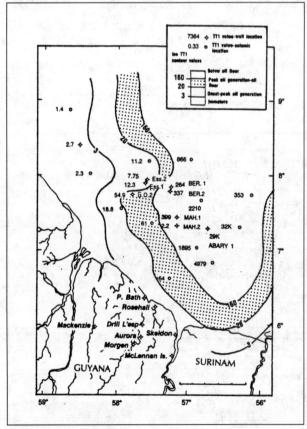

**Fig. 10-105.** Present day maturity-top source rock. From Lawrence and Coster, 1985.

These types of information are readily obtainable from seismic sections and a limited number of test wells within a basinal area. Utilization of multiple lines of investigation can lower the risk and costs for basin study and evaluation.

## MARGINAL BASINS

There is a group of basins that occurs on thinned cratonic crust. The history of these basins is similar to cratonic basins, but they persist for longer periods of time, for example, hundreds of millions of years. They rarely are completely filled with sediments. They apparently reflect a continuing process of crustal thinning, which causes subsidence and a continuously available subaqueous receptor capacity. Unconformities can terminate downdip into the basin, but the basin primarily reflects shelf sedimentary processes. These basins are ovoid to linear in shape, contain sedimentary thicknesses in excess of 10,000 meters, and may reflect differing tectonic histories.

Klemme (1983) suggests three basic patterns: (1) a simple basin without tectonic uplift on its margins, (2) a basin closed by uplift, and (3) a highly deformed linear basin filled with shelf sedimentary units (Fig. 10-106). Stratigraphic patterns reflect periods of restriction with evaporite deposition; a complex of subbasins on the shelf, on the slope, and in deeper water; and a depositional history similar to that described by Roberts and Caston (1975) (Fig. 10-21). Structural styles include those developed on the craton but also include growth faults and extensional faulting as developed on cratonic margins.

These basins are the habitat of giant and supergiant hydrocarbon accumulations. Oil is present in all stages of basin development and in all types of traps. Source potential ranges from 40,000 barrels/cubic miles in the deformed fold belts to 140,000 barrels/cubic miles in simple undeformed basins, and more than 180,000 barrels/cubic miles in deformed flank marginal basins. Oil types include waxy, paraffinic, and low-gravity, high-sulfur crudes. Reservoirs reflect shelf sedimentary processes yielding large collection areas. Unconformities are common, and onlap produces both source rocks and seals. Source rocks are various but mostly related to intrashelf and intraslope basins. Petroleum potential of these basins is related to the size of the basins and to their pattern of restriction. Total recoverable petroleum reserves can be greater than hundreds of billions of barrels (e.g., Middle East, Gulf Coast, and eastern Venezuelan basins).

Three examples of this basin type reflect the various stratigraphic response patterns and controls for hydro-

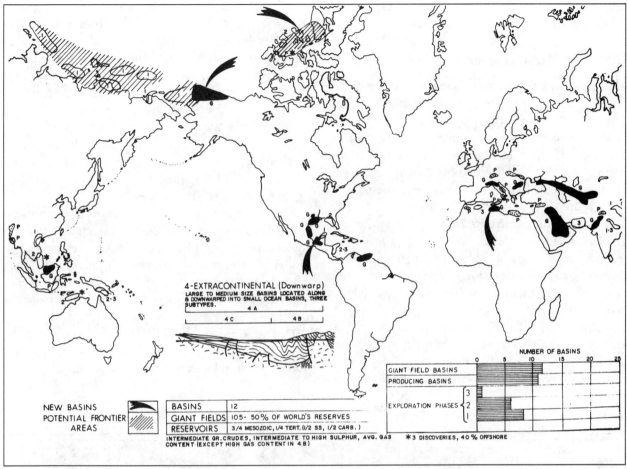

**Fig. 10–106.** Marginal cratonic basins. Personal communication. Reprinted courtesy H.D. Klemme, 1979.

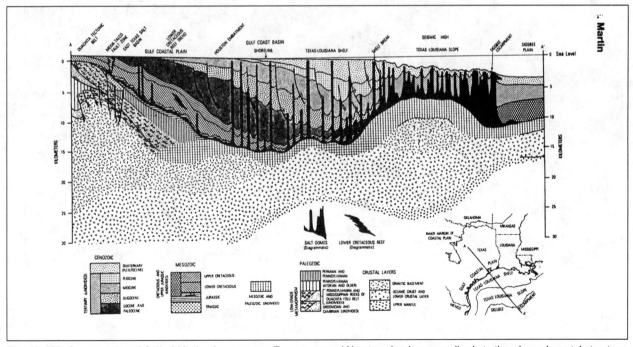

**Fig. 10–107.** Cross section of Gulf of Mexico from eastern Texas to central Yucatan showing generalized stratigraphy and crustal structure. Velocities in km/sec. Adapted from numerous sources. From Martin, 1978, reprinted by permission.

carbon accumulations. These include the Gulf Coast, Middle East, and the Prudhoe Bay basins.

## Gulf Coast Marginal Basin

Marginal basins reflect tectonic boundary conditions that may not be easily interpreted from simple paleogeographic patterns. Structural and depositional history must be more integrated over a period of more than one hundred million years (Fig. 10–107). The Gulf Coast basin reflects changes in the position of depocenters, position of the shelf edge, subsidence in the central Gulf, and patterns of restriction. The distribution of salt in the basin is not easily interpreted from present paleogeographic patterns (Martin 1978) (Fig. 10–108). A monominerallic evaporite sequence suggests deeper or deepening water in relation to a marginal sill between the basin and the ocean; where the sill was developed has not been ascertained.

The history of intrashelf basinal development can be associated with Cretaceous and Tertiary depositional centers (Winker-Edwards 1983) (Fig. 10–109). Arches between and separating depocenters probably are caus-

ally related to these depocenters. The lag between deposition and the formation of complementary uplifts suggests that unconformity patterns and history can be usefully applied to interpretation of the structural history. Local structural features can be interpreted as synsedimentary responses to the depositional history (Fig. 10–110). Depocenters are bounded by growth faults. Salt diapirism does not occur where low-density shales have been accumulated in embayments. Salt pillows and synsedimentary rim synclines produce local structural uplifts. These patterns can be related to the paleogeography and to stratigraphic responses.

Marginal basins are depocenters for both clastics and carbonates. Provenance patterns may change through time; consequently, the facies trends for reservoir units and their relation to source rocks must change. The localization of areas and structures with potentially higher concentrations of hydrocarbon reserves is the exploration objective.

Due to the presence of wide shelves with low dip, onlapping depositional sequences provide opportunity for hydrocarbon migration from thick intrashelf basinal stratigraphic units. The Cretaceous Eagleford shale

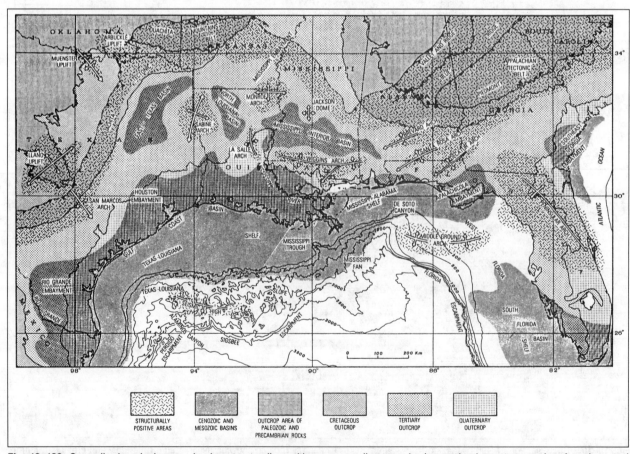

**Fig. 10–108.** Generalized geologic map showing structurally positive areas, sedimentary basins, and sub-sea topography of northern and eastern Gulf of Mexico region. Contour interval, 200 m. From Martin, 1978, reprinted by permission.

onlaps the Woodbine unconformity sandstone, and the up-dip termination of this sandstone is an area of low pressure (Fig. 10–111). Onlapping shale and fine-grained chalk provide the seal. This system resulted in the entrapment of more than ten billion barrels of recoverable oil. Similar stratigraphic and structural anomalies are present in Jurassic reservoirs, in shelf-edge Cretaceous reefs, and in unconformity traps along the updip edge of the basin.

The Gulf Coast Lower Cretaceous interval is productive in small biohermal buildups, and was deposited as regressive-transgressive sequences across the shelf. Source rocks were associated with Lower Cretaceous transgressive biogenic shale intervals, but the shelf lacked an important Cretaceous intrashelf basinal history. Clastics prograded across the shelf and prevented the development of a restricted basin bounded by shoals and a shelf edge reefal depositional system.

In the area of the Yucatan Penninsula there were differing boundary conditions. Circulation patterns in the early history of the Gulf of Mexico produced upwelling, and reefal developments both as shelf edge buildups and pinacles seaward of the shelf. The restriction of the shelf in the area of the Campeche platform by shelf edge carbonate buildups resulted in both upper Jurassic and Lower Cretaceous intrashelf depositional patterns, with basinal sequences dominated by salt and biogenic micrites.

The depositional framework was conducive to the development of source rocks, and with the periodic sea-level falls developed in the Lower Cretaceous, regressive-transgressive cycles produced periods of diagenetic alteration of shelf carbonates sequences. Salt tectonics, producing topographic uplift of platform sequences, diagentic alteration of shelf sequences, and the presence of extensive basinal areas containing source rocks, produced one of the major areas for hydrocarbon accumulations (Acevedo 1980). The Cantarell complex offshore the Campeche platform has estimated recover-

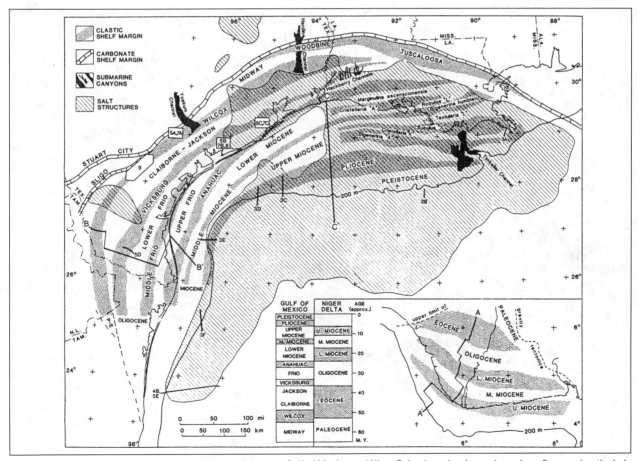

**Fig. 10–109.** Regional shelf margin trends of the northwestern Gulf of Mexico and Niger Delta, based on isopach maxima, flexures (particularly in Niger Delta), timing of maximum growth-fault activity (particularly in Texas and northeast Mexico), and stratigraphic top of geopressure (particularly in Louisiana). Where necessary, shelf margin trends are extrapolated along regional growth-fault trends (particularly in Niger Delta). In Gulf Basin, Midway and Woodbine represent stable progradation; all other clastic sequences represent unstable progradation. Ages of submarine canyons are: Yoakum and Hardin, mid-Wilcox; Hackberry, mid-Frio; Timbalier, Pleistocene. Gulf Basin exhibits major shifts in progradation rates, whereas Niger Delta exhibits fairly steady progradation. Time scales and correlations between basins are approximate. From Winker and Edwards, 1983.

able reserves of 20 billion barrels, and the Bermudez complex overlying the Jurassic salt basin has estimated recoverable reserves in excess of 1 billion barrels.

## Middle East Basin

The largest concentration of petroleum in the world is associated with the Middle East marginal basin where more than 500 billion barrels of recoverable oil are present. Major oil fields are distributed around the Arabian-Persian Gulf (Fig. 10–113). A major change from the Gulf Coast basin is the closure of the basin by uplift and thrust faulting (Fig. 10–112). The uplift changed the basin geometry, patterns of artesian flow and the pressure distribution, and the direction and pattern of migration from source rocks deposited in intrashelf basins (Fig. 10–113). The shelf stratigraphic framework is illustrated by an east-west cross section through the basin (Fig. 10–114). Intrashelf basins were developed in the Jurassic and in the Cretaceous. Basinal development was related to

shelf-edge carbonate buildups in the upper Jurassic and by subsiding shelf areas during the Lower Cretaceous. In both cases argillaceous micrites and marls were deposited. These source rocks were interrelated to structural highs on the shelf by permeable blanket carbonates and siliciclastic sands. Unconformities produced evaporite seals during onlap periods, leached porosity and permeability during lowstands of sea level, and changes in provenance patterns (Fig. 10–115).

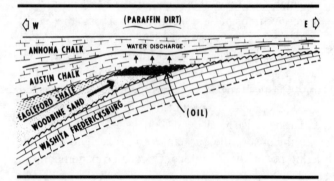

**Fig. 10–111.** Schematic profile of the East Texas field, showing local contact of the Woodbine aquifer with overlying chalk to allow escape of waters and accumulation of hydrocarbons east of the Eagle Ford shale pinchout. From Roberts, 1980, reprinted by permission.

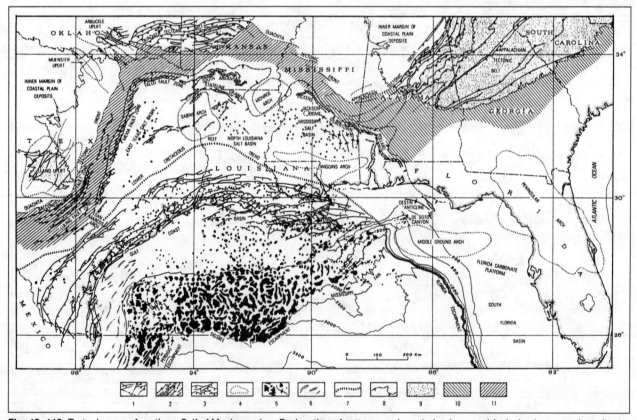

**Fig. 10–110.** Tectonic map of northern Gulf of Mexico region. Explanation of patterns and symbols: *1*, normal fault, hachures on downthrown side; *2*, reverse fault, sawteeth on overthrust plate; *3*, fault of undetermined movement; *4*, broad anticline or arch of regional extent; *5*, salt diapirs and massifs indicating relative size and shape; *6*, salt anticlines and swells (nondiapiric) showing general trend; *7*, shale domes and anticlines showing general size and trend; *8*, plutonic and volcanic rocks of Mesozoic age exclusive of basement complexes and Triassic diabase sills; *9*, updip limits of Louann salt; *10*, downdip limits of deep wells reaching rocks of Ouachita tectonic belt; *11*, uplifts of exposed Paleozoic strata and crystalline basement rocks; *12*, trend of Lower Cretaceous shelf-margin reef system; *13*, inner margin of Cretaceous and Tertiary coastal plain deposits. Scale: 1° latitude equals 110 km. From Martin, 1978, reprinted by permission.

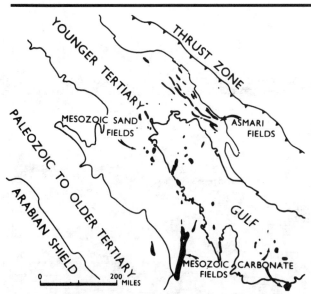

**Fig. 10–112.** Oil field distribution map. From Hull and Warman, 1968, reprinted by permission.

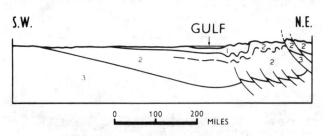

☐ YOUNGER TERTIARY

☐ MESOZOIC AND OLDER TERTIARY
(INCL. SOME PALEOZOICS)

☐ BASEMENT

**Fig. 10–113.** Arabian-Iranian Gulf, the most important oil-producing basin of this type. From Hull and Warman, 1968, reprinted by permission.

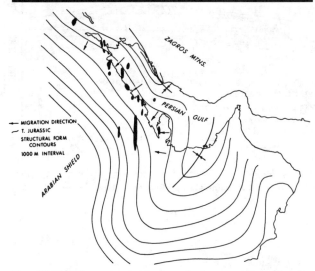

**Fig. 10–114.** Mesopotamian basin gathering area. From Momper, 1978, reprinted by permission.

A paleogeographic map for the Oxfordian to early Kimmeridgian illustrates the Jurassic depositional framework (Murris 1980) (Fig. 10–116). Large deeper water shelf basins are developed during this transgressive period, with the contemporaneous development of coastal regressive-transgressive patterns related to eustatic and depositional controls. Upper Jurassic reservoirs were developed at Ghawar and in fields adjacent to the intrashelf basin, with more than 200 billion barrels of recoverable oil from these Jurassic reservoirs. Later Jurassic, possibly Tithonian, coastal sabkha and intrashelf basinal evaporite intervals produced seals across basement faulted anticlinal structures.

The sequence unconformity history in the Lower Cretaceous was also an important control for the formation of periods of diagenetic alteration, development of carbonate buildups, and the formation of restricted intrashelf basinal patterns. Age dating of middle east events are still controversial, but the worldwide early Aptian and mid-Cenomanian unconformities may have controlled the clastic depositional sequences developed on the margin of the Arabian cratonic area (Murris 1980) (Figs. 10–117 and 10–118). A transgressive period is indicated for the middle to late Aptian, with the development of a restricted intrashelf basin (Fig. 10–119). This history is responsible for the development of Lower Cretaceous reservoirs containing tens of billions of barrels of oil in Quatar and Abu Dhabi, shelf sequences similar to those described for the Yucatan shelf of Mexico.

Similarly, the clastic reservoirs of Kuwait and the offshore Neutral Zone of Saudia Arabia and Kuwait contain more than 200 billion barrels of recoverable hydrocarbon reserves. In this northern portion of the Gulf, wave and possibly tidally dominated deltaic and shelf-deposited blanket sandstones are reservoir units (Al-Laboun 1977) (Figs. 10–120 and 10–121).

The Arabian-Persian Gulf area is the site of the largest hydrocarbon accumulations in the world. The reasons relate to the development of a broad carbonate shelf dominated by biogenic processes, but punctuated by unconformities and depositional sequences, periodically restricted, and subsequently subaerially exposed. The presence of regressive-transgressive carbonate shelf sequences, deposited across basement faulted structures with synsedimentary structural growth, and proximal to intrashelf basins with source rock potential, is a combination that contains all of the elements for supergiant hydrocarbon accumulations.

## North Slope Alaska Basin

A similar closed basinal setting was developed in the Beaufort Sea-Colville trough area of the northern Alaska

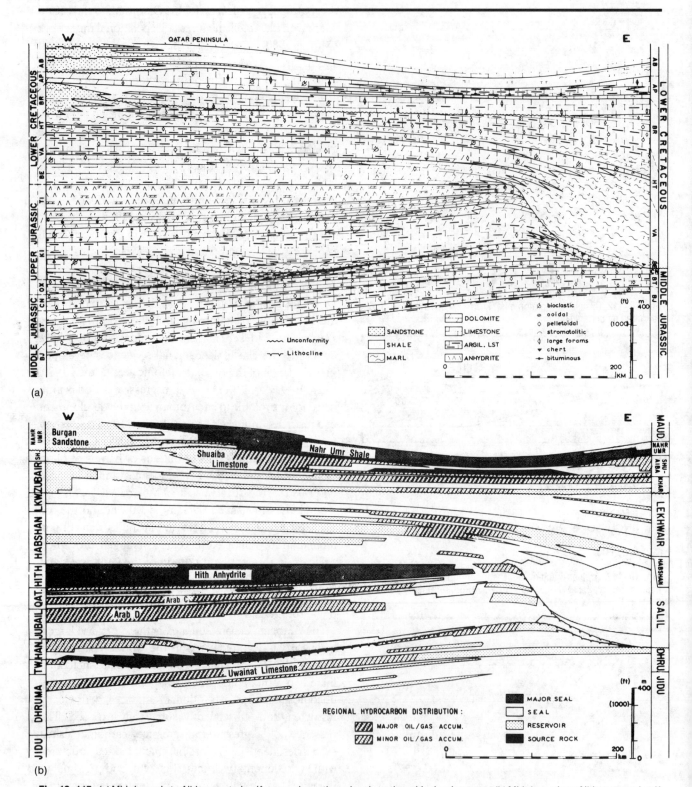

**Fig. 10–115.** *(a)* Mid-Jurassic to Albian, central gulf area: schematic regional stratigraphic development. *(b)* Mid-Jurassic to Albian, central gulf area: habitat parameters and distribution of oil and gas accumulations. From Murris, 1980, reprinted by permission.

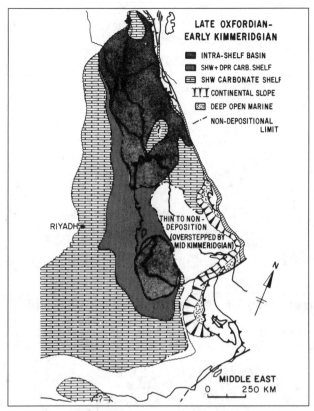

**Fig. 10–116.** Late Oxfordian to early Kimmeridgian environments of deposition. From Murris, 1980, reprinted by permission.

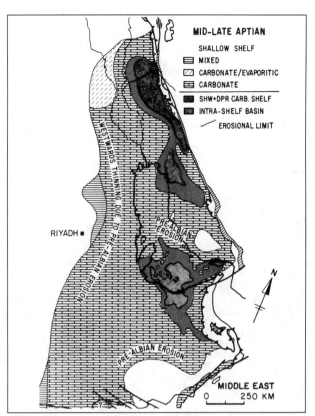

**Fig. 10–118.** Mid-Albian environments of deposition: *K,* Khuzestan sub-basin. From Murris, 1980, reprinted by permission.

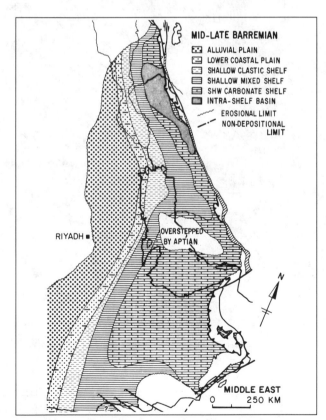

**Fig. 10–117.** Mid- to late-Barremian environments of deposition. From Murris, 1980, reprinted by permission.

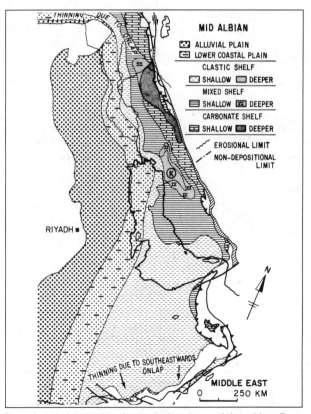

**Fig. 10–119.** Mid- to late-Aptian environments of deposition. From Murris, 1980, reprinted by permission.

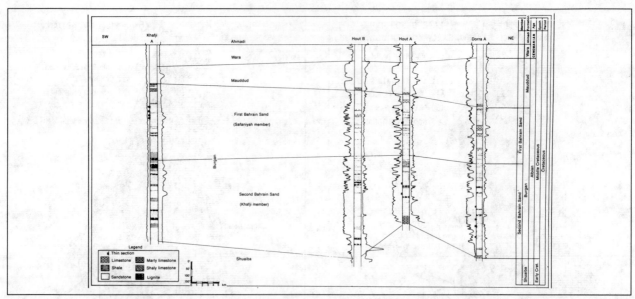

**Fig. 10–120.** Detailed stratigraphic cross section showing the Burgan formation in the Dorra-Hout-Khafji oil fields. Reprinted courtesy A. Al-Laboun.

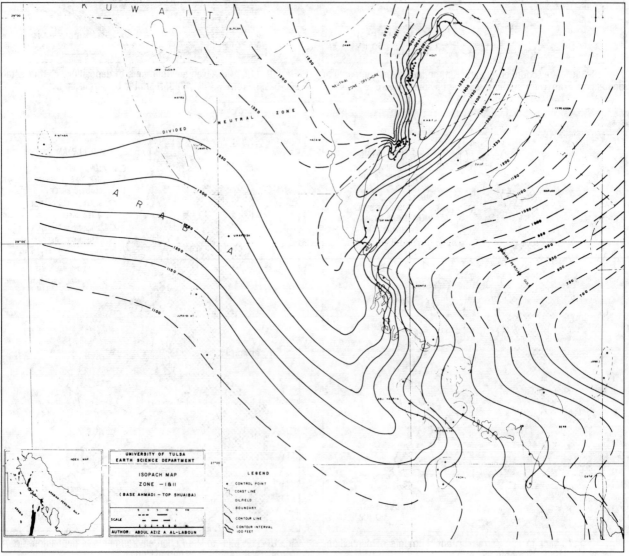

**Fig. 10–121.** Isopach map, Shuaiba and Ahmadi limestone. Reprinted courtesy A. Al-Laboun.

continental margin. The stratigraphic-structural framework indicates a large Paleozoic to Mesozoic depocenter (Fig. 10–122). The Brooks Range is a large uplifted tectonic feature that closed the southern margin of the depocenter. Paleozoic and early Mesozoic deep-water sediments, pillow basalts, and cherts in the Brooks Range indicate that an oceanic marginal basin lay to the south. Wells north of the Brooks Range penetrate continental and marine shelf sedimentary units. Unconformities occur within these shelf sedimentary sequences and provide the control for deposition. With the rise of the Brooks Range, the direction of provenance changed and shelf units were buried under a complex of fluvio-deltaic and volcanic Cretaceous sedimentary units. However, the basal Cretaceous unconformity surface extended into the basin and was onlapped by Lower Cretaceous biogenic shales.

This stratigraphic-structural history provided the framework for generation of hydrocarbons. Overpressure was developed to cause migration of fluids to the north toward the largest structures. The thinning of stratigraphic units and the convergence of unconformities produced a large collection system for hydrocarbon accumulations. The Prudhoe Bay structure was a focus for migrating fluids (Fig. 10–123) and contains more than 10 billion barrels of recoverable oil. Similar structures are presently the focus of intensive exploration and drilling activities, and additional hydrocarbon reserves have already been defined.

The basinal framework of the North Slope has been significant for the entrapment of major hydrocarbon reserves. The pattern of this framework is useful for continued exploration in the area, and the model may be applicable to other areas of the world.

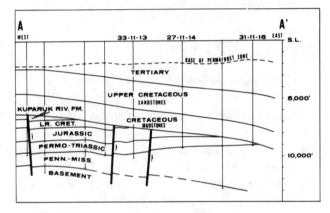

**Fig. 10–123.** Generalized cross sections of Prudhoe Bay field. After Jones and Speers, 1976, reprinted by permission.

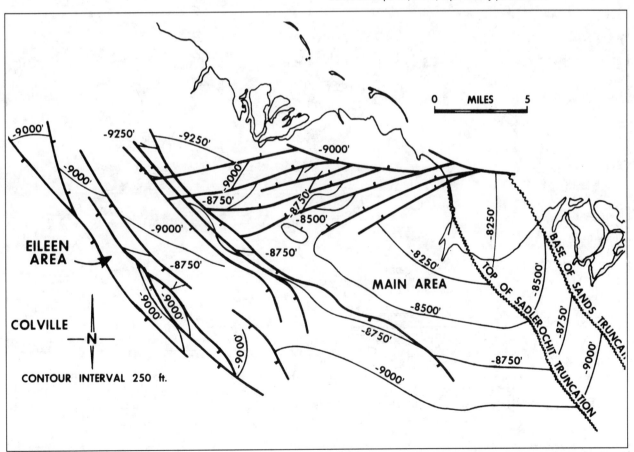

**Fig. 10–122.** Generalized structure map, top of Sadlerochit, Prudhoe Bay field. After Jones and Speers, 1976, reprinted by permission.

## Other Marginal Basinal Examples

A similar pattern occurs across the Caspian Sea in the Soviet Union, in Pakistan, in the foreland area of the Himalayas, in northeastern India, and possibly a continuation of the Beaufort Sea area to the west into the arctic ocean north of the Soviet Union. Smaller basins with billions of barrels of oil may include the eastern Venezuelan Mesozoic and lower Tertiary basin and the eastern Malaysian basin. Paleozoic, Mesozoic, and Tertiary shelf areas in the Adriatic and Sahul shelf (Bonaparte Gulf) of northwestern Australia may also be potentially of this type (personal communication, Klemme 1979) (Fig. 10–106).

## Exploration Strategies

The thickness of the basin fill, the pattern of unconformities, and the presence of large shelf structures are the principal controls for giant petroleum accumulations. The presence of blanket permeable shelf units provides the collection mechanism to focus hydrocarbon migration into low-pressure areas and structures. Migration must be from the depocenters into permeable and porous shelf sedimentary units. Pressure patterns, timing of migration, and the history of subsidence and uplift provide the controls for hydrocarbon accumulation and developing an exploration program.

## DELTAIC DEPOCENTERS

Deltas do not necessarily fit into a simple tectonic or basinal framework. They may have varying geometries and depositional patterns. Depocenters may be intrashelf or intraslope or may contribute to development of the continental rise. Sedimentary thicknesses may be more than 10,000 meters. Depositional processes control the position of the depocenter and tectonic responses. Paleogeographic and tectonic boundary conditions change from delta to delta, and these must be determined to interpret and predict stratigraphic patterns.

Deltas are not usually the site of giant and supergiant oil fields (Klemme 1983) (Fig. 10–124). Of more than 30 major deltaic depocenters, only three contain giant oil fields. Trap size appears to be the dominant control for this pattern. All aspects discussed for other basins are

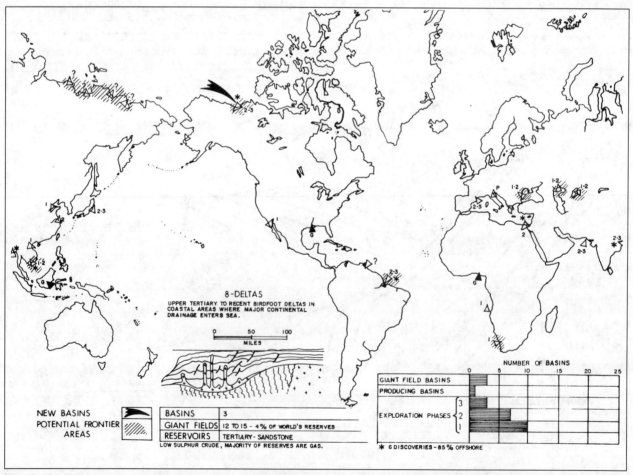

**Fig. 10–124.** Delta basins and their exploration stages. Personal communication. Reprinted courtesy H.D. Klemme, 1979.

present within the deltaic framework, including synsedimentary tectonism, overpressured biogenic shales, temperature gradients appropriate for hydrocarbon generation, and sufficient sedimentary volumes to produce large amounts of hydrocarbons. This is verified by the volume of reservoired hydrocarbons in some deltas: more than 25 billion barrels of recoverable oil from the Niger delta, 20 billion barrels from the Mississippi delta, and more than 5 billion barrels from the Mahakan Delta. Exploration and development is still underway on the Mackenzie, the Mekong, the Hwang Ho-Yellow, and the Nile deltas. Other major deltas either have no production or can be related to other basinal settings.

Tectonism is mostly the response to synsedimentary flow structures and growth faulting. Deposition is often more rapid than subsidence resulting in progradational wedges. There does not appear to be a limit on either the area or the volume of sediment deposited. Heat flow is normal but variable in relation to overpressured shales, diapiric structures, and patterns of fluid flow. Oil source rocks are developed in prodeltaic units, and gas source rocks are associated with nearshore depocenters and coastal marshes. Paraffinic oil of intermediate to high gravity is common. Gas and gas condensate are associated with both early and late periods of generation.

The largest oil fields are less than one billion barrels. Hydrocarbon productivity within basinal depocenters is high, with 190,000 barrels/cubic miles of oil and oil equivalent gas. Migration is related to tensional faults and fractures often produced by overpressured shales. Seals are associated with onlap unconformities. Sequences produce unconformities extending hundreds of meters down into basinal depocenters. Successive depocenters produce a cumulative pattern of subsidence interrelating deeper water and shelf reservoirs.

Two delta systems with giant hydrocarbon accumulations can be used as examples for basin analysis. Variations occur in terms of depositional processes, boundary conditions, and exploration strategies. Similarities are also developed, and these can be used for interpretation and prediction in other deltas.

## Niger Delta

The structural framework is a triple-point spreading center (Fig. 10–125). This has resulted in a limited area of basin fill and stacking of deposystems (Fig. 10–126). Exploration was unsuccessful for a number of years due to the absence of a time-rock correlation framework. Each fault block was complicated by shale diapirism, and the history of faulting and hydrocarbon generation and migration could not be ascertained with the available data. The structural complexity is illustrated by the pattern of faulting (Fig. 10–127). Some faults appeared to be seals, and others appeared to allow migration of hydrocarbons (Fig. 10–128). The interpretation of the depositional history from palynological correlations provided the first insight into the presence of depocenters within the deltaic system (Fig. 10–129).

With this information together with heat-flow measurements, it could be demonstrated that depocenters were the site for formation of overpressured shales and that a mechanism for hydrocarbon expulsion was thus available (Fig. 10–130). Utilizing this information, the major prolific centers of hydrocarbon accumulations were related to delta lobe depocenters (Fig. 10–131). This data

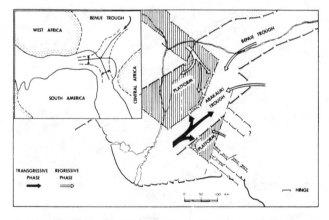

**Fig. 10-125.** Albian-Santonian evolution of the Niger delta sedimentary basin, showing rift development of the Abakaliki-Benue trough. After Weber and Daukoru, 1975.

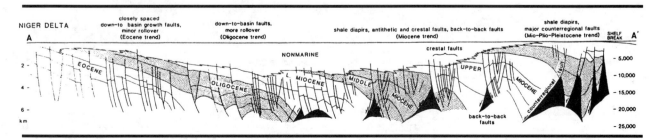

**Fig. 10-126.** Cross-section through the Niger delta depocenter illustrating extensive growth faulting occurring progressively basinward and higher in the stratigraphy as progradation continues. From Winker and Edwards, 1983.

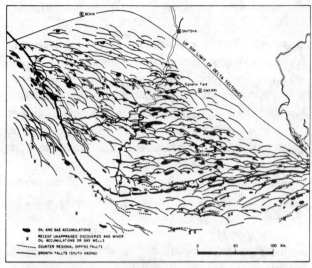

**Fig. 10-127.** Niger delta growth faults and known hydrocarbon accumulation. After Weber and Daukoru, 1975.

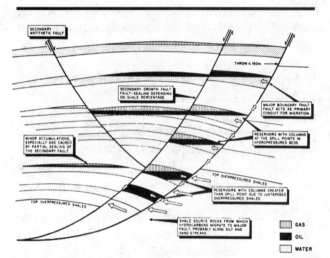

**Fig. 10-128.** Niger delta growth fault hydrocarbon migration and accumulation model. After Weber and Daukoru, 1975.

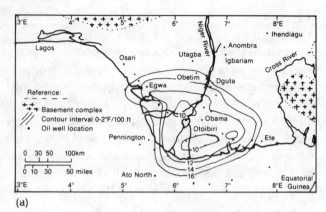

(a)

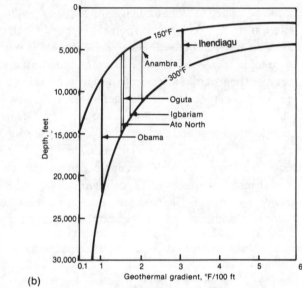

(b)

**Fig. 10-130.** Changes in geothermal gradient across the Niger delta. From Nwachukwu, 1976, reprinted by permission.

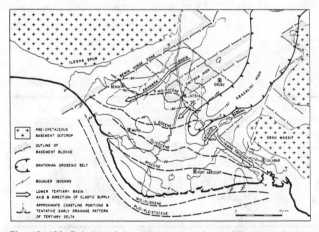

**Fig. 10-129.** Relation of depocenters to major prolific areas of production in the Niger delta. From Weber and Daukoru, 1975.

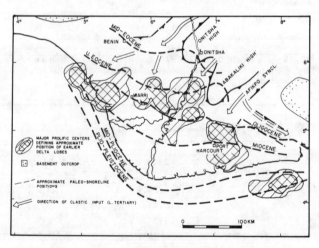

**Fig. 10-131.** From Ejedawe, 1981, reprinted by permission.

could be contoured in terms of oil-reserve density (Fig. 10–132). This information can be usefully applied, not only to understanding the causality of the hydrocarbon distribution, but also as a basis for continued exploration. Temperature maps can be used to define depocenters, overpressured shales, areas of hydrocarbon generation, and migration pathways. Also, low-density overpressured shales can be defined by seismic velocity analysis and prospective areas defined.

Exploration in the Niger delta continues into offshore areas with testing of deeper water reservoirs and evaluating unconformities for potential stratigraphic traps.

These can be more efficiently explored by the use of a combination of seismic sequence analysis, time-rock correlations, and salinity and temperature maps.

## Mississippi Delta

A similar pattern of depocenters has been established for the Mississippi delta (Fig. 10–133). The history of these depocenters for the past 20 million years is the basis for exploration (Fig. 10–134). Each depocenter has been recognized and defined by micropaleontology, including the *Harang* and *Cibicides Opima* embayments. Depositional patterns into these embayments can be examined both on a sequence and a delta lobe scale (Fig. 10–135). Similarly, the distribution of reservoir facies can be related to the depositional history, and the temperature history can be estimated from vitrinite reflectance data (Dow 1979) (Fig. 10–136). The pattern of hydrocarbon accumulations is more closely related to the vitrinite reflectance than to any specific depositional environment.

The geometry of depocenters, the pattern of overpressured shales to growth faults and diapiric structures, the pattern of reservoir units, and the pattern of hydrocarbon generation are all causally interrelated (Fig. 10–137). This provides the basis for interpretation and predicting hydrocarbon accumulations.

## Other Deltas

Exploration for giant hydrocarbon accumulations in Tertiary deltaic depocenters has been successful throughout

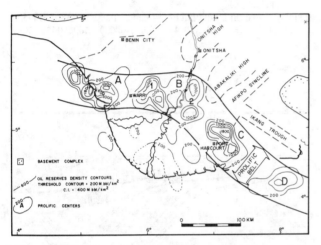

**Fig. 10–132.** Prolific oil belt, Niger delta. From Ejedawe, 1981, reprinted by permission.

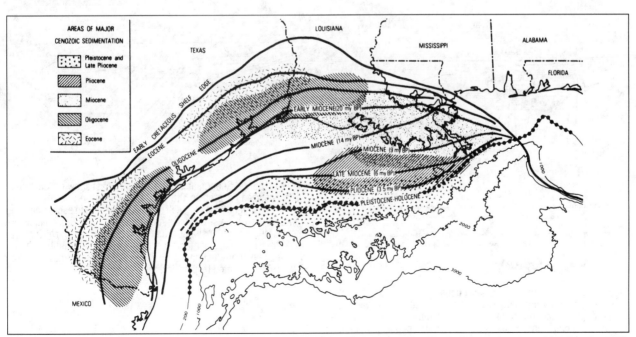

**Fig. 10–133.** Sketch map showing paleoshelf edges in Gulf Coast basin and distribution of major Tertiary depocenters. From Martin, 1978, reprinted by permission.

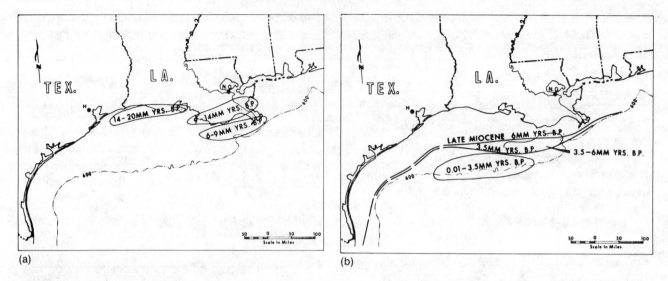

**Fig. 10-134.** (a) Position of early Pliocene and late Plio-Pleistocene depocenters based on progradation of related shelf edges. (b) Principal Miocene depocenters based on isopach maps constructed from subsurface well control. Depocenter is illustrated by closed isopach thickness. From Woodbury et al., 1973, reprinted by permission.

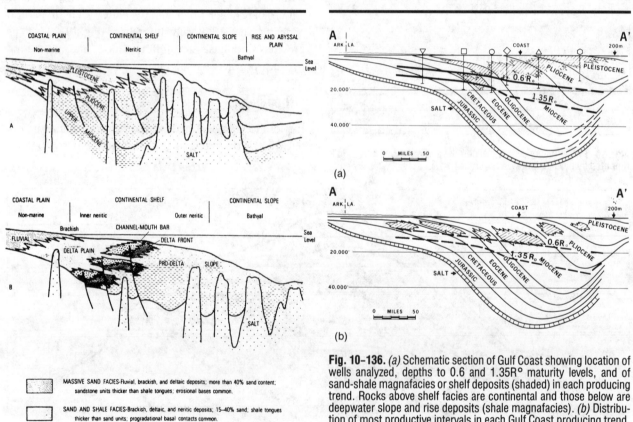

**Fig. 10-135.** Diagrammatic cross sections showing lithofacies relations in Quaternary and Tertiary sediments of Gulf Coast basin. *A,* Gross depositional model for Cenozoic sediments depicting net regression of continental-deltaic (massive sands), neritic (sandstone-shale), and bathyal (massive shale) magnafacies through Cenozoic time. *B,* Depositional model for Pleistocene sediments in Texas-Louisiana shelf region. From Martin, 1978, reprinted by permission.

**Fig. 10-136.** (a) Schematic section of Gulf Coast showing location of wells analyzed, depths to 0.6 and 1.35R° maturity levels, and of sand-shale magnafacies or shelf deposits (shaded) in each producing trend. Rocks above shelf facies are continental and those below are deepwater slope and rise deposits (shale magnafacies). (b) Distribution of most productive intervals in each Gulf Coast producing trend. Note relation of production to 0.6R° and age at reservoir rocks. From Dow, 1979, reprinted by permission.

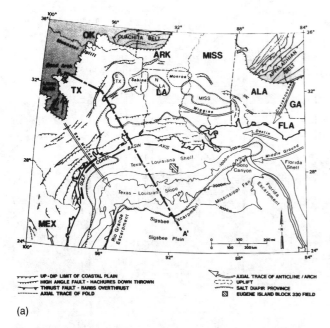

(a)

**Fig. 10-137a.** Structural-physiographic map of the northern Gulf of Mexico and adjacent land area. From Holland et al., 1980, reprinted by permission.

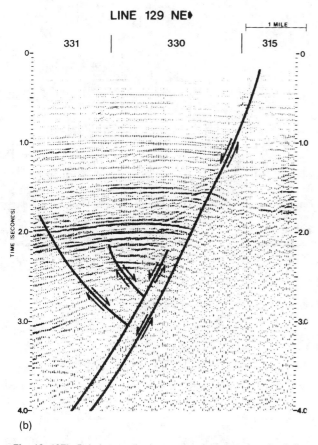

(b)

**Fig. 10-137b.** Relative amplitude processed version of a seismic line. Prominent seismic hydrocarbon indicators are associated with the oil and gas reservoirs that are traversed by the seismic line. From Holland et al., 1980, reprinted by permission.

the world, with supergiant fields in the Texas and Mexican Gulf Coasts, west Malaysia, Sabha, Sarawak and Brunei of east Malaysia, the Shengli basin in China, in the Mahakam delta of eastern Kalimantan, the McKenzie delta, and related clastics from the Beaufort Sea of Canada, and the Orinoco delta off Trinidad. New discoveries from the Irrawady, Ganges-Brahmaputra, and Indus Deltas suggests that many more deltaic depocenters will contain giant hydrocarbon accumulations. The site of many of these deltaic depocenters is associated with extensional or wrench-faulted basins. Depositional and structural controls combine in the development of hydrocarbon accumulations. The development of a single basinal theme for exploration for all deltaic systems has not been attempted, but in all instances the depositional geometry and history dominate the exploration strategy.

The Indonesian and Malaysian delta systems are dominated by wrench-fault tectonics, with both synsedimentary growth faulting and flower structures producing traps. Description of the deltaic depocenter for the Mahakam delta (Figs. 8-83 and 8-84) illustrate the complexity of both the depositional and structural histories. As in many deltas throughout the world, it is often difficult to identify source rocks, migration history, and the timing of sealing faults, but the controls indicated for both the Mississippi and Niger deltas appear to be important.

Supergiant Tertiary hydrocarbon accumulations are associated with the Orinoco delta. This basinal area is also associated with wrench faulting (Leonard 1983) (Fig. 10-138). Tertiary deltaic depocenters extend from the Gulf of Paria, across Trinidad, into the Columbus basin. The Columbus basin history illustrates the interrelation of wrench faulting and growth faulting (Leonard 1983) (Figs. 10-139 and 10-140). The depositional history is from the area of the Poui and East Queens Beach areas shows an upper Miocene to Pliocene coarsening upward interval greater than 7500 meters thick (Fig. 10-141). A composite section from the supergiant Teak field illustrates the stacking of deltaic depositional events (Bane and Chanpong 1980) (Fig. 10-142). The nature of the hydrocarbon distribution suggests the pattern of migration along faults, and indicates those faults that trap oil and those that trap gas accumulations (Fig. 10-143). These patterns suggest that the controls for hydrocarbon accumulations is the presence of faults connecting overpressured source rocks deposited in an outershelf or slope depocenter. Hydrocarbon accumulations are trapped in reservoirs by sealing faults, and are overlain by transgressive intervals related to the deltaic depositonal history. This pattern is similar to most other deltaic depocenters throughout the world.

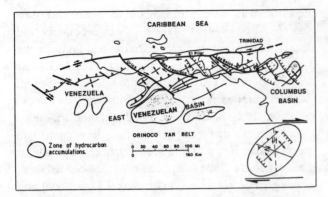

**Fig. 10-138.** Structural trends and zones of hydrocarbon accumulations in east Venezuela, Trinidad, and Columbus basin. Comparison can be made between structural trends along El Pilar fault zone and structural trends expected in an east-west wrench system, as shown in lower right. From Leonard, 1983.

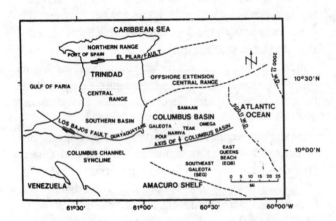

**Fig. 10-139.** Location of Columbus basin. Also shown are oil field names and geographic locations discussed in text. From Leonard, 1983.

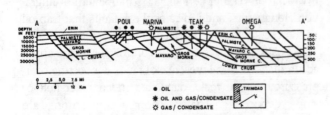

**Fig. 10-140.** Columbus basin regional cross section. Cross section demonstrates pattern of normal faulting across basin. Increasing thickness of late Pliocene-Pleistocene section (Palmiste and Erin formations and their correlatives) from is less than 10,000 ft (3,000 m) approximately 25,000 ft (7,6000 m) is also shown ($c$ = correlative). Leonard, 1983.

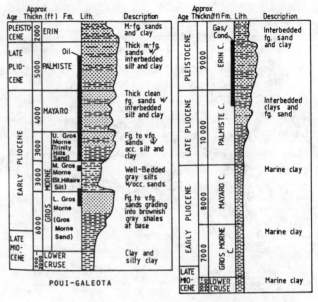

**Fig. 10-141.** Reference sections, Poui-Galeota and East Queens Beach areas. The Mayaro, Gros Morne, and lower Cruse have not been penetrated in EQB area; their presence is inferred on the basis of seismic correlations ($c$ = correlative). Leonard, 1983.

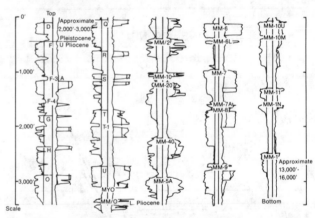

**Fig. 10-142.** Composite log section of Pliocene-Pleistocene, Teak field, Trinidad. From Bane and Chanpong, 1980, reprinted by permission.

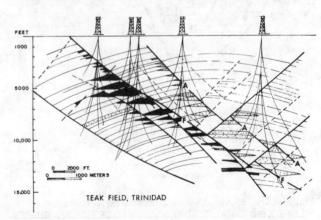

**Fig. 10-143.** Cross section, Teak field, showing oil and gas zones. From Bane and Chanpong, 1980, reprinted by permission.

## Exploration Strategies

Geophysical mapping of flow structures and growth faults is necessary to define depocenters. Sequence mapping of progradational units and onlap unconformities are needed to define reservoir patterns and their relation to source rocks. Interval velocity analysis is necessary to determine sand-shale ratios and structural history from isopach maps of time-rock depositional units. These geophysical and subsurface data can be integrated with source rock, temperature, and structural information concerning the tectonic framework of the deltaic basinal setting to establish an exploration program.

## CONVERGENT MARGIN BASINS

Convergent plate margins lead to complex tectonic responses including subsidence and formation of fore-arc trenches, back-arc extensional graben or rift basins. Wrench faulting, which involves the cratonic margin or propagation of stress through the cratonic crust, results in formation of complicated basinal patterns on the continental plate. Klemme (1983) suggested two basic tectonic settings: basins parallel to the continental plate (Type 7), and those cutting across the cratonic plate (Type 6) (Fig. 10–144). Major hydrocarbon accumulations occur in more than 30 basins around the world, but an additional 60 basins have been identified with a range of exploration phases.

These basinal settings reflect high crustal mobility and rapid changes in water depths, provenance patterns, and sedimentation rates. They produce a complicated array of response patterns. The causality of developmental changes cannot always be related to depositional processes. Geodynamic forces applied to the crustal plates result in tectonism of various types, including gravity sliding, wrench faulting, compressional uplifts, and synsedimentary extensional rhombic areas of subsidence.

These basins occur adjacent to ocean basins and can be filled with marine sedimentary detritus. Mostly these basins are developed in areas of high heat flow. Tectonics may produce deep depocenters, and synsedimentary

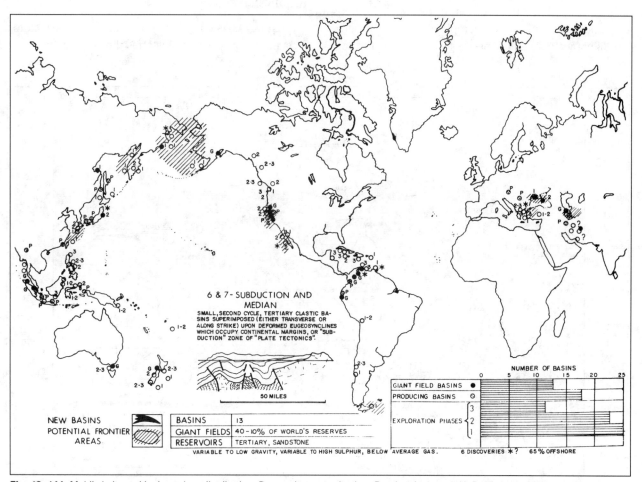

**Fig. 10–144.** Mobile belts and hydrocarbon distribution. Personal communication. Reprinted courtesy H. D. Klemme, 1974.

faulting may produce traps in the center of the basin. Relative changes in sea level lead to the development of submarine fans and transgressive onlap. Highstands of sea level lead to a reduced rate of sediment supply and accumulation of biogenic detritus. This basinal framework is conducive to preservation of organic material and the generation and entrapment of large quantities of hydrocarbons. They are the site of giant and supergiant oil fields. The Lake Maracaibo (Type 6) and the Indonesian, Cook Inlet, and California basins (Type 7) illustrate the origin of large hydrocarbon accumulations.

## Lake Maracaibo Basin

The structural history of the Maracaibo basin has been a focus of study since supergiant hydrocarbon accumulations were discovered in the 1920s. Not until the application of plate tectonic and depositional sequence concepts has it been possible to understand the causality for the accumulation of more than 40 billion barrels of recoverable hydrocarbons. Regional structural patterns with documented wrench faulting surrounding a microplate, is the the structural framework for the Cretaceous to Tertiary depositional history (Vasquez and Dickey 1972) (Fig. 10-145). The presence of a deeper water La Luna biogenic marl, within the microplate and in associated Cordillera wrench-faulted basins to the west, is the suggested source rock for Cretaceous and Tertiary hydrocarbon accumulations. (Barker et al. 1983). Wrench faults and unconformities in the Cretaceous, Paleocene, Eocene, and mid-Oligocene provided the migration pathways from the deeper part of the basin. Synsedimentary wrench and normal faulting associated with the flanks of the depocenter produced structures and traps (Gutierrez 1968). Facies and isopachous patterns were required to interpret the tectonic and depositional history of the basin (Vasquez and Dickey 1972) (Fig. 10-146). One of the principal controls for hydrocarbon migration and development of reservoir intervals was the "Eocene" unconformity, of mid-Oligocene age that truncated Eocene deltaic sequences.

Additional exploration in the central Maracaibo basin, to the west in the Llanos basin, and in the Falcon area to the east, have produced significant new discoveries, including the giant Caño Limon field on the border between Columbia and Venezuela. With an understanding of the history and magnitude of wrench faulting, additional discoveries are possible offshore Venezuela and Columbia.

## Indonesian Basins

The nature of basins and their position relative to tectonic arcs and cratonic areas illustrates the complexity of the

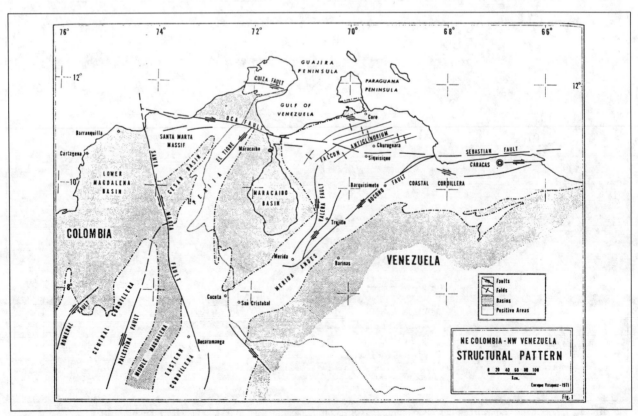

**Fig. 10-145.** Tectonic framework of western Venezuela. From Vasquez and Dickey, 1972.

setting (Barber 1985) (Fig. 10–147). Small basins contain recoverable hydrocarbons in the range of billions of barrels. A complex pattern exists of subducting plates, spreading centers, transform faults, and wrenching, but these patterns are difficult to relate causally to the formation and depositional fill of the more than 27 identifiable basins identified by Barber (1985). Many of these basins, however, reflect a depositional cycle controlled by unconformities associated with sea-level falls (Beddoes 1980) (Fig. 10–148). The principle unconformities are the post-Ypresian, mid-Oligocene, early Miocene, and the late Miocene events. Hydrocarbon accumulations have been broadly related to these sequences (Fig. 10–149).

Paleogeographic patterns at the maximum transgression for the mid-Oligocene to lower Miocene sequence are illustrated by Figure 10–150, and patterns at the maximum regression for the lower Miocene to late Miocene sequence in the early Pliocene are illustrated by Figure 10–151. The timing of these unconformity events was lacking when these figures were drawn, but the data and interpretation presented here are consistent with the events described by Vail and co-workers. The suggestion is that major progradational events were responsible for clastic reservoir development, and that sea-level rise resulted in carbonate reefs and banks to be developed during the closing stage of each sequence. Wrench faulting was superimposed on these unconformity bounded sequences, with the resulting development of traps. Stacking of sequences within each basin resulted in burial depths sufficient for hydrocarbon generation, and the faults provided the avenue for hydrocarbon migration. Discontinuous wrenching and onlapping marine shales provided the traps and the seals for the hydrocarbon accumulations. Source rocks were formed in restricted basins and reflected lacustrine, deeper water marine shale, and transgressive argillaceous micrite depositional events.

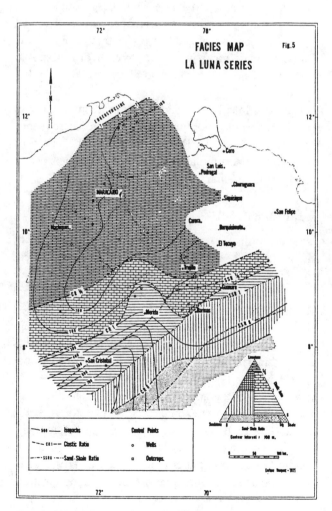

**Fig. 10-146.** Facies map, La Luna series. From Vasquez and Dickey, 1972.

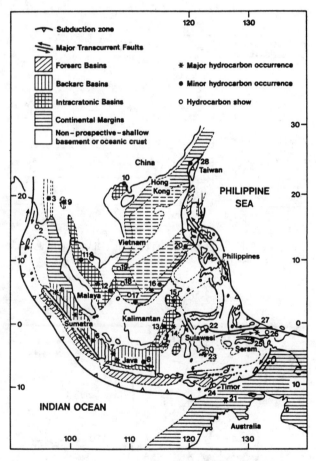

**Fig. 10-147.** Tertiary sedimentary basins, continental margins and the occurrence of hydrocarbons in Southeast Asia. Numbered occurrences: 1 = West Sumatra basin; 2 = Offshore Burma; 3 = Central Burma; 4 = North Sumatra basin; 5 = Central Sumatra basin; 6 = South sumatra basin; 7 = Sunda Strait and West Java basins; 8 = East Java basin; 9 = Fang basin; 10 = Gulf of Tongking basin; 11 = Thai basin; 12 = Malaya and West Natuna basins; 13 = Barito basin; 14 = Kutai basin; 15 = Tarakan basin; 16 = Brunei an Sabah; 17 = North West Borneo; 18 = Saigon basin; 19 = Mekong basin; 20 = Palawan; 21 = Northwest Australian Shelf; 22 = East Arm Sulawesi; 23 = Buton; 24 = Timor; 25 = Bula basin; Seram; 26 = Bituni basin; 27 = Salawati basin. From Barber, 1985.

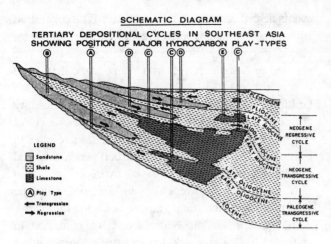

**SCHEMATIC DIAGRAM**
**TERTIARY DEPOSITIONAL CYCLES IN SOUTHEAST ASIA**
**SHOWING POSITION OF MAJOR HYDROCARBON PLAY-TYPES**

**Fig. 10-148.** Cycles of sedimentation during the Tertiary in Southeast Asia. From Beddoes, 1980.

| TERTIARY DEPOSITIONAL CYCLE | PLAY-TYPE | CUMULATIVE OIL PRODUCTION* (to Md '79) (MILLION BBLS.) | PERCENT |
|---|---|---|---|
| NEOGENE REGRESSION (LATE MIOCENE TO EARLY PLIOCENE) | Ⓔ CARBONATES | 200 ± (plus major gas production) | 1·9 % |
| | Ⓓ CLASTICS | 4200 ± (plus major gas production) | 39·3 % |
| NEOGENE TRANSGRESSION (LATE OLIGOCENE TO MIDDLE MIOCENE) | Ⓒ CARBONATES | 210 ± (plus major gas production) | 2·0 % |
| | Ⓑ CLASTICS | 5800 ± | 54·3% |
| PALEOGENE TRANSGRESSION (EOCENE TO EARLY OLIGOCENE) | Ⓐ CLASTICS | 270 ± | 2·5 % |
| | TOTALS | 10,680 ± | 100·0 % |

**Fig. 10-149.** Play-type distribution of cummulative oil production in Southeast Asia. From Beddoes, 1980.

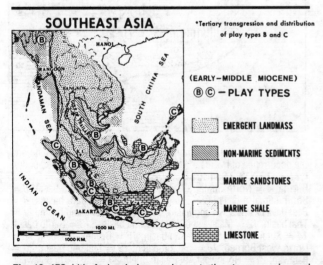

**Fig. 10-150.** Lithofacies during maximum tertiary transgression and distribution of play types *B* and *C*. From Beddoes, 1980.

## Cook Inlet Basin

Silled fore-arc basins, such as the Cook Inlet of Alaska, are other sites for major hydrocarbon accumulations. Regional structural patterns are important in defining the basinal setting (Fig. 10-153). This, together with isopach maps of the basin, suggests the positions of depocenters and possible sites for hydrocarbon accumulations (Fig. 10-154). Structures within the basin can be mapped by seismic methods, and an exploration program can be developed (Fig. 10-155).

This was the method followed for the Cook Inlet, but two critical factors were not recognized: the nature of the organic material in the depocenter and the paleogeographic and stratigraphic patterns that preceded basinal development. A cross section across the lower Cook Inlet illustrated that the principal depocenter was Mesozoic in age, not Tertiary (Fig. 10-156), and reservoir sandstones were deposited on the basal Tertiary unconformity, truncating Mesozoic source rocks (Fig. 10-157). These relationships are critical to understanding the causality of hydrocarbon accumulations and for predicting new exploration areas and strategies. These source rock, unconformity, and migration relationships are confirmed by the near identity of hydrocarbon extracts from middle Jurassic source rocks with reservoired hydrocarbons in Tertiary sediments (Magoon and Claypool 1981) (Fig. 10-158).

The Cook Inlet basin again confirms the importance of source rock, unconformity, and depositional sequence relationships as a basis for interpreting the origin of hydrocarbon accumulations and predicting their occurrence within basins.

## California Basins

The Californian foreland illustrates the importance of nonarc convergent margin basins for hydrocarbon accumulations. The structural framework is a basis for interpreting the origin of basins (Fig. 10-159). A more localized pattern reflects the strike-slip pattern developed (Fig. 10-160). The model that can be suggested is the convergent pattern of faults producing simultaneous uplifts and basinal downwarps (Fig. 10-161). The Rhombochasm is a well-developed tectonic style (Fig. 10-162). Application of this model to the Ridge basin illustrates the synsedimentary tectonic response pattern (Figs. 10-163 and 10-164). Thousands of meters of sedimentary fill can be accommodated as the faults continue to move. Marginal structural patterns reflect the convergence of strike-slip faults, and compression produces flower structures (Figs. 10-165 and 10-166).

This model can be applied on a larger scale and used to

interpret the formation of the Los Angeles basin (Fig. 10–167). Here a depocenter with more than 6,000 meters of sedimentary fill developed. Structures within the basin formed due to compression resulting from strike-slip faults that define the basinal margins. Heat flow in the basin suggests a generating depth for hydrocarbons of approximately 2,500 meters. This can be translated into cross sections showing the intervals of petroleum generation (Fig. 10–168). This can be associated with the depositional history, the patterns of mounded buildups of sandstone during periods of low sea level, and the development of onlapping stratigraphic units across unconformities, which extend down into the basin (Fig. 10–169). With these stratigraphic patterns, the only additional requirements for hydrocarbon accumulation is synsedimentary structural uplift in the central area of the basin (Fig. 10–170).

The combination of tectonics, subsidence, formation of depocenters, high heat flow, and thick overpressured

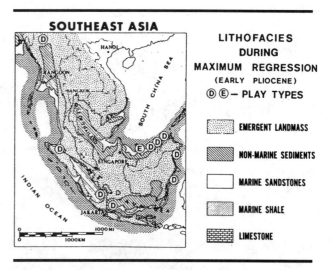

**Fig. 10–151.** Lithofacies during maximum regression. From Beddoes, 1980.

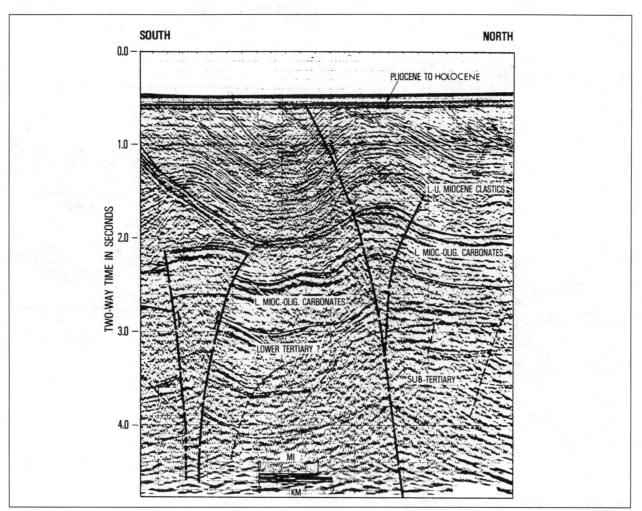

**Fig. 10–152.** Seismic refraction line across small displacement, covergent wrench fault in Palawan shelf of South China Sea (lat. 11°N, long. 119°E). Zone has steeply dipping, upthrust profiles that merge at depth into single, subvertical strand. Narrow horst at left margin of line closely resembles strike-slip zone and demonstrates fallibility of structural-stype identifications made with structural criteria from a single seismic profile. From Harding, 1985, reprinted by permission.

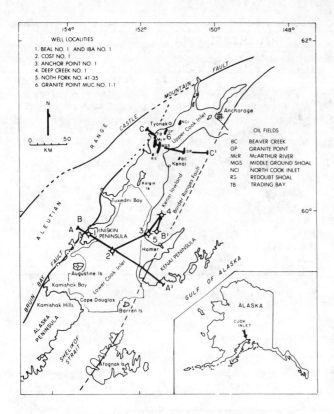

**Fig. 10-153.** Cook Inlet area indicating geographic names; well locations; cross sections *A-A', B-B', C-C'*; and oil fields. From Magoon and Claypool, 1981, reprinted by permission.

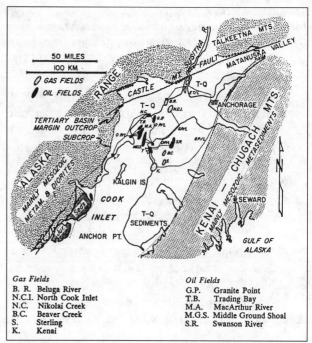

**Fig. 10-155.** Outline map of Cook Inlet basin showing major oil and gas fields. After Kirschner and Lyon, 1973.

*Gas Fields*
B. R. Beluga River
N.C.I. North Cook Inlet
N.C. Nikolai Creek
B.C. Beaver Creek
S. Sterling
K. Kenai

*Oil Fields*
G.P. Granite Point
T.B. Trading Bay
M.A. MacArthur River
M.G.S. Middle Ground Shoal
S.R. Swanson River

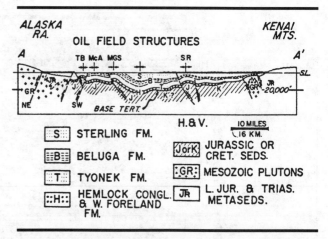

**Fig. 10-156.** Tectonic map and cross section of Cook Inlet. From Kirschner and Lyon, 1973.

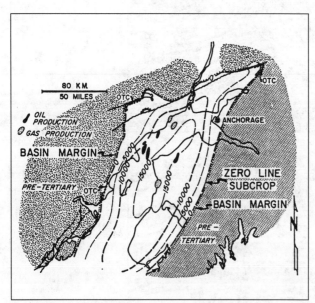

**Fig. 10-154.** Generalized isopach map of Kenai Group. The Kenai Group (Neogene) includes a maximum of more than 15,000 ft (4,575 m) and roughly 18,000 cu mi (75,000 cu km) of sedimentary rocks. All oil and gas production in basin is from these strata. From Kirschner and Lyon, 1973, reprinted by permission.

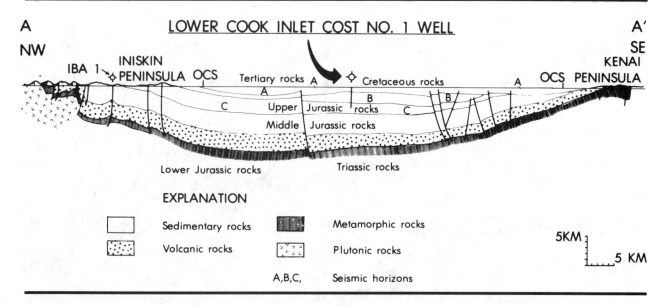

## LOWER COOK INLET COST NO. 1 WELL

**Fig. 10–157.** Lower Cook Inlet cost No. 1 well placed on cross section *AA'* (Fig. 10–153). *OCS* refers to offshore shale boundary of Fig. 10–154. From Magoon and Claypool, 1981, reprinted by permission.

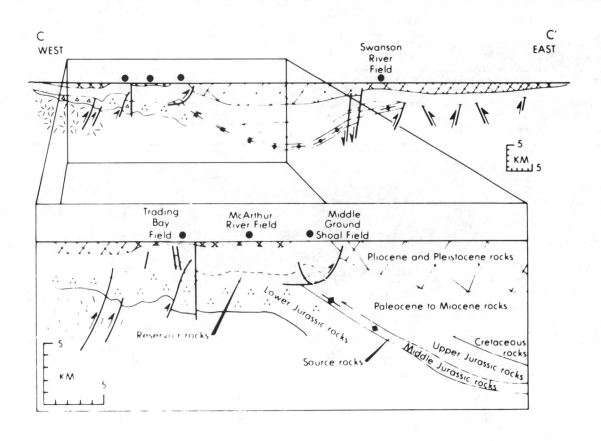

**Fig. 10–158.** Cross section *CC* transects upper Cook Inlet. Detail of west flank depicts relation of Middle Jurassic oil source bed in Tertiary reservoir rocks. From Magoon and Claypool, 1981, reprinted by permission.

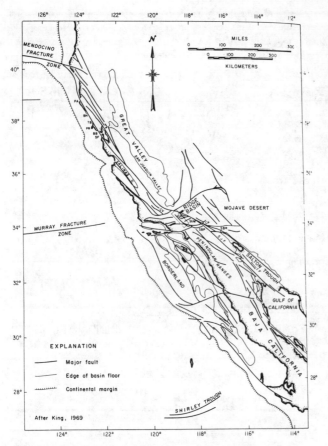

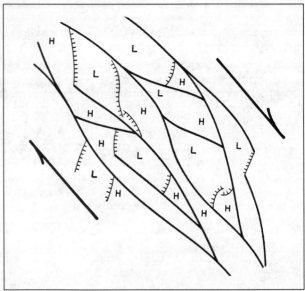

**Fig. 10–161.** Sketch map of region in a transform regime, showing pull-apart basins and tipped fault wedges where right-slip faults converge or diverge. From Crowell, 1974.

**Fig. 10–159.** Major faults and present sedimentation sites of California and Baja California. Geographic localities from north to south: *PA* = Point Arena; *BH* = Bodega Head; *TB* = Tomales Bay; *PR* = Point Reyes; *BB* = Bolinas Bay; *SF* = San Francisco; *CV* = Cuyama Valley; *SGM* = San Gabriel Mountains; *CP* = Cajon Pass; *SBR* = San Bernardino Range; and *SBP* = San Bernardino Plain. Faults: *SGF* = San Gabriel fault; *SFJ* = San Jacinto fault; and *IF* = Imperial fault. From Crowell, 1974.

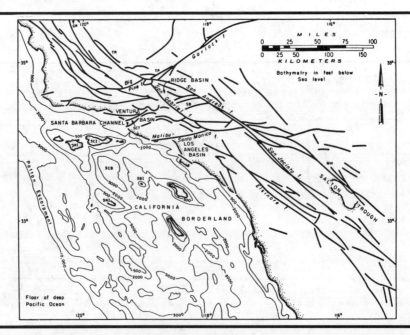

**Fig. 10–160.** Map of southern California and California Borderland showing major onshore faults. Abbreviations: *TP* = Tejon Pass; *SB* = Soledad basin; *SCT* = Santa Clara trough; *SRI* = Santa Rosa Island; *SCI* = Santa Cruz Island; *SCB* = Santa Cruz basin; *SBI* = Santa Barbara Island; *SNI* = San Nicholas Island; *GR* = Gabilan Range; *TR* = Temblor Range; and *MH* = Mecca Hills. From Crowell, 1974.

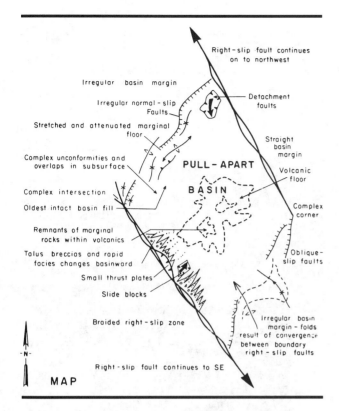

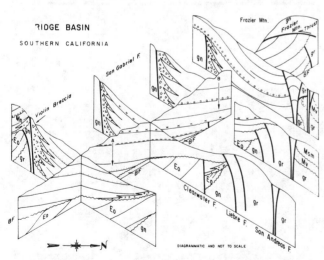

**Fig. 10-163.** Isometric sketch of Ridge basin, southern California (diagrammatic and not to scale). Strata of Ridge basin are not labeled. Stratigraphic section at *A* is overlapped at *B*. Symbols: *BF* = basin floor; *gn* = gneiss; *gr* = granitic rocks; *Eo* = mainly Eocene; *Mv* = Miocene volcanic rocks; *Msm* = Miocene Santa Margarita formation; and *Mm* = Miocene Modelo formation. From Crowell, 1974.

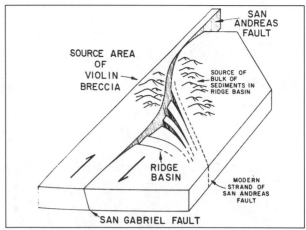

**Fig. 10-164.** Block diagram illustrating origin of Ridge basin at a curve in San Andreas fault. During Pliocene Epoch San Gabriel fault was probably main San Andreas strand. From Crowell, 1974.

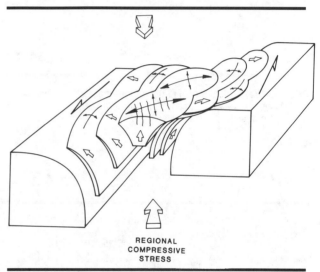

**Fig. 10-165.** Convergent left-lateral strike-slip fault causing upthrust blocks parallel to P direction. Personal communication. Courtesy P. Link, 1983.

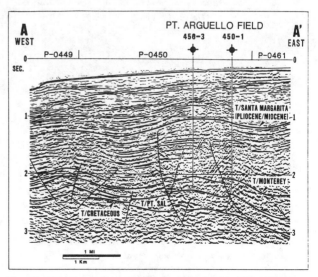

**Fig. 10-166.** Time-migrated seismic section passing near Chevron-Phillips *P-0450-3* and through Chevron-Phillips *P-0450-1.* From Crain, Mero and Patterson, reprinted by permission, 1985.

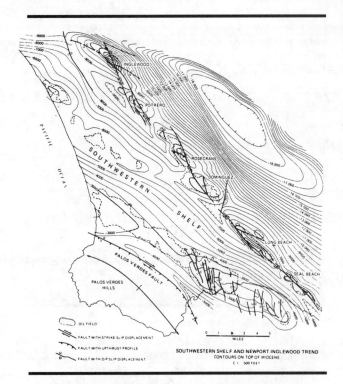

Fig. 10–167. Structure map of the northwestern and central part of the Newport-Inglewood trend, Los Angeles basin. From Harding, 1984, reprinted by permission.

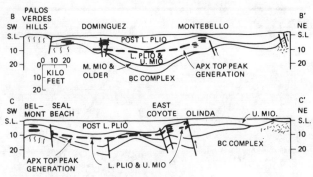

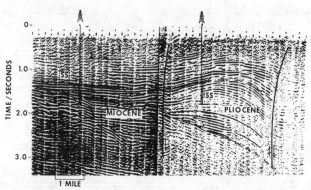

Fig. 10–168. Diagrammatic northeast-southwest cross section, Los Angeles basin, showing oil field and approximate top peak oil generation. From Jones, 1980, reprinted by permission.

Fig. 10–169. Mounded onlap-fill seismic facies, Pliocene example. Although located in structurally complex area, sandstone-rich mound is believed to be depositional in origin. From Sangree et al., 1977, reprinted by permission.

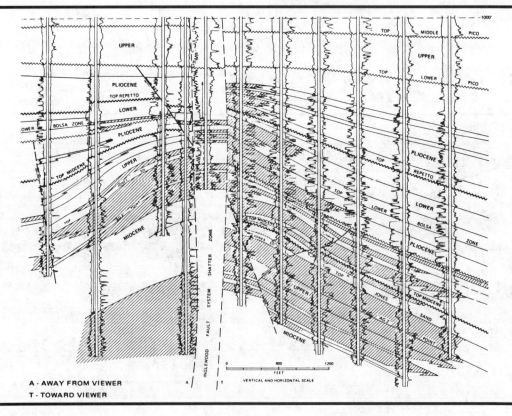

Fig. 10–170. Well log cross section illustrating wrench-fault tectonism. From Harding, 1973, reprinted by permission.

clastics can only lead to petroleum generation, migration, and hydrocarbon accumulations. This basin has generated more than 5 million barrels/cubic miles of sedimentary fill, and more than 5 billion barrels of recoverable hydrocarbons were reservoired. Again, it is the combination of controls that produces the giant and supergiant hydrocarbon accumulations. These controls can be recognized from the stratigraphic record, and patterns can be predicted from seismic, subsurface, and surface information.

## REFERENCES FOR FURTHER STUDY — CHAPTER 10

Bally, A.W., P.L. Bender, T.R. McGetchin, and R.I. Walcott, eds., 1980, Dynamics of plate interiors: Am. Geophys. Union, and Geol. Soc. Amer., Geodynamics Series, v. 1, 162 p.

Harding, T.P. and J.D. Lowell, 1979, Structural styles, their plate tectonic habitats and hydrocarbons traps in petroleum provinces: Am. Assoc. Petrol. Geol. Bull., v. 63, p. 1016–1058

Howell, D.G., ed., 1985, Tectonostratigraphic Terranes of the Circum-Pacific Region: Houston, Texas, The Circum-Pacific Council for Energy and Mineral Resources, 581 p.

Lowell, J.D., 1985, Structural Styles in Petroleum Exploration: Tulsa, Oklahoma, Oil & Gas Consultants Internat., Inc., 460 p.

Stanley, D.J., and G.T. Moore, eds., 1983, The Shelfbreak Critical Interface on Continental Margins: Tulsa, OK, Soc. Econ. Paleont. and Mineral., Spec. Pub. no. 33, 467 p.

Watkins, J.S., L. Montadent, and D.W. Dickerson, eds., 1982, Geological and Geophysical Investigation of Continental Margins: Am. Assoc. Petrol. Geol. Mem., no. 29, 472 p.

Ziegler, P.A., 1982, Geological Atlas of Western and Central Europe: Amsterdam, Holland, Elsevier Scientific Pub. Co., 130 p.

Acevedo, J.A., 1980, Giant fields of the southern zone—Mexico *in* Halbouty, M.T., ed., Giant Oil and Gas Fields of the Decade 1968-1978: Am. Assoc. Petrol. Geol. Mem., no. 30, p. 339-385.

Ackoff, R.L., S.K. Gupta, and J.S. Minas, 1962, Scientific Method: Optimizing Applied Research Decisions: New York, John Wiley & Sons, 464 p.

Adams, J.E., and M.L. Rhodes, 1960, Dolomitization by seepage refluxion: Am. Assoc. Petrol. Geol. Bull., v. 44, p. 1912-1920.

Agassiz, L., 1842, On glaciers and the evidence of their having once existed in Scotland, Ireland, and England: Geol. Soc. London Proc., v. 3, p. 327-332.

Agassiz, L., 1896, Geological sketches (2nd Series): Boston, Houghton, Mifflin and Co., 229 p.

Ager, D.V., 1981, The Nature of the Stratigraphic Record, 2nd ed.: New York, John Wiley & Sons, 122 p.

Al-Laboun, A., 1977, The depositional framework of the Wara-Mauddud-Burgan formations in Northeastern Arabia: MS. thesis, Univ. of Tulsa, 95 p.

Albritton, C.C., Jr., ed., 1963, The Fabric of Geology: Standford, California, Freeman, Cooper, & Co., 374 p.

Albritton, C.C., Jr., 1980, The Abyss of Time: San Francisco, CA, Freeman, Cooper & Co., 251 p.

Allen, G.P., D. Laurier, and J. Thouvenin, 1974, Etude sedimentologique du Delta de la Mahakam, Notes et Memoires, no. 15: Paris, TOTAL Compagnie Française des Petroles, 156 p.

Allen, J.R.L., 1964, Studies in fluvial sedimentation: Six cyclothems from the lower Old Red Sandstone, Anglo-Welsh Basin: Sedimentology, v. 3, p. 163-198.

Allen, J.R.L., 1965a, A review of the origin and characteristics of Recent alluvial sediments: Sedimentology, v. 5, p. 89-191.

Allen, J.R.L., 1965b, Coastal geomorphology of eastern Nigeria: beach ridge barrier islands and vegetated tidal flats: Geol. Mijnbouw, v. 44, p. 1-21.

Allen, J.R.L., 1967, Depth indicators of clastic sequences: Mar. Geol., v. 5, p. 429-446.

Allen, J.R.L., 1968, Current Ripples: Their Relation to Pattern of Water and Sediment Motion: Amsterdam, North-Holland Publ. Co., 433 p.

Allen, J.R.L., 1970, Sediments of the Modern Niger Delta: a summary review, *in* Morgan, J.P., and R.H. Shaver, eds., Deltaic Sedimentation Modern and Ancient: Soc. Econ. Paleont. Mineral. Spec. Publ., no. 15, p. 138-151.

Allen, J.R.L., 1980, Sand waves: a model of origin and internal structure: Sediment. Geol., v. 26, p. 281-328.

Allen, J.R.L., 1984, Sedimentary Structures, Their Character and Physical Basis: New York, Elsevier, 1256 p.

Allen, P.A., and P. Homewood, 1984, Evolution and mechanics of a Miocene tidal sandwave: Sedimentology., v. 31, p. 63-81.

Almgren, A.A., 1978, Timing of Tertiary submarine canyons and marine cycles of deposition in the southern Sacramento Valley, California, *in* Stanley, D.J., and G. Kelling, eds., Submarine Canyon and Fan Sedimentation: Stroudsburg, PA, Dowden, Hutchinson, and Ross, p. 276-291.

Amsden, T.W., 1962, Silurian and Early Devonian carbonate rocks of Oklahoma: Am. Assoc. Petrol. Geol. Bull., v. 46, p. 1502-1519.

Anderson, R.Y., W.E. Dean, Jr., D. W. Kirkland, and H.I. Snider, 1972, Permian Castile varved evaporite sequence, west Texas and New Mexico: Geol. Soc. Am. Bull., v. 83, p. 53-86.

Anderson, R.Y., and D.W. Kirkland, 1960, Origin, varves, and cycles of Jurassic Todilto Formation, New Mexico: Am. Assoc. Petrol. Geol. Bull., v. 44, p. 37-52.

Anderton, R., 1976, Tidal-shelf sedimentation: an example from the Scottish Dolradian: Sedimentology, v. 23, p. 429-458.

Andressen, M.J., 1962, Paleodrainage patterns: their mapping from subsurface data and their paleogeographic value: Am. Assoc. Petrol. Geol. Bull., v. 46, p. 398-405.

Apostel, L., 1961, Towards the formal study of models in the nonformal sciences, *in* Freudenthal, H., ed., The Concept and Role of the Model in Mathematics and Natural and Social Sciences: Dordrecht, Holland, D. Reidel Publishing Co., p. 1-37.

Arthur, K.R., D.R. Cole, G.G.L. Henderson, and D.W. Kushnir, 1982, Geology of the Hibernia discovery, *in* Halbouty, M.T., ed., The Deliberate Search for the Subtle Trap: Am. Assoc. Petrol. Geol. Mem., no. 32, p. 181-196.

Arthur, M.A., and R.E. Garrison, eds., 1986, Milankovitch cycles through geologic time: Paleoceanography, v. 1, p. 369-586.

Aubouin, J., 1959, Place des Hellenides parmi les edifices structuraux de la Mediterranée orientale (2 thesé) Paris, (1958): Ann. Geol. Pays Helleniques, v. 10, p. 487-525.

Aubouin, J. 1965, Geosynclines: Developments in Geotectonics: Amsterdam, Elsevier, 335 p.

Auldridge, L., 1979, Worldwide production: Oil & Gas Jour., v. 77, no. 53, p. 67-122.

Baker, D.R., 1962, Organic geochemistry of the Cherokee Group in southeastern Kansas and northeastern Oklahoma: Am. Assoc. Petrol. Geol. Bull., v. 45, p. 1621-1642.

Bally, A.W., 1980, Basins and subsidence—a summary, *in* Bally, A.W., P.L. Bender, T.R. McGetchin, and R.I. Walcott, eds., Dynamics of Plate Interiors: Am. Geophys. Union and Geol. Soc. Am. Geodynam. Ser., v. 1, p. 5-20.

Bally, A.W., 1987, Seismic sequences as applied to basin-fill analysis: Taranaki Basin, New Zealand, *in* Bally, A.W., ed, Atlas of Seismic Stratigraphy: Am. Assoc. Petrol. Geol. Stud. in Geol., no. 27, p. 53-71.

Bane, S.C., and R.R. Chanpong, 1980, Geology and development of the Teak Oil Field, Trinidad, West Indies, *in* Halbouty, M.T., ed., Giant Oil and Gas Fields of the Decade 1968-1978: Am. Assoc. Petrol. Geol. Mem., no. 30, p. 387-398.

Banks, N.L., 1973, Tide-dominated offshore sedimentation, Lower Cambrian, north Norway: Sedimentology, v. 20, p. 213-228.

Barber, A.J., 1985, The relationship between the tectonic evolution of southeast Asia and hydrocarbon occurrences, *in* Howell, D.G., ed., Tectonostratigraphic Terranes of the Circum-Pacific Region: Houston, The Circum-Pacific Council for Energy and Mineral Resources, p. 523-528.

Barker, C., 1972, Aquathermal pressuring—role of temperature in development of abnormal-pressure zones: Am. Assoc. Petrol. Geol. Bull., v. 45, p. 2068-2071.

Barker, C., 1978, Primary migration—the importance of water-mineral-organic matter interactions in the source rock: Am. Assoc. Petrol. Geol. Cont. Educ. Course Notes, no. 8, p. D-1–D-19.

Barker, C., 1979, Organic geochemistry in petroleum exploration: Am. Assoc. Petrol. Geol. Cont. Educ. Course Notes, no. 10, p. 19-32.

Barker, C., H. Bockmeulen, and P.A. Dickey, 1983, Geology and geochemistry of crude oil, Bolivar Coastal field, Venezuela: Am. Assoc. Petrol. Geol. Bull., v. 67, p. 242-270.

Barrell, J., 1906, Relative geological importance of continental, littoral, and marine sedimentation: Jour. Geol., v. 14, p. 316-356, 430-457, 524-568.

Basu, D.N., A. Banerjee, and D.M. Tamhane, 1980, Source area, migration trends of oil and gas in Bombay offshore basin India: Am. Assoc. Petrol. Geol. Bull., v. 64, p. 209-220.

Bates, R.L., and J.A. Jackson, eds., 1959, Glossary of Geology, 3rd. ed.: Alexandria, Va., Amer. Geol. Inst., 788 p.

Beddoes, L.R., 1980, Hydrocarbon plays in Tertiary basins of southeast Asia (preprint): Offshore Southeast Asia Conference, SEAPEX, February 26-29.

Belderson, R.H., 1986, Offshore tidal and non-tidal sandridges and sheets: Difference in morphology and hydrodynamic setting, in Knight, R.J., and J.R.McLean, eds., Shelf Sands and Sandstones: Can. Soc. Petrol. Geol. Mem., no. 11, p. 293-302.

Belderson, R.H., M.A. Johnson, and N.H. Kenyon, 1982, Bedforms, in Stride, A.H., ed., Offshore Tidal Sands, Process and Deposits: London, Chapman & Hall, p. 27-57.

Bengston, C.A., 1982, Structural and stratigraphic uses of dip profiles in petroleum exploration, in Halbouty, M.T., ed.: Am. Assoc. Petrol. Geol. Mem., no. 32, p. 31-46.

Bennison, A.P., 1972, Pennsylvanian Rocks of the Tulsa Area, in Bennison, A.P., ed., Tulsa's Physical Environment: Tulsa Geol. Soc. Dig., v. 37, p. 49-66.

Berg, O.R., 1982, Seismic detection and evaluation of delta and turbidite sequences: their application to the exploration for the subtle trap, in Halbouty, M.T., ed., The Deliberate Search for the Subtle Trap: Am. Assoc. Petrol. Geol. Mem., no. 32, p. 57-75.

Berg, R.R., 1975, Depositional environment of Upper Cretaceous Sussex sandstone House Creek Field, Wyoming: Am. Assoc. Petrol. Geol. Bull. v. 59, p. 2099-2110.

Bernard, H.A., R.J. LeBlanc, and C.F. Major, 1962, Recent and Pleistocene geology of southeast Texas, in Rainwater, E.H., and R.P. Zingula, eds., Geology of the Gulf Coast and Central Texas: Houston, Houston Geol. Soc., p. 175-224.

Berry, W.B.N., 1970, On teaching stratigraphic paleontology, in Yockelson, E.L., ed., Proc. N. Am. Paleont. Conv., part 1, v. 1: Lawrence, KS, Allen Press, p. 12-22.

Biddle, K.T., and D.R. Seeley, 1983, Structure of a subduction complex, in Bally, A.W., ed., Seismic Expression of Structural Styles, v. 3, Tectonics of Compressional Processes: Am. Assoc. Petrol. Geol. Stud. Geol. Ser., no. 15, p. 129-134.

Bissell, H.J., 1964, Ely, Arcturus, and Park City Groups (Pennsylvanian-Permian) in eastern Nevada and western Utah: Am. Assoc. Petrol. Geol. Bull., v. 48, p. 565-636.

Bluck, B.J., 1964, Sedimentation of an alluvial fan in southern Nevada: Jour. Sediment. Petrol., v. 34, p. 395-400.

Bois, C., P. Bouché, and R. Pelet, 1982, Global geological history and the distribution of hydrocarbon reserves: Am. Assoc. Petrol. Geol. Bull., v. 66, p. 1248-1271.

Bonham, L.C., 1980, Migration of hydrocarbons in compacting basins, in Roberts, W.H., III, and R.J. Cordell, eds., Problems of Petroleum Migration: Am. Assoc. Petrol. Geol. Stud. Geol., no. 10, p. 69-88.

Bott, M.H.P., 1979, Subsidence mechanisms and passive continental margins: in Watkins, J.S, L. Montadert, and P.W. Dickerson, eds., Geological and Geophysical Investigations of Continental Margins: Am. Assoc. Petrol. Geol. Mem., no. 29, p. 3-10.

Boucot, A.J., 1962, Discussion: Hunton Group (Silurian and Devonian) and related strata in Oklahoma: Am. Assoc. Petrol. Geol. Bull., v. 46, p. 1528-1532.

Bouma, A.H., 1962, Sedimentology of some flysch deposits: Amsterdam, Elsevier, 168 p.

Bourcart, J., 1964, Les sables de la Mediterranée occidentale, in Bouma, A.H., and A. Brouwer, eds., Turbidites, Developments in Sedimentology: New York, Elsevier, v. 3, p. 148-155.

Bova, J.A., and J.F. Read, 1987, Incipiently drowned facies within a cyclic peritidal ramp sequence, early Ordovician Chepultepec interval, Virginia Appalachians: Geol. Soc. Am. Bull., v. 98, p. 714-727.

Bowen, N. L., 1928, The evolution of igneous rocks: Princeton, NJ, Princeton Univ. Press. 332 p.

Bradley, J.S., 1975, Abnormal formation pressure: Am. Assoc. Petrol. Geol. Bull., v. 59, p. 957-973.

Braitsch, O., 1971, Salt Deposits—Their Origin and Composition: New York, Springer-Verlag, 297 p.

Bretsky, P.W., and D.M. Lorenz, 1970, Adaptive response to environmental stability—a unifying concept in paleoecology, in Yockelson, E.L., ed., Proc. N. Am. Paleontol. Conv., part A, v. 1: Lawrence, KS, Allen Press, p. 522-550.

Briggs, L.I., 1958, Evaporite facies: Jour. Sediment. Petrol., v. 28, no. 1, p. 46-56.

Brink, A.H., 1974, Petroleum geology of Gabon basin: Am. Assoc. Petrol. Geol. Bull., v. 58, p. 216-235.

Brinkmann, R., 1929, Statistisch-biostratigraphische Untersuchungen an mittelJurassischen Ammoniten über Artbegriff und Stammesentwicklung, Abh.,: Gesellschaft der Wissenschaften zu Gottingen, Math-Phys. Klasse, N.F. l3, no. 3, p. 1-249.

Bristow, C.S., 1987, Bramaputra River channel migration and deposition, in Ethridge, F.G., R. M. Flores, and M.D. Harvey, eds., Recent Developments in Fluvial Sedimentology: Soc. Econ. Paleont. Mineral. Spec. Publ., no. 39, p. 63-74.

Brongersma-Sanders, M., and P. Groen, 1970, Wind and water depth and their bearing on the circulation in evaporite basins: N. Ohio Geol. Soc. Third Symp. on Salt Trans., v. 1, p. 3-7.

Bronowski, J., and B. Mazlish, 1960, The Western Intellectual Tradition: New York, Harper & Row, 512 p.

Brown, L.F., Jr., 1984, Seismic stratigraphic analysis of depositional systems: Austin, TX, 110 p.

Brown, L.F., Jr., and W. L. Fisher, 1977, Seismic-stratigraphic interpretation of depositional systems: examples from Brazilian rift and pull-apart basins, in Payton, C.E., ed., Seismic stratigraphy—applications to hydrocarbon exploration: Am. Assoc. Petrol. Geol. Mem., no. 26, p. 213-248.

Brown, L.F., Jr., and W. L. Fisher, 1980, Seismic Stratigraphic Interpretation and Petroleum Exploration: Am. Assoc. Petrol. Geol. Short Course Notes, no. 16, 192 p.

Brown, R.W., 1962, Paleocene flora of the Rocky Mountain and Great Plains: U.S. Geol. Surv. Prof. Pap., no. 375, 119 p.

Bruce, C.H., 1973, Pressured shale and related sediment deformation: mechanism for development of regional contemporaneous faults: Am. Assoc. Petrol. Geol. Bull., v. 55, p. 878-886.

Bubb, J.N., and W.G. Hatlelid, 1977, Seismic recognition of carbonate buildups, in Payton, C.E., ed., Seismic Stratigraphy: Applications to Hydrocarbon Exploration: Am. Assoc. Petrol. Geol. Mem., no. 20, p. 186-204.

Bucher, W. H., 1936, The concept of natural law in geology: Ohio J. Sci., v. 36, p. 183-194.

Buffon, 1778, Histoire naturelle des époques de la nature: Histoire Naturelle, Général et Particulière, supplément V.

Bull, W.B., 1972, Recognition of alluvial-fan deposits in the stratigraphic record, in Rigby, J.K., and W.K. Hamblin, eds., Recognition of Ancient Sedimentary Environments: Soc. Econ. Paleon. Mineral. Spec. Publ., no. 16, p. 63-83.

Burk, C.A., and C.L. Drake, eds., 1974, The Geology of

Continental Margins: New York, Springer-Verlag, p. 3–10.

Busch, D.A., 1953, The significance of deltaic subsurface exploration: Tulsa Geol. Soc. Dig., v. 21, p. 71–80.

Busch, D.A., 1973, Genetic units in delta prospecting: Am. Assoc. Petrol. Geol. Bull., v. 55, p. 1137–1154.

Busch, D.A., 1974, Stratigraphic Traps in Sandstone—Exploration Techniques: Am. Assoc. Petrol. Geol. Mem., no. 21, 194 p.

Campbell, C.V., 1976, Reservoir Geometry of a fluvial sheet sandstone: Am. Assoc. Petrol. Geol. Bull., v. 60, p. 1009–1019.

Campos, C.W.M., K. Muira, and L.A.N. Reis, 1975, The east Brazilian continental margin and petroleum prospects: Ninth World Petroleum Congress (preprint), Panel Discussion, v. 2, p. 71–81.

Canby, T.Y, 1983, Satellites that serve us: National Geographic, v. 164, no. 3, p. 281–335.

Cant, D.F., and F.J. Hein, 1986, Depositional sequences in ancient shelf sediments—some contrasts in style, in Knight, R.J., and J.R. McLean, eds., Shelf Sands and Sandstones: Can. Soc. Petrol. Geol. Mem., no. 11, p. 303–312.

Cavaroc, V.V., and J.C. Ferm, 1968, Siliceous spiculites as shoreline indicators in deltaic sequences: Geol. Soc. Am. Bull., v. 79, p. 263–272.

Chamberlin, T.C., 1897, The method of the multiple working hypothesis: Jour. Geol., v. 5, p. 837–848.

Chang-shu, Y., and S. Jia-song, 1988, Tidal sand ridges on the east China Sea shelf, in de Boer, P.L., A. van Gelder, and S.D. Nio, eds., Tide-Influenced Sedimentary Environments and Facies: Dordrecht, Holland, D. Reidel Publ. Co., p. 23–38.

Chapin, C.E., 1985, Two-stage Laramide orogeny in southwestern United States: tectonics and sedimentation (abst.): Am. Assoc. Petrol. Geol. Bull., v. 69, p. 2044.

Chatfield, J., 1972, Case history of Red Wash Field, Uintah County, Utah., in King, R.E., ed., Stratigraphic Oil & Gas Fields—Classification, Exploration Methods, and Case Histories: Am. Assoc. Petrol. Geol. Mem., no. 16, p. 342–353.

Chenoweth, P.A., 1967, Unconformity analysis: Am. Assoc. Petrol. Geol. Bull., v. 51, p. 4–27.

Chien, N., 1961, The braided stream of the lower Yellow River: Sci. Sinica, v. 10, p. 734–754.

Chowdary, L.R., 1975, Reversal of basement-block motions in Cambay Bay, India, and its importance in petroleum exploration: Am. Assoc. Petrol. Geol. Bull., v. 59, p. 85–96.

Clark, T.H., and C.W. Stearns, 1968, Geological Evolution of North America, 2nd ed.: New York, Ronald Press, 570 p.

Clayton, R.W., and R.P. Comer, 1983, A tomographic analysis of mantle heterogeneities from body wave travel times (abst.): EOS, Trans. Am. Geophys. Union, v. 64, p. 776.

Clifton, H.E., 1976, Wave-formed sedimentary structures: a conceptual model, in Davis, R.A., Jr., and R.L. Ethington, eds., Beach and Nearshore Sedimentation: Soc. Econ. Paleont. Mineral. Spec. Pub., no. 24, p. 126–148.

Clifton, H.E., R.E. Hunter, and R.L. Phillips, 1971, Depositional structures and processes in nonbarred high-energy near-shore: Jour. Sediment. Petrol., v. 41, p. 651–670.

Coleman, J.M., 1969, Brahmaputra River channel processes and sedimentation: Sediment. Geol., v. 3, p. 129–239.

Coleman, J.M., and S.M. Gagliano, 1964, Cyclic sedimentation in the Mississippi Delta Plain: Gulf Coast Assoc. Geol. Soc. Trans., v. 14, p. 67–80.

Coleman, J.M., and S.M. Gagliano, 1965, Sedimentary structures—Mississippi River deltaic plains: Primary Sedimentary Structures and Their Hydrodynamic Interpretation—A Symposium: Soc. Econ. Paleont. Mineral. Spec. Publ., no. 12, p. 133–148.

Coleman, J.M., and S.M. Gagliano, 1969, Sedimentary structures: Mississippi River deltaic plain: Recent and Ancient Deltaic Sediments: A Comparison: Baton Rouge, Louisiana State Univ., 205 p.

Coleman, J.M., and D.B. Prior, 1980, Deltaic sand bodies: Am. Assoc. Petrol. Geol. Cont. Educ. Short Course Notes, no. 15, 171 p.

Coleman, J.M., and L.D. Wright, 1973, Variability of modern river deltas: Gulf Coast Assoc. Geol. Soc. Trans., v. 23, p. 33–36.

Coleman, J.M., and L.D. Wright, 1975, Modern river deltas: variability of processes and sand bodies, in Broussard, M.L., ed., Deltas, Models for Exploration: Houston, Houston Geol. Soc., p. 99–149.

Collins, A.G, 1975, Geochemistry of Oilfield Waters: Developments in Petroleum Science: New York, Elsevier, v. 1, 496 p.

Collinson, J.D., 1986, Deserts, in Reading, H.G., ed., Sedimentary Environments and Facies, 2nd ed.: Oxford, Blackwell Scientific Publications, 615 p.

Connolly, J.R., and G.C. Ferm, 1971, Permian-Triassic sedimentation patterns, Sydney Basin, Australia: Am. Assoc. Petrol. Geol. Bull. v. 55, p. 2018–2032.

Conrey, B.L., 1959, Sedimentary history of the early Pliocene in the Los Angeles basin, California (unpubl. Ph.D. dissertation): Univ. Southern California, 273 p.

Conrey, B.L., 1967, Early Pliocene sedimentary history of the Los Angeles Basin, California: Calif. Div. Mines and Geol. Spec. Rep., no. 93, 63 p.

Courtillot, V., J. Besse, 1987, Magnetic field reversals, polar wander, and core-mantle coupling: Science, v. 237, p. 1140–1147.

Crain, W.E., W.E. Mero, and D. Patterson, 1985, Geology of the Point Arguello discovery: Am. Assoc. Petrol. Geol. Bull., v. 69, p. 537–545.

Creager, J.S., and R.W. Sternberg, 1972, Some specific problems in understanding bottom sediment distribution and dispersal on the continental shelf, in Swift, D.J.P., D.B. Duane, and O.H. Pilkey, eds., Shelf Sediment Transport—Process and Pattern: Stroudsburg, PA, Dowden, Hutchinson, and Ross, p. 347–362.

Crowell, J.C., 1974, Origin of Late Cenozoic basin in southern California, in Dickinson, W.R., ed., Tectonics and Sedimentation: Soc. Econ. Paleont. Mineral. Mem., no. 22, p. 190–204.

Crowell, J.C., 1975, The San Gabriel fault and Ridge basin, southern California, in Crowell, J.C., ed., San Andreas Fault in Southern California: Calif. Div. Mines and Geol. Spec. Rep., v. 118, p. 208–233.

Curray, J.R., 1960, Sediments and history of Holocene transgression, Continental Shelf, northwest Gulf of Mexico, in Shepard, F.P., F.B. Phleger, and T.H. van Andel, eds., Recent Sediments, Northwest Gulf of Mexico: Am. Assoc. Petrol. Geol., v. 44, p. 221–266.

Curray, J.R., 1964, Transgressions and regressions, in Miller, R.L., ed., Papers in Marine Geology: New York, McMillan Co., p. 175–203.

Curray, J.R., 1965, Late Quaternary history, continental shelves of the United States, in Wright, H.E., and D.G. Frey, eds. The Quaternary of the United States: Princeton, NJ, Princeton Univ. Press, p. 723–735.

Curray, J.R., 1969, Shore zone sand bodies: barriers, cheniers, and beach ridges, in Stanley, D.J., ed., The New Concepts of Continental Margin Sedimentation: Washington, DC, Am. Geol. Inst. Short Course Lecture Notes, Philadelphia, p. JC11-1—JC11-18.

Curray, J.R., and D.G. Moore, 1971, Growth of the Bengal deep-sea fan and denudation in the Himalayas: Geol. Soc. Am. Bull, v. 82, p. 563–572.

Dailly, G.C., 1975, Some remarks on regression and transgression in deltaic sediments, in Yorath, C.T., E.R. Parker, and D.J. Glass, eds., Canada's Continental Margins and Offshore Petroleum Exploration: Can. Soc. Petrol. Geol. Mem., no. 4, p. 791–820.

Dailly, G.C., 1982, Slope readjustment during sedimentation on continental margins, in Watkins, J.S., and C.L. Drake, eds., Studies in Continental Margin Geology: Am. Assoc. Petrol. Geol. Mem., no. 34, p. 593–610.

Damuth, J.E., R.O. Kawsmann, R.D. Flood, R.H. Belderson, and M.A. Gorini, 1983, Age relationships of distributary channel on Amazon deep-sea fan: Implications for fan growth pattern: Geology, v. 11, p. 47–473.

Dana, J.D., 1873, On some results of the earth's contraction from cooling, including a discussion of the origin of mountains and the nature of the earth's interior: Am. Jour. Sci., ser. 3, v. 5, p. 423–443; v. 6, p. 6–14, 104–115, 161–172.

Dapples, E.C., 1955, General lithofacies relationship of St. Peter sandstone and Simpson Group: Am. Assoc. Petrol. Geol. Bull., v. 39, p. 444–467.

Darwin, C., 1859, On the Origin of Species: London, John Murray, 502 p.

Dauzacker, M.D., H. Schaller, A.C.M. Castro, Jr., M.M. de Souza, 1985, Geology of Brazil's Atlantic margin basins: Oil & Gas Jour., v. 83, no. 9, p. 142–144.

Davies, P.J., J. Marshall, and D. Hopley, 1985, Relationships between reef growth and sea level in the Great Barrier Reef: Fifth Int. Coral Reef Symp. Proc., Tahiti, v. 3, p. 95–103.

Davis, R.A., 1983, Depositional Systems: A Genetic Approach to Sedimentary Geology: Englewood Cliffs, NJ, Prentice Hall, 669 p.

de Boer, P.L., A. van Gelder, and S. D. Nio, eds., 1988, Tide-influenced Sedimentary Environments and Facies: Dordrecht, Holland, D. Reidel Pub. Co., 539 p.

de Maeyer, Ph., and S. Wartel, 1988, Relation between superficial sediment grainsize and morphological features, in de Boer, P.L., A. van Gelder, and S.D. Nio, eds., Tide-Influenced Sedimentary Environments and Facies: Dordrecht, Holland, D. Reidel Publish. Co., p. 91–100.

Demaison, G.J., and G.T. Moore, 1980, Anoxic environments and oil source bed genesis: Am. Assoc. Petrol. Geol. Bull., v. 64, p. 1179–1209.

DeMatharel, M., P. Lehmann, and T. Oki, 1980, Geology of the Bekapai field, in Halbouty, M.T., ed., Giant Oil and Gas Fields of the Decade 1968-1978: Am. Assoc. Petrol. Geol. Mem., no. 30, p. 459–470.

Denney, C.S., 1967, Fans and pediments: Am. Jour. Sci., v. 265, p. 81–105.

Deutsch, K.W., 1948-1949, Some notes on research on the role of models in the natural and social sciences: Synthese, v. 7, p. 506–533.

Dewey, J.F., and J.M. Bird, 1970, Mountain belts and the new global tectonics: Jour. Geophys. Res., v. 75, p. 2625–2647.

Dickinson, K.A., 1968, Petrology of Buckner Formation in adjacent parts of Texas, Louisiana, and Arkansas: Jour. Sediment. Petrol., v. 38, p. 555–567.

Dickinson, W.R., 1974, Plate tectonics and sedimentation, in Dickinson, W.R., ed., Tectonics and Sedimentation: Soc. Econ. Paleont. Mineral. Spec. Publ., no. 22, p. 1–27.

Dietz, R.S., 1961, Continent and ocean basin evolution by spreading of the sea floor: Nature, v. 190, p. 854–857.

Dow, W.G., 1979, Petroleum source beds on continental slopes and rises, in Watkins, J.S., L. Montadert, and P.W. Dickerson, eds., Geological and Geophysical Investigations of Continental Margins: Am. Assoc. Petrol. Geol. Mem., no 29, p. 423–442.

Droz, L., and G. Bellarche, 1985, Rhone deep-sea fan morphostructure and growth pattern: Am. Assoc. Petrol. Geol. Bull., v. 69, p. 460–479.

Dutton, S.P., 1982, Pennsylvanian fan-delta and carbonate deposition, Mobeetie field, Texas Panhandle: Am. Assoc. Petrol. Geol. Bull., v. 66, p. 389–407.

Dziewonski, A.M., and D.L. Anderson, 1984, Seismic tomography of the earth's interior: Am. Sci., v. 72, p. 483–494.

Dziewonski, A.M., and J.H. Woodhouse, 1987, Global images of the earth's interior: Science, v. 236, p. 37–48.

Eicher, D.L., 1968, Geologic Time: Englewood Cliffs, NJ, Prentice-Hall, 150 p.

Ejedawe, J.E., 1981, Patterns of incidence of oil reservoirs in Niger Delta basin: Am. Assoc. Petrol. Geol. Bull., v. 65, p. 1574–1585.

Eldredge, N., and S. J. Gould, 1972, Punctuated equilibria: an alternative to phyletic gradualism, in Schopf, T.J.M., ed., Models in Paleobiology: San Francisco, Freeman,

Cooper, & Co. p. 82–115.

Elias, G.K., 1963, Habitat of Pennsylvanian algal biotherms, Four Corners Area: Farmington, NM, 4th Four Corners Geol. Soc. Field Conf., p. 185–203.

Elliott, R.E., 1968, A classification of subaqueous sedimentary structures based upon rheology and kinematical properties: Sedimentology, v. 5, p. 193–209.

Emery, K.O., 1960, The continental shelves: Sci. Amer., v. 221, p. 106–122.

Emery, K.O., 1969, Relict sediments on continental shelves of the world: Am. Assoc. Petrol. Geol. Bull., v. 52, p. 445–464.

Enos, P., 1969, Anatomy of a flysch: Jour. Sediment. Petrol., v. 39, p. 680–723.

Erickson, D.B., M. Ewing, and B.C. Heezen, 1952, Turbidity currents and sediments in the North Atlantic: Am. Assoc. Petrol. Geol. Bull., v. 36, p. 489–511.

Ethridge, F.G., and W.A Wescott, 1984, Tectonic setting, recognition and hydrocarbon reservoir potential of fan-delta deposits, in Koster, E.H., and R.J. Steel, eds., Sedimentology of Gravels and Conglomerates: Can. Soc. Petrol. Geol. Mem., no. 10, p. 217–236.

Evans, C.R., D.K. Melvor, and K. Magara, 1975, Organic matter, compaction history, and hydrocarbon occurrence—MacKenzie Delta, Canada: Ninth World Petrol. Congr., v. 3, no. 2, p. 149–157.

Evans, G., 1965, Intertidal flat sediments and their environments of deposition in the wash: Geol. Soc. London Quart. Jour., v. 121, p. 209–245.

Evans, G., 1975, Intertidal flat deposits of the wash, western margin of the North Sea, in Ginsburg, R.N., ed., Tidal Deposits: New York, Springer-Verlag, p. 13–20.

Evans, W.E., 1970, Imbricate linear sandstone bodies of the Viking formation in Doddsland-Hoosir area of southwestern Saskatchewan: Am. Assoc. Petrol. Geol. Bull., v. 54, p. 469–480.

Falt, L.M, 1986, Tectonic control on sedimentation in the Statfjord formation, Statfjord Field, N. North Sea (abst.): Canberra, Australia, 12th Intern. Sediment. Congr., p. 101.

Falvey, D.A., 1974, The development of continental margins in plate tectonic theory: Aust. Petrol. Explor. Assoc. Jour., p. 95–106.

Faul, H., 1966, Ages of Planets and Stars: New York, McGraw-Hill, 109 p.

Fischer, A.G., 1975, Tidal deposits Dachstein Limestone of the North Alpine Triassic, in Ginsburg, R.N., ed., Tidal Deposits: New York, Springer-Verlag, p. 235–242.

Fischer, A. G., 1981, Climatic oscillations in the biosphere: in Nitecki, M.H., ed., Biotic Crises in Ecological and Evolutionary Time: New York, Academic Press, p. 103–131.

Fischer, A.G., and M.A. Arthur, 1977, Secular variations in the pelagic realm, in Cook, H. E., and P. Enos, eds., Deep-water Carbonate Environments: Soc. Econ. Paleont. Mineral. Spec. Publ., no. 25, p. 19–50.

Fisher, J.J., 1968, Barrier island formation: discussion: Geol. Soc. Am. Bull., v. 79, p. 1421–1424.

Fisher, R.V., 1983, Flow transformation in sediment gravity flows: Geology, v. 11, p. 273–274.

Fisher, W.L., 1969, Facies characterization of Gulf Coast Basin delta systems with some Holocene analogues: Gulf Coast Assoc. Geol. Soc. Trans., v. 19, p. 239–261.

Fisher, W.L., and J.H. McGowen, 1969, Depositional systems in Wilcox Group (Eocene) of Texas and their relation to occurrence of oil and gas: Am. Assoc. Petrol. Geol. Bull., v. 53, p. 30–34.

Fisk, H.N., 1961, Bar-finger sands of the Mississippi Delta, in Peterson, J.A., and J.C. Osmond, eds., Geometry of Sandstone Bodies: Am. Assoc. Petrol. Geol. Mem., no. 3, p. 29–52.

Fleming, R.H., 1938, Tides and tidal currents in the Gulf of Panama: Jour. Mar. Res., v. 1, p. 192–206.

Flores, R.M., and C.L. Pillmore, 1987, Tectonic control on alluvial paleoarchitecture of the Cretaceous and Tertiary Raton basin, Colorado and New Mexico, in Ethridge, F.G., R.M. Flores, and M.D. Harvey, eds., Recent Developments in Fluvial Sedimentology: Soc. Econ. Paleont. Mineral. Spec. Publ., no. 39, p. 311–320.

Forgotson, J.M., Jr., A.T. Statler, and M. David, 1966, Influence of regional tectonics and local structure on deposition of Morrowan formation in west Anadarko basin: Am. Assoc. Petrol. Geol. Bull., v. 50, p. 518–532.

Frank, J.R., S. Cluff, and J.M. Bauman, 1982, Painter reservoir, east Painter reservoir and Clear Creek fields, Uinta County, Wyoming, in Powers, R.B., ed., Geologic Studies of the Cordilleran Thrust Belt: Denver, Rocky Mountain Assoc. Geol., p. 601–611.

Frankel, H., 1979, Why drift theory was accepted with the confirmation of Harry Hess's concept of sea-floor spreading, in Schneer, C.J., Two Hundred Years of Geology in America: Hanover, NH, Univ. of New Hampshire Press, p. 337–353.

Freeman, W.E., and G.S. Visher, 1975, Stratigraphic analysis of the Navajo sandstone: Jour. Sediment. Petrol., v. 45, no. 3, p. 651–668.

Frey, R.W., 1971, Ichnology—the study of fossil and recent Lebensspuren, in Trace Fossils: Louisiana State Univ. School of Geosciences Misc. Publ., no. 71–1, p. 91–125.

Frey, R.W., 1975, The Study of Trace Fossils. A Study of Principles, Problems, and Procedures in Ichnology: New York, Springer-Verlag, 562 p.

Friedman, G.M., and J.E. Sanders, 1978, Principles of Sedimentology: New York, John Wiley & Sons, 792 p.

Frostick, L.E., and I. Reid, eds., 1987, Desert Sediments Ancient and Modern: London Geological Society, Special Publication no. 35: Palo Alto, CA, Blackwell Scientific Publications, 401 p.

Füchtbauer, H., 1974, Sediments and Sedimentary Rocks 1: Stuttgart, Schweizerbart'sche Verlagsbuchhandlung, 464 p.

Gagliano, S.M., and W. McIntire, 1968, Reports on the Mekong River Delta: Louisiana State Univ. Coastal Studies Inst. Tech. Rep., no. 57, 143 p.

Galloway, W.E., 1968, Depositional systems of the Lower Wilcox Group, North-Central Gulf Coast Basin: Gulf Coast Assoc. Geol. Soc. Trans., v. 18, p. 275–289.

Galloway, W.E., 1975, Process framework for describing the morphologic and stratigraphic evolution of deltaic depositional systems, in Broussard, M.L., ed., Deltas, Models for Exploration: Houston, Houston Geol. Soc., p. 87–98.

Galloway, W.E., 1976, Sediments and stratigraphic framework of the Copper River fan delta: Alaska Jour. Sediment. Petrol., v. 46, p. 726–737.

Galloway, W.E., and L.F. Brown, 1973, Depositional systems and shelf-slope relations on cratonic basin margin, uppermost Pennsylvanian of north-central Texas: Am. Assoc. Petrol. Geol. Bull., v. 66, p. 649–688.

Galloway, W.R., M.S. Yancey, and A.P. Whipple, 1977, Seismic stratigraphic model of depositional platform, Eastern Anadarko Basin, Oklahoma, in Seismic Stratigraphy—Application to Hydrocarbon Exploration: Am. Assoc. Petrol. Geol. Mem., no. 26, p. 439–450.

Garrels, R.M., and F.T. Mackenzie, 1971, Evolution of Sedimentary Rocks: New York, W.W. Norton and Co., 397 p.

Germanov. A.L., 1963, Role of organic substances in the formation of hydrothermal sulfide deposits: Int. Geol. Rev., v. 5, p. 379–394.

Ghigone, J.I., and G. de Andrade, 1970, General geology and major oil fields of Reconcavo basin, Brazil, in Geology of Giant Petroleum Fields: Am. Assoc. Petrol. Geol. Mem., no. 14, p. 337–358.

Gilbert, G.K., 1914, The Transportation of Debris by Running Water: U.S. Geol. Surv. Prof. Pap., no. 86, 263 p.

Gilbert, G.K., 1886, The inculcation of scientfic method by example, with an illustration drawn from the Quaternary geology of Utah: Am. Jour. Sci., ser. 3, v. 31, p. 284–299.

Gilluly, J., 1963, The scientific philosophy of G. K. Gilbert, in Albritton, C.C., Jr., ed., The Fabric of Geology: Stanford, CA, Freeman, Cooper, & Co., p. 218–224.

Ginsburg, R.N., 1971, Landward movement of carbonate mud—new model for regressive cycles in carbonates: Am. Assoc. Petrol. Geol. Bull., v. 55, p. 340.

Ginsburg, R.N., ed., 1975, Tidal Deposits: A Casebook of Recent Examples and Fossil Counterparts: New York, Springer-Verlag, 428 p.

Ginsburg, R.N., and L.S. Hardie, 1975, Tidal and storm deposits, Northeastern Andros Island, Bahamas, in Ginsburg, R.N., ed., Tidal Deposits: New York, Springer-Verlag, p. 201–208.

Ginsburg, R.N., and N.P. James, 1974, Holocene carbonate sediments of continental shelves, in Burk, C.A., and C.L. Drake, eds., The Geology of Continental Margins: New York, Springer-Verlag, p. 137–155.

Glennie, K.W., 1970, Desert sedimentary environments: Developments in Sedimentology: New York, Elsevier Pubishing Co., no. 14, 222 p.

Gloe, C.S., 1960, The geology of the Latrobe Valley coalfields: Australian Institute of Mining and Metallurgy, Proc. no. 194, p. 57–125.

Gödel, Kurt, 1931, Monatshefte fur Mathamatik und Physik, v. 38: New York, Springer-Verlag.

Goetz, J.L., W.J. Prins, and J.F. Logar, 1977, Reservoir delineation by wireline techniques (preprint): Jakarta, Indonesia, 6th Annu. Conv. Indones. Petrol. Assoc.

Goldhammer, R.K., 1988, Superimposed platform carbonate cycles eustatic response of an aggradational carbonate buildup, Middle Triassic of the Dolomites (abst.): Am. Assoc. Petrol. Geol. Bull., v. 72, p. 190.

Gorsline, D.S., and K.O. Emery, 1959, Turbidity current deposits in San Pedro and Santa Monica basins off southern California: Geol. Soc. Am. Bull., v. 70, p. 279–290.

Gould, H.R., 1970, The Mississippi Delta complex, in Morgan, J.P., and R.H. Shaver, eds., Deltaic Sedimentation Modern and Ancient: Soc. Econ. Paleont. Mineral. Spec. Publ., no 15, p. 3–30.

Gould, H.R., and E. McFarlan, Jr., 1959, Geologic history of the Chenier Plain southwestern Louisiana: Gulf Coast Assoc. Geol. Soc. Trans., v. 9, p. 261–270.

Gould, H.R., and J.P. Morgan, 1962, Coastal Louisiana swamps and marshlands, in Rainwater, E.H., and R.P. Zingula, eds., Geology of the Gulf Coast and Central Texas and Guidebook of Excursions: 1962 Annual Meeting, Houston Geol. Soc., p. 287–341.

Gould, S.J., 1985, The paradox of the first tier: an agenda for paleobiology: Paleobiology, v. 11, p. 2–12.

Gould, S.J., 1986, Evolution and the triumph of homology, or why history matters: Am. Sci., v. 74, p. 60–69.

Gould, S.J., and N. Eldredge, 1977, Punctuated equilbria: the tempo and mode of evolution reconsidered: Paleobiology, v. 3, p. 115–151.

Grant, A.C., K.D. McAlpine, and J.A. Wade, 1986, The continental margin of eastern Canada: geological framework and petroleum potential, in Halbouty, M.T., ed., Future Petroleum Provinces of the World: Am. Assoc. Petrol. Geol. Mem., no. 40, p. 177–205.

Gregory, J.L., 1966, A Lower Oligocene delta in the subsurface of southeastern Texas, in Deltas in Their Geologic Framework: Houston, Houston Geol. Soc., p. 213–227.

Gries, R.R., and R.C. Dyer, eds., 1987, Seismic Exploration of the Rocky Mountain Region: Denver, CO, Rocky Mountain Assoc. Geologists, 300 p.

Grieve, R.A.F., 1982, The geological record of impacts, in Silver, L.T., and P.H. Schultz, eds., Geological Implications of Impacts of Large Asteroids and Comets on the Earth: Geol. Soc. Am. Spec. Pap., no. 190, p. 25–37.

Griffith, E.G., 1966, Geology of Saber Bar, Lagoon, Logan and Wold Counties, Colorado: Am. Assoc. Petrol. Geol. Bull., v. 50, p. 2112–2118.

Griffith, L.S., M.G. Pitcher, and G.W. Rice, 1969, Analysis of a Lower Cretaceous reef complex, in Friedman, G.M., ed., Depositional Environments in Carbonate Rocks: Soc. Econ. Paleont. Mineral. Spec. Publ., no. 14, p. 120–138.

Griggs, G.B., and L.D. Kulm, 1970, Physiography of Cascadia deep-sea channel: Geol. Soc. Am. Bull., v. 81, p. 82–94.

Grotzinger, J.P., 1986, Cyclicity and paleoenvironmental dynamics, Rocknest platform, northwest Canada: Geol. Soc. Am. Bull., v. 97, p. 1208–1231.

Grow, J.A., R.E. Mattlock, and J. Schlee, 1979, Multichannel seismic depth section and internal velocities over outer continental shelf and upper continental slope between Cape Hatteras and Cape Cod, in Watkinson, J.S., L. Montadent, and A.W. Dickerson, eds., Geological and Geophysical Investigations of Continental Margins: Am. Assoc. Petrol. Geol. Mem., no. 29, p. 65–83.

Guettard, J.E., 1751, Memoire et carte mineralogique sur la nature & l'Angleterre: Paris, Memoires de l'Academie Royale des Sciences, v. for 1746, p. 363–392.

Gutierrez, F.J., 1968, A subsurface study of the Bachaquero Interfield area, Bolivar coastal fields, Lake Maracaibo, Venezuela (unpubl. M.S. thesis): Univ. of Tulsa, 70 p.

Haber, F.C., 1959, The Age of the World: Moses to Darwin: Baltimore, MD, The Johns Hopkins Press, 303 p.

Hack, J.T., 1960, Interpretation of erosional topography in humid temperate regions: Am. Jour. Sci., v. 258-A, p. 80–97.

Hagan, G.M., and B.W. Logan, 1975, Prograding tidal-flat sequences: Hutchinson Embayment, Shark Bay, Western Australia, in Ginsburg, R.N., ed., Tidal Deposits: New York, Springer-Verlag, p. 215–222.

Halbouty, M.T., and T.D. Barber, 1961, Port Acres and Port Arthur fields, Jefferson County, Texas: Gulf Coast Assoc. Geol. Soc. Trans., v. 11, p. 225–234.

Halbouty, M.T., A.A. Meyerhoff, R.E. King, R.H. Dott, Sr., H.D. Klemme, and T. Shabad, 1970, World's giant oil and gas fields, geologic factors affecting their formation, Part I, giant oil and gas fields, in Halbouty, M.T., ed., Geology of Giant Petroleum Fields: Am. Assoc. Petrol. Geol. Mem., no. 14, p. 502–528.

Hale, L.A., 1959, Frontier formations—Coalville, Utah, and nearby areas of Wyoming and Colorado: Early Cretaceous Rocks of Wyoming, 17th Ann. Wyo. Geol. Assoc. Field Conf., Casper, Wyoming, p. 211–220.

Hamilton, W.S., Jr., 1979, Tectonics of the Indonesian Region: U.S. Geol. Surv. Prof. Paper, no. 1078, 345 p.

Hamilton, W.S., Jr., and C.P. Cameron, 1986, Facies relationships and depositional environments of lower Tuscaloosa formation reservoir sandstones in McComb field area, southwest Mississippi (abst.): Am. Assoc. Petrol. Geol. Bull., v. 70, p. 598–599.

Hampton, M.A., 1972, The role of subaqueous debris flows in generating turbidity currents: Jour. Sediment. Petrol., v. 42, p. 775–793.

Haq, B.U., J. Hardenbol, and P.R. Vail, 1987, Chronology of fluctuating sea levels since the Triassic: Science, v. 235, p. 1156–1167.

Haq, B.U., and F.W. van Eysinga, 1987, Geological Time Table; 4th rev. ed.: New York, Elsevier Science Publ. Co., Inc., chart.

Harbaugh, J.W., and D.F. Merriam, 1968, Computer Applications in Stratigraphic Analysis: New York, John Wiley & Sons, 282 p.

Harding, T.P., 1973, Newport-Inglewood trend, California—an example of wrenching style of deformation: Am. Assoc. Petrol. Geol. Bull., v. 57, p. 97–116.

Harding, T.P., 1984, Graben hydrocarbon occurrences and structural style: Am. Assoc. Petrol. Geol. Bull., v. 68, p. 333–362.

Harding, T.P., 1985, Seismic characteristics and identification of negative flower structures, positive flower structures, and positive structural inversion: Am. Assoc. Petrol. Geol. Bull., v. 69, p. 582–600.

Harding, T.P., and J.D. Lowell, 1979, Structural styles, their plate-tectonic habitats, and hydrocarbon traps in petroleum provinces: Am. Assoc. Petrol. Geol. Bull., v. 63, p. 1016–1058.

Harland, W.B., A.V. Cox, P.C. Llewellyn, C.A.G. Picton, A.G. Smith, and R. Walters, 1982, A geologic time scale: Cambridge, Engl., Cambridge Univ. Press, 131 p.

Harms, J.C., 1975, Stratification and sequences in prograding shoreline deposits: Soc. Econ. Paleont. Mineral. Short Course no. 2, p. 81–102.

Harms, J.C. and F.A. Exum, 1966, Stratigraphic traps in a valley fill, western Nebraska: Am. Assoc. Petrol. Geol.

Bull., v. 50, p. 2119–2149.

Harms, J.C., and W.J. McMichael, 1983, Sedimentology of the Brae oilfield area: North Sea Jour. Petrol. Geol., v. 5, p. 437–439.

Harms, J.C., and P. Tackenberg, 1972, Seismic signatures of sedimentation models: Geophysics, v. 37, p. 45–58.

Harms, J.C., P. Tackenberg, E. Pickles, and R.E. Pollock, 1981, The Brae oilfield area, in Illing, L.W., and G.D. Hobson, eds., Petroleum Geology of the Continental Shelf of Northwest Europe London: Heyden, p. 352–357.

Hays, J.D., J. Imbrie, and J.J. Shackleton, 1976, Variations in the Earth's orbit: pacemaker of the ice ages: Science, v. 194, p. 1121–1132.

Heckel, P.H., 1972, Recognition of Ancient shallow marine environments, in Rigby, J.K., and W.K. Hamblin, eds., Recognition of Ancient Sedimentary Environments: Soc. Econ. Paleont. Mineral. Spec. Publ., no. 16, p. 226–286.

Heckel, P.H., 1986, Sea-level curve for Pennsylvanian eustatic marine transgressive-regressive depositional cycles along mid-continent outcrop belt, North America: Geology, v. 14, p. 330–334.

Hedburg, H.D., 1961, The stratigraphic panorama: Geol. Soc. Am. Bull., v. 72, p. 499–517.

Hedburg, H.D., ed., 1972, An International Guide to Stratigraphic Classification, Terminology, and Usage: Lethaia, v. 5, p. 283–295, p. 297–323.

Hedburg, H.D., 1974, Relation of methane generation to under-compacted shales, shale diapirs, and mud volcanoes: Am. Assoc. Petrol. Geol. Bull., v. 58, p. 661–673.

Hedburg, H.D., 1980, Methane generation and petroleum migration, in Roberts, W.H., III, and R.J. Cordell, eds., Problems of Petroleum Migration: Am. Assoc. Petrol. Geol. Stud. in Geol. no. 10, p. 179–206.

Heezen, B.C., and C.D. Hollister, 1971, The Face of the Deep: New York, Oxford Univ. Press, 659 p.

Heisenberg, W., 1927, Uber den anschaulichen Inhalt der quanten-theoretischen Kinematik and Mechanik: Berlin, Zeitschrift fur Physik, v. 43, 197 p.

Henbest, L.G., 1952, Significance of evolutionary explosions in geologic time: Jour. Paleont., v. 26, p. 298–394.

Heritier, F.E., P. Lossel, and E. Wathne, 1979, Frigg field—large submarine-fan trap in lower Eocene rocks of North Sea graben: Am. Assoc. Petrol. Geol. Bull., v. 63, p. 1999–2020.

Heritier, F.E., P. Lossel, and E. Wathne, 1980, Frigg Field: large submarine fan trap in Lower Eocene rocks of the Viking Graben, North Sea, in Halbouty, M.T., ed., Giant Oil and Gas Fields of the Decade: 1968–1978: Am. Assoc. Petrol. Geol. Mem., no. 30, p. 59–80.

Herschel, J.F.W., 1830, A Preliminary Discourse on the Study of Natural Philosophy: London, Longman, 372 p.

Hitchon, B., 1969, Fluid flow in the Western Canada sedimentary basin, part 1: effect of topography: Water Resources Res., v. 5, no. 1, p. 186–195.

Hitchon, B., 1974, Application of geochemistry to the search for crude oil and natural gas, in Levinson, A.A., ed., Introduction to Exploration Geochemistry: Calgary, Applied Publ. Ltd., p. 509–545.

Hite, R.J., 1970, Shelf carbonate sedimentation controlled by salinity in the Paradox basin, southeast Utah: Northern Ohio Geol. Soc. Third Symp. on Salt Trans., p. 48–66.

Hobday, D.K., 1978, Fluvial deposits of the Ecca and Beaufort groups in the eastern Karoo Basin, South Africa, in Miall, A.D., ed., Fluvial Sedimentology: Can. Soc. Petrol. Geol. Mem., no. 5, p. 413–431.

Hobson, J., M.L. Fowler, and E.H. Beaumont, 1982, Depositional and statistical exploration models of Upper Cretaceous off-shore sandstone complex, Sussex member, House Creek field, Wyoming: Am. Assoc. Petrol. Geol. Bull., v. 66, p. 689–707.

Holland, D.S., W.E. Nunan, D.R. Lammlein, and R.L. Wordhaus, 1980, Eugene Island Block 330 Field, offshore Louisiana, in Halbouty, M.T., ed., Giant Oil and Gas Fields of the Decade: 1968–1978: Am. Assoc. Petrol. Geol. Mem., no. 30, p. 253–280.

Hollenshead, C.T., and R.L. Pritchard, 1961, Geometry of producing Mesa Verde sandstones, San Juan Basin, in Peterson, J.A., and J.C. Osmond, eds., Geometry of Sandstone Bodies: Tulsa, Am. Assoc. Petrol. Geol., p. 98–118.

Holmgren, D.A., J.D. Moody, and H.H. Emmerich, 1975, The structural setting for giant oil and gas fields: Ninth World Petrol. Congr. Proc., v. 2, p. 45–54.

Houbolt, J.J.H.C., 1968, Recent sediments in the Southern Bight of the North Sea: Geol. Mijnbouw, v. 47, p. 245–273.

Houston Geological Society, 1965, Deltas in Their Geological Framework, 251 p.

Howard, J.D., 1972, Trace fossils as criteria for recognizing shorelines in the stratigraphic record, in Rigby, J.K., and W.K. Hamblin, eds., Recognition of Ancient Sedimentary Environments: Soc. Econ. Paleont. Mineral. Spec. Publ., no. 16, p. 215–225.

Hoyt, J.H., 1967, Barrier island formation: Geol. Soc. Am. Bull., v. 78, p. 1125–1136.

Hoyt, W.V., 1959, Erosional channel in the middle Wilcox near Yoakum, Lavaca Co., Texas: Trans. Gulf Coast Assoc. Geol. Soc., v. 9, p. 41–50.

Hubbard, R.J., J. Pape, and D.G. Roberts, 1985, Depositional sequence mapping as a technique to establish tectonic and stratigraphic framework and evaluate hydrocarbon potential on a passive continental margin, in Berg, O.R., and D.G. Woolverton, eds., Seismic Stratigraphy II: An Integrated Approach to Hydrocarbon Exploration: Am. Assoc. Petrol. Geol. Mem., no. 39, p. 79–92.

Hubbert, M.K., 1937, Theory of scale models as applied to the study of geologic structures: Geol. Soc. Am. Bull., v. 48, p. 1459–1520.

Hull, C.E., and H.R. Warman, 1968, Asmari oilfields of Iran: Am. Assoc. Petrol. Geol. Bull., v. 52, p. 428–437.

Hunt, J.M., 1976, Geochemistry of Petroleum: Am. Assoc. Petrol. Geol. Cont. Educ. Lecture Note Ser., 10 p.

Hutton, James, 1788, Theory of the earth; or an investigation of the laws observable in the composition, dissolution, and restoration of land upon the globe: Royal Soc. Edinburgh Trans., v. 1, p. 209–304.

Huxley, A.L., 1963, Literature & Science: London, Chatto & Windus, 99 p.

Huxley, J., 1961, The Humanist Frame: New York, Harper & Row, 432 p.

Imbrie, J., and J. Buchanan, 1965, Sedimentary structures in modern carbonate sands of the Bahamas, in Middleton, G.V., ed., Primary Sedimentary Structures and Their Hydrodynamic Interpretation: Soc. Econ. Paleont. Mineral. Spec. Publ., no. 12, p. 149–172.

Imbrie, J., and E.G. Purdy, 1962, Classification of modern Bahamian carbonate sediments, in Ham, W.E., ed., Classification of Carbonate Rocks: Am. Assoc. Petrol. Geol. Mem., no. 1, p. 253–272.

Ingle, J.C., Jr., 1966, The movement of beach sand, in Developments in Sedimentology: New York, Elsevier, 221 p.

Irwin, M.L., 1965, General theory of epeiric clear water sedimentation: Am. Assoc. Petrol. Geol. Bull., v. 49, p. 445–459.

Isacks, B., J. Oliver, and L.R. Sykes, 1968, Seismology and the new global tectonics: Jour. Geophys. Res., v. 73, p. 5855–5899.

Johnson, H.D., 1977, Shallow marine sand bar sequences, an example from late Precambrian of north Norway: Sedimentology, v. 24, p. 245–270.

Johnson, H.D., 1978, Shallow siliciclastic seas, in Reading, H.G., ed. Sedimentary Environments and Facies: New York, Elsevier, p. 207–258.

Johnson, M.A., N.H. Kenyon, R.H. Belderson, and A.H. Stride, 1982, Sand transport, in Stride, A.H., ed., Offshore Tidal Sands Processes and Deposits: London, Chapman and Hall, p. 58–94.

Jones, C.L., 1972, Permian Basin potash deposits, southwestern United States, in Richter-Bernburg, G., ed., Geology of Saline Deposits: Hanover Symp., UNESCO Proc., p. 191–201.

Jones, H.P., and R.G. Speers, 1976, Permo-Triassic reservoirs of Prudhoe Bay Field, North Slope, Alaska, in Braustein, J., ed., North American Oil and Gas Fields: Am. Assoc. Petrol. Geol. Mem., no. 24, p. 23–50.

Jones, R., 1962, Advancement in exploration techniques using log derived data factors: Wyom. Geol. Assoc. 17th Annu. Field Conf. Guidebook, p. 268–272.

Jones, R.W., 1980, Some mass balance and geologic constraints on migration mechanisms, in Roberts, W.H., III, and R.J. Cordell, eds., Physical and Chemical Constraints on Petroleum Migration: Am. Assoc. Petrol. Geol. Cont. Educ. Course Note Ser., no. 8, p. A–1—A–43.

Kanes, W.H., 1970, Facies and Development of the Colorado River Delta in Texas, in Morgan, J.P., and R.H. Shaver, eds., Deltaic Sedimentation: Modern and Ancient: Soc. Econ. Paleont. Mineral. Spec. Publ., no. 15, p. 78–106.

Kay, M., 1947, Geosynclinal nomenclature and the craton: Am. Assoc. Petrol. Geol. Bull., v. 31, p. 1289–1293.

Kay, M., 1951, North American Geosynclines: Geol. Soc. Am. Mem., no. 48, 143 p.

Kendall, M.G., and W.R. Buckland, 1960, A Dictionary of Statistical Terms: Edinburg, Oliver and Boyd Ltd., 107 p.

Kenyon, N.H., 1970, Sand ribbons of European tidal seas: Mar. Geol., v. 9, p. 25–39.

Kerr, R.A., 1987, Milankovitch climate cycles through the ages: Science, v. 235, p. 973–974.

Kerr, R.A., 1987, Capturing El Niño in models: Science, v. 238, p. 1507–1508.

Kerr, R.W., 1984, Developing a big picture of earth's mantle: Science, v. 255, p. 702–703.

Kiefer, W.R., 1965, Stratigraphy and the geologic history of the uppermost Cretaceous, Paleocene, and Lower Eocene in the Wind River basin, Wyoming: U.S. Geol. Surv. Prof. Pap., no. 495A, p. A1–77.

King, R.H., 1947, Sedimentation in Permian Castile Sea: Am. Assoc. Petrol. Geol. Bull., v. 31, p. 470–477.

Kingston, D.R., C.P. Dishroom, and P.A. Williams, 1983, Hydrocarbon plays and global basin classification: Am. Assoc. Petrol. Geol. Bull., v. 67, p. 2175–2193.

Kirschner, C.E., and C.A. Lyon, 1973, Stratigraphic and tectonic development of Cook Inlet petroleum province, in Pitcher, M.G., ed., Arctic Geology: Am. Assoc. Petrol. Geol. Mem., no. 19, p. 396–407.

Klein, G.deV., 1971, A sedimentary model for determining paleotidal range: Geol. Soc. Am. Bull. v. 82, p. 2585–2592.

Klein, G.deV., 1975a, Tidalites in the Eureka Quartzite (Ordovican), Eastern California and Nevada, in Ginsburg, R.N., ed., Tidal Deposits: New York, Springer-Verlag, p. 145–152.

Klein, G.deV., 1975b, Sedimentary tectonics in the southwest Pacific marginal basin based on Leg 3 Deep Sea Drilling Project cores from the south Fiji, Hebrides, and Coral Sea basin: Geol. Soc. Am. Bull., v. 86, p. 1012–1018.

Klein, G.deV., 1976, Holocene Tidal Sedimentation: Stroudsburg, PA, Dowden, Hutchinson, and Ross, 423 p.

Klein, G.deV., 1985, Sandstone Depositional Models in Petroleum Exploration: Boston, Int. Human Resources Develop. Corp., 209 p.

Kleispehn, K.L., R.J. Steel, E. Johannessen, and A. Netland, 1984, Conglomeratic fan-delta sequences, late Carboniferous—early Permian, western Spitsbergen, in Koster, E.H., and R.J. Steel, eds., Sedimentology of Gravels and Conglomerates: Can. Soc. Petrol. Geol. Mem., no. 10, p. 279–294.

Klemme, H.D., 1975, Giant oil fields related to their geologic setting—a possible guide to exploration: Can. Petrol. Geol. Bull., v. 23, no. 1, p. 30–66.

Klemme, H.D., 1983, Field size distribution related to basin characteristics: Oil & Gas Jour., v. 81, no. 2, p. 168–178.

Knight, R.J., and D.L. McLean, eds., 1986, Shelf Sands and Sandstones: Can. Soc. of Petrol. Geol. Mem., no. 11, 347 p.

Kochel, R.C., and R.A. Johnson, 1984, Geomorphology

and sedimentology of humid temperate alluvial fans, central Virginia, *in* Koster, E.H., and R.J. Steel, eds., Sedimentology of Gravels and Conglomerates: Can. Soc. Petrol. Eng. Mem., no. 10, p. 109–122.

Kolb, C.R., and J.R. Van Lopik, 1966, Depositional environments of the Mississippi River deltaic plain—southeastern Louisiana, *in* Deltas in Their Geologic Framework: Houston, Houston Geol. Soc., p. 17–62.

Kolmer, J.R., 1973, A wave tank analysis of the beach foreshore grain size distribution: Jour. Sediment. Petrol., v. 43, p. 200–204.

Komar, P.D., 1972, Relative significance of head and body spill from a channelized turbidity current: Geol. Soc. Am. Bull., v. 83, p. 1151–1156.

Kraft, J.C., 1971, Sedimentary facies patterns and geologic history of a Holocene marine transgression: Geol. Soc. Am. Bull., v. 82, p. 2131–2158.

Kraft, J.C., and E.A. Allen, 1975, A transgressive sequence of Late Holocene epoch tidal environmental lithosomes along the Delaware Coast, *in* Ginsburg, R.N., ed., Tidal Deposits: New York, Springer-Verlag, p. 39–46.

Kraft, J.C., M.J. Chrzastowski, D.F. Belknap, M.A. Toscano, and C.H. Fletcher III, 1987, The transgressive barrier-lagoon coast of Delaware: morphostratigraphy, sedimentary sequences and responses to relative rise in sea level, *in* Nummedal, D., O.H Pilkey, and J.D. Howard, eds., Sea-level Fluctuation and Coastal Evolution: Soc. Econ. Paleont. Mineral. Spec. Publ. no. 41, p. 129–143.

Kraft, J.C., and C.J. John, 1979, Lateral and vertical facies relations of transgressive: Am. Assoc. Petrol. Geol. Bull., v. 63, p. 2145–2163.

Kreisa, R.D., and R.J. Moiola, 1986, Sigmoidal tidal bundles and other tide-generated sedimentary structures of the Curtis formation, Utah: Geol. Soc. Am. Bull., v. 97, p. 381–387.

Krinsley, D.H., and J.C. Doornkamp, 1973, Atlas of Quartz Sand Surface Textures: Cambridge, Cambridge Univ. Press, 91 p.

Krumbein, W.C., and F.A. Graybill, 1965, An Introduction to Statistical Models in Geology: New York, McGraw Hill Book Co., 475 p.

Krumbein, W.C., and W.G. Libby, 1957, Applications of moments to vertical variability maps of stratigraphic units: Am. Assoc. Petrol. Geol. Bull., v. 41, p. 197–211.

Krumbein, W.C., and L.L. Sloss, 1951, Stratigraphy and Sedimentation: San Francisco, W.H. Freeman and Co., 497 p.

Krumbein, W.C., and L.L. Sloss, 1963, Stratigraphy and Sedimentation, 2nd edition: San Francisco, W.H. Freeman and Co., 660 p.

Krumme, G.W., 1981, Stratigraphic significance of limestones of the Marmaton Group (Pennsylvanian, Des Moinesian) in Eastern Oklahoma: Okla. Geol. Surv. Bull., no. 131, 67 p.

Kuenen, Ph.H., 1950, Marine Geology: New York, John Wiley & Sons, 568 p.

Kuenen, Ph.H., A. Faure-Muret, M. Lanteaume, and F. Fallot, 1957, Observations sur le flysch des Alpes Maritimes Françaises et Italiennes: Bull. Soc. Geol. France, v. 7, p. 11–26.

Kuenen, Ph.H., and C.I. Migliorini, 1950, Turbidity currents as a cause of graded bedding: Jour. Geol., v. 58, p. 91–126.

Kuipers, A., 1961, Model and insight, *in* Freudenthal, H., ed., The Concept and the Role of the Model in Mathematical and Natural and Social Sciences: Dordrecht, Holland, D. Reidel Publ. Co., p. 125–132.

Kumar, N., and J.E. Sanders, 1975, Inlet sequence formed by the migration of Fire Island Inlet, Long Island, New York, *in* Ginsburg, R.N., ed., Tidal Deposits: New York, Springer-Verlag, p. 75–84.

Ladd, H.S., 1944, Reefs and other bioherms: Nat. Res. Council Div. Geol. and Geogr. Annu. Rep., no. 4, app. K, p. 26–29.

Ladd, H.S., 1950, Recent Reefs: Am. Assoc. Petrol. Geol.

Bull., v. 34, p. 203–214.

Lande, R., 1985, Expected time for random genetic drift of a population between stable phenotypic states: Proc. Nat. Acad. Sci. U.S.A., v. 82, p. 7641–7645.

Lanteaume, M., B. Beaudoin, and R. Campredon, 1967, Figures sedimentares du flysch "gres d'Annot" du synclinal de Peira-Cava: Paris, Centre National de la Recherche Scientifique, 97 p.

Larsonneur, C., 1975, Tidal deposits, Mont Saint-Michel Bay, France, *in* Ginsburg, R.N., ed., Tidal Deposits: New York, Springer-Verlag, p. 21–30.

Lawrence, D.A., and B.P.J. Williams, 1987, Evolution of drainage systems in response to Acadian deformation: the Devonian Battery Point formation, eastern Canada, *in* Ethridge, F.G., R.M. Flores, and M.D. Harvey, eds., Recent Developments in Fluvial Sedimentology: Soc. Econ. Paleont. Mineral. Spec. Publ., no. 39, p. 287–299.

Lawrence, S., and P. Coster, 1985, Petroleum potential of off-shore Guyana: Oil & Gas Jour., v.83, p. 67–74.

Legun, A.S., and B.R. Rust, 1982, The Upper Carboniferous Clifton formation of northern New Bruswick: coal bearing deposits of a semi-arid alluvial plain: Can. Jour. Earth Sci., v. 19, p. 1775–1785.

Leonard, R., 1984, Geology and hydrocarbon accumulations, Columbus basin, offshore Trinidad: Oil & Gas Jour., v. 82, no. 15, p. 132–138.

Levorsen, A. I., 1960, Paleogeographic Maps: San Francisco, W.H. Freeman and Co., 174 p.

Lewin, R., 1986, Punctuated equilibrium is now old hat: Science, v. 231, p. 672–673.

Lewis, T.R., 1983, The process of formation of ocean crust: Science, v. 220, no. 4593, p. 151–157.

Lewontin, R.C., 1963, Models, mathematics, and metaphors, *in* Synthese 15: Dordrecht, Holland, D. Reidel Publ. Co., p. 222–244.

Leythaeuser, D., R.G. Schaefer, and M. Radke, 1987, On the primary migration of petroleum: 12th World Petrol. Congr. Proc., v. 2, Exploration: New York, John Wiley & Sons Ltd., p. 227–236.

Lindseth, R.O., 1981, Stratigraphic traps with synthetic sonic logs, Short course notes: Midland, TX, Soc. Explor. Geophys.

Lindsey, J.P., 1982, Applied seismic stratigraphic interpretation, Short course notes: Houston, GeoQuest International Inc.

Link, M.H., and R.H. Osborne, 1978, Lacustrine facies in the Pliocene Ridge basin Group Ridge basin, California, *in* Matter, A., and M.E. Tucker, eds., Modern and Ancient Lake Sediments: Int. Assoc. Sediment Spec. Publ., no. 2: Oxford, Engl., Blackwell Scientific Publ., p. 169–187.

Lockman-Balk, C., 1970, Upper Cambrian faunal patterns on the craton: Geol. Soc. Am. Bull., v. 81, p. 3197–3224.

Lohsee, John, 1972, A Historical Introduction to the Philosophy of Science: London, Oxford University Press, 218 p.

Lopatin, N.V., 1971, Temperature and geologic time as factors in coalification (in Russian): Akad. Nauk. SSSR Izv. Ser. Geol., no. 3, p. 95–106.

Lowe, D.R., 1976, Grain flow and grain flow deposits: Jour. Sediment. Petrol., v. 46, p. 188–199.

Lowe, D.R., 1976, Subaqueous liquified and fluidized sediment flows and their deposits: Sedimentology, v. 22, p. 157–204.

Lowe, D.R., 1979, Sediment gravity flows, their classification and some problems of application to natural flow and deposits, *in* Doyle, L.J., and O.H. Pilkey, eds., Geology of Continental Slopes: Soc. Econ. Paleont. Mineral. Spec. Publ., no. 27, p. 75–82.

Lowe, D.R., 1982, Sediment gravity flows II: Depositional models with special reference to the deposits of high-density turbidity currents: Jour. Sediment. Petrol., v. 52, p. 279–298.

Lowell, J.D., 1985, Structural Styles in Petroleum Exploration: Tulsa, OK, Oil & Gas Consultants Internat., Inc., 460 p.

Lowman, S.W., 1947, Fundamental research in sedimentology: Am. Assoc. Petrol. Geol. Bull., v. 31, p. 501–512.

Lowman, S.W., 1949, Sedimentary facies in Gulf Coast: Am. Assoc. Petrol. Geol. Bull., v. 33, p. 1939–1997.

Lozo, F.E., and F.L. Strickland, 1956, Stratigraphic notes on the outcrop of the Basal Cretaceous, Central Texas: Gulf Coast Assoc. Geol. Soc. Trans., v. 6, p. 67–78.

Lucas, P.T., and J.M. Drexler, 1976, Altamont-Bluebell—a major, naturally fractured stratigraphic trap, Uinta Basin, Utah, in Braunstein, J., ed., North American Oil and Gas Fields: Am. Assoc. Petrol. Geol. Mem., no. 24, p. 121–135.

Lucia, F.J., 1972, Recognition of evaporite-carbonate shore-line sedimentation, in Rigby, J.K., and W.K. Hamblin, eds., Recognition of Ancient Sedimentary Environments: Soc. Econ. Paleontol. Mineral. Spec. Publ., no. 16, p. 160–191.

Lumsden, D.N., E.D. Pittman, and R.S. Buchanan, 1971, Sedimentation and petrology of Spiro and Foster Sands (Pennsylvanian), McAlester Basin, Oklahoma: Am. Assoc. Petrol. Geol. Bull., v. 55, p. 254–266.

Lyday, J.R., 1978, Controls on late Cretaceous-Paleocene sedimentation, Wyoming (unpub. MS thesis): Univ. Tulsa, 98 p.

Lyell, Sir Charles, 1830, Principles of Geology, being an attempt to explain the former changes of the earth's surface, by reference to causes now in operation, v. 1: London, John Murray, 511 p.

Mackin, J.H., 1948, Concept of the graded river: Geol. Soc. Am. Bull., v. 59, p. 463–512.

Mackin, J.H., 1963, Methods of investigation in geology, in Albritton, C.C., Jr., ed., The Fabric of Geology: Stanford, CA, Freeman, Cooper & Co., p. 135–163.

MacPhearson, B.A., 1978, Sedimentation and trapping mechanism in Upper Miocene Stevens and older turbidite fans of southeastern San Joaquin Valley, California: Am. Assoc. Petrol. Geol. Bull., v. 62, p. 2243–2274.

Magoon, L.R., and G.B. Claypool, 1981, Petroleum geology of Cook Inlet Basin—an exploration model: Am. Assoc. Petrol. Geol. Bull., v. 65, p. 1043–1061.

Mallory, U.S., 1970, Biostratigraphy—a major basis for paleontological correlation, in Yockelson, E.L., ed., Proc. North American Paleontol. Conv., part A, v. 1: Lawrence, KS, Allen Press.

Mallory, W.W., 1972, Geologic Atlas of the Rocky Mountain Region: Denver, Rocky Mountain Assoc. Geol., 331 p. 553–565.

Martin, R., 1966, Paleogeomorphology and its application to exploration for oil and gas (with examples from Western Canada): Am. Assoc. Petrol. Geol. Bull., v. 50, p. 2277–2311.

Martin, R.G., 1978, Northern and Eastern Gulf of Mexico continental margin: stratigraphic and structural framework, in Bouma, A.H., G.T. Moore, and J.M. Coleman, Framework, Facies and Oil Trapping Characteristics of the Upper Continental Margin: Am. Assoc. Petrol. Geol. Stud. Geol., no. 7, p. 21–42.

Masroua, L.F., 1973, Pattern of Pressures in the Morrow sands of northwestern Oklahoma (unpubl. M.S. Thesis): Univ. of Tulsa, 78 p.

Matthews, R.D., and G.C. Egleson, 1974, Origin and implication of a mid-basin potash facies in the Salina Salt of Michigan: Northern Ohio Geol. Soc. Fourth Symp. on Salt Trans., v. 1, p. 15–34.

Mayr, E., 1951, Bearing of some biological data on geology: Geol. Soc. Am. Bull., v. 62, p. 537–546.

McCall, P.L., and M.J.S. Tenesz, eds., 1982, Animal Sediment Relations: The Biogenic Alteration of Sediments: New York, Plenum Press, 336 p.

McCave, I.N., 1972, Transport and escape of fine-grained sediment from shelf areas, in Swift, D.J.P., D.B. Duane, and D.H. Pilkey, eds., Shelf Sediment Transport—Process and Pattern: Stroudsburg, PA, Dowden, Hutchinson, and Ross, p. 225–248.

McConnell, C.L., 1985, Salinity and temperature anomalies over structural oil fields, Carter County, Oklahoma: Am. Assoc. Petrol. Geol. Bull., v. 69, p. 781–787.

McCoy, A.W., 1934, An interpretation of local, structural development in mid-Continent areas associated with deposits of petroleum, in Wrather, W.E., and F.A. Lahee, eds., Problems of Petroleum Geology: Tulsa, OK, Am. Assoc. Petrol. Geol., p. 581–627.

McIver, R.D., 1967, Composition of kerogen—clue to its role in the origin of petroleum: 7th World Petrol. Congr. Proc., v. 2, p. 25–36.

McKenzie, D., 1978, Some remarks on the development of sedimentary basins: Earth Planet. Sci. Letters., v. 40, p. 25–32.

McPherson, J.G., G. Shanmugam, and R.J. Moiola, 1987, Fan-deltas and braid delta: Varieties of coarse-grained deltas: Geol. Soc. Am., v. 99, no. 3, p. 331–340.

Matter, A., and M.E. Tucker, 1978, Modern and ancient lake sediments: an introduction, in Matter, A., and M.E. Tucker, eds., Modern and Ancient Lake Sediments: Int. Assoc. Sediment. Spec. Publ., no. 2, Oxford, Engl., Blackwell Sci. Publ., p. 1–6.

Meadows, P., 1957, Models, systems and science: Am. Soc. Rev., v. 22, p. 3–9.

Meissner, F.F., 1978, Petroleum Geology of the Bakken Formation, Williston Basin, North Dakota and Montana: Montana Geol. Soc., 24th Ann. Conf., Billings, Mont., p. 207–227.

Melton, F.A., 1968, Regional unconformities of flatlands: Am. Assoc. Petrol. Geol. Bull., v. 52, p. 313–321.

Menard, H.W. 1983, Insular erosion, isostasy, and subsidence: Science, v. 220, p. 913–918.

Merriam, D.F., and J.W. Harbaugh, 1964, Trend Surface Analysis of Regional and Residual Components of Geologic Structures in Kansas: Kans. Geol. Surv. Spec. Distr. Publ., no. 11, 27 p.

Meyer, H.J., and H.W. McGee, 1985, Oil and gas fields accompanied by geothermal anomalies in Rocky Mountain region: Am. Assoc. Petrol. Geol. Bull., v. 69, p. 933–945.

Middleton, G.V., 1967, Experiments on density and turbidity currents, III: deposition of sediment: Can. Jour. Earth Sci., v. 4, p. 475–505.

Middleton, G.V., ed., 1965, Primary sedimentary structures and their hydrodynamic interpretation: Soc. Econ. Paleont. Mineral. Spec. Publ., no. 12, 265 p.

Middleton, G.V., and M.A. Hampton, 1973, Sediment gravity flows mechanics of flow and deposition, in Middleton, G.V., and A.H. Bouma, eds., Turbidites and Deep Water Sedimentation: Soc. Econ. Paleont. Mineral. Short Course Notes, Anaheim, CA, 157 p.

Middleton, G.V., and M.A. Hampton, 1976, Subaqueous sediment transport and deposition by sediment gravity flows, in Stanley, D.J., and D.J.P. Swift, eds., Marine Sediment Transport and Environmental Management: New York, John Wiley, p. 197–218.

Mill, J.S., 1865, System of logic (vol. I): London, Longmans, Green, 480 p.

Miller, R.L., 1954, A model for the analysis of environments of sedimentation: Jour. Geol., v. 62, p. 108–113.

Miller, R.V., and G.S. Visher, 1974, Environmental significance of shale properties: Am. Assoc. Petrol. Geol. Bull., v. 58, p. 93–94.

Mitchum, R.M., Jr., and P.R. Vail, 1977, Seismic stratigraphy and global changes of sea level, part 7: seismic stratigraphic interpretation procedure, in Payton, C.E., ed., Seismic Stratigraphy—Applications to Hydrocarbon Exploration: Am. Assoc. Petrol. Geol. Mem., no. 26, p. 135–143.

Mitchum, R.M., Jr., P.R. Vail, and J.B. Sangree, 1977, Seismic stratigraphy and global changes of sea level, part 6: stratigraphic interpretation of seismic reflection patterns in depositional sequences, in Payton, C.E., ed., Seismic Stratigraphy—Applications to Hydrocarbon Exploration: Am. Assoc. Petrol. Geol. Mem., no. 26, p. 117–134.

Mitchum, R.M., Jr., P.R. Vail, and S. Thompson III, 1977, Seismic stratigraphy and global changes of sea level, part 2: the depositional sequence as a basic unit for stratigraphic analysis, in Payton, C.E., ed., Seismic Stratigraphy—Applications to Hydrocarbon Exploration: Am. Assoc. Petrol. Geol. Mem., no. 26, p. 53–62.

Momper, J.A., 1978, Oil migration limitation suggested by geological and geochemical consideration, *in* Physical and Chemical Constraints on Petroleum Migration: Am. Assoc. Petrol. Geol. Cont. Educ. Short Course Notes, no. 8, p. B1–B60.

Moore, D.G., 1966, Structure, litho-orogenic units, and postorogenic basin fill by reflection profiling—California continental borderland: San Diego, CA, U.S. Navy Electronics Lab., 151 p.

Moore, G.T., and D.O. Asquith, 1971, Delta term and concept: Geol. Soc. Am. Bull., v. 82, p. 2563–2568.

Moore, G.T., G.W. Starke, L.C. Bonham, and H.O. Woodbury, 1978, Physiography, stratigraphy, and sedimentational patterns: *in* Bouma, A.H., G.T. Moore, and J.M. Coleman, eds., Framework, Facies, and Oil-Trapping Characteristics of the Upper Continental Margin: Am. Assoc. Petrol. Geol. Stud. in Geol., no. 7, p. 155–192.

Morgan, J.P., ed., 1970, Deltaic Sedimentation, Modern and Ancient: Soc. Econ. Paleont. Mineral. Spec. Publ., no. 15, 312 p.

Morton, R.A., and W.A. Price, 1987, Late Quaternary sea-level fluctuations and sedimentary phases of the Texas coastal plain and shelf, *in* Nummedal, D., O.H. Pilkey, and J.D. Howard, eds., Sea-Level Fluctuations and Coastal Evolution: Soc. Econ. Paleont. Mineral. Spec. Publ., no. 41, p. 181–198.

Mudge, M.R., 1970, Origin of the disturbed belt in Northwestern Montana: Geol. Soc. Am. Bull., v. 80, p. 1021–1049.

Murris, R.J., 1980, Middle East: stratigraphic evolution and oil habitat: Am. Assoc. Petrol. Geol. Bull., v. 64, p. 597–618.

Mutti, E., 1974, Examples of ancient deep-sea fan deposits from circum-Mediterranean geosynclines, *in* Dott, R.H., Jr., and R.H. Shaver, eds., Modern and Ancient Geosynclinal Sedimentation: Soc. Econ. Paleont. Mineral. Spec. Publ., no. 19, p. 92–105.

Nagel, E., P. Suppes, and A. Tarski, 1962, International Congress for Logic, Methodology and Philosophy of Science: Proceedings of the 1960 International Congress, Stanford Univ. Press, 661 p.

Nanz, R.N., 1954, Genesis of Oligocene santstone reservoir, Seeligson field, Jim Wells and Kleberg Counties: Am. Assoc. Petrol. Geol. Bull., v. 38, no. 1, p. 96–117.

Neidell, N., and E. Poggiagliolmi, 1977, Stratigraphic modeling of interpretation, geophysical principles and techniques, *in* Payton, C.E., ed., Seismic Stratigraphy—Applications to Hydrocarbon Exploration: Am. Assoc. Petrol. Geol. Mem., no. 26, p. 389–416.

Nelson, C.H., 1976, Late Pleistocene and Holocene depositional trends, processes, and history of Astoria deep-sea fan, northeast Pacific: Mar. Geol., v. 20, p. 129–173.

Nelson, C.H., and L.D. Kulm, 1973, Submarine fans and channels: *in* Middleton, G.V., and A.W. Bouma, eds., Turbidites and Deep Water Sedimentation: Anaheim, CA, Soc. Econ. Paleont. Mineral. Pacific Sect. Short Course, p. 39–78.

Nelson, C.H., L.D. Kulm, P.R. Carlson, and J.R. Duncan, 1970, Development of the Astoria Canyon—fan physiography and comparison with similar systems: Mar. Geol., v. 8, no. 3–4, p. 259–291.

Nelson, C.H., and T.H. Nilsen, 1974, Depositional trends of Modern and ancient deep-sea fans, *in* Dott, R.H., and R.H. Shaver, eds., Modern and Ancient Geosynclinal Sedimentation: Soc. Econ. Paleont. Mineral. Spec. Publ., no. 19, p. 69–91.

Nemec, W., and R.J. Steel, eds., 1988, Fan Deltas, Sedimentology and Tectonic Settings: Bishopbriggs, Scotland, Blackie and Son Ltd., 464 p.

Newell, N.D., 1967, Paraconformities, *in* Teichert, C., and E.L. Yochelson, eds., Essays in Paleontology and Stratigraphy. R.C. Moore Commemorative Volume: Univ. Kans. Dept. Geol. Spec. Publ., no. 2, p. 349–367.

Newman, C.M., 1985, Neo-Darwinian evolution implies punctuated equilibria: Nature (London), v. 315, p. 400.

Nwachukwu, A., 1976, Approximate geothermal gradients in Niger Delta sedimentary basin: Am. Assoc. Petrol. Geol. Bull., v. 60, p. 1073–1077.

Odin, G.S., ed., 1982, Numerical Dating in Stratigraphy: New York: Wiley Interscience, New York, part I, II, 1040 p.

Oertel, G.G., and J.D. Howard, 1972, Water circulation and sedimentation at estuary entrances on the Georgia Coast, *in* Swift, D.J.P., D.B. Duane, and O.H. Pilkey, eds., Shelf Sediment Transport: Stroudsburg, PA, Dowden, Hutchinson, and Ross, p. 411–427.

Ollier, C., 1986, Advances in continental sedimentation (address): Canberra, Australia, 12th Int. Sediment. Congr.

Olsen, P.E., 1986, A 4 million-year lake record of early Mesozoic orbital climatic forcing: Science, v. 234, p. 842–848.

Oomkens, E., 1974, Lithofacies relations in the Quarternary Niger Delta complex: Sedimentology, v. 21, p. 195–221.

Orr, R.D., J.R. Johnston, and E.M. Manko, 1977, Lower Cretaceous geology and heavy-oil potential of the Lloydminster area: Can. Soc. Petrol. Geol. Bull., v. 25, p. 1187–1221.

Ottmann, R.D., P.L. Keyes, and M.A. Ziegler, 1976, Jay field, Florida—A Jurassic stratigraphic trap Braunstein, J., ed., North American Oil and Gas Fields: Am. Assoc. Petrol. Geol. Mem., no. 24, p. 276–286.

Otvos, E.G., 1970, Development and migration of barrier islands, Northern Gulf of Mexico: Geol. Soc. Am. Bull., v. 81, p. 341–348.

Palmer, A. R., 1983, The decade of North American geology 1983 time scale: Geology, v. 11, p. 503–504.

Palmer, J.J., and A.J. Scott, 1984, Stacked shoreline and shelf sandstones of La Ventana tongue (Campanian) northwestern New Mexico: Am. Assoc. Petrol. Geol. Bull., v. 68, p. 74–91.

Penland, S., R. Boyd, and J.R. Suter, 1988, Transgressive depositional system of the Mississippi Delta plain: a model for barrier shoreline and shelf sand development: Jour. Sediment. Petrol., v. 58, p. 932–949.

Pennington, W.D., 1983, Role of shallow phase changes in the subduction of oceanic crust: Science, v. 220, p. 1045–1047.

Pettijohn, F.J., 1975, Sedimentary Rocks, 3rd ed.: New York, Harper & Row, 628 p.

Pettijohn, F.J., P.E. Potter, and R. Siever, 1973, Sand and Sandstone: New York, Springer-Verlag, 618 p.

Philippi, G.T., 1965, On the depth, time and mechanism of petroleum generation: Geochim. Cosmochim. Acta, v. 29, p. 1021–1049.

Phleger, F.B., 1969, A modern evaporite deposit in Mexico: Am. Assoc. Petrol. Geol. Bull., v. 53, p. 824–829.

Picard, M.D., and L.R. High, 1981, Physical stratigraphy of ancient lacustrine deposits, *in* Ethridge, F.G., and R.M. Flores, eds., Recent and Ancient Nonmarine Depositional Environments: Models for Exploration: Soc. Econ. Paleont. Mineral. Spec. Publ., no. 31, p. 233–260.

Pilkey, O.H., S.D. Locker, and W.J. Cleary, 1980, Comparison of sand-layer geometry on flat floors of Modern depositional basins: Am. Assoc. Petrol. Geol. Bull., v. 64, p. 841–856.

Pitman, W.C. III, 1978, Relationship between eustacy and stratigraphic sequences of passive margins: Geol. Soc. Am. Bull., v. 80, no. 9, p. 1389–1403.

Pitman, W.C. III, 1979, The effect of eustatic sea level changes on stratigraphic sequences at Atlantic margins, *in* Watkins, J.S., L. Montadent, and D.W. Dickerson, eds., Geological and Geophysical Investigation of Continental Margins: Am. Assoc. Petrol. Geol. Mem., no. 29, p. 330–365.

Pitman, W.C. III, and J.R. Heirtzler, 1966, Magnetic anomalies over the Pacific-Antarctic ridge: Science, v. 154, p. 1164–1170.

Pittendrigh, C.S., 1958, Adaptation, natural selection, and behavior, *in* Roe, A., and G.G. Simpson, Behavior and Evolution: New Haven, CT, Yale Univ. Press, 557 p.

Postma, H., 1967, Sediment transportation and sedimentation in the estuarine environment, in Lauff, G.F., ed., Estuaries: Washington, DC, Am. Assoc. Advance. Sci., 737 p.

Potsma, G., 1986, Classificatin for sediment gravity-flow deposits based on flow conditions during sedimentation: Geology, v. 14, p. 291–294.

Potter, P.E., 1959, Facies model conference: Science, v. 129, p. 1292–1294.

Potter, P.E., 1962, Sand body shape and map patterns of Pennsylvanian sandstones in Illinois: Illinois Geol. Surv. Rep. Invest., no. 339, 35 p.

Potter, P.E., 1963, Late Paleozoic sandstones of the Illinois Basin: Illinois Geol. Surv. Rep. Invest., no. 217, 92 p.

Potter, P.E., and F.J. Pettijohn, 1965, Paleocurrents and Basin Analysis, 1st ed.: New York, Springer-Verlag, 296 p.

Powell, T.G., 1987, Depositional controls on source rock character and crude composition: 12th World Petrol. Congr. Proc., v. 2, Exploration: New York, John Wiley & Sons, Ltd., p. 31–42.

Pryor, W.A., and E.J. Amaral, 1971, Large scale cross stratification in the St. Peter sandstone: Geol. Soc. America Bull., v. 82, p. 239–243.

Purdy, E.G., 1974, Reef configurations: cause and effect, in Laporte, L.F., ed., Reefs in Time and Space: Soc. Econ. Paleont. Mineral. Spec. Publ., v. 18, p. 9–76.

Putnam, P.E., 1982, Fluvial channel sandstones within upper Manville (Albian) of Lloydminster area, Canada—geometry, petrography, and paleogeographic implications: Am. Assoc. Petrol. Geol. Bull., v. 66, p. 436–459.

Putnam, P.E., 1983, Fluvial deposits and hydrocarbon accumulations: examples from the Lloydminster area, Canada, in Collinson, J.D., and J. Lewin, eds., Modern and Ancient Fluvial Systems: Int. Assoc. Sediment. Spec. Publ., no. 6, Oxford, Engl., Blackwell Sci. Publ., p. 517–532.

Qiu, Y., P. Xue, and J. Xiao, 1987, Fluvial sandstone bodies as hydrocarbon reservoirs in lake basins, in Ethridge, F.G., R.M. Flores, and M.D. Harvey, eds., Recent Developments in Fluvial Sedimentology: Soc. Econ. Paleont. Mineral. Spec. Publ., no. 39, p. 329–342.

Raaf, J.F.M. de, J.R. Boersma, and A. van Gelder, 1977, Wave generated structures and sequences from a shallow marine succession. Lower Carboniferous, County Cork, Ireland: Sedimentology, v. 4, p. 1–52.

Rabinowitz, P.D., M.F. Coffin, and D. Falvey, 1983, The separation of Madagascar and Africa: Science, v. 220, p. 67–69.

Rahmani, R.A., 1988, Nearshore sedimentation of a late Cretaceous epicontinental sea, Drumheller, Alberta, Canada, in de Boer, P.L., A. van Gelder, and S.D. Nio, eds., Tide-influenced Sedimentary Environments and Facies: Dordrecht, Holland, D. Reidel Publishing Co., p. 433–471.

Rampino, M.R., and R.R. Stothers, 1984, Terrestrial mass extinctions, cometary impacts, and the sun's motion perpendicular to the galactic plane: Nature, v. 308, p. 709–11.

Rampino, M.R., and R.R. Stothers, 1984, Geological rhythms and cometary impacts: Science, v. 226, p. 1427–1431.

Rampino, M.R., and R.R. Stothers, 1988, Flood basalt lt volcanism during the past 250 million years: Science, v. 241, p. 663–668.

Ramsayer, G.R., 1979, Seismic stratigraphy, a fundamental exploration tool: Offshore Tecnology Conference Pap., no. 3568.

Rascoe, B., Jr., and D.L. Baars, 1972, The Permian system, in Mallory, W.W., ed., Geologic Atlas of the Rocky Mountain Region: Denver, Rocky Mountain Assoc. Geol., p. 143–165.

Raup, D.M., 1985, Mathematical models of cladogenesis: Paleobiology, v. 11, p. 42–52.

Raup, D.M., and J.J. Sepkoski, 1984, Periodicity of extinction in the geologic past: Proc. Natl. Acad. Sci. U.S.A., v. 81, p. 801–805.

Read, J.F., 1973, Carbonate cycles, Pillara Formation (Devonian Canning basin, Western Australia: Can. Soc. Petrol. Geol. Bull., v. 21, p. 38–51.

Read, J.F., 1975, Tidal flat facies in carbonate cycles, Pillara formation (Devonian) Canning basin, Western Australia, in Ginsburg, R.N., ed., Tidal Deposits: New York, Springer-Verlag, p. 251–256.

Read, J.F., J.P. Grotzinger, J.A. Bova, and W.F. Koerschner, 1986, Models for generation of carbonate cycles: Geology, v. 14, p. 107–110.

Reading, H.G., ed., 1986, Sedimentary Environments and Facies, 2nd ed.: New York, Elsevier, 625 p.

Redfield, A.C., 1958, The influence of the continental shelf on the tides of the Atlantic Coast of the United States: Jour. Mar. Res., v. 17, p. 432–448.

Reineck, H.E., 1976a, Layered sediments of tidal flats, beaches and shelf bottoms, in Lauff, G.H., ed., Estuaries: Washington, DC, Amer. Assoc. Adv. Sci., p. 191–206.

Reineck, H.E., 1976b, Sediment transportation and sedimentation in the estuarine environment, in Lauff, G.F., ed., Estuaries: Washington, DC, Amer. Assoc. Adv. Sci., p. 158–179.

Reineck, H.E., 1972, Tidal flats, in Rigby, J.K., and W.K. Hamblim, eds. Recognition of Ancient Sedimentary Environments: Soc. Econ. Paleont. Mineral. Spec. Publ., no. 6, p. 146–159.

Reineck, H.E., 1975, German North Sea tidal flats, in Ginsburg, R.N., ed., Tidal Deposits: New York, Springer-Verlag, p. 5–12.

Reineck, H.E., and I.B. Singh, 1980, Depositional Sedimentary Environments: New York, Springer-Verlag, 549 p.

Ricci-Lucchi, F., A. Colella, G.G. Ori, and F. Ogliani, 1981, Pliocene fan deltas of the intra-Appenninic basin, Bolgna in Ricci Lucchi, F., ed., Excursion Guidebook with Contributions on Sedimentology of Some Italian Basins: Int. Assoc. Sediment., 2nd European Meeting, Bologna, Excursion no. 4, p. 81–162.

Rich, J.L., 1934, Problems on the origin, migration, and accumulation of oil, in Wrather, W.E., and F.A. Lahee, eds., Problems of Petroleum Geology: Tulsa, OK, Am. Assoc. Petrol. Geol., p. 337–345.

Rich, J.R., 1951, Probable fondo origin of Marcellus, Ohio, New Albany, Chattanooga Bituminous shales: Am. Assoc. Petrol. Geol. Bull., v. 35, p. 2017–2040.

Richards, F.A., 1957, Oxygen in the ocean, in Hedgpeth, J.W., ed., Treatise on Marine Ecology and Paleoecology: Geol. Soc. Am. Mem., no. 67, v. 1, p. 185–238.

Richter-Bernburg, G., 1972, Saline deposits in Germany: a review and general introduction to the excursions, in Geology of Saline Deposits: Hanover Symp., UNESCO Proc., p. 275–287.

Riggs, S.R., 1984, Paleoceanogrphic model of Neogene phosphorite deposition, U.S. Atlantic continental margin: Science, v. 233, no. 4632, p. 123–131.

Rizzini, A., 1975, Sedimentary sequences of Lower Devonian sediments (Uan Caza formation) South Tunisia, in Ginsburg, R.N., ed., Tidal Deposits: New York, Springer-Verlag, p. 187–195.

Roberts, D.G., and V.N.D. Caston, 1975, Petroleum potential of the deep Atlantic Ocean: Ninth World Petrol. Congr. Proc., v. 2, p. 281–298.

Roberts, W.H., III, 1980, Design and function of oil and gas traps: in Roberts, W.H., III, and R.J. Cordell, eds., Problems of Petroleum Migration: Am. Assoc. Petrol. Geol. Stud. in Geol., no. 10, p. 217–240.

Rodriguiz, J., and R.C. Gutschick, 1970, Late Devonian and Early Mississippian ichnofacies from western Montana and northern Utah: Geol. Jour. Spec. Iss., no. 3, p. 407–438.

Root, M.R., 1980, Zenith field—a significant Dakota (Muddy) sandstone discovery: Oil & Gas Jour., v. 78, no. 34, p. 184–186.

Rosenblueth, A., and Wiener, N., 1944–1945, The role of models in science: Phil. Sci., v. 11–12, p. 316–321.

Ross, C.A., 1974, Paleogeography and provinciality, in Paleogeography Provinces and Provinciality: Soc. Econ.

Paleont. Mineral. Spec. Publ., no. 21, p. 1–17.

Sabins, F.F., Jr., 1963, Anatomy of stratigraphic traps, Bisti oilfield, New Mexico: Am. Assoc. Petrol. Geol. Bull., v. 47, p. 193–228.

Sagoe, K.-M., and G.S. Visher, 1977, Population breaks in grain-size distributions of sand—a theoretical model: Jour. Sediment. Petrol., v. 47, p. 285–310.

Sales, J.K., 1968, Crustal mechanism of Cordilleran foreland deformation: a regional and scale approach: Am. Assoc. Petrol. Geol. Bull., v. 52, p. 2016–2144.

Sangree, J.B., D.C. Waylett, D.E. Frazier, G.B. Amery, and W.J. Fennessy, 1978, Recognition of continental-slope seismic facies, offshore Texas-Louisiana, in Bouma, A.H., and G.T. Moore, eds., Framework, Facies, and Oil-Trapping Characteristics of the Upper Continental Margin: Am. Assoc. Petrol. Geol. Stud. Geol., no. 7, p. 87–116.

Sangree, J.B., and J.M. Widmier, 1977, Seismic stratigraphy and global changes of sea level, part 9: Seismic interpretation of clastic depositional facies, in Payton, C.E., ed., Seismic Stratigraphy—Applications to Hydrocarbon Exploration: Am. Assoc. Petrol. Geol. Mem., no. 26, p. 165–184.

Scheidegger, A. E. 1961, Theoretical Geomorphology: Englewood Cliffs, NJ, Prentice Hall Inc., 333 p.

Schlumberger Well Log Services, 1981, Dipmeter Interpretation, v. 1, Fundamentals: New York, Schlumberger Well Log Services, 61 p.

Schmalz, R.F., 1969, Deep-sea evaporite deposition: a genetic model: Am. Assoc. Petrol. Geol. Bull., v. 53, no. 4, p. 798–823.

Schneer, C.J., ed., 1979, Two Hundred Years of Geology in America: Hanover, NH, Univ. of New Hampshire Press, 385 p.

Schneider, J.F., 1975, Recent tidal deposits, Abu Dhabi UAE, Arabian Gulf, in Ginsburg, R.N., ed., Tidal Deposits: New York, Springer-Verlag, p. 91–94.

Schoch, R.M., 1986, Phylogeny Reconstruction in Paleotology: New York, Van Nostrand Reinhold Co., 353 p.

Scholl, D.W., and M.S. Marlow, 1974, Sedimentary sequences in Modern Pacific trenches and the deformed circumpacific eugeosyncline, in Dott, R.H., Jr., and R.H. Shaver, eds., Modern and Ancient Geosynclinal Sedimentation: Soc. Econ. Paleont. Mineral. Mem., no. 19, p. 193–211.

Scholl, D.W., and M. Stuiver, 1967, Recent submergence of South Florida: a comparison with adjacent coasts and other eustatic data: Geol. Soc. Am. Bull., v. 78, p. 437–454.

Schuchert, C., 1923, Sites and natures of the North-American geosynclines: Am. Assoc. Petrol. Geol. Bull., v. 34, p. 151–260.

Schumm, S.A., 1977, The Fluvial System: New York, John Wiley & Sons, 338 p.

Schwab, F.L., 1976, Modern and ancient sedimentary basins comparative accumulation rates: Geology, v. 4, p. 723–727.

Schwartz, M.L., 1971, The multiple causality of barrier islands: Jour. Geol., v. 79, p. 91–94.

Schwarzacher, W., 1975, Sedimentation Models and Quantitative Stratigraphy: New York, Elsevier, 382 p.

Scott, J.D., 1970, Cherokee sandstones of a portion of Noble County, Oklahoma (unpub. MS thesis): Univ. Oklahoma.

Scruton, P.G., 1953, Deposition of evaporites: Am. Assoc. Petrol. Geol. Bull., v. 37, p. 2498–2512.

Sears, J.D., C.B. Hunt, and T.A. Hendricks, 1941, Transgressive and regressive Cretaceous deposits in Southern San Juan Basin, New Mexico: U.S. Geol. Surv. Prof. Pap., no. 193F, p. 101–121.

Seeley, D.R., P.R. Vail, and G.G. Wallin, 1974, Trench slope model, in The Geology of Continental Margins: New York, Springer-Verlag, p. 249–260.

Seeley, D.R., and W.R. Dickenson, 1977, Structure and stratigraphy of forearc basins, in Geology of Continental Margins: Am. Assoc. Petrol. Geol. Cont. Educ. Short Course Notes, no. 5, p. 101–121.

Seilacher, A., 1964, Biogenic sedimentary structures, in Imbrie, J., and N. Newell, eds., Approaches to Paleoecology: New York, John Wiley & Sons, p. 296–316.

Seilacher, A., 1967, Bathymetry of trace fossils: Mar. Geol., v. 5, no. 516, p. 413–428.

Sepkoski, J. John, Jr., and M.L. Hulver, 1985, An atlas of Phanerozoic diversity diagrams, in Valentine, J.W., ed., Phanerozoic Diversity Patterns: Princeton, NJ, Princeton Univ. Press; and San Francisco, Pacific Div. Am. Assoc. Adv. Sci., p. 11–40.

Shanmugan, G., and R.J. Moiola, 1982, Eustatic control of turbidites and winnowed turbidites: Geology, v. 10, p. 231–235.

Shannon, J.P., Jr., 1962, Hunton Group (Silurian-Devonian) and related strata in Oklahoma: Am. Assoc. Petrol. Geol. Bull., v. 46, no. 1, p. 1–29.

Shaw, A.B., 1964, Time in Stratigraphy: New York, McGraw-Hill, 365 p.

Shearman, D.J., 1966, Origin of marine evaporites by diagenesis: Inst. Mining Metal. Trans., v. 75, p. 208–215.

Sheldon, R.P., 1963, Physical stratigraphy and mineral resources of Permian rocks in Western Wyoming: U.S. Geol. Surv. Prof. Pap., no. 313B, p. 49–273.

Sheldon, R.P., 1964, Paleolatitudinal and paleogeographic distribution of phosphate: U.S. Geol. Surv. Prof. Pap., no. 501C, p. 106–113.

Shepard, D.P., and R.F. Dill, 1966, Submarine Canyons and Other Sea Valleys: Chicago, Rand McNally, 381 p.

Shepard, F.P., 1948, Submarine geology: New York, Harper and Brothers, 348 p.

Shepard, F.P., N.F. Marshall, P.A. McLoughlin, and G.G. Sullivan, 1979, Currents in submarine canyons and other sea valleys: Am. Assoc. Petrol. Geol. Stud. Geol., no. 8, 173 p.

Shepard , F.P., and D.G. Moore, 1954, Sedimentary environments differentiated by coarse fraction studies: Am. Assoc. Petrol. Geol. Bull., v. 38, p. 1792–1802.

Shepard, F.P., F.B. Phleger, and Tj.H. van Andel, eds., 1960, Recent Sediments, Northwest Gulf of Mexico: Tulsa, OK, Am. Assoc. Petrol. Geol., 394 p.

Sheriff, R.E., 1977, Limitation of resolution of seismic reflection and geologic detail derivable from them, in Payton, C.E., ed., Seismic Stratigraphy—Applications to Hydrocarbon Exploration: Am. Assoc. Petrol. Geol. Mem., no. 26, p. 3–14.

Shinn, E.A., R.M. Lloyd, and R.N. Ginsburg, 1969, Anatomy of a modern carbonate tidal flat, Andros Island, Bahamas: Jour. Sediment. Petrol., v. 39, p. 1202–1228.

Silver, B.A., and R.G. Todd, 1969, Permian cyclic strata northern Midland and Delaware basins, west Texas and southern New Mexico: Am. Assoc. Petrol. Geol. Bull., v. 53, no. 11, p. 2223–2251.

Simpson, G.G., 1963, Historical science, in Albritton, C.C., ed., The Fabric of Geology: Stanford, CA, Freeman, Cooper, & Co., p. 13–24.

Sleep, N.H., 1971, Thermal effects of the formation of Atlantic continental margins by continental break-up: Geophysical Journ. Royal Astronomical Soc., v. 24, p. 325–350.

Sloss, L.L., 1949, Physical Geology, lecture notes: Northwestern Univ.

Sloss, L.L., 1962, Stratigraphic models in exploration: Am. Assoc. Petrol. Geol. Bull., v. 46, no. 7, p. 1050–1057.

Sloss, L.L., 1963, Sequences in the cratonic interior North America: Geol. Soc. Am. Bull., v. 74, p. 93.

Sloss, L.L., 1972, Synchrony of Phanerozoic sedimentary-tectonic events of the North American and Russian platform: Int. Geol. Conf., 24th session, sect. 6, p. 24–32.

Sloss, L.L., E.C. Dapples, and W.C. Krumbein, 1960, Lithofacies Maps—An Atlas of the United States and Southern Canada: New York, John Wiley & Sons, 108 p.

Smith, A.E., 1966, Modern deltas: comparison maps, in Shirley, M.L.. ed., Deltas in their Geologic Framework: Houston, Houston Geol. Soc., p. 234–251.

Smith, M.K., 1969, Developments in seismic processing for

geologic interpretation (abst.): Am. Assoc. Petrol. Geol. Bull., v. 53, p. 742.

Smith, William, 1815a, A delineation of strata of England and Wales, with part of Scotland: London, J. Cary, 16 maps.

Smith, William, 1815b, A memoir to the map and delineation of the strata of England and Wales with part of Scotland: London, J. Carey, 51 p.

Southard, J.B., and D.A. Cacchione, 1972, Experiments on bottom sediment movement by breaking internal waves, in Swift, D.J.P., D.B. Duane, and O.H. Pilkey, eds., Shelf Sediment Transport: Stroudsburg, PA, Dowden, Hutchinson, and Ross, p. 83–98.

Spieker, E.M., 1949, Sedimentary facies and associated diastrophism in the Upper Cretaceous of central and eastern Utah: Geol. Soc. Am. Mem., no. 39, pp. 55–81.

Stanley, D.J., 1961, Etudes sédimentologiques des grès d'Annot et de leurs équivalents latéraux: Inst. Franç. Pétrole, Ref. 6821, Société des Editions Technip, Paris, 158 p.

Stanley, D.J., 1974, Modern flysch sedimentation in a Mediterranean island arc setting, in Dott, R.J., Jr., and R.H. Shaver, eds., Modern and Ancient Geosynclinal Sedimentation: Soc. Econ. Paleont. Mineral. Spec. Publ., no. 19, p. 240–259.

Stanley, D.J., and A.H. Bouma, 1964, Methodology and paleogeographic interpretation of flysch formations: a summary of studies in the Maritime Alps, in Bouma, A.H., and A. Brouwer, eds., Turbidites; Developments in Sedimentology, v. 3: New York, Elsevier, p. 34–64.

Stanley, D.J., and G. Kelling, eds., 1978, Sedimentation in Submarine Canyons, Fans, and Trenches: Stroudsburg, PA, Dowden, Hutchinson, and Ross, 1978, 345 p.

Stanley, S.M., 1979, Macroevolution: Pattern and Process: San Francisco, W. H. Freeman, 332 p.

Stanley, S.M., 1985, Rates of evolution: Paleobiology, v. 11, p. 13–26.

Stearns, D.W., W.R. Sacrison, and R.C. Hanson, 1975, Structural history of southwestern Wyoming as evidenced from outcrop and seismic: Rocky Mountain Assoc. Petrol. Geol. Field Trip Guide, p. 9–20.

Stille, H., 1913, Evolution and Revolutionen in der Erdgeschichte: Berlin, Borntraeger, 32 p.

Stille, H., 1936, Wege and ergebnisse der geolgisch-tektonischen Forschung: Festschr. 2: Jahre K.-Wilhelm Gesellsch., Ford, Wiss., Bd. 2., p. 77–97.

Stille, H., 1940, Einfuhrung in den Bau Nordamerikasz: Berlin, Borntraeger, 717 p.

Stone, D.S., 1969, Wrench faulting and Rocky Mountain tectonics: Mountain Geol., v. 6, no. 2, p. 67–79.

Stride, A.H., ed., 1982, Offshore Tidal Sands—Processes and Deposits: New York, Chapman and Hall, 213 p.

Sturm, M., and A. Matter, 1978, Turbidites and varves in Lake Brienz (Switzerland): deposition of clastic detritus by density currents, in Matter, A., and M.E. Tucker, eds., Modern and Ancient Lake Sediments: Int. Assoc. Sediment. Spec. Publ., no. 2, p. 147–168.

Suppes, P., 1962, Models of data, in Nagel, E., P. Suppes, and A. Tarski, eds., Logic, Methodology and Philosophy of Science: Proc. 1960 Int. Congr., Stanford, CA, Stanford Univ. Press, p. 250–261.

Swann, D.H., 1964, Late Mississippian rhythmic sediments of Mississippi Valley: Am. Assoc. Petrol. Geol. Bull., v. 48, p. 638–658.

Swift, D.J.P., 1967, Shoreface erosion and transgressive stratigraphy: Jour. Geol., v. 76, p. 444–456.

Swift, D.J.P., 1970, Quarternary shelves and the return to grade: Mar. Geol., v. 8, p. 5–30.

Swift, D.J.P., 1976, Continental shelf sedimentation, in Fairbridge, R., ed., Encyclopedia of Sedimentology: Stroudsburg, PA, Dowden, Hutchinson, and Ross, p. 190–196.

Swift, D.J.P., D.J. Stanley, and J.R. Curray, 1979, Relict sediments on continental shelves: a reconsideration: Jour. Geol., v. 79, p. 322–346.

Ten Haaf, E., 1959, Graded beds of northern Apennines (unpubl. Ph.D. dissertation): Groningen, Netherlands,

State Univ., 102 p.

Termier, P., 1902, Quatar, coupes a travers les Alpes Franco-Italiennes: Bull. Soc. Geol. France, v. 2, p. 411–432.

Terwindt, J.H.J., 1988, Paleo-tidal reconstructions of inshore tidal depositional environments, in de Boer, P.L., A. van Gelder, and S.D. Nio, eds., Tide-influenced Sedimentary Environments and Facies: Dordrecht, Holland, D. Reidel Publishing Co., p. 233–264.

Thayer, C.W., 1974, Marine paleocology in Upper Devonian of New York: Lethaia, v. 7, no. 2, p. 121–155.

Thom, B.G., P.S. Roy, A.D. Short, J. Hudson, and R.A. Davis, Jr., 1986, Modern coastal and estuarine environments of deposition in southeastern Australia: Guide for Excursion 4A: 12th Int. Sediment. Congr., Canberra, Australia, Dept. of Geography, Univ. of Sydney, Australia, p. 81–87.

Thomas, G.E., 1962, Grouping of carbonate rocks into textural and porosity units for mapping purposes, in Ham, W.E., ed., Classification of Carbonate Rocks: Am. Assoc. Petrol. Geol. Mem., no. 1, p. 193–223.

Thomas, G.E., 1971, Continental plate tectonics—southwest Wyoming, in Symposium on Wyoming Tectonics and their Significance: Wyoming Geol. Assoc. Guidebook, no. 23, p. 103–123.

Tillman, R.W., and C.T. Siemers, eds., 1984, Silciclastic Shelf Sediments: Soc. Econ. Paleont. Mineral. Spec. Publ., no. 34, 268 p.

Tissot, B., B. Durand, J. Espitalié, and J. Combaz, 1974, Influence of the nature and diagenesis of organic matter in the formation of petroleum: Am. Assoc. Petrol. Geol. Bull., v. 58, p. 499–506.

Tissot, B.P., R. Pelet, and Ph. Ungerer, 1987, Thermal history of sedimentary basins, maturation indices, and kinetics of oil and and gas generation: Am. Assoc. Petrol. Geol. Bull., v. 71, no. 12, p. 1445–1446.

Tissot, B.P., and D.H. Welte, 1984, Petroleum Formation and Occurrence: Berlin, Springer-Verlag, 699 p.

Todd, R.G., and R.M. Mitchum, Jr., 1977, Identification of Upper Triassic, Jurassic, and Lower Cretaceous seismic sequences in the Gulf of Mexico and offshore West Africa, in Payton, C.E., ed., Seismic Stratigraphy—Applications to Hydrocarbon Exploration: Am. Assoc. Petrol. Geol. Mem., no. 26, p. 145–164.

Toth, J., 1980, Cross-formational gravity-flow of groundwater a mechanim of the transport and accumulation of petroleum (the generalized hydraulic theory of petroleum migration, in Roberts, W.H., III, and R.J. Cordell, eds., Problems of Petroleum Migration: Am. Assoc. Petrol. Geol. Stud. Geol. Ser., no. 10, p. 121–168.

Trask, P.D., 1953, Chemical studies of sediments of the western Gulf of Mexico: Mass. Inst. Technol. Woods Hole Oceanogr. Inst., Pap. Phys. Oceanogr. Meterol., v. 12, p. 47–120.

Trask, P.D., and H.W. Patnode, 1942, Source Beds of Petroleum: Tulsa, OK, Am. Assoc. Petrol. Geol., 566 p.

Triplehorn, D.M., 1966, Morphology, internal structure, and origin of glauconite pellets: Sedimentology, v. 6, p. 247–266.

Vail, P.R., 1987, Seismic stratigraphy interpretation using sequence stratigraphy, part 1: seismic stratigraphy interpretation procedure, in Bally, A.W., ed., Atlas of Seismic Stratigraphy: Am. Assoc. Petrol. Geol. Stud. Geol. Ser., v. no. 1, no. 27, p. 1–10.

Vail, P.R. and J. Hardenbol, 1979, Sea-level changes during the Tertiary: Oceanus, v. 22, p. 71–79.

Vail, P.R., R.M. Mitchum, Jr., J.B. Sangree, and S. Thompson III, 1975, Stratigraphic framework and eustatic cycles from seismic stratigraphic analysis: Am. Assoc. Petrol. Geol. and Soc. Econ. Paleont. Mineral. Annu. Meeting, v. 2, p. 76–77.

Vail, P.R., R.M. Mitchum, Jr., and S. Thompson III, 1977a, Seismic stratigraphy and global changes of sea level, part 3: relative changes of sea level from coastal onlap, in Payton, C.E., ed., Seismic stratigraphy—Applicatons to hydrocarbon exploration: Am. Assoc. Petrol. Geol. Mem., no. 26, p. 63–97.

Vail, P.R., R.M. Mitchum, and S. Thompson, 1977b, Seismic

stratigraphy and global changes of sea-level, part 4: global cycles of relative changes of sea-level, *in* Payton, C.E., ed., Seismic Stratigraphy—Applications to Hydrocarbon Exploration: Am. Assoc. Petrol. Geol. Mem., no. 26, p. 83–98.

Vail, P.R., R.G. Todd, and J.B. Sangree, 1976, Stratigraphic interpretation of seismic data: short course notes: Tulsa, OK, Am. Assoc. Petrol. Geol.

Vail, P.R., R.G. Todd, and J.B. Sangree, 1977, Chronostratigraphic significance of seismic reflections, *in* Payton, C.E., ed., Seismic Stratigraphy—Applications to Hydrocarbon Exploration: Amer. Assoc. Petrol. Geol. Mem., no. 26, p. 99–116.

Vail, P.R., and R.O. Wilbur, 1966, Onlap key to worldwide unconformities and depositional cycles: Am. Assoc. Petrol. Geol. Annu. Meeting (St. Louis), p. 638–639.

Valentine, J.W., 1973, Evolutionary Paleoecology of the Marine Biosphere: Englewood Cliffs, NJ, Prentice-Hall.

Valentine, J.W., and E.M. Moores, 1972, Global tectonics and the fossil record: Jour. Geol., v. 80, no. 2, p. 167–184.

Valentine, J.W., and E.M. Moores, 1974, Plate tectonics and the history of life in the oceans: Sci. Am., v. 230, no. 4, p. 80–89.

Valyasko, M.G., 1968, Scientific works in the field of geochemistry and the genesis of salt deposits in the U.S.S.R., *in* Richter-Bernburg, G., ed., Geology of Saline Deposits: Hannover Symp., UNESCO Proc., p. 289–311.

Van Hinte, J.E., 1976a, A Jurassic time scale: Am. Assoc. Petrol. Geol. Bull., v. 60, p. 489–497.

Van Hinte, J.E., 1976b, A Cretaceous time scale: Am. Assoc. Petrol. Geol. Bull., v. 60, p. 498–516.

Van Straaten, L.M.J.U., 1965, Coastal barrier deposits in south and north Holland in particular areas around Scheveningen and Ijmuiden: Mededel. Geol. Stichtung., v. 17, p. 41–75.

Van Wagoner, J.C., R.M. Mitchum, Jr., H.W. Posamentier, and P.R. Vail, 1987, Seismic stratigraphy interpretation using sequence stratigraphy, part 2: key definitions of sequence stratigraphy, *in* Bally, A.W., ed., Atlas of Seismic Stratigraphy: Am. Assoc. Petrol. Geol. Stud. Geol. Ser., v. 1, no. 27, p. 11–14.

Vasquez, E., and P.A. Dickey, 1972, Major faulting in northwestern Venezuela and its relation to global tectonics, *in* Petzall, C., ed., VI Caribbean Geological Conf. Trans.: Margarita, Venezuela, p. 191–202.

Vedros, S., and G.S. Visher, 1978, The Red Oak sandstone— a submarine fan deposit, *in* Stanley, D.J., and G. Kelling, eds., Sedimentation in Submarine Canyons, Fans, and Trenches: Stroudsburg, PA, Dowden, Hutchinson & Ross, Stroudsburg, p. 292–310.

Verdier, A.C., T. Oki, and A. Suardy, 1980, Geology of the Handil field (east Kalimantan, Indonesia), *in* Halbouty, M.T., ed., Giant Oil and Gas Fields of the Decade 1968–1978: Am. Assoc. Petrol. Geol. Mem., no. 30, p. 399–422.

Vest, E.L., Jr., 1970, Oil fields of Pennsylvanian-Permian horse-shoe atoll, West Texas, *in* Halbouty, M. ed., Geology of Giant Petroleum Fields: Am. Assoc. Petrol. Geol. Mem., no. 14, p. 185–203.

Vilas, F., A. Sopena, L. Rey, A. Ramos, M.A. Nombela, and A. Arche, 1988, The Corrubedo tial inlet, Galicia, NW Spain sedimentary processes and facies, *in* de Boer, P.L., A. van Gelder and S.D. Nio, eds., Tide-influenced Sedimentary Environments and Facies: Dordrecht, Holland, D. Reidel Publishing Co., p. 183–200.

Vincellette, R.R., and E. Chittum, 1981, Exploration for oil accumulations in Entrada sandstone, San Juan Basin, New Mexico: Am. Assoc. Petrol. Geol. Bull., v. 65, no. 12, p. 2546–2570.

Vine, F.J., 1963, Spreading of the ocean floor: new evidence: Science, v. 154, p. 1405–1415.

Vine, F.J., and P.M. Mathews, 1966, Magnetic anomalies over ocean ridges: Nature, v. 199, p. 947–949.

Visher, G.S., 1963, The use of the vertical profile in environmental reconstruction (abst.): Am. Assoc. Petrol. Geol. Bull., v. 47, p. 374.

Visher, G.S., 1965a, The use of the vertical profile in environmental reconstruction: Am. Assoc. Petrol. Geol. Bull., v. 49, p. 41–61.

Visher, G.S., 1965b, Fluvial processes as interpreted from ancient and Recent fluvial deposits, *in* Middleton, G.V., ed., Primary Sedimentary Structures and their Hydrodynamic Interpretation: Soc. Econ. Paleont. Mineral. Spec. Publ., no. 12, p. 116–132.

Visher, G.S., 1968, Guidebook to the geology of the Bluejacket-Bartlesville sandstone, Oklahoma: Oklahoma City Geol. Soc., 72 p.

Visher, G.S., 1969, Grain-size distribution and depositional processes: Jour. Sediment. Petrol., v. 39, no. 3, p. 1074–1106.

Visher, G.S., 1970, Environmental criteria—their use and misuse: Gulf Coast Assoc. Geol. Soc. Trans., 22nd Annu. Conv., p. 67–72.

Visher, G.S., 1972, Physical characteristics of fluvial deposits, *in* Rigby, J.K., and W.K. Hamblin, eds., Recognition of Ancient Sedimentary Environments: Soc. Econ. Paleont. Mineral. Spec. Publ., no. 16, p. 84–97.

Visher, G.S., 1978a, Size-frequency studies—description, significance, and application, *in* Fairbridge, R.W., ed., Encyclopedia of Earth Sciences, Vol. VI-A: Stroudsburg, PA, Dowden, Hutchinson, and Ross, p. 370–374.

Visher, G.S., 1978b, Textural interpretation of physical aspects of gravity flows (preprint): 10th Int. Congr. Sediment. (Jerusalem).

Visher, G. S., 1984, Exploration Stratigraphy: Tulsa, OK, PennWell Publishing Co., 334 p.

Visher, G.S., and W.E. Freeman, 1976, Stratigraphic analysis of the Navajo sandstone: reply to discussion: Jour. Sediment. Petrol., v. 46, p. 491–497.

Visher, G.S., and J.D. Howard, 1974, Dynamic relationship between hydraulics and sedimentation in the Altamaha Estuary: Jour. Sediment. Petrol., v. 44, no. 2, p. 502–521.

Visher, G.S., and N.J. Hyne, 1973, Dynamic continental shelf model (abst.): Estuary–Shelf Symp. Proc. (Bordeaux, France).

Visher, G.S., B.S. Saitta, and R.S. Phares, 1971, Pennsylvanian delta patterns and petroleum occurrences in Eastern Oklahoma: Am. Assoc. Petrol. Geol. Bull., v. 55, p. 1206–1230.

Visser, M.J., 1980, Neap-spring cycles reflected in Holocene subtidal largescale bedform deposits: a preliminary note: Geology, v. 8, p. 543–546.

Vrba, E.S., 1980, Evolution, species and fossils how does life evolve?: S. Afr. Jour. Sci., v. 76, p. 61–84.

Waldrop, M.M., 1988, After the fall: Science, v. 239, p. 977.

Walker, K. R. and L. Laporte, 1970, Congruent fossil communities from Ordovician and Devonian carbonates of New York: Jour. Paleont., v. 44, p. 928–944.

Walker, T.R., and J.C. Harms, 1972, Eolian origin of flagstone beds, Lyons sandstone (Permian), type area, Boulder, CO, Mountain Geol., v. 9, nos. 2–3, p. 279–288.

Walther, Johannes, 1893–1894, Einleitung in die Geologie als historische Wissenschaft. Beofachtungen icher die Bildung der Gesteine und ihrer organiachen Einschlüsse: Jena, Gustav Fischer, 1055 p.

Wanless, H.R., A.R. Campos, J.C. Horne, R.C. Trescott, R.S. Vail, J.R. Baroffio, and D.E. Orlopp, 1970, Late Paleozoic deltas in the Central and Eastern United States, *in* Morgan, J.P., ed., Deltaic Sedimentation, Modern and Ancient: Soc. Econ. Paleont. Mineral. Spec. Publ., no. 15, p. 215–245.

Waples, D.W., 1980, Time and temperature in petroleum formation Application of Lopatin's method to petroleum exploration: Am. Assoc. Petrol. Geol. Bull., v. 64, p. 916–926.

Ward, P.D., and P.W. Signor, III, 1985, Evolutionary patterns of Jurassic and Cretaceous ammonites; an analysis of clade shape, *in* Valentine, J.W., ed., Phanerozoic Diversity Patterns, Profiles in Macroevolution: New Jersey, Princeton Univ. Press., and San Francisco, Pacific Division, Amer. Assoc. for the Advancement of Science, p. 399–418.

Warner, G.A., 1982, Source and time of generation of hydrocarbons in the Fossil Basin, western Wyoming thrust belt, *in* Powers, R.B., ed., Geologic Studies in the Cordilleran Thrust Belt: Rocky Mountain Assoc. Geol., p. 805–815.

Watney, W.L., 1984, Recognition of favorable reservoir trends in upper Pennsylvanian cyclic carbonates in western Kansas, *in* Hyne, N.J., ed., Limestones of the Mid-continent: Tulsa Geol. Soc. Spec. Pub., no. 2, p. 201–245.

Webb, D.J., 1976, A model of continental-shelf resonances: Deep-Sea Res., v. 23, p. 1–15.

Weber, K.J., 1971, Sedimentologic aspects of oil fields in the Niger Delta: Geol. Mijnbouw, v. 50, p. 559–576.

Weber, K.J., and E. Daukoru, 1975, Petroleum geology of the Niger Delta: Ninth World Petrol. Congr. Proc., v. 2, p. 209–221.

Weimer, R.J., 1960, Upper Cretaceous stratigraphy of the Rocky Mountain area: Am. Assoc. Petrol. Geol. Bull., v. 44, p. 1–20.

Weimer, R.J., and J.D. Haun, 1960, Guide to the Geology of Colorado: Rocky Mountain Assoc. Geol., 310 p.

Welder, F.A., 1959, Processes of Deltaic Sedimentation in the Lower Mississippi River: Louisiana State Univ. Coastal Studies Inst. Tech. Rep., no. 12, 90 p.

Weller, J.M., 1960, Stratigraphic Principles and Practice: New York, Harper and Row, 725 p.

Wells, N.A., and J.A. Dorr, Jr., 1987, A reconnaissance of sedimentation on the Kosi alluvial fan of India, *in* Ethridge, F.G., R.M. Flores, and M.D. Harvey, eds. Recent Developments in Fluvial Sedimentology: Soc. Econ. Paleont. Mineral. Spec. Publ., no. 39, p. 51–62.

Wermund, E.G., and W.A. Jenkins, Jr., 1970, Recognition of deltas by fitting trend surfaces to upper Pennsylvanian sandstones in North-Central Texas, *in* Morgan, J.P., ed., Deltaic Sedimentation, Modern and Ancient: Soc. Econ. Paleont. Mineral. Spec. Publ., no. 15, p. 256–269.

Wescott, W.A., and F.G. Ethridge, 1980, Fan-delta sedimentology and tectonic setting—Yallahs fan delta, southeast Jamaica: Am. Assoc. Petrol. Geol. Bull., v. 64, p. 374–399.

Wezel, F.C., D. Savelli, M. Bellagamba, M. Tramontana, and R. Bartole, 1981, Plio-Quaternary depositional style of sedimentary basins along insular Tyrrhenian margins, *in* Wezel, F.C., ed., Sedimentary Basins of Mediterranean Margins: CNR Italian Project of Oceanography, p. 239–269.

Wheeler, H.E., 1963, Post-Sauk and Pre-Absaroka Paleozoic stratigraphic patterns in North America: Am. Assoc. Petrol. Geol. Bull., v. 47, p. 1497–1526.

Whewell, W., 1859, Novum Organon Renvatum: London, John W. Parker & Son.

Whitehead, A. N., 1962, The Function of Reason, 3rd ed.: Beacon Hill, Beacon Hill Press, 90 p.

Whitten, E.H.T., 1964, Process-response models in geology: Geol. Soc. Am. Bull., v. 75, p. 455–464.

Wiener, Norbert, 1961, Cybernetics, or control and communication in the animal and the machine, 2nd ed.: New York, M.I.T. Press and John Wiley, 212 p.

Williams, J.J., D.C. Connor, and R.E. Peterson, 1975, The Piper oil-field: a fault-block structure with Upper Jurassic beach-bar reservoir sands: Am. Assoc. Petrol. Geol. Bull., v. 59, p. 1581–1601.

Williams, W.D., and J.S. Dixon, 1985, Seismic interpretation of the Wyoming overthrust belt, *in* Greis, R.R., and R.C. Dyer, eds., Seismic Exploration of the Rocky Mountain Region: Rocky Mountain Assoc. Geol. and Denver Geophys. Soc., p. 13–22.

Wilson, I.G., 1973, Ergs: Sediment. Geol., v. 10, p. 77–106.

Wilson, J.B., 1986, Faunas of tidal current and wave-dominated continental shelves and their use in the recognition of storm deposits, *in* Knight, R.J., and J.R. McLean, eds., Shelf Sands and Sandstones: Can. Soc. Petrol. Geol. Mem., no. 11, p. 313–326.

Wilson, J.L., 1975, Carbonate Facies in Geologic History: New York, Springer-Verlag, 471 p.

Winker, C.D., and M.B. Edwards, 1983, Unstable progradational clastic shelf margins, *in* Stanley, D.J., and G.T.

Moore, eds., The Shelfbreak Critical Interface on Continental Margins: Soc. Econ. Paleont. Mineral. Spec. Publ., no. 33, p. 139–157.

Wise, D.U., 1963, Keystone faulting and gravity sliding driven by basement uplift of Owl Creek Mountains, Wyoming: Am. Assoc. Petrol. Geol. Bull., v. 47, no. 4, p. 586–598.

Wise, D.U., 1974, Continental margins, freeboard and the volume of continents and oceans through time *in* Burk, C.A., and C.L. Drake, eds., The Geology of Continental Margins: New York, SpringerVerlag, p. 45–58.

Wolman, M.G., and J.P. Miller, 1960, Magnitude and frequency of forces in geomorphic processes: Jour. Geol., v. 68, no. 1, p. 54–74.

Woodbury, H.O., T.B. Murray, P.J. Packford, and W.H. Akers, 1973, Pliocene and Pleistocene depocenters, outer continental shelf, Louisiana and Texas: Am. Assoc. Petrol. Geol. Bull., v. 57, p. 2428–2439.

Wright, L.D., and J.M. Coleman, 1973, Variations in morphology of major river deltas as functions of ocean waves, land river discharge regimes: Am. Assoc. Petrol. Geol. Bull., v. 57, p. 370–398.

Wright, L.D., and J.R. Coleman, 1974, Mississippi River mouth processes, effluent dynamics, and morphologic development: Jour. Geol., v. 82, no. 6, p. 751–778.

Wright, S., 1988, Surfaces of selective value revisted (notes and comments): Am. Natural., v. 131, no. 1, p. 115–123.

Wylie, P.J., 1987, Magma genesis and plate tectonics, Union Lecture no. UL4: Vancouver, Canada, Int. Union Geod. Geophys.

Young, F.G., and R.A. Rahmani, 1974, Bioturbation structures in clastic rocks, *in* Shawa, M.S., ed., Use of Sedimentary Structures for Recognition of Clastic Environments: Calgary, Canada, Can. Soc. Petrol. Geol., p. 40–51.

Zaaza, M.W., 1974, Stratigraphic distribution of giant petroleum fields as controlled by worldwide correlative unconformities and onlap-offlap stratigraphic sequences (unpub. MS thesis): Univ. Tulsa, 126 p.

Zaaza, M.W., and G.S. Visher, 1975, Worldwide distribution of giant oil and gas fields as controlled by stratigraphic sequences: Am. Assoc. Petrol. Geol. Annu. Meeting (Dallas), p. 84.

Ziegler, P.A., 1978, North-western Europe: tectonics and basin development: Geol. Mijnbouw, v.57, p. 589–626.

Ziegler, P.A., 1982, Geological Atlas of Western and Central Europe: Amsterdam, Holland, Elsevier Sci. Publ. Co., 130 p.

Ziegler, P.A., 1983, Inverted basins in the Alpine foreland, *in* Bally, A.W., ed., Seismic Expression of Seismic Styles: Am. Assoc. Petrol. Geol. Stud. Geol., no. 15, part III, p. 3.3–3.12.